NEUTRINO PHYSICS

Thoroughly revised and updated, this new edition presents a coherent and comprehensive overview of modern neutrino physics.

The book covers all the major areas of current interest, with chapters discussing the intrinsic properties of neutrinos, the theory of the interaction of neutrinos with matter, experimental investigations of the weak interaction in neutrino processes, the theory and supporting experiment for the basic properties of the interaction of neutrinos with fermions, and on astrophysics and cosmology. Several chapters have been completely rewritten.

This edition presents new data on solar neutrinos and an update of the results of searches for double beta decay. It also contains a new chapter on direct measurements of the neutrino mass, with high precision data from experiments at Fermilab and CERN, and at the Kamiokande Laboratory in Japan.

An essential reference text for particle physicists, nuclear physicists and astrophysicists.

KLAUS WINTER studied physics at the University of Hamburg and finished his graduate studies at the College de France (Paris) with a PhD degree at the Sorbonne. He then joined CERN (Geneva) and has led several experiments in particle physics, including a measurement of the π^0 lifetime, giving the first confirmation of the color property of quarks, the first experimental confirmation of the discovery of CP violation in K^0 decay, and a precise test of the $\Delta S/\Delta Q$ rule. Since 1975 he has performed a series of neutrino experiments at CERN, with special emphasis on the study of neutral current phenomena. He is Professor of Physics at the Humboldt University (Berlin) and has been Regents Professor at the University of California and Guest Professor at the College de France.

T0189036

CAMBRIDGE MONOGRAPHS ON PARTICLE PHYSICS, NUCLEAR
PHYSICS AND COSMOLOGY: 14

General editors: T. Ericson, P. V. Landshoff

NEUTRINO PHYSICS

Second edition

Edited by Klaus Winter
CERN, Geneva

CAMBRIDGE
UNIVERSITY PRESS

CAMBRIDGE UNIVERSITY PRESS
Cambridge, New York, Melbourne, Madrid, Cape Town, Singapore, São Paulo, Delhi

Cambridge University Press
The Edinburgh Building, Cambridge CB2 8RU, UK

Published in the United States of America by Cambridge University Press, New York

www.cambridge.org
Information on this title: www.cambridge.org/9780521650038

First published 1991
Second edition 2000
This digitally printed version 2008

A catalogue record for this publication is available from the British Library

ISBN 978-0-521-65003-8 hardback
ISBN 978-0-521-08405-5 paperback

Contents

Preface

This book is an introduction to the theory and the experiments of neutrinos and their interactions with matter. It is intended to give a comprehensive overview of the current knowledge in neutrino physics and to discuss the prospects of further developments for the next decade.

The book was written for experimental physicists and graduate students working in the fields of particle physics, nuclear physics, and astrophysics. Theoretical physicists may find it useful to consult this book for experimental information and for orientation on theoretical issues.

The amount of material has grown tremendously since Princeton University Press published the book by James S. Allen in 1958. Following the first series of experiments with neutrino beams at accelerators in the 1960s, the late Carlo Franzinetti had made plans for a book on neutrino physics, but it was postponed so that it could include the results from a second and third generation of experiments at the 400–500 GeV accelerators at CERN and at the Fermi National Laboratory. These results are now available. The weak bosons $W^{\pm}$ and Z^0 have been discovered at CERN in proton–antiproton collisions, and their mass values have been determined. Recently, precise determinations of the number of fermion families with light neutrinos and of the partial decay width of the Z^0 were performed at Stanford (SLC) and at CERN (LEP). Therefore the time for a new book on neutrino physics seems well chosen.

I decided to organize the material as a reference collection of essays and I approached those physicists who had made important contributions to the field to write them. Special attention was given to areas of growing importance, such as solar neutrinos, supernova neutrinos, nucleon structure, and tests of quantum chromodynamics at the first high-energy electron–proton collider HERA at DESY (Hamburg), and precision tests of the Standard Model of the electroweak interactions. The study of nuclear β-decay and its contribution to elucidating the structure of the weak interaction has been omitted as several excellent books are available on this subject. Other topics such as elastic neutrino scattering, resonance production, and charm particle production by neutrinos have also been omitted to keep the work down to one volume.

Chapter 1 is devoted to the history of the subject. Instead of preparing new text, I have chosen to reprint the famous essay by Wolfgang Pauli, "On the Earlier and More Recent History of the Neutrino" written in 1957, following the detection of the neutrino as a free particle by Reines and Cowan. It is the first publication of this essay in the English language. This chapter sets out the main new ideas and discoveries about neutrinos, and seeing the original papers will, I believe, provide a special thrill for the reader.

Chapter 2 deals with the intrinsic or static properties of neutrinos. Today we can summarize all the evidence by saying neutrinos are massless and electrically neutral. However, more subtle questions arise in studying their particle-antiparticle properties, for example, in the search for neutrinoless double β-decay and in their electromagnetic interactions. Today we know that there are only three fermion families with light neutrinos in nature. Do these neutrino flavors mix? This question is deeply related to that of a nonzero mass of neutrinos.

The basic elements of the standard electroweak theory are introduced in Chapter 3, on the theory of the interaction of neutrinos with matter. This chapter also shows how the framework for describing the experimental results is derived, and raises open questions about the high-energy behavior of the electroweak theory. One section is devoted to coherent effects in neutrino propagation through matter.

Chapter 4 reviews the experimental investigations of the weak interaction in neutrino processes. The topics of interest are the space–time structure of the charged weak current, the flavor structure of the charged weak quark current, and the structure of the neutral weak current, all of which are important to the electroweak theory. Attention is also given to the observation and the study of the properties of the weak bosons $W^{\pm}$ and Z^0 and to the evidence for the gauge nature of the electroweak current.

Chapter 5 proceeds on the assumption that the basic properties of the interaction of neutrinos with fermions are known, and neutrinos are used as probes to study the parton structure of the nucleon. The particular advantages of neutrinos are compared with deep inelastic electron and muon scattering. The experimental results on total and differential cross sections are examined, along with methods for extracting the structure functions. The results are confronted with the theory of quantum chromodynamics. They will provide a reference for future work on the inverse reactions at HERA and for further tests of the validity of quantum chromodynamics.

Neutrino physics began as a branch of nuclear physics, went on to become an important field of particle physics, and has recently developed fundamental applications to astrophysics. Solar neutrinos, supernova neutrinos and their first detection, and the possible role of massive neutrinos in cosmology are treated in Chapter 6.

A second volume is planned with details on neutrino detectors and neutrino beams.

I am grateful to all the friends and colleagues who have helped me clarify many of the topics covered in this book. I owe special thanks to my collaborators in CHARM and CHARM II scientific publications.

I am dedicating this book to my wife Krisztina.

Klaus Winter
Geneva, 5 May 1991

Preface to the second edition

Looking today, roughly ten years after this book was written, at the status of Neutrino Physics, it is amazing how much progress has been achieved. Some of it is truly fundamental, the discovery of neutrino flavor oscillation. Of course, according to our rules, it has still to be confirmed. It provides a first outlook into physics beyond the Standard Model. The Standard Model itself has now reached a status of maturity, after confirmation by measurements with a precision which seemed unthinkable a decade ago. It is now generally believed that its underlying symmetry pattern cannot be accidental and that it will be incorporated into a future Grand Unified Theory.

The structure of this book has met, I am told, with general approval. It has therefore not been modified. Most chapters have been updated, some had to be completely rewritten. Because of the original approach that the book is written by scientists who have themselves made important contributions, some new authors appear.

Looking at the open questions, details of the neutrino mixing and neutrino masses, and of CP violation in neutrino reactions, the particle-antiparticle properties of neutrinos, the problem of the detection of relic neutrinos, their density in space and their contribution to the energy density of the universe, their flavor composition and neutrino-antineutrino asymmetry, and the electromagnetic properties of neutrinos, I feel assured that Neutrino Physics will continue to develop and to become an even more central part of elementary particle physics.

I am grateful to the authors of the first and the second edition and to many colleagues who have again helped me to clarify the topics covered in this book.

It is a special pleasure to thank my wife Krisztina who has always given me her full support; I am gratefully dedicating this book to her.

Klaus Winter
Geneva, 3 November 1998

Contributors

Altarelli, Guido	CERN, Geneva, Switzerland, and Università dì Roma III, Italy.
Bilenky, Samoil M.	Joint Institute for Nuclear Research, Dubna, Russia.
Blondel, Alain	Ecole Politechnique, Paris, France and CERN, Geneva, Switzerland.
Caravaglios, Francesco	CERN, Geneva, Switzerland.
Denegri, Daniel	CERN, Geneva, Switzerland, and Centre d'Etudes Nucleaires, Saclay, France.
Diemoz, Marcella	Istituto Nazionale di Fisica Nucleare, Sezione di Roma, Italy.
Einsweiler, Ken	University of California, Berkeley, CA, USA.
Ferroni, Fernando	Istituto Nazionale di Fisica Nucleare, Sezione di Roma, Italy.
Holzschuh, Eugen	Physics Institute, University of Zürich, Switzerland.
Kayser, Boris	Lawrence Berkeley Laboratory, Berkeley, CA, and National Science Foundation, Washington, D.C., USA.
Kirsten, Till	Max-Planck-Institut für Kernphysik, Heidelberg, Germany.
Klapdor-Kleingrothaus, Hans Volker	Max-Planck-Institut für Kernphysik, Heidelberg, Germany.
Kleinknecht, Konrad	Institut für Physik, Johannes Gutenberg-Universität, Mainz, Germany.
Kolb, Edward W.	University of Chicago and NASA/ Fermilab Astrophysics Center, Batavia, IL, USA.
Longo, Egidio	Istituto Nazionale di Fisica Nucleare, Sezione di Roma, Italy.
Maiani, Luciano	Università dì Roma "La Sapienza," Italy and CERN, Geneva, Switzerland.

Mangano, Michelangelo	Università dì Roma III and CERN, Geneva, Switzerland.
Marciano, William J.	Brookhaven National Laboratory, Upton, NY, USA.
Martinelli, Guido	Istituto Nazionale di Fisica Nucleare, Sezione di Roma, Italy.
Rolandi, Luciano	CERN, Geneva, Switzerland.
Schramm, David N.	University of Chicago and NASA/Fermilab Astrophysics Center, Batavia, IL, USA.
Shandarin, Sergei F.	Institute for Physical Problems, Moscow, USSR.
Sirlin, Alberto	Department of Physics, New York University, NY, USA.
Totsuka, Yoji	Institute for Cosmic Ray Research, University of Tokyo, Japan.
Turner, Michael S.	University of Chicago and NASA/Fermilab Astrophysics Center, Batavia, IL, USA.
Winter, Klaus	CERN, Geneva, Switzerland and Humboldt University, Berlin, Germany.
Wolfenstein, Lincoln	Carnegie Mellon University, Pittsburgh, PA, USA.

1

History

1.1 On the earlier and more recent history of the neutrino

WOLFGANG PAULI, 1957*

1 Problems concerning the interpretation of the continuous energy spectrum of beta rays

The continuous energy spectrum of beta rays discovered by J. Chadwick in 1914 [CHA 14] immediately posed difficult problems with respect to its theoretical interpretation. Was it directly due to the primary electrons emitted from the radioactive nucleus or was it to be attributed to secondary processes? The first hypothesis, which proved to be the correct one, was advocated by C. D. Ellis [ELL 22a], the second one by L. Meitner [MEI 22]. Meitner appealed to the fact that nuclei possess discrete energy states, as was known from alpha and gamma rays. She focused attention on the discrete energies of electrons, which had also been observed for many beta-radioactive nuclei. Ellis interpreted them as electrons being ejected from the outer shells by inner conversion of monochromatic nuclear gamma rays and assigned them to the observed X-ray lines. According to Meitner's theory, however, at least one of the electrons of discrete energy should be a genuine primary electron from the nucleus, which, in a secondary process, could then emit from the outer shells more electrons with smaller energies.[1] However, this postulated primary electron of discrete energy was never detected. Moreover, there are beta-radioactive nuclei, like RaE, that do not emit gamma rays and for which the electrons with discrete energies are missing altogether. In the polemic that arose between Ellis and Meitner, Ellis summarized [ELL 22b] his point of view in the following way:

> The theory of Miss Meitner is a very interesting attempt to provide a simple explanation of β-decay. The experimental facts, however, do not fit the framework of this theory and there is every indication that the simple analogy between α- and β-decay cannot be maintained. The β-decay is a considerably more complicated process and the general suggestions I made in this context appear to me to require the least constraint.

---

* Translation by Gabriele Zacek (CERN, Geneva), of "Zur älteren und neueren Geschichte des Neutrinos," published in Wolfgang Pauli, *Physik und Erkenntnistheorie*, pp. 156–80; Friedr, Vieweg, & Sohn, Braunschweig/Wiesbaden, 1984.

[1] In a later work [MEI 25] Meitner has proven experimentally that the γ-rays, contrary to an earlier opinion of Ellis, were emitted by the nucleus, which is generated *after* the emission of the α- or β-particle.

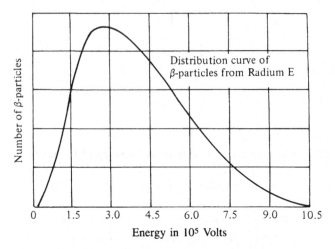

Fig. 1   Continuous beta spectrum of RaE.

This statement obviously did not bring researchers any closer to an answer to the question of how to interpret the continuous beta spectrum, and opinion remained divided on whether the spectrum was of primary origin (Ellis) or whether an initially discrete energy did broaden into a continuum by subsequent secondary processes (Meitner). This dispute finally came to an end in an experiment: the *measurement of the absolute heat in the absorption of beta electrons*. It was known from counting experiments that *one* electron is emitted from the nucleus per decay. In subsequent secondary processes, the heat measured in the calorimeter per decay should correspond to the upper limit of the beta spectrum; in the primary process, however, it should correspond to its mean energy. Ellis and W. A. Wooster [ELL 27] performed the measurement on RaE. The result for each decay, converted to Volts, was a heat of

$$344\,000 \text{ Volts} \pm 10\%$$

which corresponded well to the mean energy of the beta spectrum (Fig. 1). The upper boundary of the beta spectrum, however, would correspond to about 1 million Volts, which was completely excluded by the experiments. Ellis stressed that his experiment still left open the possibility of restoring the energy balance by a continuous gamma spectrum that would not have been absorbed in the calorimeter and would have escaped observation.

Meitner was not yet convinced by this experiment and immediately decided to repeat it with an improved apparatus. W. Orthmann, a collaborator of Nernst, designed a special differential calorimeter for this purpose. This calorimeter made it possible to repeat the heat measurement of the beta electrons from RaE with increased precision. The outcome,

$$337\,000 \text{ Volts} \pm 6\%$$

confirmed the result from Ellis and Wooster.

Moreover, in special experiments using ionization tubes, Meitner [MEI 30] proved that the continuous gamma spectrum postulated by Ellis was not present. Following these experimental results, there remained only two theoretical possibilities for the *interpretation of the continuous beta spectrum*:

1 The conservation of energy holds only statistically in this particular interaction, which gives rise to beta radioactivity.
2 The conservation of energy holds strictly in each primary process; however, an additional, very penetrating radiation is emitted together with the electrons, which consists of *new, neutral particles*.

The first possibility was supported by Bohr, the second one by myself. Before treating the history of these further questions, which was finally settled in favor of the second possibility, we must explain how our ideas about nuclear structure developed.

2 Neutrino and nuclear structure

Following Rutherford's first experiments on artificially induced transformations of nuclei, it was generally accepted that nuclei consist of protons and electrons. Rutherford himself discussed nuclear structure in this way in his famous Bakerian Lecture [RUT 20]. Among other things, the lecture presented the hypothesis of the existence of a nucleus with charge 0 and its eventual properties. Soon it became known (compare, e.g., [CLA 21]) that Rutherford had proposed the name *neutron* for these new hypothetical particles. He thought of them as a combination of protons and electrons of nuclear dimensions. Consequently, he urged his laboratory to perform experiments looking for these neutrons in hydrogen discharges, which of course had to remain fruitless.

The idea that the nuclei were made up of protons and electrons was eventually dismissed, albeit reluctantly. The decisive blow came from the quantum and wave mechanics theory advanced in 1927. According to this theory, there are two sorts of particles, the antisymmetric fermions and the symmetric bosons. Composite particles are fermions or bosons with the number of their constitutive fermions odd or even. An equivalent argument also holds for the spin, with fermions always possessing half a unit and bosons always an entire unit of spin. Since it was soon found that electrons and protons are fermions, the idea that they alone were the building blocks of all nuclei led to the conclusion that the parity of the charge number should determine the symmetry character of the nuclei. This conclusion was not confirmed by experience. The first counterexample was the "nitrogen anomaly," as we called it then. Using the band spectra, R. Kronig [KRO 28] and W. Heitler and G. Herzberg [HEI 29] showed that nitrogen with a charge number 7 and mass number 14 has spin 1 and Bose statistics. Similar cases followed, such as Li 6 (charge 3, mass 6) and the deuteron (charge 1, mass 2); both also had spin 1 and Bose statistics. Thus it was shown that the symmetry character of the nuclei was determined by the parity of the mass number and not by the parity of the charge number.

Using the idea of a new particle, I tried to combine this problem of the spin and statistics of nuclei with the problem of the continuous beta spectrum, without abandoning the conservation of energy. In December 1930, when the heavy neutron had not yet been discovered experimentally, I sent a letter on this topic to a meeting of physicists in Tübingen, where Geiger and Meitner in particular were present.[2]

> Public letter to the group of the Radioactives at the district society meeting in Tübingen:

> Physikalisches Institut   Zürich, 4. Dec. 1930
> der Eidg. Technischen Hochschule   Gloriastr.

> Zürich

> Dear Radioactive Ladies and Gentlemen,

> As the bearer of these lines, to whom I graciously ask you to listen, will explain to you in more detail, how because of the "wrong" statistics of the N and ^{6}Li nuclei and the continuous β-spectrum, I have hit upon a desperate remedy to save the "exchange theorem"[3] of statistics and the law of conservation of energy. Namely, the possibility that there could exist in the nuclei electrically neutral particles, that I wish to call neutrons, which have spin $\frac{1}{2}$ and obey the exclusion principle and which further differ from light quanta in that they do not travel with the velocity of light. The mass of the neutrons should be of the same order of magnitude as the electron mass and in any event not larger than 0.01 proton masses. – The continuous β-spectrum would then become understandable by the assumption that in β-decay, a neutron is emitted in addition to the electron such that the sum of the energies of the neutron and electron is constant. Now the question that has to be dealt with is which forces act on the neutrons? The most likely model for the neutron seems to me, because of wave mechanical reasons (the details are known by the bearer of these lines), that the neutron at rest is a magnetic dipole of a certain moment μ. The experiments seem to require that the effect of the ionization of such a neutron cannot be larger than that of a γ-ray and then μ should not be larger than $e * 10^{-3}$ cm.

---

[2] I am indebted to Mrs. Meitner for keeping a copy of this letter and for leaving it to me.
[3] This reads: exclusion principle (Fermi statistics) and half-integer spin for an odd number of particles; Bose statistics and integer spin for an even number of particles.

For the moment, however, I do not dare to publish anything on this idea and I put to you, dear Radioactives, the question of what the situation would be if one such neutron were detected experimentally, if it would have a penetrating power similar to, or about 10 times larger than, a γ-ray.

I admit that on a first look my way out might seem to be unlikely, since one would certainly have seen the neutrons by now if they existed. But nothing ventured nothing gained, and the seriousness of the matter with the continuous β-spectrum is illustrated by a quotation of my honored predecessor in office, Mr. Debey, who recently told me in Brussels: "Oh, it is best not to think about it, like the new taxes." Therefore one should earnestly discuss each way of salvation. – So, dear Radioactives, examine and judge it. – Unfortunately I cannot appear in Tübingen personally, since I am indispensable here in Zürich because of a ball on the night of 6/7 December. – With my best regards to you, and also to Mr. Back, your humble servant,

W. Pauli

You see how modest the numbers were that I still had in mind at that time. To tell the truth, the penetration power of these particles, which today are called neutrinos, is about 100 light-years of Pb instead of 10 cm; compared with the gamma rays the factor is 10^{16} to 10^{17} instead of 10, the rest mass and the magnetic moment theoretically are 0, and the experimental upper limits are 0.002 electron masses and 10^{-9} Bohr magnetons [COW 57a].

I soon received a reply to my letter from Geiger, who had discussed my question with the others in Tübingen, especially with Meitner. Unfortunately, I do not have this reply any more. I recall, however, that his answer was positive and encouraging: From the experimental point of view, my new particles would indeed be possible.

Because of the empirical nuclear masses, I had quickly abandoned the idea that the neutral particles emitted in beta decay were at the same time constituents of the nuclei.

In a talk I gave on the occasion of a meeting of the American Physical Society in Pasadena in June 1931, I reported for the first time on my idea of new, very penetrating neutral particles in beta decay. I no longer believed that they made up the building blocks of the nucleus and hence did not call them neutrons any more. In fact, I used no special name for them. The matter still seemed to me to be quite uncertain, however, and I did not have my talk printed. In the same year, 1931, I traveled from America to Rome, where a large international congress on nuclear physics was to take place in October. There I met Fermi, who immediately expressed a lively interest in my idea and a very positive attitude toward my new neutral particles, as well as Bohr, who on the contrary advocated his idea of the statistical conservation of energy in beta decay. A little later he published this idea in his Faraday lecture [BOH 32]. To give you an impression of his ideas at that time, I quote the following section ([BOH 32], p. 383).

At the present stage of atomic theory, however, we may say that we have
no argument, either empirical or theoretical, for upholding the energy
principle in the case of β-ray disintegrations, and are even led to compli-
cations and difficulties in trying to do so. Of course, a radical departure
from this principle would imply strange consequences, in case such a
process could be reversed. Indeed, if, in a collision process, an electron
could attach itself to a nucleus with loss of its mechanical individuality,
and subsequently be recreated as a β-ray, we should find that the energy
of this β-ray would generally differ from that of the original electron.
Still, just as the account of those aspects of atomic constitution essential
for the explanation of the ordinary physical and chemical properties of
matter implies a renunciation of the classical idea of causality, the features of atomic stability, still deeper-lying, responsible for the existence
and the properties of atomic nuclei, may force us to renounce the very
idea of energy balance. I shall not enter further into such speculations
and their possible bearing on the much debated question of the source of
stellar energy. I have touched upon them here mainly to emphasize that
in atomic theory, notwithstanding all the recent progress, we must still he
prepared for new surprises.

Concerning the more general possibility of surprises in those interactions that we
today call "weak," Bohr should maintain his point in another respect. However, his
idea that there was only a statistical conservation of energy in these interactions
seemed unacceptable to both Fermi and me. We had many private discussions on
this topic in Rome in 1931, and I saw no theoretical reason to consider the law of the
conservation of energy as less certain than, for example, the law of the conservation
of electric charge. From an empirical point of view, it seemed to me decisive, whether
the beta spectra of electrons showed a sharp upper limit or whether they showed a
Poisson distribution dropping off toward infinity. In the first case, in my opinion,
my idea of new particles would be established.[4] At that time the question was not yet
decided experimentally, but Ellis, who was also present in Rome, already had plans
to take this experimental problem up once more.

In the following year, Chadwick discovered the long-searched-for neutron with
charge number 0 and mass number 1 through the bombardment of lighter nuclei
with alpha particles. My new particle emitted in beta decay was thereupon called
neutrino by Fermi in talks in Rome, to distinguish it from the heavier neutron,[5] and
this Italian name was soon commonly adopted. Then the new idea about nuclear
structure rapidly took shape, with the nuclei consisting of protons and neutrons,
which we today call "nucleons," and which are both fermions with spin $\frac{1}{2}$. Various
authors came to this idea independently; in Italy it was advocated by Majorana, who
was supported by Fermi.

---

[4] For the theoretical interpretation of the upper limit of the spectrum, see also Ellis and Mott [ELL 33].
[5] I owe this information to Mr. E. Amaldi.

Thus at the Solvay meeting on atomic nuclei in Brussels in October 1933, where Joliot and Chadwick, among others, reported on their experimental discovery of positron decay and of the neutron and Heisenberg reported on the structure of the nucleus, a general clarification took place. Also, Fermi and Bohr were again present. It was now evident that, on the basis of this conception of nuclear structure, the neutrinos, as they were now called, had to be fermions in order to conserve statistics in beta decay. Furthermore, Ellis reported on new experiments carried out by his student W. J. Henderson [HEN 34], which established the sharp upper limit of the beta spectrum and consolidated its interpretation.

In view of the new circumstances, my earlier precaution of delaying publication now seemed to me unnecessary.

Following Heisenberg's lecture, I communicated my ideas on the neutrino (as it now was called) in the discussion, which also was printed in the report of the conference [PAU 34] and is reproduced here:

> The difficulty connected with the existence of the continuous spectra of beta rays arises, as one knows, from the fact that the mean lifetimes of the nuclei that emit these rays and also of the resulting daughter nuclei, have well determined values. Thus one necessarily concludes that the state, as well as the energy and the mass of the nucleus, which is left over after the expulsion of the β-particles, are also well determined. I do not want to elaborate on the efforts one could use to avoid this conclusion, but I think in accordance with Mr. Bohr, that one will always encounter unsurmountable difficulties in the explanation of the experimental facts.
>
> In the context of these ideas, two interpretations of the experiments are suggested. The one that is defended by Mr. Bohr admits that the laws of energy and momentum conservation are violated if one deals with a nuclear process where light particles play an essential role. This hypothesis seems to me unsatisfactory, not even plausible. First, the electric charge is conserved in the process and I do not see why the conservation of charge should be more fundamental than the conservation of energy and momentum. Furthermore, it is precisely the kinematic relations that govern various properties of the β-spectra (the existence of an upper limit and the connection to the γ-spectra, Heisenberg's criterion of stability). If the conservation laws should not hold, one would obviously have to conclude from these relations that β-decay is always accompanied by a loss and never by a gain in energy; this conclusion implies an irreversibility of this process with respect to time, which seems to me not to be acceptable at all.
>
> In June 1931 on the occasion of a conference in Pasadena I proposed the following interpretation: The conservation laws remain valid, since the emission of the β-particles is accompanied by a very penetrating radiation of neutral particles, which has not been observed up to now. The sum of the energies of the β-particle and the neutral particle (or the neutral particles, since one does not know whether there is only one or

whether there are several), which are emitted by the nucleus in a single process equals the energy which corresponds to the upper limit of the β-spectra. It goes without saying that we admit for all elementary processes not only the conservation of energy but also the conservation of momentum, of angular momentum and of the type of statistics.

As for the properties of these neutral particles, the atomic weights of the radioactive elements in particular teach us that their mass cannot exceed the mass of the electron by a lot. To distinguish them from the heavy neutrons Mr. Fermi has suggested the name "neutrino." It is possible that the rest mass of the neutrinos equals zero, so that they have to propagate, like the photons, with the speed of light. In any case their penetrating power exceeds many times that of photons of the same energy. It seems to me admissible that the neutrinos have spin $\frac{1}{2}$ and that they obey Fermi statistics, even though experience does not provide us with any direct proof of this hypothesis. We do not know anything about the interaction of the neutrinos with other matter particles and with photons: The hypothesis that they possess a magnetic moment, as I have proposed earlier (Dirac's theory foresees the possibility of the existence of neutral magnetic particles), does not seem to me established at all.

In connection with these ideas, the experimental study of the momentum balance in β-decays is a problem of utmost importance; one can predict that the difficulties will be great because of the smallness of the recoil energy of the nucleus.

The difficulty with recoil measurements referred to above was not overcome until quite recently.

Subsequently, Chadwick reported on the first unsuccessful efforts to experimentally detect an absorption of neutrinos, which yielded an upper limit on the magnetic moment of the neutrino of 0.001 magnetons. Bohr's opposition had weakened considerably since his Faraday lecture. Having become very cautious about claiming the invalidity of the conservation of energy, he restricted himself to his much more general statement that nobody knew which surprises still were in store for us in this field. By the way, only as late as 1936 [BOH 36] he accepted entirely the validity of the conservation of energy in beta decay and the neutrino, even though Fermi's theory had already been successfully developed by then.

3 Formulation of a theory of beta decay

Soon afterward, stimulated by the discussions at the Solvay conference, Fermi developed his theory of beta decay [FER 33, 34]. Part of Fermi's conclusions concerning the shape of the beta spectrum and the inference about the rest mass of the neutrino were drawn at the same time and independently by F. Perrin [PER 33], who was also present at the Solvay conference. For this, a complete theory of the interaction is not necessary if one restricts oneself to the so-called allowed transitions, where the nonrelativistic approximation for the nucleons in the nucleus is sufficient. Apart from corrections, which only become important for larger nuclear charges due to the Coulomb interaction between the nucleus and the electron, the shape of the beta spectrum for these transitions is entirely determined by the statistical weight factor $\rho(E_e)$ of the density of states in phase space. This factor, depending very sensitively on the value of the rest mass m_ν of the neutrinos, is given by

$$\rho(E_e)\, dE_e = p_e^2\, dp_e\, p_\nu^2\, \frac{dp_\nu}{dE_\nu} = p_e E_e p_\nu E_\nu\, dE_e. \tag{1}$$

Here, the natural units $\hbar = c = 1$ are adopted, the indices e, ν refer to electron and neutrino, respectively, and the energy E is related to the momentum through the relation $E^2 = p^2 + m^2$, such that $dE/dp = p/E$.

If ΔE is the energy difference of the nucleus in the initial and final state of the decay, the law of energy conservation requires

$$E_\nu = \Delta E - E_e. \tag{2}$$

Since m_ν is the minimum energy of the neutrino, the upper limit E_0 of the electron energy of the spectrum is

$$E_0 = \Delta E - m_\nu. \tag{3}$$

Thus,

$$E_\nu = E_0 - E_e + m_\nu \tag{4}$$

and

$$\rho(E_e)\, dE_e = p_e E_e (E_0 - E_e + m_\nu) \sqrt{(E_0 - E_e)(E_0 - E_e + 2m_\nu)}\, dE_e. \tag{5}$$

In the case $m_\nu \neq 0$, the behavior of (5) in the vicinity of the upper limit E_0, namely, for $E_0 - E_e \ll m_\nu$, is completely different from the behavior for $m_\nu = 0$; that is,

$$\rho(E_e)\, dE_e = p_e E_e (E_0 - E_e)^2\, dE_e, \quad \text{for } m_\nu = 0. \tag{6}$$

In comparison with the empirical shape of the spectrum, Fermi and Perrin had already inferred $m_\nu = 0$ in 1933.

In accordance with the same principles, the most precise estimate of the upper limit on the rest mass of the neutrino m_ν is derived from the precise measurements of the beta spectra of tritium (H_3) by L. M. Langer and R. J. D. Moffat [LAN 52].[6] The result is found in the discussions of L. Friedman and Smith [FRI 58a], J. J. Sakurai

---

[6] Besides the statistical factor of ρ an additional correction had to be taken into account here for $m_\nu \neq 0$, which was noted for the first time by J. R. Pruett [PRU 48]. The correction depends on a factor, which in general can lie between -1 and $+1$. For the general expression of this factor, see E. P. Enz [ENZ 57]. For the type of interaction assumed today, however, this factor is equal to zero.

[SAK 58a], and L. Friedman [FRI 58b]

$$m_\nu < 250 \, \text{eV} = 0.002 m_e.$$

Thus in what follows we always assume $m_\nu = 0$.

The Kurie plot of allowed transitions shows that (besides a factor $F(Z, E_e)$, i.e., the Coulomb correction) the statistical density $\rho(E_e)$ alone determines the shape of the beta spectrum. The experimental technique had to be refined before this result could be established.[7] In the Kurie plot,

$$\sqrt{N(E_e)/F(Z, E_e) p_e E_e} = K(x) \tag{7}$$

is plotted as a function of

$$x = (E_e - m_e)/(\Delta E - m_e) \tag{8}$$

where $N(E_e) \, dE_e$ is the number of electrons emitted per second and integrated over all directions.

For $m_\nu = 0$ the theory yields

$$K(x) = 1 - x. \tag{9}$$

Figure 2 shows a typical example of the linear character of the Kurie plot. On the basis of Fermi's theory of beta decay from 1933 and its generalizations, further conclusions can be drawn from the empirical result that, for allowed transitions, already the statistical weight factor alone determines the shape of the beta spectrum. Fermi had devoted all of his attention to the formalism of quantum electrodynamics developed by Heisenberg and myself, where the fields are represented as sums of space-time-dependent creation and absorption operators, and soon had reformulated them more elegantly in his own contributions. Immediately after the congress in Brussels, he began to develop a theory of beta decay as an example of an application of these field quantization methods in as close connection to quantum electrodynamics as possible. For the energy of the interaction per cm^3, he thus made the ansatz of a sum of products of the components of *four* different spinor fields (corresponding to two nucleons and two leptons, respectively) at the same space-time point. It is possible that this *local character of the Fermi interaction* will have to be refined later, but in any case it has proved to be an extremely good approximation. The entire expression describing the density of the interaction energy has to be a relativistic invariant, which, moreover, strictly obeys the law of conservation of electric charge. There are five typical possibilities, depending on whether the scalar products used are of two scalars (S), two pseudoscalars (P), two

---

[7] An example for a forbidden transition is the beta decay of RaE, which has played such an important role in the history of the interpretation of the continuous electron spectrum. The shape of the RaE spectrum is determined not only by the factor ρ, the density of states, which even today still makes an interesting object for study [BUH 58a,b].

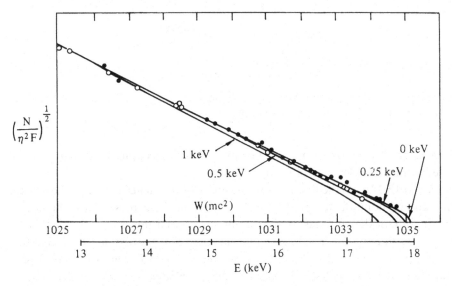

Fig. 2 Kurie plot of the tritium spectrum.

vectors (V), two pseudo- or axialvectors (A), or two antisymmetric tensors (T). By analogy with quantum electrodynamics, Fermi chose the V type in particular.

Initially, each of these types seemed to result in only *one* constant. However, this is based on special assumptions. One of them, as illustrated in the next paragraph, is the conservation of a leptonic charge, which up to now has withstood all the tests. The other one is the assumption of an invariance under spatial reflection and unchanged electrical charge ("parity"). In the last paragraph we will see that, surprisingly, this assumption did *not* prove to be correct. Thus, in the case of the "Fermi interaction," the most general expression that corresponds to the five types contains 10 arbitrary constants. However, in nature one special case is realized (see Section 5), so that finally only *one* quotient of coupling constants still remained undefined.

For the following discussion, we note, first, that in the nonrelativistic approximation the pseudoscalar type P makes no contribution to nucleons. To obtain information about the type P, it is necessary to consider "forbidden" transitions, for which this nonrelativistic approximation vanishes, while here we confine ourselves to "allowed" beta decay transitions in the nonrelativistic approximation and consequently omit the case P.

According to the selection rules for the angular momentum J of the nuclei, these transitions divide into two classes:

$$S, V \quad \begin{array}{c} \Delta J = 0 \\ (0 \rightarrow 0 \text{ allowed}) \end{array} \qquad \text{Fermi (F)} \tag{10a}$$

$$T, A \quad \begin{array}{c} \Delta J = 0, \pm 1 \\ (0 \rightarrow 0 \text{ forbidden}) \end{array} \qquad \text{Gamow–Teller (GT).} \tag{10b}$$

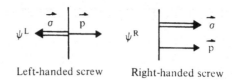

Left-handed screw   Right-handed screw

Fig. 3   Relative direction of spin σ and momentum p, for states characterized by ψ^L and ψ^R of a free Dirac particle with zero rest mass.

There are both pure Fermi and pure Gamow–Teller transitions, while in the general case both matrix elements differ from zero.

Fierz [FIE 37] was the first to draw the important conclusion that in the general case an additional factor of $(1 \pm bm_e/E_e)$ arises in the expression for the energy distribution of the beta spectrum, and, moreover, that this is only the case where S, V or T, A are mixed. The linearity of the Kurie plot showed, however, that to a good approximation these "Fierzterms" should be zero. This leads to the conclusion *that cases S and V and cases T and A cannot both be present at the same time.*[8]

B. Stech and J. H. D. Jensen [STE 55] have related this result to a formal transformation property of the density of the interaction energy, which proved to be successful and suitable for generalization when parity violation was later discovered. To illustrate this, we have to introduce the 4×4 matrix denoted by γ_5. This matrix has two eigenvalues $+1$ and two eigenvalues -1, such that

$$(1 + \gamma_5)/2 \equiv a^L, \quad (1 - \gamma_5)/2 \equiv a^R \tag{11}$$

are projection operators. The letters L and R refer to left and right and justify themselves by the fact that the corresponding spinor components

$$\psi^L = a^L \psi; \quad \psi^R = a^R \psi \tag{12}$$

refer to states with spin σ and momentum p (i.e., direction of motion) either antiparallel or parallel (Fig. 3).

These states are identical to the stationary states of a free particle only in the case of a particle with rest mass 0, like the neutrino, while for the electron the mass term in the Dirac equation couples the L and R components. However, for electron energies that are large compared to their rest mass[9] one can still talk more or less about L and R states in the case of a free particle.

The original "Stech–Jensen transformation" now corresponds to the fact that one has to multiply the L-component of the electron and of the neutrino at the same time by $+1$ and the R-component by -1, which according to (11), (12) is equivalent to

$$\psi' = \gamma_5 \psi. \tag{13}$$

---

[8] In this form the conclusion is correct only if invariance with respect to time reversal holds, which seems to be fulfilled in nature.
[9] We always use the natural units $\hbar = c = 1$.

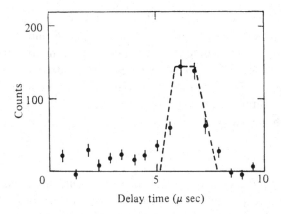

Fig. 4   Time of flight of the recoil Cl atoms from reaction (15).

The five interaction types than divide into two classes:

$$S, T, P \quad \text{and} \quad V, A \tag{14a,b}$$

where, according to the transformation under discussion, *the first one is multiplied by* -1, *the second one by* $+1$. Stech and Jensen postulated that after the transformation the entire density of the interaction energy should show either one or the other of the two properties, which excludes *a mixing of both classes*. At the same time, this guarantees the vanishing of the "Fierzterms" and has also proved successful.

As a further theoretical conclusion derived from the linearity of the Kurie plot, note that S. Kusaka [KUS 41] could exclude a value of $\frac{3}{2}$ for the spin of the neutrino and thus establish the value of $\frac{1}{2}$ for the neutrino spin assumed by Fermi.

In addition, the validity of momentum conservation in the emission of neutrinos could be experimentally verified apart from the deeper problem concerning the type of interaction. Particularly transparent is the experimental setup of G. W. Rodeback and J. S. Allen [ROD 52] [see reprint in this chapter], which uses the K-capture reaction in ^{37}Ar:

$$^{37}\text{Ar} + e^- \rightarrow {}^{37}\text{Cl} + \nu \ (K\text{-capture}). \tag{15}$$

The recoil of the Auger electrons can be neglected, so that the momentum of the neutrinos manifests itself only in the recoil of the Cl atoms. Experimentally, their recoil energy has to be determined from their time of flight and agrees with the calculated value under the assumption $m_\nu = 0$ (compare also the test of momentum conservation in the usual beta decay by C. W. Sherwin [SHE 51]; see Fig. 4).

4 Experimental detection of the absorption of free neutrinos:
Conservation of a leptonic charge, neutrino, and antineutrino

Despite the success of Fermi's theory of the interaction underlying beta decay and based on the neutrino hypothesis, various physicists still did not believe the neutrino itself was real. The absorption of the free neutrino was still missing. Detection became feasible when it was suggested that uranium reactors be used as neutrino sources, which emit on the order of 10^{20} neutrinos per second. For the following discussion it is important to note that, practically speaking, one has to deal here exclusively with negatron (e^-) decay, whereas for positron decay (e^+) no such sources are available. According to today's convention, the neutrino accompanying the e^- is called antineutrino with the symbol $\bar{\nu}$. The negatron decay then corresponds to the reaction

$$n \rightarrow p + e^- + \bar{\nu} \tag{16}$$

which takes place for free neutrons and for neutrons bound in nuclei. It follows that, theoretically, the absorption of an antineutrino by a proton is possible, which turns into a neutron and a positron. This process is derived from

$$p + \bar{\nu} \rightarrow n + e^+ \tag{17}$$

the common e^- decay (16), by inverting a subprocess, where instead of e^- absorption, the emission of e^+ occurs now.

The enormous technical difficulties in the experimental detection of this reaction arising from the smallness of the reaction cross section were finally overcome by F. Reines and C. L. Cowan, Jr. [COW 56a; REI 56]. A "giant amplifier" had to be built, which made it possible to detect the neutrons and positrons created according to (17) by the absorption of antineutrinos emitted from the uranium reactor. After a finite time of flight, the neutrons were absorbed by Cd nuclei, followed by the emission of gamma radiation, while the positrons became visible as gamma radiation following annihilation with negatrons. A delayed coincidence allowed detection of both gamma rays. Figure 5 illustrates the experimental setup. In the first publication of the experimental result (1956), the measured cross section was quoted as

$$Q = 6.3 \times 10^{-44}\,\text{cm}^2 \pm 25\% \text{ per neutrino}$$

resulting in 2.88 ± 0.22 collisions per hour, owing to the large neutrino flux provided by the reactor.

To compare the measured absorption cross section with the theoretical one, the energy spectrum of the electrons emitted in the fission processes will have to be determined experimentally. Moreover, the theoretical value for the absorption (assuming an empirically given emission probability) still depends on the assumption of the special type of the interaction, owing to a factor that can take a value between 1 and 2. This was discussed more precisely by C.P. Enz [ENZ 58]. The value of the absorption cross section recently published by Reines and Cowan [REI 58; CAR 58],

$$Q = (6.7 \pm 1.5) \times 10^{-43}\,\text{cm}^2 \text{ per fission}$$

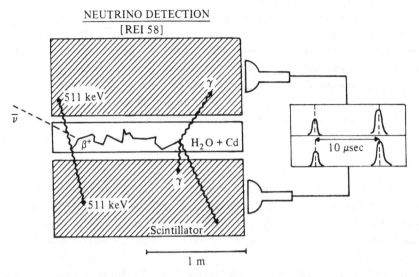

NEUTRINO DETECTION
[REI 58]

511 keV

$\bar{\nu}$

β^+

γ

H₂O + Cd

511 keV

γ

Scintillator

10 μsec

1 m

Fig. 5 Experimental setup for the detection of antineutrinos according to [REI 58]

(using the measured electron spectrum from the fission products) agrees with the theoretical value following from the presently adopted two-component model of the neutrino (see Section 5)

$$Q = (6.0 \pm 1) \times 10^{-43} \text{ cm}^2 \text{ per fission.}$$

We have already encountered the question of whether there are two mirror-reflected images of the neutrino, namely, besides the antineutrino $\bar{\nu}$, which is emitted together with the e^-, the neutrino ν which is emitted in the case of bound protons together with the e^+ according to

$$p \rightarrow n + e^+ + \nu. \tag{16a}$$

Following this idea, the reaction (17) should be impossible with ν instead of $\bar{\nu}$. However, this cannot be tested experimentally in this way since no reactors with positron emission exist as neutrino sources. But one can consider the inverse process to the reaction (15)

$$^{37}\text{Cl} + \bar{\nu} \rightarrow ^{37}\text{Ar} + e^- \tag{18}$$

by taking antineutrinos $\bar{\nu}$ instead of ν, which corresponds to

$$n + \bar{\nu} \rightarrow p + e^- \tag{18a}$$

that is, to neutrons bound in the ^{37}Cl nucleus. This reaction should be impossible if the idea of the two mirror-reflected images of the neutrino holds.

One can formulate this in a more transparent way with the help of a *"leptonic charge," the sum of which must be conserved in all possible reactions.* Although it has nothing to do with the electromagnetic charge, the leptonic charge, like the electromagnetic one, can assume both signs. A common sign for the leptonic charge for all leptons is purely conventional, whereas the sign of the ratio of the leptonic and electromagnetic charge has to be determined experimentally for different particles. For example, it is not a matter of convention whether the muon (μ-meson) μ^+ and e^+; and likewise μ^- and e^- have the same or the opposite leptonic charge. The values of the leptonic charge for e, μ, and ν now adopted are

$$+1 \text{ for } \mu^-, e^-, \nu, \quad -1 \text{ for } \mu^+, e^+, \bar{\nu}. \tag{19}$$

We shall come back to the muon once more. For heavy particles (baryons) like n and p and for bosons like π, the leptonic charge shall be equal to zero. One sees that this assignment *together with the assumption of a conservation law for leptonic charge allows the reactions (16), (16a), (17) and forbids the reactions (18) or (18a)*.

Reaction (18) was investigated by R. Davis [DAV 55, 56] with a negative result and an upper limit of $0.9 \times 10^{-45} \text{cm}^2$ for the reaction cross section. The experimental precision, which was limited because of the cosmic ray background, is not very relevant for theoretical purposes. In any case, the largest value theoretically possible for this cross section is $2.6 \times 10^{-45} \text{cm}^2$, and one can also imagine theories where this maximal value is multiplied by some factor between 0 and 1.

The combination of the reactions (18a) and (16) would give rise to the *emission of two electrons e without the emission of neutrinos* and the simultaneous transformation of two neutrons into two protons. This reaction, obviously violating the conservation of leptonic charge and known as "double beta decay," has often been looked for in vain. The most precise negative result known corresponds to the absence of a transition $^{150}\text{Nd} (Z = 60) \rightarrow {}^{150}\text{Sm} (Z = 62)$ [COW 56b, 57b]. The lifetime of Nd turned out to be greater than 4.4×10^{18} years. In this case, however, the theoretical estimate is uncertain, since unknown matrix elements enter the picture. The largest acceptable value for the half-life of Nd is 4×10^{15} years, which could, however, be pushed up to 1.9×10^{18} years.

To summarize, we can say that the quantitative empirical confirmation of a fundamental law such as the law of the conservation of leptonic charge indeed leaves much to be desired. On the other hand, all known experiments are in agreement with the assumption of this conservation law. Thus in what follows we take the latter as granted.

5 Violation of parity: Law of weak interaction

In the wake of a critical discussion of the mirror symmetry in weak interactions two years ago, a new development has emerged in this new field of physics, according to which the neutrino properties treated in this lecture only constitute a special case. The so-called θ-τ puzzle of the K-meson decay stimulated T. D. Lee and C. N. Yang

[LEE 56] to investigate more closely the empirical evidence, in weak interactions, of the validity of both charge symmetry *C* (generally *C* is understood to exchange particles and antiparticles) and the spatial reflection *P* (derived from parity; where by definition the sign of the charge does not change). They found the existing evidence unsatisfactory and specified the experiments required to check it. To the great surprise of many physicists, among whom I also count myself, the first performance of some of these experiments gave the result published in January 1957, that in beta decay [WU 57] (aligned nuclear spins of ^{60}Co), as well as in the creation and decay of μ-mesons [GAR 57] (cyclotron), [FRI 57] (photographic plates), the symmetry operations *C* and *P* *cannot* be fulfilled individually. Concerning the principal importance of symmetry questions, I can refer to another paper of mine, where the three categories of interactions – strong, electromagnetic, and weak – are also illustrated in more detail ([PAU 58]; there is also further literature on experiments until the end of 1957.) Therefore I will only make the following brief remarks here. Besides the symmetry operations *C* and *P*, there still exists the time reversal *T* (by definition, without a change of sign of charge). The so-called *CPT* theorem guarantees that, under the very general assumptions of invariance with respect to the continuous Lorenz group, invariance with respect to the product of the three discrete operations *C*, *P*, and *T* (in any sequence) can already be inferred. Moreover, the experiments performed up to now (i.e., September 1958 – [GOL 58a]) have proved *compatible with the symmetry T or with the equivalent symmetry operation CP.*

At this point I feel it is appropriate to relate Bohr's warning mentioned earlier concerning the separate violation of the symmetries *C* or *P*, namely, that one has "to be prepared for surprises" in the weak interactions (as they are called today). While his special, and later abandoned, idea of violation of the law of energy conservation in these interactions would have affected the *continuous* group of space and time translations (contained in the inhomogeneous Lorentz group), our real surprise refers to the reduction of symmetry in the *discrete* groups of reflections in weak interactions. This surprise would certainly not have turned up if *all* laws of nature would *only* show the weaker symmetry *CP* or *T*. Therefore, one can also say that the problem is to understand *why the strong and the electromagnetic interactions individually show the higher symmetries C or P.* This problem is still unsolved. Although in the case of electromagnetism the higher mirror symmetry could be related to the special type of interaction, the situation is more difficult in the strong interactions. Moreover, here the question arises which has to be decided empirically, whether this higher mirror symmetry is really present in *all* strong interactions or only in pion–nucleon and nucleon–nucleon interactions. The answer to these questions must be left to the future.

The weaker mirror symmetry of the weak interaction is not restricted to the neutrino and thus cannot be attributed to special neutrino properties alone. For example, it is established for certain in the decay of the neutral hyperon Λ^0 into a proton *p* and a negative pion π^-. For the neutrino there exists a particular possibility, which was already mentioned in Section 3: the so-called *two-component model*. According to this model, *only the two R components* or *only the two L components* shall be present in nature.[10] Following the first work of Lee and Yang, different authors independently proposed [SAL 57; LEE 57; LAN 57] to apply this model to the neutrino. Indeed, then the free neutrino already possesses only the mirror symmetry, here denoted with *CP* or *T*, where spatial reflections (reversal of the direction of motion with respect to the spin direction) are at the same time connected to transitions from neutrino to antineutrino. *Up to now, this two-component model of the neutrino has proved successful in all experimental results.* For some time, I looked at this particular model with a certain skepticism [SAL 57; LEE 57; LAN 57], since it seemed to me that the special role of the neutrino was emphasized too strongly. It turned out, however, that by further developing the ideas of Stech and Jensen (see Section 3), the model allowed an interesting generalization for the form of the interaction energy for all weak interactions. Initially the experiments studying electron polarization in beta decay, as well as the angular distribution of electrons for polarized nuclear spin, were compatible with the following alternative: either one only has *an A and V interaction together with an L model of the neutrinos* or *only an S and T interaction together with an R model of the neutrino.*

From the two-component model one could first decide among the two possibilities in μ-decay

$$\mu \rightarrow e + \nu + \nu \text{ (or } \mu \rightarrow e + \bar{\nu} + \bar{\nu}) \quad \text{and} \quad \mu \rightarrow e + \nu + \bar{\nu} \qquad (20)$$

in favor of the second possibility. Only for this type does the shape of the electron spectrum (so-called Michel parameter, $\rho = \frac{3}{4}$) agree with experience. Furthermore, the measurement of the spatial direction of flight of the electrons with respect to the direction of motion of the μ-mesons created in the process

$$\pi^+ \rightarrow \mu^+ + \nu \quad \text{or} \quad \pi^- \rightarrow \mu^- + \bar{\nu} \qquad (21)$$

demonstrated that, according to the two-component model, the only *interactions* left over for the process (20) had to be of the *type V and A* and of the same strength. Concerning the μ-meson, we note in addition that the weak interaction also has to act between (p,n) and (μ,ν), as is shown by the capture of μ-mesons of nuclei.

For a long time, the search for the reaction

$$\pi^+ \rightarrow e^+ + \nu \quad \text{or} \quad \pi^- \rightarrow e^- + \bar{\nu} \qquad (22)$$

[10] For particles with rest mass 0 this "two-component" theory was mentioned for the first time by H. Weyl [WEY 29]. In my article, "Prinzipien der Wellenmechanik" ("Principle of Wave Mechanic") [PAU 33], see especially p. 226, this theory is critically discussed. This took place *before* the formation of Dirac's hole theory, so that the mirror symmetries of the model (*CP* or *T*) remained unnoticed in the transition from particle to antiparticle.

has remained fruitless and it was not until recently that its existence was successfully proven [FAZ 58; IMP 58]. On older negative experiments and the theoretical estimation, see, for example, [LOK 55] and [AND 57]. However, it is still somewhat premature to address the quantitative question of the relative abundances of the two reactions (22) and (21) by a comparison of experiment and theory. The order of magnitude of the cross-sectional ratio of electron to meson decay modes of the pion is 10^{-5} to 10^{-4}. It was more difficult to decide between the alternatives S, T versus V, A interaction in beta decay. Here the long delay has been due to wrongly analyzed recoil measurements in ^{6}He. The first *correct hint pointing to the alternative* (V, A) was obtained from the angular correlation of electron and neutrino [HER 57] investigated in ^{37}Ar recoil experiments. Furthermore, in agreement with this was the outcome of an elegant experiment performed by M. Goldhaber, L. Grodzins, and A. W. Sunyar [GOL 58b], which by observing the sense of circular polarization of gamma rays emitted from the inner atomic shells after electron capture directly allowed them to infer the helicity of the emitted neutrinos by means of resonant scattering of the gamma rays on daughter nuclei. *The experiment on* ^{152}*Eu gave an L neutrino.* Together with the results from other experiments already mentioned, this corresponds to the *alternative* (V, A). Further confirmations [GOL 58a] soon followed (from new recoil experiments on ^{6}He, among others), so that the (V, A) alternative can now be regarded as well established.

On the basis of the Stech–Jensen transformation and the two-component model of the neutrino, the following postulate suggests itself for the theoretical interpretation: *The Hamiltonian of each weak four-fermion interaction shall "universally" contain either only R or only L components of the involved fermions.*[11] Equivalent to this postulate is the formulation that in the transformation $\psi' = \gamma_5\psi$ the density of the interaction energy for each particle separately should "universally" remain unchanged or change its sign.[12] The Stech–Jensen transformation referred to a pair of particles simultaneously, while the two-component model of the neutrino is equivalent to the validity of the result of the transformation for the neutrino alone. The *postulate of the extended Stech–Jensen transformation now under discussion is therefore a generalization of the two-component model of the neutrino.* As can easily be seen, this postulate leads to the only possible law of interaction (which is automatically *CP* and *T* invariant):

$$[\bar{\psi}_1\gamma_\mu(1 \pm \gamma_5)\psi_2][\bar{\psi}_3\gamma_\mu(1 \pm \gamma_5)\psi_4] \equiv [\bar{\psi}_1\gamma_\mu(1 \pm \gamma_5)\psi_4][\bar{\psi}_3\gamma_\mu(1 \pm \gamma_5)\psi_2]. \quad (23)$$

The identity of both expressions is a purely algebraic one. The sign of γ_5 has to be the same "universally." Its choice depends on the convention of what is considered a particle and an antiparticle. Here, as usual, $\bar{\psi} = \psi^*\gamma_4$ and ψ^* denotes the hermitean

---

[11] This should also include the notion that the Hamiltonian should not explicitly contain the derivations of these spinor components. For particles with nonvanishing rest mass, the R components can be expressed by the first derivatives of the L components, and vice versa.

[12] This postulate or an equivalent one was proposed independently by various authors [SUD 58; SAK 58b; FEY 58].

conjugate operator of ψ. The coupling constant is not explicitly written in (23). The postulate in the form used here does *not* require the equality of the coupling constants for the interaction of different particles. The postulate of the "universal" weak R and L interaction would in general *require the equality of the strength of the V and A interaction. In this form, however, this is not empirically correct for nucleons in beta decay.* The empirical result can now be summarized as follows.[13] The interaction for beta decay is

$$\frac{1}{\sqrt{2}}C[\bar{p}\gamma_\mu(1+\lambda\gamma_5)n][\bar{e}\gamma_\mu(1+\gamma_5)\nu] + \text{herm. conj.} \qquad (24)$$

In this more general ansatz of the interaction, both CP and T invariance are equivalent to the statement that the constant λ is real, which is well supported experimentally [GOL 58a]. The numerical values of the constants are

$$\lambda = 1.25 \pm 0.04 \quad C = (1.410 \pm 0.009)10^{-49}\, \text{erg cm}^{-3}.$$

In μ-meson decay, the ansatz

$$\frac{1}{\sqrt{2}}C[\nu\gamma_\mu(1+\gamma_5)\mu][e\gamma_\mu(1+\gamma_5)\nu] + \text{herm. conj.}$$

$$\equiv \frac{1}{\sqrt{2}}C[\nu\gamma_\mu(1+\gamma_5)\nu][e\gamma_\mu(1+\gamma_5)\mu] + \text{herm. conj.} \qquad (25)$$

is sufficient. *The two constants C in nucleon and muon decay are empirically equal to good approximation.*

To interpret the deviation of the constant λ from 1, Feynman and Gell-Mann[14] have suggested an interesting hypothesis. The term $[\bar{p}\gamma_\mu(1+\lambda\gamma_5)n]$ in (24) shall be replaced by the corresponding component of the total isospin current, including the π-meson, such that the law of interaction now reads

$$C\left\{\frac{1}{\sqrt{2}}(\bar{p}\gamma_\mu(1+\gamma_5)n) - \left(\pi_0\frac{\partial\pi^*}{\partial x_\mu} - \pi^*\frac{\partial\pi_0}{\partial x_\mu}\right)\right\}$$

$$* [\bar{e}\gamma_\mu(1+\gamma_5)\nu] + \text{herm. conj.} \qquad (24a)$$

and that the postulate of the "universal" weak L interaction is restored. Here the field $\pi_0(x)$ corresponds to the neutral, the (complex) field $\pi(x)$ to the charged π-meson. For an explanation of λ, the concept of "renormalization of coupling constants" is applied. The conservation of the total isospin in (strong) pion–nucleon coupling takes care that only the coupling constant of the axial (A) part of the interaction is modified while the V coupling constant remains unchanged.

---

[13] See [GOL 58a]. Here, especially, a new measurement of the half-life of the free neutron of 11.7 ± 0.3 min has been made use of. It was undertaken by the Russian authors A. N. Sosnovskij, P. E. Spivak, Yu. A. Prokofiev, I. E. Kutikov, and Yu. P. Dobrynin.

[14] See [SUD 58; SAK 58b; FEY 58]. Compare also [GER 56] and further [GEL 58], where possible experimental tests of the new ansatz are discussed.

One must add that only a calculation of λ from further empirical data of pion–nucleon interaction would transform the still incomplete formal scheme of "renormalization" into a real theory. For the moment, however, such a calculation does not exist. The proposed direct interaction of pions with electrons and neutrinos gives rise to the possibility of experimental verification that one has to wait for.

We have followed the history of the neutrino along part of its way and we have seen how the original ideas and conceptions were later justified. Now it seems that a point has been reached where the physics of the neutrino joins the more general field of elementary particle physics. Today each of these particles is still described by its own field and each type of interaction by its own coupling constant. What, for example, is the meaning of the small numerical value of the constants of the Fermi interaction having the dimensions of a cross section, when compared with other atomic cross sections? The next step, namely, to overcome the phenomenological physics of individual fields and coupling constants, in favor of a unified conception, is supposedly much more difficult than everything else accomplished up to now.

References

[AND 57]   H. A. Anderson and C. M. G. Lattes, *Nuovo Cimento* 6 (1957) 1356.
[BOH 32]   N. Bohr, Faraday Lecture, Chemistry and the Quantum Theory of Atomic Constitution, *J. Chem. Soc.* (London) (1932) 349–84.
[BOH 36]   N. Bohr, *Nature* 138 (1936) 25.
[BUH 58a] W. Bühring and J. Heintze, *Phys. Rev. Lett.* 1 (1958) 177.
[BUH 58b] W. Bühring and J. Heintze, *Z. Physik.* 153 (1958) 237.
[CAR 58]   R. E. Carter, F. Reines, J. J. Wagner, and W. E. Wyman, *Phys. Rev.* 113 (1958) 280.
[CHA 14]   J. Chadwick, *Verh. d. deutschen Phys. Ges.* 16 (1914) 383.
[CLA 21]   J. L. Classon, *Phil. Mag.* 42 (1921) 596.
[COW 56a] C. L. Cowan, Jr., F. Reines, F. B. Harrison, H. Kruse, and A. D. Guire, *Science* 124 (1956) 103.
[COW 56b] C. L. Cowan, Jr., F. B. Harrison, L. M. Langer, and F. Reines, *Nuovo Cimento* 3 (1956) 649.
[COW 57a] C. L. Cowan, Jr. and F. Reines, *Phys. Rev.* 107 (1957) 528.
[COW 57b] C. L. Cowan, Jr., F. B. Harrison, L. M. Langer, and F. Reines, *Phys. Rev.* 106 (1957) 825 (L).
[DAV 55]   R. Davis, *Phys. Rev.* 97 (1955) 766.
[DAV 56]   R. Davis, *Bull. American Phys. Soc.*, Washington Meeting (1956), 219.
[ELL 22a] C. D. Ellis, *Proc. Roy. Soc.* (A), 101 (1922) 1.
[ELL 22b] C. D. Ellis, *Z. Physik.* 10 (1922) 303.
[ELL 27]   C. D. Ellis and W. A. Wooster, *Proc. Roy. Soc.* (A) 117 (1927) 109.
[ELL 33]   C. D. Ellis and N. E. Mott, *Proc. Roy. Soc.* (A) 141 (1933) 502.
[ENZ 57]   C. P. Enz, *Nuovo Cimento* 6 (1957) 250.
[ENZ 58]   C. P. Enz, *Helv. Phys. Acta* 31 (1958) 69.
[FAZ 58]   T. Fazzini, G. Fidecaro, A. W. Merrison, H. Paul, and A. V. Tollestrup, *Phys. Rev. Lett.* 1 (1958) 247.
[FER 33]   E. Fermi, *Ricercha Scient.* 2 (1933) 12.
[FER 34]   E. Fermi, *Z. Physik.* 88 (1934) 161.
[FEY 58]   R. Feynman and M. Gell-Mann, *Phys. Rev.* 109 (1958) 193.
[FIE 37]   M. Fierz, *Z. Physik.* 104 (1937) 553.

[FRI 57]   J. L. Friedman and V. L. Telegdi, *Phys. Rev.* 105 (1957) 1681.
[FRI 58a]   L. Friedman and L. G. Smith, *Phys. Rev.* 109 (1958) 2214.
[FRI 58b]   L. Friedman, *Phys. Rev. Lett*, 1 (1958) 101.
[GAR 57]   R. L. Garwin, L. M. Ledermann, and M. Weinrich, *Phys. Rev.* 105 (1957) 1415.
[GEL 58]   M. Gell-Mann, *Phys. Rev.* 111 (1958) 362.
[GER 56]   S. S. Gershtein and Ia. B. Zel'dovich, *JETP* 2 (1956) 576.
[GOL 58a]   M. Goldhaber, in *Proc. Eighth Int'l. Conference on High-Energy Physics*, Geneva (1958) 233.
[GOL 58b]   M. Goldhaber, L. Grodzin, and A. W. Sunyar, *Phys. Rev.* 109 (1958) 1015.
[HEI 29]   W. Heitler and G. Herzberg, *Naturw.* 17 (1929) 673.
[HEN 34]   W. J. Henderson, *Proc. Roy. Soc.* (A) 147 (1934) 572.
[HER 57]   W. B. Herrmannsfeld, D. R. Maxson, P. Stähelin, and J. S. Allen, *Phys. Rev.* 107 (1957) 641 (L).
[IMP 58]   G. Impeduglia, P. Plano, A. Prodell, N. Samios, M. Schwartz, and J. Steinberger, *Phys. Rev. Lett.* 1 (1958) 249.
[KRO 28]   R. Kronig, *Naturw.* 16 (1928) 335.
[KUS 41]   S. Kusaka, *Phys. Rev.* 60 (1941) 61.
[LAN 52]   L. M. Langer and R. J. D. Moffat, *Phys. Rev.* 88 (1952) 689.
[LAN 57]   L. Landau, *Nucl. Phys.* 3 (1957) 127.
[LEE 56]   T. D. Lee and C. N. Yang, *Phys. Rev.* 104 (1956) 254.
[LEE 57]   T. D. Lee and C. N. Yang, *Phys. Rev.* 105 (1957) 1671.
[LOK 55]   S. Lokanathan and J. Steinberger, *Nuovo Cimento* 2, Supl. 151 (1955).
[MEI 22]   L. Meitner, *Z. Physik.* 9, 101, and 145 (1922); 11, 35 (1922).
[MEI 25]   L. Meitner, *Z. Physik.* 34 (1925) 807.
[MEI 30]   L. Meitner and W. Orthmann, *Z. Physik.* 60 (1930) 143.
[PAU 33]   W. Pauli, *Handbuch der Physik.* Berlin (1933).
[PAU 34]   W. Pauli, in *Rapp. Septième Conseil Phys. Solvay, Brussels 1933*, Gautier-Villars, Paris (1934).
[PAU 58]   W. Pauli, *Experentia* 14/1 (1958) 1.
[PER 33]   F. Perrin, *Comptes rendues* 197 (1933) 1624.
[PRU 48]   J. R. Pruett, *Phys. Rev.* 73 (1948) 1219.
[REI 56]   F. Reines and C. L. Cowan, Jr., *Nature* 178 (1956) 446.
[REI 58]   F. Reines and C. L. Cowan, Jr., *Phys. Rev.* 113 (1958) 273.
[ROD 52]   G. W. Rodeback and J. S. Allen, *Phys. Rev.* 86 (1952) 446.
[RUT 20]   E. Rutherford, *Proc. Roy. Soc.* (A) 97 (1920) 374.
[SAK 58a]   J. J. Sakurai, *Phys. Rev. Lett.* 1 (1958) 40.
[SAK 58b]   J. J. Sakurai, *Nuovo Cimento* 7 (1958) 649.
[SAL 57]   A. Salam, *Nuovo Cimento* 5 (1957) 229.
[SHE 51]   C. W. Sherwin, *Phys. Rev.* 82 (1951) 82.
[STE 55]   B. Stech and J. H. D. Jensen. *Z. Physik.* 141 (1955) 175 and 403.
[SUD 58]   E. C. G. Sudarshan and R. E. Marshak, *Phys. Rev.* 109 (1958) 1860.
[WEY 29]   H. Weyl, *Z. Physik* 56 (1929) 330.
[WU 57]   C. S. Wu, E. Ambler, R. W. Hayward, D. D. Hoppes, and R. P. Hudson, *Phys. Rev.* 105 (1957) 1413.

1.2 Inverse β process

B. PONTECORVO, 1946*

1 Introduction

The Fermi theory of the β disintegration is not yet in a final stage; not only detailed problems are to be solved, but also the fundamental assumption – the neutrino hypothesis – has not yet been definitely proven. I will recall briefly the main experimental facts which have led Pauli to propose the neutrino hypothesis.

1 In a β disintegration, the atomic nucleus Z changes by one unit, while the mass number does not change.
2 The β-spectrum is continuous, while the parent and the daughter states correspond to well-defined energy values of the nuclei Z and $Z \pm 1$.
3 The difference in energy between the initial and final states involved in a β transition is equal to the *upper* limit of the continuous spectrum.

We see that the fundamental facts can be reconciled only with one of the following alternative assumptions:

1 The law of the conservation of the energy does not hold in a single β process.
2 The law of the conservation of the energy is valid, but a new hypothetical particle, undetectable in any calorimetric measurement – the neutrino – is emitted together with a β-particle in a β transition in such a way that the energy available in such a transition is shared between the electron and the neutrino. This suggestion was made by Pauli, and on this basis Fermi has built a consistent quantitative theory of the β disintegration. In addition to the difficulties already mentioned, the assumption 2 removes some difficulties connected with the conservation of the spin and of the type of statistics not covered here.

The main neutrino properties follow "by definition" and are zero charge, spin $\frac{1}{2}$, and Fermi's statistics.

The problem of the β disintegration has been attacked experimentally in many ways:

β-spectroscopy. This is the study of the form of the spectrum, the relationship between the energy release and the probability of disintegration, the ratio of positron to electron emission in cases where both electrons and positrons can be emitted, the ratio of the number of the K-capture transitions to positron transitions.

---

* This section is a reprint of B. Pontecorvo, Report PD-205 of the National Research Council of Canada, Division of Atomic Energy, Chalk River, Ontario, November 13, 1946.

Neutron decay. This fundamental β transition, the transformation of a free neutron into a proton, has not yet been detected. Plans for its detection, as well as for the study of the angular distribution of the proton and electron emitted, have been made in several laboratories in the United States and in the Chalk River Laboratory.

Experiments on the recoils of nuclei in a β-ray disintegration. Several authors have attempted experiments of this type. The common feature of all these experiments is that the magnitude of the recoil energy of the nucleus, having undergone a β-decay process, is examined in the light of the laws of the conservation of energy and momentum. The most significant results were obtained by Allen, who studied the recoil of a nucleus having undergone a K electron capture, and by Jacobsen and Kofoed-Hansen, who deduced from their experiments that neutrinos and electrons are emitted prevalently in the same direction. It should be noticed that experiments of this type, while of fundamental significance in the understanding of the β process, cannot bring decisive *direct* evidence on the basic assumption of the existence of the neutrino. This statement can be understood if we keep in mind that recoil experiments are interpreted on the basis of the laws of the conservation of the energy and momentum in individual β processes, i.e., on the basis of the alternative 2, which, in effect, corresponds essentially to the assumption of the existence of the neutrino.

Direct proof of the existence of the neutrino must, consequently, be based on experiments, the interpretation of which does not require the law of conservation of energy, i.e., on experiments in which some characteristic process produced by *free neutrinos* (a process produced by neutrinos after they have been emitted in a β disintegration) is observed.

2 Inverse β process

It is clear that inverse β transformations produced by neutrinos are processes of this type and certainly can be produced by neutrinos, if neutrinos exist at all. They consist of the concomitant absorption of a neutrino and emission of a β-particle (positron or negatron) by a nucleus. It is obvious, on thermodynamical grounds, that such processes must have an extremely low yield since their inverse, the β process, is so unlikely. It has been currently stated in the literature that an inverse β process produced by neutrinos cannot be observed, due to the low yield. As it will be shown below, this statement seems to be too drastic. The object of this note is to show that the experimental observation of an inverse β process produced by neutrinos is not out of the question with the modern experimental facilities, and to suggest a method which might make an experimental observation feasible.

For completeness, we will mention also some inverse β processes produced by particles other than a neutrino; an inverse β process, more generally, can be defined as the transformation of a neutron into a proton, or vice versa, produced artificially by bombardment with neutrinos, electrons, or γ-rays. These processes are

1 Absorption of negative β-particle ($β$) with emission of a neutrino

$$(\nu)\beta^- + Z \rightarrow \nu + (Z - 1).$$

2 Absorption of a neutrino with emission of a β-particle:

$$\nu + Z \rightarrow \beta^- + (Z + 1); \quad \nu + Z \rightarrow \beta^+ + (Z - 1).$$

3 Absorption of a neutrino accompanied by a K electron capture

$$\nu + Z + \beta^-(K) \rightarrow Z - 1.$$

4 Processes involved by γ-radiation

$$\gamma + Z \rightarrow \gamma + \beta^- + (Z + 1)$$
$$\gamma + Z \rightarrow \gamma + \beta^+ + (Z - 1).$$

3 Proposed method

It is true that the actual β transition involved, i.e., the *actual emission* of a β-particle in processes 2 and 3 and the emission of X-radiation in process 3, is certainly not detectable in practice. However, the nucleus of charge $Z \pm 1$, which is produced in any of the reactions indicated above, may be (and generally will be) radioactive, with a decay period well-known (see, for example, Seaborg's table of radioelements). *Consequently, the radioactivity of the produced nucleus may be looked for as proof of the inverse β process.*

The essential point in this method is that radioactive atoms produced by an inverse β-ray process have different chemical properties from the irradiated atoms. Consequently, it may be possible to concentrate the radioactive atoms of known period from a very large irradiated volume. In the case of electron irradiation, the effective volume irradiated may be of the order of cubic centimeters; in the case of γ-ray irradiation, the volume may be of the order of a liter, and for neutrino irradiation, the volume is limited only by practical consideration and may be as high as 1 cubic meter. Elements to be considered for irradiation must be selected according to a compromise between their desirable properties, which are:

1 The material to be irradiated must not be too expensive, since large volumes are involved.
2 The nucleus produced in inverse β transformation must be radioactive with a period of at least one day, because of the long time involved in the separation.
3 The separation of the radioactive atoms from the irradiated material must be relatively simple. If a chemical separation is involved, it is necessary that the addition of only a few grams of a nonisotopic carrier, per hundred liters of

material treated, gives an efficient separation. Isotopic carriers must be used only in the last phase of the separation. An electrochemical separation is another possibility presenting some advantages because of the absence of carriers. If the nucleus formed in the inverse β process is a rare gas, the separation can be obtained by physical methods, again without a carrier, for example, by boiling the material irradiated. This is the most promising method, according to Dr. O. Frisch and the writer.

4 The maximum energy of the β-rays emitted by the radioelement produced must be very small; i.e., the difference in mass of the element Z and $Z \pm 1$ must be small. This is so because the probability of an inverse β process increases rapidly with the energy of the particle emitted, as will be explained. Of course, the requirement that the mass of Z is close to the mass of $Z \pm 1$ is not important if the bombarding particles have an energy much higher than the difference in the masses of Z and $Z \pm 1$. While γ-rays or electrons produced by betatrons or synchrotrons may easily satisfy this condition, strong sources of high-energy neutrinos are not available, so that the requirement is of importance in a neutrino experiment.

5 The background (i.e., the production of element $Z \pm 1$ by other causes than the inverse β process) must be as small as possible.

4 An example

There are several elements which can be used for neutrino radiation in the suggested investigation. Chlorine and bromine, for example, fulfill reasonably well the desired conditions. The reactions of interest would be

$$\nu + {}^{37}Cl \rightarrow \beta^- + {}^{37}A \qquad \nu + {}^{79,81}Br \rightarrow \beta^- + {}^{79,81}Kr$$
$$ {}^{37}A \rightarrow {}^{37}Cl \qquad\qquad {}^{79,81}Kr \rightarrow {}^{79,81}Br$$

(34 days; K capture)   (34 hrs; emission of positrons of 0.4 MeV).

The experiment with chlorine, for example, would consist in irradiating with neutrinos a large volume of chlorine or carbon tetrachloride for a time of the order of one month, and extracting the radioactive ${}^{37}Ar$ from such volume by boiling. The radioactive argon would be introduced inside a small counter; the counting efficiency is close to 100 percent, because of the high Auger electron yield. Conditions 1, 2, 3, 4, are reasonably fulfilled in this example. It can be shown also that condition 5, implying a relatively low background, is fulfilled.

Causes other than inverse β processes capable of producing the radioelement looked for are:

(n, p) **Processes and nuclear explosions.**   The production of background by (n, p) process against the nucleus bombarded is zero, if the particular inverse β process selected involves the emission of a negatron rather than the emission of a positron. This is the case in the inverse β process which would produce ${}^{37}Ar$ from

^{37}Cl. Similar arguments show that "cosmic-ray stars" cannot produce a direct background of ^{37}Ar from ^{37}Cl. As for (n, p) processes in impurities, the fact that ^{37}K does not exist in nature rules out this possibility.

(n, γ) **Process.** This effect can produce background only through impurities. In principle, at least, it can be reduced by addition of neutron-absorbing material. In the case considered, ^{37}Ar could be produced by absorption of neutrons in ^{36}Ar present to an extent of 0.3 percent in natural argon still present as contamination. It is estimated that $(n, 2n)$ effects, again through impurities, would not produce high background.

(p, n) **Effects.** These effects are estimated to be very small. They would arise from cosmic rays, and are consequently independent of the neutrino strength used. They could be investigated in a blank experiment.

5 Cross sections

If W is the mass difference between the two atoms involved in the inverse transition, E_p is the energy of the impinging particle, E is the energy of the emitted particle, we have $E = E_p - W$. We will see that the cross section σ_{inv} for the inverse β process increases rapidly with E, so that there is advantage in having a small W, at least for an energy of the primary particle smaller than 10 MeV.

Fierz and Bethe first gave a theoretical value for the cross sections of an inverse β process. A general dimensional argument given by Bethe and Peierls will be given here. This argument permits the estimate of the order of magnitude of σ_{inv} by using only the empirical knowledge of the β-ray lifetimes.

On thermodynamical grounds, the cross section σ_{inv} of an inverse β process produced by neutrino must be given by a formula of the type $\sigma = K/\tau \, \text{cm}^2$, where $1/\tau$ is the probability per unit time of a β-disintegration involving energy E, and K is a constant of proportionality having the dimensions of $\text{cm}^2 \times \text{sec}$. The largest possible length involved is the wavelength of the impinging neutrino, and the longest time involved is that length divided by c. Thus we can write the above formula in the form

$$\sigma_{\text{inv}} \leq \bar{\lambda}^2 \times \frac{\bar{\lambda}^2}{c} \times \frac{1}{\tau}$$

which has a quite clear physical meaning. From the above formula, we can recognize immediately that the cross section will increase with the energy of the impinging particle if $1/\tau$ increases with a power of E bigger than E^3. Now $1/\tau$, according to our knowledge of the β-disintegration, increases about as E^5 for energy on the order of 1 MeV. For very high energies, the dependence of $1/\tau$ on the energy is not known. It might be considerably higher. The Konopinski and Uhlenbeck modification of Fermi theory would give a dependence $\alpha \, E^7$. We can conclude that the cross section for an inverse β process produced by neutrinos with emission of a β-particle increases with a high power of the energy of the bombarding neutrino.

For $E = 5$ MeV, τ might be as small as 0.1 sec; $\bar{\lambda}^2$ and λ/c are respectively, of the order of 10^{-21} cm^2 and 10^{-22} sec so that σ_{inv}, for neutrinos of 5 MeV, may be of the order of 10^{-42} cm^2. The evaluation is more complicated when many levels participate in the process, because of the uncertain dependence of the matrix elements on the excitation energy. Assuming, for example, that 1 m^3 of CCl$_4$ is used for the experiment, the number of nuclei of ^{37}Cl is about 10^{28}, and the number of disintegrations N per second of ^{37}Ar produced at saturation in such volume is $N = $ neutrino flux $\times \sigma_{inv} \times 10^{28} \sim$ neutrino flux $\times 10^{-14}$. The effect might be detected if N is of the order of 1, requiring a neutrino flux of the order of 10^{14} neutrinos per cm^2/sec. Such a value of the neutrino flux, though extremely high, is not too far from what could be obtained with present-day facilities.

6 Sources

The neutrino flux from the sun is of the order of 10^{10} neutrinos/cm^2/sec. The neutrinos emitted by the sun, however, are not very energetic. The use of high-intensity piles permits two possible strong neutrino sources:

1 The neutrino source is the pile itself, *during operation*. In this case, neutrinos must be utilized beyond the usual pile shield. The advantage of such an arrangement is the possibility of using high-energy neutrinos emitted by all the very short-period fission fragments. Probably this is the most convenient neutrino source.

2 The neutrino source is the "hot" uranium metal extracted from a pile, or the fission fragment concentrate from "hot" uranium metal. In this case, neutrinos can be utilized near to the surface of the source, but the high-energy neutrinos emitted by the short-period fragments are not present.

In the case of the investigation of inverse β processes produced by electrons of γ-rays of high energy, the best source is a betatron or a synchrotron.

1.3 Neutrino recoils following the capture of orbital electrons in ^{37}Ar

GEORGE W. RODEBACK AND JAMES S. ALLEN, 1952*

1 Introduction

The measurement of the energy of the recoil nucleus associated with orbital electron capture in an isotope in gaseous form should provide an answer to the question of

* George W. Rodeback and James S. Allen, Department of Physics, University of Illinois, Urbana, Illinois. Reprinted from *Phys. Rev.*, May 15, 1952, Vol. 86, No. 4, pages 446–50. (Received 21 January 1952.) This investigation was supported jointly by the AEC and ONR.

whether or not single neutrinos are emitted in this type of radioactive decay, as was pointed out by Crane [CRA 48]. This experiment was an attempt at such a measurement in which radioactive ^{37}Ar was used.

The isotope ^{37}Ar has properties which make it well suited for a recoil energy determination by a time-of-flight measurement. First of all, the expected recoil velocity results in a time-of-flight of the order of a few microseconds for a drift distance of several centimeters. In addition, the excited atom, following orbital electron capture, returns to its ground state primarily by the emission of one or more Auger electrons of less than 3000 eV energy. The detection of these low energy electrons with an electron multiplier provides a means of initiating the time measurement. Since the low-velocity recoil atoms which emit the Auger electrons are either singly or multiply ionized, they can be accelerated through an electric field to an energy which renders them easily detectable with an electron multiplier.

The primary experimental requirements for this measurement were to continuously maintain a gaseous source of constant and suitable strength and to record the data in a reliable fashion for long periods of time. The geometry finally employed proved to be a compromise between good velocity resolution and high coincidence counting rates.

2 Properties of ^{37}Ar

Assuming the emission of a single neutrino, the orbital electron capture disintegration of ^{37}Ar is represented by the equation

$$^{37}\text{Ar} + e_{K,L} \rightarrow {}^{37}\text{Cl} + \nu + Q,$$

where $e_{K,L}$ is the captured orbital electron, ν the emitted neutrino, and Q is the disintegration energy. If the neutrino has zero rest mass, the Q of the above reaction is given by the $^{37}\text{Ar} - {}^{37}\text{Cl}$ mass difference. A value of 816 ± 4 keV for this mass difference has been obtained from a recent $^{37}\text{Cl}(p,n)^{37}\text{Ar}$ threshold measurement [RIC 50] together with an $n-p$ mass difference [TAS 49] of 782 ± 1 keV. If the entire disintegration energy is carried away by the single neutrino and the recoiling nucleus and if we assume that linear momentum is conserved between the neutrino and the recoiling nucleus, the energy of the nuclear recoil should be 9.67 ± 0.08 eV corresponding to a velocity of 0.711 ± 0.004 cm/μsec. The contribution of the binding energy of the orbital electron to the reaction energy is negligible, and has been omitted in the above computations.

^{37}Ar decays entirely by orbital electron capture with a half-life of 34 days [WEI 44]. About 93 percent of the disintegrations result from K capture with about 90 percent of this fraction resulting in emission of Auger electrons; the remainder is K X-ray emission. The other 7 percent of the total number of disintegrations result from L capture, as was first reported by Pontecorvo *et al.* [PON 49]. Using the

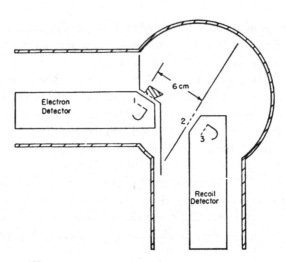

Fig. 1   Schematic of ^{37}Ar time-of-flight apparatus. The effective source volume is indicated by the shaded trapezoidal cross section in front of the grid 1. The recoil ^{37}Cl ions resulting from a disintegration within the source volume traverse a field free path to grid 2 and then enter the ion counter after an acceleration through a potential difference of

proportional counter technique, this group measured the spectrum of energies due to the emission of Auger electrons and X-rays [KIR 48] and confirmed the existence of Auger electron energies of about 2400, 2600, and 200 eV corresponding to $K-L^2$, $K-LM$, and $L-M^2$ converted electrons. These energies are in agreement with the values computed from the known critical absorption wavelengths of chlorine [COM 35].

According to Morrison and Schiff [MOR 40], about 0.05 percent of the disintegrations should result from radiative orbital electron capture. In this radiative capture process the available energy is shared between a neutrino, a γ-quantum, and the recoiling nucleus. An almost continuous spread of recoil momenta should result from this type of disintegration. However, this effect was not observed in the present experiment since the expected counting rate was much smaller than the chance coincidence rate. Maeder and Preiswerk [MAE 51] have recently shown that radiative electron capture does occur in ^{55}Fe.

3 Recoil chamber and method of recording data

Figure 1 shows a schematic cross section of the chamber in which the time-of-flight measurements were made. During the run the total pressure of the gases in the chamber, including that of the ^{37}Ar, was maintained at about 10^{-5} mm Hg. This corresponds to a mean free path of about 500 cm for argon atoms. The shaded trapizoidal cross section indicates the effective source volume which was defined by baffles and also by the region seen simultaneously by both detectors. The necessary baffles, shields, and wire gauze grids (denoted by dotted lines) were all maintained at ground potential with the exception of grid 3, which was maintained at approximately −4500 volts with respect to ground. A delayed coincidence was recorded for an ^{37}Ar disintegration occurring within the source volume when the resulting Auger electron passed through grid 1 into the electron detector and when, in addition, the ionized chlorine atom traversed the field free distance to grid 2 and was counted by the recoil detector. Both detectors were Allen-type electron multipliers [ALL 47].

Delayed coincidences of the recoil detector output with respect to the electron detector output were recorded by a 20-channel delayed-coincidence circuit. The outputs of the two multipliers were fed through two identical channels consisting of preamplifiers, linear pulse amplifiers, and discriminators. The final output of the electron detector channel emerged from a pulse shaper circuit and initiated the delayed-coincidence circuit. The final output of the recoil detector channel also was shaped and fed into the multichannel coincidence circuit. Time calibration of the entire delayed-coincidence circuit, including measurement of resolving times and total delays for individual channels, was accomplished by the introduction at the preamplifier inputs of two pulses separated by a precisely known time interval.

The 20 delayed-coincidence channels follow consecutively in time and usually the adjacent channels overlap by as much as 10 percent of the time width of an individual channel. During the period of a run the total counts for each channel are computed on the basis of its resolving time and the measured singles counts from both detectors. The true counts are given by the difference between the total coincidence counts and the calculated counts. What is plotted, however, is the number of true counts per channel of unit time width (i.e., true counts divided by resolving time) as a function of the total elapsed time to the midpoint of a coincidence channel. The plot then represents an experimental determination of the differential time distribution of the recoils coming from the source volume.

The relative statistical accuracy in the number of true counts can be defined as the ratio of true counts to the probable statistical error in the observed number of counts. An expression for this ratio in terms of the constants of the counting arrangement can be obtained. Assuming an extended source which emits monoenergetic recoils, it can be shown that this ratio is proportional to $(\tau T)^{\frac{1}{2}}$ where τ is the resolving time of a channel and T is the total elapsed counting time. Therefore, when the details of the time-of-flight spectrum are desired (implying that the resolving time be shorter than the time spread expected for the recoils coming from the extended source volume), increased resolution is obtained at the expense of the statistical accuracy, and this loss can be compensated for only by increasing the counting time. For most of the experiments to be described below, the source strength was adjusted to produce a true to chance ratio greater than unity. The statistical accuracy in the number of coincidences per channel then was determined by the length of the observation period.

4 Results

Figure 2 shows several time-of-flight distributions obtained for a geometry similar to that of Fig. 1 with a mean traversal distance for the recoil ions of about 5 cm. The conditions for each run were similar, except that the resolving times of the coincidence channels were successively decreased, thus yielding increasingly greater detail of the distribution. The statistical accuracy was made nearly the same for each plot by controlling the time duration of each run.

The peak at about 7 μsec is the result of the recoils originating in the source volume. The abrupt cessation of the time-of-flight distribution at about 9 μsec, with no further coincidences indicated out to at least 35 μsec, indicates the absence of recoils with velocities less than the expected value of 0.711 cm/μsec. This sharp cutoff, together with the fact that the distribution has the general shape expected for monoenergetic recoils from the source volume, is interpreted as experimental verification of the unique energy of the recoil atoms. This explanation assumes that the zero and short time counts of the distribution can be satisfactorily explained.

Following the runs shown in Fig. 2, the apparatus was altered slightly to give a better defined source volume. Figure 3 shows the resulting distribution. The dashed curve is the predicted distribution based on the shape and location of the source volume and the assumed value of recoil velocity which gives the best fit.

The far edge of the measured distribution was used to calculate an experimental value for the recoil velocity based on the time for a singly ionized ^{37}Cl atom to traverse the maximum field-free distance from the source volume to grid 2 of Fig. 1. The following considerations entered into the calculation of the delay time to be used:

1 The ideal time-of-flight distribution is modified due to the thermal velocities of the ^{37}Ar atoms in the source. The root-mean-square velocity at room temperature is about 0.04 cm/μsec and results in a spreading at the base of the ideal distribution. The expected spread is very similar to the observed far edge of the distribution in Fig. 3. Therefore, the intercept of the dotted curve at 7.8 μsec is taken as the observed time-of-flight corresponding to the maximum recoil distance. However, further corrections must be made to this value based on the following:

2 The time required for singly ionized ^{37}Cl atom to be accelerated between grids 2 and 3 (Fig. 1) and thence to travel at constant velocity to the sensitive region of the first dynode is calculated to be 0.3 μsec.

3 The field between grids 2 and 3 (Fig. 1) was found to have penetrated the supposedly field-free region between the source volume and grid 2. This field distribution was subsequently measured, and calculations based on this show that the traversal time in this region for a singly ionized atom is decreased by 1.0 μsec. For doubly ionized atoms this is 1.4 μsec. With the above corrections

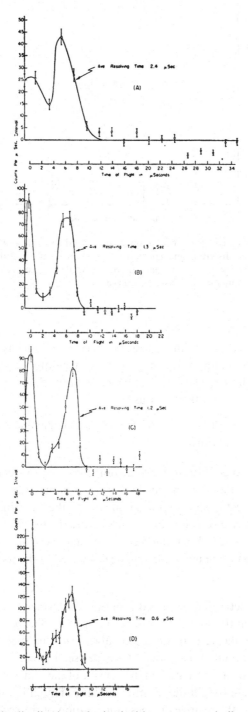

Fig. 2 Time-of-flight distributions obtained with a geometry similar to that of Fig. 1. The mean time-of-flight path was 5 cm. The average channel width of the time recorder was progressively decreased in order to yield increasingly greater details of the distribution.

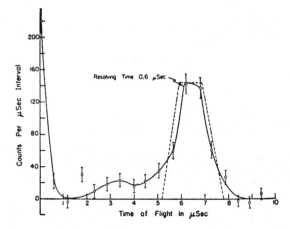

Fig. 3   Time-of-flight distribution obtained with improved definition of the source volume. The dashed curve is the distribution expected for monoenergetic recoils coming from the source volume. The tail of the solid curve in the region of 8 μsec is due to the thermal velocities of the ^{37}Ar atoms in the gaseous source.

for a singly ionized atom, the corrected maximum time is $7.8 - 0.3 + 1.0 = 8.5$ μsec. The maximum distance is 6.0 cm, resulting in a recoil velocity determination of 0.71 ± 0.06 cm/μsec, which is in excellent agreement with the expected value of 0.711 ± 0.004 cm/μsec.

The following two additional corrections are negligible compared to the uncertainties of this experiment:

1 Time of flight of Auger electrons from source volume to electron detector.
2 The momentum of the recoiling nucleus is equal in magnitude and opposite in direction to the vector sum of the momenta of the neutrino and the Auger electron, since the electron emission occurs before the recoil has moved through an appreciable distance. When the recoil atom and electron are at 90° as in the present arrangement, the maximum change expected in the recoil momentum is 0.2 percent.

Based on the geometry and estimated counting efficiencies, the probability of an ^{37}Ar disintegration in the source volume resulting in a coincidence count is about 10^{-5}. Calculations indicate that during the above run the specific activity in the source volume was about 200 disintegrations per sec per cm^3.

By imposing a retarding potential on the left side of the grid 2 in Fig. 1 and noting the effect on the time-of-flight distribution as this potential is varied, it should be possible to determine the relative degrees of ionization of the recoil atoms. Due to instrumental difficulties it was possible only to verify that a retarding potential of 16.5 volts eliminated the distribution of delayed counts from the source volume.

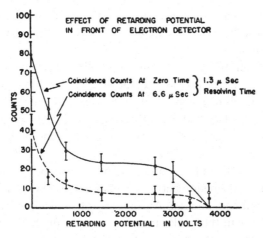

Fig. 4 Curves showing the effect of retarding potentials on the electrons entering the electron counter.

5 Origin of zero and short time counts

A series of measurements were made in an attempt to explain the origin of the zero time peak and the other coincidences occurring at times shorter than those of the main time-of-flight peak. Zero time coincidences were recorded for two separate runs made under identical conditions except that the first was made with no ion accelerating field in front of the ion detector (first dynode at ground) while the second run was made under the normal conditions for a time-of-flight measurement. It was found that there were 20 percent as many zero time coincidences without the accelerating field as with it. Therefore, at least 80 percent of the zero time coincidences are to be associated with the presence of the 4500-volt difference of potential between grids 2 and 3. It is likely that some of the coincidences measured without the field can be detected only under this condition and may be due to X-ray–electron or electron–electron coincidences resulting from cascade Auger processes in the ^{37}Cl atom.

A series of runs were made to determine the effect of an electron-retarding potential in front of the electron detector. Grid 1 together with the first dynode of the electron detector (Fig. 1) was maintained at a negative potential with respect to the grounded source volume baffle. Figure 4 shows the results of these runs where the zero time and 6.6-µsec (at maximum of delayed peak) coincidences are plotted as a function of the retarding potential at the electron detector. For each of these runs the total number of ion detector counts was maintained at the same value.

On the basis of the above results the following conclusions are made concerning a normal run:

1 The electrons counted by the electron detector for the zero time and delayed coincidences have nearly identical energy spectra.

2 The delayed coincidences at 6.6 μsec, when the electron-retarding potential is greater than 2800 volts, probably are due to X-ray–ion coincidences originating in the source volume.

3 No appreciable number of zero time coincidences originate in the source volume. Probably the best evidence for this is the fact that as mentioned in Section 4 a retarding potential of 16.5 volts eliminated the delayed coincidences at 6.6 μsec. This indicates the absence of high-energy positive ions, which would be necessary for an essentially zero time of flight over a path of 5 cm.

4 Most of the zero and short time counts are believed to be due to disintegrations occurring in or near the region between grids 2 and 3. The coincidence would be between the recoil ion and an Auger electron which undergoes an elastic scattering process with the baffle structure surrounding the source volume and subsequently enters the electron detector. The acceleration of electrons by the field between grids 2 and 3 will aid this process. This is indicated by Fig. 4, where appreciable zero time counts are recorded for electron energies above the maximum Auger electron energy.

5 A few zero time counts may originate from those secondary electrons emitted at the first dynode of the ion detector which escape outwards and after accleration between grids 3 and 2 are elastically scattered into the electron detector.

6 According to the above explanations, a number of delayed coincidences involving scattered Auger electrons are to be expected from disintegrations occurring in the region between the source volume and grid 2. From solid angle considerations a peak in the number of coincidences should occur near the source volume, and going toward grid 2, this number should at first fall to a minimum and finally increase rapidly as grid 2 is approached. The general shape of the time-of-flight distribution (Fig. 3) below 5 μsec seems to agree with these predictions.

6 Method of providing source

The ^{37}Ar used for these measurements was initially obtained from the Oak Ridge reactor, where it was prepared by neutron bombardment of ^{40}Ca. The radioisotope arrived mixed with a small amount of air. Preceding a run, the oxygen and nitrogen were removed from approximately 1 millicurie of the 30 millicurie source by exposure to outgassed metallic calcium heated to 565°C. The remaining gas was then introduced into the high-pressure side of a three-stage diffusion and booster pump combination after the fore pump had been sealed off from the system. The resulting pressure was a few microns of Hg. The calcium purifier was now operated at 300°C and adequately performed the functions of the usual fore pump and, in addition, removed the impurities from the ^{37}Ar. With this arrangement the recoil chamber could be kept evacuated to less than 10^{-6} mm of Hg.

During a run, the ^{37}Ar in the reservoir at the rough vacuum side of the pumps was allowed to leak into the recoil chamber through a needle valve at a rate giving a suitable counting rate for the equipment. A steady state was very quickly reached and the recycling process could be steadily maintained for over 12 hours. The best source used resulted in a total pressure of about 10^{-5} mm of Hg in the recoil chamber. A relative measure of the source strength present in the reservoir was continuously provided by a monitor. This monitor was a thin mica window Geiger counter which was exposed to the gaseous source and responded to the small percentage of ^{37}Cl recoils emitting ^{37}Ar radiation. The high absorptivity of charcoal at liquid air temperatures was utilized for storage of the ^{37}Ar between runs. All but a very small fraction of the source in the reservoir could be collected and sealed off in a cooled tube containing outgassed charcoal powder. This made it possible to use the same source sample many times.

7 Conclusions

The results of this experiment indicate that for most of the ^{37}Ar orbital electron capture disintegrations, the missing energy of a disintegration is shared between the recoil nucleus and a single neutrino. Linear momentum is shown to be conserved between the recoil nucleus and a single neutrino. Additional experiments will be necessary to further clarify the origin of the zero time and short time coincidence counts, and further refinements in the time-of-flight method should yield a more accurate value of the recoil velocity. Future investigations based on the techniques of this experiment should reveal the details of the processes which occur when the electronic levels of the excited ^{37}Cl atom return to the ground state.

References

[ALL 47] J. S. Allen, *Rev. Sci. Instr.* 18 (1947) 739.

[COM 35] A. H. Compton and S. K. Allison, *X-Rays in Theory and Experiment*, D. Van Nostrand (1935).

[CRA 48] H. R. Crane, *Revs. Modern Phys.* 20 (1948) 195.

[KIR 48] Kirkwood, Pontecorvo, and Hanna, *Phys. Rev.* 74 (1948) 497.

[MAE 51] D. Maeder and P. Preiswerk, *Phys. Rev.* 84 (1951) 595.

[MOR 40] P. Morrison and L. I. Schiff, *Phys. Rev.* 58 (1940) 24.

[PON 49] Pontecorvo, Kirkwood and Hanna, *Phys. Rev.* 75 (1949) 982. [Translation: *Soviet Phys. JETP* 10 (1960) 1236.]

[RIC 50] Richards, Smith, and Brown, *Phys. Rev.* 80 (1950), 524.

[TAS 49] Taschek, Jarvis, Argo, and Hemmendinger, *Phys. Rev.* 75 (1949) 1268.

[WEI 44] Weimer, Kurbatov, and Pool, *Phys. Rev.* 66 (1944) 209.

1.4 Detection of the free neutrino: A confirmation

C. L. COWAN, JR., F. REINES, F. B. HARRISON,

H. W. KRUSE AND A. D. MCGUIRE, 1956*

A tentative identification of the free neutrino was made in an experiment performed at Hanford [REI 53a,b] in 1953. In that work the reaction

$$\nu_- + p^+ \rightarrow \beta^+ + n^0 \tag{1}$$

was employed wherein the intense neutrino flux from fission-fragment decay in a large reactor was incident on a detector containing many target protons in a hydrogenous liquid scintillator. The reaction products were detected as a delayed pulse pair; the first pulse being due to the slowing down and annihilation of the positron and the second to capture of the moderated neutron in cadmium dissolved in the scintillator. To identify the observed signal as neutrino-induced, the energies of the two pulses, their time-delay spectrum, the dependence of the signal rate on reactor power, and its magnitude as compared with the predicted rate were used. The calculated effectiveness of the shielding employed, together with neutron measurements made with emulsions external to the shield, seemed to rule out reactor neutrons and gamma radiation as the cause of the signal. Although a high background was experienced due both to the reactor and to cosmic radiation, it was felt that an identification of the free neutrino had probably been made.

1 Design of the experiment

To carry this work to a more definitive conclusion, a second experiment was designed,[1] and the equipment was taken to the Savannah River Plant of the U.S. Atomic Energy Commission, where the present work was done.[2] This work confirms the results obtained at Hanford and so verifies the neutrino hypothesis suggested by Pauli [PAU 33] and incorporated in a quantitative theory of beta decay by Fermi [FER 34].

* C. L. Cowan, Jr., F. Reines, F. B. Harrison, H. W. Kruse, and A. D. McGuire. The authors are on the staff of the University of California, Los Alamos Scientific Laboratory, Los Alamos, N.M. Distributed by University of California. Los Alamos Scientific Laboratory. Reprinted from *Science*. July 20, 1956, Vol. 124, No. 3212, pages 103–4.

[1] C. L. Cowan, Jr. and F. Reines, invited paper, American Physical Society, New York Meeting, Jan. 1954; the results of the present work were presented in a post deadline paper, American Physical Society, New Haven Meeting, June, 1956.

[2] We wish to thank the many people at the Los Alamos Scientific Laboratory who assisted in the preparation of the experiment and to mention especially A. R. Ronzio, C. W. Johnstone and A. Brousseau for their help in the chemical and electronic problems. M. P. Warren and R. Jones were invaluable members of the group during both the preparation and field phase of the problem. We also wish to thank the E. I. du Pont de Nemours Company and their personnel at the Savannah River Plant for their constant cooperation and assistance during our stay at the reactor. This work was performed under the auspices of the U.S. Atomic Energy Commission.

In this experiment, a detailed check of each term of Eq. (1) was made using a detector consisting of a multiple-layer (club-sandwich) arrangement of scintillation counters and target tanks. This arrangement permits the observation of prompt spatial coincidences characteristic of positron annihilation radiation and of the multiple gamma ray burst due to neutron capture in cadmium as well as the delayed coincidences described in the first paragraph.

The three "bread" layers of the sandwich are scintillation detectors consisting of rectangular steel tanks containing a purified triethylbenzene solution of terphenyl and POPOP[3] in a chamber 2 feet thick, 6 feet 3 inches long, and 4 feet 6 inches wide. The tops and bottoms of these chambers are thin to low-energy gamma radiation. The tank interiors are painted white, and the solutions in the chambers are viewed by 110 5-inch Dumont photomultiplier tubes connected in parallel in each tank. The energy resolution of the detectors for gamma rays of 0.5 MeV is about 15 percent half-width at half-height.

The two "meat" layers of the sandwich serve as targets and consist of polyethylene boxes 3 inches thick and 6 feet 3 inches by 4 feet 6 inches on edge containing a water solution of cadmium chloride. This provides two essentially independent "triad" detectors, the central scintillation detector being common to both triads. The detector was completely enclosed by a paraffin and lead shield and was located in an underground room of the reactor building which provides excellent shielding from both the reactor neutrons and gamma rays and from cosmic rays.

The signals from a bank of preamplifiers connected to the scintillation tanks were transmitted via coaxial lines to an electronic analyzing system in a trailer van parked outside the reactor building. Two independent sets of equipment were used to analyze and record the operation of the two triad detectors. Linear amplifiers fed the signals to pulse-height selection gates and coincidence circuits. When the required pulse amplitudes and coincidences (prompt and delayed) were satisfied, the sweeps of two triple-beam oscilloscopes were triggered, and the pulses from the complete event were recorded photographically. The three beams of both oscilloscopes recorded signals from their respective scintillation tanks independently. The oscilloscopes were thus operated in parallel but with different gains in order to cover the requisite pulse-amplitude range. All amplifier pulses were stored in long low-distortion delay lines awaiting electronic decision prior to this acceptance.

Manual analysis of the photographic record of an event then yielded the energy deposited in each tank of a triad by both the first and second pulses and the time delay between the pulses. Using this system, various conditions could be placed on the pulses of the pair comprising an acceptable event. For example, acceptance of events with short time delays (over ranges up to 17 microseconds, depending on the cadmium concentration used) resulted in optimum signal-to-background ratios, while analysis of those events with longer time delays yielded relevant accidental

---

[3] Triethylbenzene scintillator, studied first in connection with the Hanford experiment in the search for higher proton densities, was purified by methods developed in collaboration with A. R. Ronzio; POPOP, a scintillation spectrum shifter, was developed by F. N. Hayes.

background rates. Spectral analyses of pulses comprising events with short time delays were also made and compared with those with long delays.

This method of analysis was also employed to require various types of energy deposition in the two tanks of a triad. For instance, the second pulse of an event could be required to deposit at least a given energy in each tank, and in addition, maximum and minimum limits could be placed on the total energy of the pulse. Application of criteria such as these assisted in discriminating between events satisfying the physical aspects of a neutrino capture and the various backgrounds experienced. Simultaneous presentation of the three tank outputs on the three beams of the oscilloscopes also permitted rejection of pseudo events due to penetrating cosmic rays, thus utilizing the two triads as shields for one another.

The varying rates observed by changing the response of the system assisted in ascertaining that the gamma rays observed did indeed arise in the target tanks. The efficiency of the system was calibrated in each case by the use of dissolved copper-64 positron source in the target tanks and by using a plutonium–beryllium neutron source. The neutron calibrations utilized the 4.2 MeV gamma ray emitted by the source as the first pulse of a delayed pair, the second being due to capture of the associated neutron in the cadmium. In addition, secondary calibrations were performed each week using the cosmic ray penetration pile-up peak [REI 54] and standardized pulsers to check for drift in the apparatus. Standard pulses were recorded each day on the oscilloscope cameras to maintain a constant film calibration. Running counts were made of all single and prompt coincidence rates relevant during the experiment as checks for drift or changes in background. Long-term stability of the equipment was easily maintained, and the results of the two independent triad detectors agreed well throughout the experiment.

2 Experimental results

Using this equipment near one of the reactors at the Savanah River Plant, the following results were obtained bearing on the reactions expressed by Eq. (1).

1. A reactor-power-dependent signal was observed which was (within 5 percent) in agreement with a cross section for reaction 1 of 6.3×10^{-44} cm^2. The predicted cross section[4] for the reaction, however, is uncertain by ±25 percent. In one set of runs, the neutrino signal rate was 0.56 ± 0.06 count per hour, and with changed requirements it was 2.88 ± 0.22 counts per hour. The total running time, including reactor downtime, was 1371 hours. The signal-to-background ratio associated with the higher signal rate quoted was about 3 to 1. The neutrino signal was greater than 20 times the accidental background associated with the reactor.

---

[4] This value for the predicted cross section is calculated from the decay of the neutron as observed by J. M. Robson [*Phys. Rev.* 83, 349 (1951)] and the spectrum of beta radiation from fission fragments as measured by C. O. Muehlhause at Brookhaven National Laboratory. We are indebted to Muehlhause for communication of his results in advance of publication.

2. A signal rate produced by reaction 1 must be a linear function of the number of protons provided as targets for the neutrinos. This was tested by diluting the light water solution in a target tank with a heavy water solution to yield a resultant proton density of one-half of normal. The neutron detection efficiency measured using the plutonium–beryllium source was essentially unchanged. The reactor signal fell to one-half of its former rate.

3. Reaction 1 states that the first pulse of a delayed pair observed must be due to the annihilation radiation of a positron in the target tank. This would produce a $\frac{1}{2}$-MeV gamma ray entering each detector tank of the triad simultaneously after some degradation in the water target. Events were thus chosen which satisfied these time and spatial conditions. Analysis of the pulse-amplitude spectra of these gamma rays associated with short time-delay events yielded spectra which matched that produced by the dissolved copper-64 source, having a peak at about 0.3 MeV. Spectra obtained for the first pulse of events with long delays (accidental events) were, on the other hand, monotonically decreasing with energy, as was the background spectrum producing the accidental events.

A differential absorption measurement was made using first a $\frac{3}{16}$-inch and then a $\frac{3}{8}$-inch-thick lead sheet between the target tank and one scintillation tank of a triad. The measured neutron detection efficiency was changed to about 70 percent of its former value in the first case and to about 45 percent in the second. The reactor signal rate fell sharply, however, as required for events with first pulse gamma rays of 0.5 MeV originating in the target tank.

4. The second pulse of the delayed pair signal observed was identified as being due to the capture of a neutron by cadmium in the water target. In addition to the prompt spatial coincidence required and the total-energy limits of 3 to 11 MeV imposed on a pulse for acceptance, analysis of the time-delay spectrum yielded excellent agreement with that expected for the cadmium concentration used in the target water [REI 54]. Doubling of the cadmium concentration produced the expected shift in the time-delay spectrum without increasing the signal rate. Removal of the cadmium from the target water resulted in disappearance of the reactor signal.

5. As it is possible for a fast neutron or energetic gamma ray entering the detector from the outside to produce pseudo events with many of the characteristics of true neutrino captures, the observed reactor signal was tested for these effects. A strong americium–beryllium neutron source was used outside the detector shield to produce pseudo signals. Tests of the pseudo signal with the lead sheet described in paragraph 3 resulted in a negligible drop in rate beyond that accounted for by the lowered neutron detection efficiency mentioned in paragraph 3, in contrast with the strong response of the reactor signal. The spectrum of first pulse amplitude of the neutron-produced signal with short time delays fell monotonically with increasing energy, in contrast with the characteristic spectra obtained with both the reactor signal and the dissolved copper-64 positron source.

The results of the heavy water dilution measurement described in paragraph 2 also militate against reactor-produced neutrons or gamma rays as the agent producing the signal observed.

Finally, a gross shielding experiment was performed in which the detector shield was augmented by bags of sawdust saturated with water. When stacked, the density of the added shield was 0.5 grams per cubic centimeter, its minimum thickness was 30 inches, and its average thickness was about 40 inches. This absorber would reduce the signal caused by neutrons to about one-tenth of its former rate, depending somewhat upon the direction of the incoming neutrons, and would produce a similar decrease in a signal caused by gamma rays. No decrease was observed in the reactor signal within the statistical fluctuations quoted in paragraph 1.

References

[FER 34]   E. Fermi, *Z. Physik*. 88 (1934) 161.
[PAU 33]   W. Pauli, *Handbuch der Physik*. Berlin (1933).
[REI 53a]   F. Reines and C. L. Cowan, Jr., *Phys. Rev*. 90 (1953) 492.
[REI 53b]   F. Reines and C. L. Cowan, Jr., *Phys. Rev*. 92 (1953) 830.
[REI 54]   F. Reines *et al*., *Rev. Sci. Instr*. 25 (1954) 1061.

1.5 Electron and muon neutrinos

B. PONTECORVO, 1960*

1 Introduction

Bethe and Peierls [BET 34] in 1934 were the first to estimate the cross section for production of β-particles in the collisions of free neutrinos with nuclei at energies of the order of 1 MeV. As is well known, the cross section turned out to be of the order of 10^{-44} cm^2 and for this reason effects due to free neutrinos were for a long time considered to be unobservable. Subsequently it was shown by the author and by Alvarez [PON 46; ALV 49] that such experiments are quite feasible and only recently Reines and Cowan, and also Davis, successfully performed experiments in which free antineutrinos from reactors were used. These experiments proved that neutrinos can be observed and are therefore "real," that they are two-component neutrinos [REI 53a, 59], and also that the neutrino and antineutrino are different particles [DAV 52, 59].

The aim of the present work is to emphasize the possibility of solving certain physical problems by studying effects due to free neutrinos not heretofore discussed. The corresponding experiments may turn out to be not feasible today; however, it seems to us that a discussion of them is no more premature than the discussion in its time of the antineutrino experiments from reactors.

* B. Pontecorvo, Joint Institute for Nuclear Research. Reprinted from *Soviet Physics JETP* (1960) Vol. 37(10), No. 6. pages 1236–40. Translated by A. M. Bincer from *J. Exptl. Theoret. Phys.* (U.S.S.R.), December 1959, Vol. 37, pages 1751–7. (Submitted to *JETP* July 9, 1959.)

The question discussed is the possibility of deciding, in principle, whether the neutrino emitted in the $\pi \to \mu$ decay (ν_μ) and the neutrino emitted in β decay (ν_e) are identical particles or not.

2 Reactions due to neutrinos

All slow processes known to us are apparently due to the interactions of the following fermion pairs:

$$(e^+\nu_e), (\mu^+\nu_\mu), (p\bar{n}), (p\bar{\Lambda})$$
$$(e^-\bar{\nu}_e), (\mu^-\bar{\nu}_\mu), (\bar{p}n), (\bar{p}\Lambda). \tag{1}$$

Any pair of particles may interact with itself or with another pair and, according to the Markov–Sakata–Okun' scheme, of all the strange particles only the Λ hyperon is included in the "strange" pair. In the language of the universal interactions theory [SAK 46; MAR 56; OKU 58a,b; SUD 57; FEY 58c], this scheme implies that the current J^+ entering into the weak interactions Lagrangian consists of four terms

$$J^+ = J(e^+\nu_e) + J(\mu^+\nu_\mu) + J(p\bar{n}) + J(p\bar{\Lambda}) \tag{2}$$

each of which corresponds to one of the above-mentioned pairs.

Some of the processes that may be induced by free neutrinos are listed below using the assumption that the Markov–Sakata–Okun' scheme and the universal interactions theory are valid.

The identity of ν_e and ν_μ is an open question and is discussed in the next section. Neither theoretical nor experimental arguments exist for the assertion that ν_e and ν_μ are identical particles. Therefore below, as well as in the above expression for the current, we use the notation $(e^+\nu_e)$, $(\mu^+\nu_\mu)$ instead of the conventional $(e^+\nu)$, $(\mu^+\nu)$.

We consider here collision of neutrinos with real targets, i.e., with negative electrons, protons, and nuclei (A). Among the processes listed, only 1, 2, 3, and 10 have been previously discussed in the literature. In the present work we restrict ourselves for the majority of the processes to brief comments inserted in Table 1. We discuss in more detail only those processes that have a bearing on the question of the distinguishability of the ν_e and ν_μ particles.

3 Are ν_e and ν_μ identical particles?

The upper limit on the mass of the neutral leptons emitted in μ decay, the magnitude of the Michel parameter ρ, and theoretical considerations permit one to conclude that the neutral leptons in μ decay have either vanishing or very small mass and are not identical. On this basis the decay of μ mesons is usually described by the scheme $\mu \to e + \nu + \bar{\nu}$. However, it is easy to see that the totality of experimental and theoretical information only implies that the two neutral leptons in μ decay must not be identical, and does not necessarily require that they be particle and antiparticle.

Table 1. *Some reactions induced by free neutrinos on real targets*

Process no.	Reaction	Remarks
1	$\bar{\nu}_e + p \to e^+ + n$	In the study of this process [REI 53a; REI 59], free neutral leptons were observed for the first time. The experiment confirmed the two component nature of the neutrino.
2	$\bar{\nu}_e + Cl^{37} \to Ar^{37} + e^-$	The absence of this process [PON 46; ALV 49; DAV 52, 59] proved that ν_e and $\bar{\nu}_e$ are distinct.
3	$\nu_e + Cl^{37} \to Ar^{37} + e^-$	A study of this process could be of interest for astrophysics, in particular for measuring the neutrino flux from the sun [CAM 58].
4	$\nu_e + A \to \pi^+ + e^- + A,$ $\bar{\nu}_e + A \to \pi^- + e^+ + A$	Inverse $\pi - e$ decay in the field of a nucleus. π^+ mesons are produced by ν_e, π^- mesons are produced by $\bar{\nu}_e$
5	$\bar{\nu}_e + e^- \to \pi^- + \pi^0$	
6	$\bar{\nu}_e + p \to \Lambda + e^+,$ $\bar{\nu}_e + A \to$ hyperfragment $+ e^+$	Only $\bar{\nu}$ (and not ν) can produce strange particles.
7	$\bar{\nu}_e + n \to \Sigma^- + e^+$	This process is possible only in nuclear matter.
8	$\nu_e + A \to K^+ + e^- + A,$ $\bar{\nu}_e + A \to K^- + e^+ + A$	See process (4).
9	$\bar{\nu}_e + e^- \to K^- + K^0$	See process (5).
10	$\bar{\nu}_e + e^- \to \bar{\nu}_e + e^-,$ $\nu_e + e^- \to \nu_e + e^-$	Neutrino scattering by electrons is predicted by the universal theory of weak interactions [FEY 58a].
11	$\nu_e + A \to \nu_e + e^+ + e^- + A,$ $\bar{\nu}_e + A \to \bar{\nu}_e + e^+ + e^- + A$	e^+e^- pair production in the field of a nucleus.[a] This process is the inverse of the lepton electron bremsstrahlung described in [PON 59a,b].
12	$\bar{\nu}_e + e^- \to \bar{\nu}_\mu + \mu^-,$ $\nu_e + e^- \to \nu_\mu + \mu^-$	Inverse μ decay. Forbidden if $\nu_e \neq \nu_\mu$.
13	$\bar{\nu}_e + A \to \bar{\nu}_\mu + e^+ + \mu^- + A,$ $\nu_e + A \to \nu_\mu + e^- + \mu^+ + A$	$\mu - e$ pair creation in the field of a nucleus.
14	$\bar{\nu}_e + p \to \nu^+ + n,$ $\bar{\nu}_\mu + p \to e^+ + n$	Inverse μ capture. Forbidden if $\nu_e \neq \nu_\mu$
15	$\nu_\mu + A \to \pi^+ + \mu^- + A,$ $\bar{\nu}_\mu + A \to \pi^- + \mu^+ + A$	Inverse $\pi - \mu$ decay in the field of a nucleus.
16	$\bar{\nu}_\mu + p \to A + \mu^+,$ $\bar{\nu}_\mu + p \to A + e^+,$ $\bar{\nu}_\mu + A \to$ hyperfragment $+ \mu^+$	Forbidden if $\nu_e \neq \nu_\mu$.
17	$\nu_\mu + A \to \mu^- + K^+ + A,$ $\bar{\nu}_\mu + A \to \nu^+ + K^- + A,$	
18	$\nu_\mu + A \to \nu_\mu + \mu^+ + \mu^- + A,$ $\bar{\nu}_\mu + A \to \bar{\nu}_\mu + \mu^+ + \mu^- + A$	Neutrino scattering by μ meson in the field of a nucleus.
19	$\nu_\mu + e^- \to \nu_e + \mu^-,$ $\bar{\nu}_\mu + e^- \to \bar{\nu}_e + \mu^-$	Inverse μ decay. Forbidden if $\nu_e \neq \nu_\mu$.
20	$\nu_\mu + A \to A + \mu^- + e^+ + \nu_e,$ $\bar{\nu}_\mu + A \to A + \mu^+ + e^- + \bar{\nu}_e$	$\mu - e$ pair creation in the field of a nucleus.
21	$\nu_\mu + e^- \to \nu_\mu + e^-$	If $\nu_e \neq \nu_\mu$ this reaction is possible only as a second-order process.

[a] This process was recently studied theoretically by Ya. B. Smorodinskiĭ and Chou Kuang Chao.

The possible existence of two pairs of neutrinos was already considered by Oneda and Pati [ONE 59]. At first sight, the hypothesis of two types of neutrinos – electron neutrino (ν_e, $\bar{\nu}_e$) and muon neutrino (ν, $\bar{\nu}_\mu$) – may seem an unnecessary complication. However, there are arguments that make the hypothesis of distinct electron and muon neutrinos attractive.

The absence in nature of a number of processes of the type $\mu + p \rightarrow e + p$, $\mu \rightarrow e + e + e$, etc. shows that in each of the currents in the weak interactions Lagrangian apparently only pairs consisting of one charged and one neutral particle are permitted (see Eqs. (1) and (2)). The existence of only a "charged" current, as was pointed out by Gell-Mann and Feynman [FEY 58a], could be very naturally explained if there existed in nature a charged vector boson B, coupled to various fermions with "intermediate" strength; then all weak interaction processes known to us would be due to an interaction of second order in the "intermediate" coupling constant. As was shown by Feinberg, and also by Gell-Mann and Feynman [FEI 58; FEY 58b], the nonlocality in the μ–e decay process corresponding to an intermediate vector boson would imply a probability for decay through the channel $\mu \rightarrow e + \gamma$ in contradiction with experimental data [BER 59].

However, it is not difficult to see that even if the B meson existed, the probability in question would be zero[1] (as is fully consistent with experiment) if the electron and muon neutrinos are different. Thus, the fact that the weak interactions current in the Lagrangian is "charged" could be very well explained on the hypothesis of an intermediate boson only provided ν_e and ν_μ are different. Besides this argument it seems to us the hypothesis of two different types of neutrinos, unable to annihilate each other,[2] is attractive from the point of view of symmetry and systematics of particles and also could help us understand the difference in the nature of the muon and electron.

It follows from the above that experimental information on the identity of ν_e and ν_μ is of paramount importance. One possibility consists in measuring the helicity of the μ meson. If only one kind of neutrino–antineutrino pair exists in nature, then the *V-A* interaction requires that the helicity of the μ^- meson be positive. Should this helicity be found experimentally to be negative, it would serve as a strong argument in favor of the existence of two kinds of neutrinos; the μ^+ decay could be described according to the scheme $\mu^+ \rightarrow e^+ + \nu_e + \nu_\mu$.

The experiments of Love *et al.* [LOV 59] however show that the helicity of the μ^- meson is apparently positive. Therefore the question of the existence of two kinds of neutrinos in nature remains open. The positive helicity of the μ^- meson indicates, however, that if two kinds of neutrino–antineutrino pairs do exist in nature, then the weak interactions should be described as in (1) and the decay of the μ^+ meson should proceed according to the scheme $\mu^+ \rightarrow e^+ + \nu_e + \bar{\nu}_\mu$. Here, as

---

[1] The $\mu \rightarrow e + \gamma$ process is also possible in the absence of the B meson in higher order of perturbation theory as long as there is only one kind of neutrino–antineutrino pair, whereas it is absolutely forbidden if $\nu_e \neq \nu_\mu$.

[2] We note, in particular, that if ν_e, $\bar{\nu}_e$ and $\bar{\nu}_\mu$ are different, the muonium system ($\mu^+ e^-$) cannot make transitions [PON 57, 58] into the antimuonium system ($\mu^- e^+$) in any order of approximation.

usual, the electron neutrino is defined as the particle emitted with the positron in β-decay. Its helicity has been determined experimentally and was found to be negative [GOL 58] (the helicity of $\bar{\nu}_e$ is, of course, the opposite). As regards ν_μ and $\bar{\nu}_\mu$: these particles are defined as having respectively negative and positive helicities. The decay of the π^+ meson is thus described by $\pi^+ \to \mu^+ + \nu_\mu$. This notation was used in the preceding section.

There remains one more fundamental possibility for settling the question of whether ν_e and ν_μ are different particles, and this is discussed in the following section.

4 Proposed experiment for detecting the difference between ν_e and ν_μ

The method discussed below is in essence analogous to that used to determine whether the neutrino and antineutrino (we are referring here to ν_e and $\bar{\nu}_e$) are identical [PON 46; DAV 52, 59] or whether the K^0 meson and $\bar{K}^0$ meson are identical [BAL 55, 56]. In these cases particle and antiparticle were proved to be distinct when transitions, whose matrix elements would be nonvanishing if the particle and antiparticle were identical, were not observed experimentally. For example, the absence of the process $\bar{\nu}_e + {}^{37}Cl \to {}^{37}Ar + \beta^-$ proves that ν_e and $\bar{\nu}_e$ are distinct since the process $\nu_e + {}^{37}Cl \to {}^{37}Ar + \beta^-$ without any doubt does occur.

In our case we are concerned not with the already settled problem about the distinction between neutrino and antineutrino but rather with the distinction between ν_e and ν_μ (or $\bar{\nu}_e$ and $\bar{\nu}_\mu$). If ν_e and ν_μ are different then it is already known which reactions should produce ν_e and $\bar{\nu}_e$ and should not produce ν_μ and $\bar{\nu}_\mu$ (and vice versa).

To settle the question, it is necessary to ascertain experimentally whether a beam of $\bar{\nu}_\mu$ is capable of inducing transitions that can definitely be induced by $\bar{\nu}_e$. From the experimental point of view a beam of muon neutrinos is more attractive than a beam of electron neutrinos for the following reasons. The usual intense sources of electron neutrinos are radioactive isotopes. Their very nature makes them incapable of emitting high-energy neutrinos. A good source of muon neutrinos is the $\pi - \mu$ decay in which the neutrinos are produced with high energies. It would be of interest to use a high-energy antineutrino, say $\gg 100\,\text{MeV}$, since the cross section for neutrino-induced processes grows rapidly with energy. However, at very high energies the intensity of generation of muon neutrinos is reduced due to the relativistic increase in the lifetime of the π mesons and therefore we shall discuss an experiment for a neutrino with energy $< 100\,\text{MeV}$.

Now as an example let us consider the reactions (see Table 1, processes 1–21)

$$\bar{\nu}_\mu + p \to \mu^+ + n, \tag{a}$$
$$\bar{\nu}_\mu + p \to e^+ + n. \tag{b}$$

The reaction (b), if ν_e and ν_μ are identical, was successfully observed by Reines and Cowan [REI 53, 59], and if $\nu_e \neq \nu_\mu$, the reaction is unobservable. The reaction

(a) is a threshold reaction and therefore can never be observed for ν_μ energies < 100 MeV. The problem is to determine the cross section for reaction (b). When the neutrons from reaction (b) are in the energy region where their detection is possible with good efficiency inside a large scintillation counter containing cadmium, the method of Reines and Cowan is fully applicable. When the event caused by reaction (b) takes place, two pulses will appear in the scintillation counter, one corresponding to the release of the positron energy (the neutron receives a small share of the energy) and the other, delayed with respect to the first one, corresponding to the release of the photon energy from the neutron capture in cadmium. To detect the reaction (b), a Reines and Cowan type scintillation counter may be placed in a beam of muon antineutrinos incapable of inducing reaction (a) (for energy reasons) and containing a negligibly small admixture of electron antineutrinos which could cause the "trivial" reaction $\bar{\nu}_e + p \to e^+ + n$.

To clarify the experimental conditions, let us discuss the production of neutral leptons of various kinds in cyclic accelerators where protons attain an energy of, say, 700 MeV. The radioactive elements that are produced in the target and in other parts of the accelerator are sources of ν_e and, to a smaller degree, of $\bar{\nu}_e$ of low energy ($\lesssim 10$ MeV). These electron neutrinos will not give rise to an appreciable background because (a) their energy is low and it is easy to discriminate against them by an analysis of the corresponding pulses from the scintillator; and (b) the cross section for the reaction $\bar{\nu}_e + p \to e^+ + n$ is proportional to the square of the energy of the incident antineutrinos and is relatively small at low energies.

Pions of both signs will be produced at the target of the synchrocyclotron. They will give rise to neutral leptons according to the scheme

$$(1)\ \pi^+ \to \mu^+ + \nu_\mu, \quad (2)\ \mu^+ \to e^+ + \nu_e + \bar{\nu}_\mu,$$

$$(3)\ \pi^- \to \mu^- + \bar{\nu}_\mu, \quad (4)\ \mu^- \to e^- + \bar{\nu}_e + \nu_\mu,$$

$$(5)\ \mu^- + \text{nucleus} \to \nu_\mu.$$

The admixtures of ν_e and ν_μ in the beam are harmless since it is already known that the neutrino (both ν_μ and ν_e) is incapable of inducing the reaction under study. It is easy to see that "harmful" $\bar{\nu}_e$ come only from the decay 4 of μ^- mesons. However μ^- mesons stopped in a material of high atomic number (it is not difficult to take care of the possibility of mesons stopping in light materials) practically do not decay, and μ-meson decay in flight may be ignored since the mean free path for μ-meson decay is measured in hundreds of meters and it is reasonable to place the detector for reaction (b) at a distance of ≈ 10 m from the target.

It is thus possible to achieve a beam of $\bar{\nu}_\mu$ particles with practically no admixture of $\bar{\nu}_e$. Furthermore, the $\bar{\nu}_\mu$ from reaction (2) (stopped μ^+ mesons) will have an average energy of ~ 35 MeV, and although the $\bar{\nu}_\mu$ from reaction (3) may have a significantly higher energy (decay in flight), their intensity will in general be low.[3]

The number of $\bar{\nu}_\mu$ produced in reaction (2) may be close to the number of π^+ mesons produced in the target and therefore the number of $\bar{\nu}_\mu$ that present-day machines are capable of producing could be equal to 10^{12} sec^{-1}. At the present time models of new accelerators are being discussed which would be capable in principle of a proton intensity larger by three orders of magnitude. Thus one may think that a flux $\Phi = 10^8 \bar{\nu}_\mu/\text{cm}^2$ sec at a distance of 10 m from the target may be realistic in the not too distant future. The cross section for reaction (b) was estimated using perturbation theory and found equal to 2×10^{-41} cm^2 if $\bar{\nu}_e \equiv \bar{\nu}_\mu$, for $\bar{\nu}_e$ energy equal to 35 Mev. If one were to use a scintillation counter of the Reines and Cowan type (1–2 m), then the number of events (for $\nu_e \equiv \bar{\nu}_\mu$) will be ~ 1 per hour ($\Phi \sim 10^8$ cm^{-2} sec^{-1}) for a registration efficiency equal to unity. As was recently shown by Reines and Cowan [REI 53, 59], the efficiency may exceed 0.5. Technically, the registration of one of the events under consideration is less difficult than in the Reines and Cowan experiment since the energy of the emitted β^+ particles is high. Thus the feasibility of the experiment depends on the size of the background; this is difficult to estimate a priori. One can only say that unfortunately the signal-to-background ratio should be considerably lower than in the Reines and Cowan experiment. Attention is called to the fact that the $\bar{\nu}_\mu$ from the reaction (2) are emitted isotropically, in contrast to the neutrons produced in the target. This makes it possible to reduce the difficulties due to the background from the accelerator by placing the $\bar{\nu}_\mu$ detector at an angle of $\gtrsim 90°$ with respect to the direction of the high energy protons incident on the target.

To sum up, one could say that an experiment to establish the identity of ν_e and ν_μ, although very difficult, should be seriously considered in the planning of new accelerators. In particular, the problem of shielding of the $\bar{\nu}_\mu$ detector from radiation should be looked to in the very first stages of design.

In conclusion, the author takes great pleasure in thanking Chou Kuang Chao, L. B. Okun', and Ya. A. Smorodinskiĭ for numerous discussions, and also E. M. Lipmanov, who kindly showed us, prior to publication, the paper in which arguments are presented in favor of the hypothesis of two kinds of neutrino pairs.

References

[ALV 49]   L. W. Alvarez, UCRL Report 328 (1949).
[BAL 55]   Balandin, Balashov, Zhukov, Pontecorvo, and Selivanov, *JETP* 29 (1955) 265.
[BAL 56]   Balandin, Balashov, Zhukov, Pontecorvo, and Selivanov, *Soviet Phys. JETP* 1 (1956) 98.
[BER 59]   Berley, Lee and Bardon, *Phys. Rev. Lett.* 2 (1959) 357.

[3] The $\bar{\nu}_\mu$ flux from the $\pi^- \to \mu^-$ decay in flight may be appreciable only if the high-energy proton beam falls on an external target located several meters away from the massive parts (shield, etc.) of the accelerator.

[BET 34]   H. A. Bethe and R. Peierls, *Nature* 133 (1934) 532.
[CAM 58]   A. G. W. Cameron, *Annual Review of Nuclear Science* (1958) 306. Chapter 1
[DAV 52]   R. Davis, *Phys. Rev.* 86 (1952) 976.
[DAV 59]   R. Davis, *Bull. Am. Phys. Soc.*, Washington meeting (1959).
[FEI 58]   G. Feinberg, *Phys. Rev.* 110 (1958) 1482.
[FEY 58a]   R. Feynman and M. Gell-Mann, *Phys. Rev.* 109 (1958) 193.
[FEY 58b]   R. Feynman and M. Gell-Mann, *Phys. Rev.* 109 (1958) 1860.
[FEY 58c]   R. P. Feynman and M. Gell-Mann, *Proc. Eighth Int'l. Conf. on High-Energy Physics*, Geneva (1958).
[GOL 58]   M. Goldhaber, L. Grodzin, and A. W. Sunyar, *Phys. Rev.* 109 (1958) 1015.
[LOV 59]   Love, Marder, Nadelhaft, Siegel, and Taylor, *Phys. Rev. Lett.* 1 (1958) 107.
[MAR 56]   M. A. Markov, *Proc. Sixth Int'l. Conf. on High-Energy Physics*, Rochester (1956).
[OKU 58a]   L. B. Okun', *JETP* 34 (1958) 469.
[OKU 58b]   L. B. Okun', *Soviet Phys. JETP.* 7 (1958) 322.
[ONE 59]   S. Oneda and J. C. Pati, *Phys. Rev. Lett.* 2 (1959) 125.
[POM 46]   B. Pontecorvo, Report PD-205 (1946).
[PON 57]   B. Pontecorvo, *JETP* 33 (1957) 549.
[PON 58]   B. Pontecorvo, *JETP* 6 (1958) 429.
[PON 59a]   B. Pontecorvo, *JETP* 36 (1959) 1615.
[PON 59b]   B. Pontecorvo, *Soviet Phys. JETP.* 9 (1959) 1148.
[REI 53]   F. Reines and C. L. Cowan, Jr., *Phys. Rev.* 90 (1953) 492.
[REI 59]   F. Reines and C. L. Cowan, Jr., *Phys. Rev.* 113 (1959) 273.
[SAK 46]   S. Sakata, *Prog. Theor. Phys.* 16 (1946) 686.
[SUD 57]   E. C. G. Sudarshan and R. E. Marshak, *Proc. Conf. on Mesons and Newly Discovered Particles*, Padua-Venice (1957).

1.6 Feasibility of using high-energy neutrinos to study the weak interactions

M. SCHWARTZ, 1960*

For many years, the question of how to investigate the behavior of the weak interactions at high energies has been one of considerable interest. It is the purpose of this note to show that experiments pointed in this direction, though not quite feasible with presently existing equipment, are within the capabilities of present technology and should be possible within the next decade.

We propose the use of high-energy neutrinos as a probe to investigate the weak interactions.

A natural source of high-energy neutrinos are high-energy pions. Such pions will produce neutrinos whose laboratory energy will range with equal probability from zero to 45 percent of the pion energy, and whose direction will tend very much toward the pion direction. For example, 1 BeV/c pions will emit neutrinos with an average energy of ~ 220 MeV in such a way that $\sim\frac{1}{2}$ of the neutrinos will fall within a cone of half-angle $7°$. For orientation purposes, the mean decay distance for such a pion would be 50 meters.

The best-known source of pions is a proton accelerator where the beam is allowed to impinge on a target. Let us assume that we have available a 3 BeV

* M. Schwartz, Columbia University. Reprinted from *Phys. Rev. Lett.* (1960) Vol. 4, No. 6, pages 306–7. (Submitted February 23, 1960.)

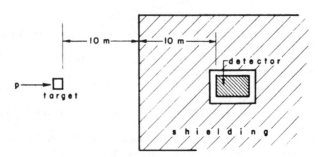

Fig. 1   Proposed experimental arrangement.

proton beam and 10 000 kilograms of material for sensing a neutrino interaction. We may then estimate the proton flux necessary to produce one interaction per hour with a cross section of σ cm^2. To do this, let us consider the simple setup shown in Fig. 1. Let I be the number of incident protons per unit time, and let, say, $I/10$ charged pions with energy ≥ 2 BeV be produced at the target. These pions emerge in a cone of about 45° half-angle, or in about 2 steradians of solid angle. We now let them travel for a distance of 10 meters before hitting a 10-meter shielding wall in front of the detector. Approximately 10 percent of the pions will decay with an average neutrino energy of about 400 MeV. Each square centimeter of detector subtends a solid angle of $\frac{1}{4} \times 10^{-6}$ steradian. Hence, the high-energy neutrino flux at the detector is $(\frac{1}{10}I)(\frac{1}{4} \times 10^{-6})(\frac{1}{10})(\frac{1}{2}) \cong 1 \times 10^{-9}I$. If there are 10000 kilograms of detector present, the number of events per unit time is given by

$$N \sim (10^7)(6 \times 10^{23})(10^{-9}I)\sigma = 6 \times 10^{21}I\sigma.$$

For an intensity of $I = 5 \times 10^{12}$ protons/sec $= 1.8 \times 10^{16}$ protons/hr, the high-energy neutrino flux is ~ 5000 neutrinos sec^{-1} cm^{-2}. With a cross section $\sigma \sim 10^{-38}$ cm^2, the number of counts is $N \sim 1$ per hour in 10 000 kg of detector. The estimate here given is for neutrinos from high-energy pions. There is, as a matter of fact, a much greater flux of lower-energy neutrinos from lower-energy pions. However, because the neutrino cross section decreases rapidly with decreasing energy, the rate is not likely to be improved by more than a factor of two.

This estimate places the experiment outside the capabilities of existing machines by one or two orders of magnitude. Optimistic estimates for accelerators which are currently under construction, namely, the 3-BeV machine at Princeton and the 10-BeV machine at Argonne, indicate that the experiments may be barely feasible in the near future. However, for really quantitative experiments it will be necessary to use high-intensity machines such as the FFAG machine proposed by MURA or the 10-BeV linear proton accelerator discussed by Blewett at Brookhaven. In these machines, one hopes to attain a beam intensity of the order of 10^{15} protons/sec at an energy of about 10 BeV.

The higher energy of the primary beam of protons makes the experiment easier because of the increased multiplicity of pions, the more concentrated forward distribution of the pions, and the increased cross section for neutrino reactions. Balanced against these is the fact that the percentage of higher-energy pions that decay in 10 meters is smaller. The net result is likely to give a counting rate per primary proton that probably increases more than linearly with the primary proton energy.

Thus, a high-intensity 10-BeV proton machine with a beam intensity $\sim 10^{15}$ protons/sec may give a counting rate of more than 10^3 per hour, using the experimental setup described above. If that proves to be the case, it is perhaps desirable to have magnetic lenses to analyze and focus the pions so as to obtain more monoenergetic neutrino beams.

I would like to express my gratitude to Dr. T. D. Lee and Dr. C. N. Yang for many stimulating discussions which led to the above proposal.

Note added in proof. The author's attention has been called to a somewhat related paper which has just appeared: B. Pontecorvo, *J. Exptl. Theoret. Phys.* (U.S.S.R.) 37, 1751 (1959). [This paper is reprinted in this volume.]

1.7 Observation of high-energy neutrino reactions and the existence of two kinds of neutrinos

G. DANBY, J.-M. GAILLARD, K. GOULIANOS, L. M. LEDERMAN,

N. MISTRY, M. SCHWARTZ, AND J. STEINBERGER, 1962*

In the course of an experiment at the Brookhaven AGS, we have observed the interaction of high-energy neutrinos with matter. These neutrinos were produced primarily as the result of the decay of the pion:

$$\pi^{\pm} \rightarrow \mu^{\pm} + (\nu/\bar{\nu}). \tag{1}$$

It is the purpose of this letter to report some of the results of this experiment including (1) demonstration that the neutrinos we have used produce μ mesons but do not produce electrons, and hence are very likely different from the neutrinos involved in β-decay and (2) approximate cross sections.

1 Behavior of cross section as a function of energy

The Fermi theory of weak interactions, which works well at low energies, implies a cross section for weak interactions which increases as phase space. Calculation indicates that weak interacting cross sections should be in the neighborhood of

* G. Danby, J.-M. Gaillard, K. Goulianos, L. M. Lederman, N. Mistry, M. Schwartz, and J. Steinberger, Columbia University and Brookhaven National Laboratory. Reprinted from *Phys. Rev. Lett.*, July 1, 1962, Vol. 9, No. 1, pages 36–44. (Received 15 June 1962.)

10^{-38} cm^2 at about 1 BeV. Lee and Yang [LEE 60b] first calculated the detailed cross sections for

$$\nu + n \rightarrow p + e^-$$
$$\bar{\nu} + p \rightarrow n + e^+ \tag{2}$$

$$\nu + n \rightarrow p + \mu^-$$
$$\bar{\nu} + p \rightarrow n + \mu^+ \tag{3}$$

using the vector form factor deduced from electron scattering results and assuming the axial vector form factor to be the same as the vector form factor. Subsequent work has been done by Yamaguchi [YAM 60] and Cabibbo and Gatto [CAB 60]. These calculations have been used as standards for comparison with experiments.

2 Unitarity and the absence of the decay $\mu \rightarrow e + \gamma$

A major difficulty of the Fermi theory at high energies is the necessity that it break down before the cross section reaches $\pi\lambda^2$, violating unitarity. This breakdown must occur below 300 BeV in the center of mass. This difficulty may be avoided if an intermediate boson mediates the weak interactions. Feinberg [FEI 58] pointed out, however, that such a boson implies a branching ratio $(\mu \rightarrow e + \gamma)/(\mu \rightarrow e + \nu + \bar{\nu})$ of the order of 10^{-4}, unless the neutrinos associated with muons are different from those associated with electrons.[1] Lee and Yang[2] have subsequently noted that any general mechanism which would preserve unitarity should lead to a $\mu \rightarrow e + \gamma$ branching ratio not too different from the above. Inasmuch as the branching ratio is measured to be $\lesssim 10^{-8}$ [BAR 62; FRA 62], the hypothesis that the two neutrinos may be different has found some favor. It is expected that if there is only one type of neutrino, then neutrino interactions should produce muons and electrons in equal abundance. In the event that there are two neutrinos, there is no reason to expect any electrons at all.

The feasibility of doing neutrino experiments at accelerators was proposed independently by Pontecorvo [PON 59] and Schwartz [SCH 60]. It was shown that the fluxes of neutrinos available from accelerators should produce of the order of several events per day per 10 tons of detector.

The essential scheme of the experiment is as follows: A neutrino "beam" is generated by decay in flight of pions according to reaction (1). The pions are produced by 15-BeV protons striking a beryllium target at one end of a 10-ft.-long straight section. The resulting entire flux of particles moving in the general direction of the detector strikes a 13.5-m-thick iron shield wall at a distance of 21 m from the target. Neutrino interactions are observed in a 10-ton aluminum spark chamber located behind this shield.

---

[1] Several authors have discussed this possibility. Some of the earlier viewpoints are given by [KON 53, 58; KAW 57; NIS 57; SCH 57; BLU 59; ONE 59; LEE 60b].
[2] T. D. Lee and C. N. Yang (private communications). See also [LEE 60a].

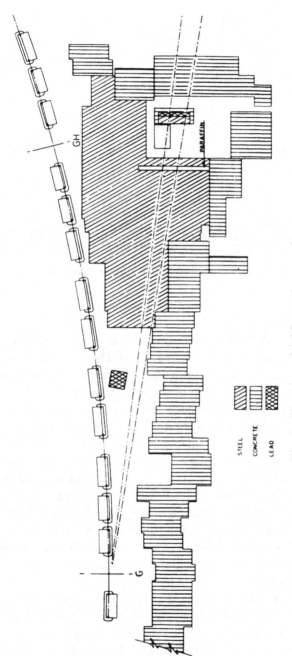

Fig. 1   Plan view of AGS neutrino experiment.

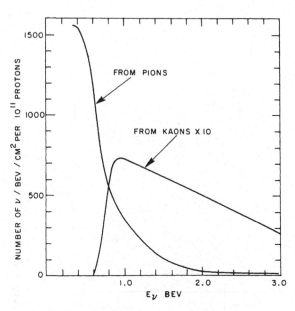

Fig. 2   Energy spectrum of neutrinos expected in the arrangement of Fig. 1 for 15-BeV protons on Be.

The line of flight of the beam from target to detector makes an angle of 7.5° with respect to the internal proton direction (see Fig. 1). The operating energy of 15 BeV is chosen to keep the muons penetrating the shield to a tolerable level.

The number and energy spectrum of neutrinos from reaction (1) can be rather well calculated, on the basis of measured pion-production rates [BAK 61] and the geometry. The expected neutrino flux from π decay is shown in Fig. 2. Also shown is an estimate of neutrinos from the decay $K^{\pm} \rightarrow \mu^{\pm} + \nu(\bar{\nu})$. Various checks were performed to compare the targeting efficiency (fraction of circulating beam that interacts in the target) during the neutrino run with the efficiency during the beam survey run. (We believe this efficiency to be close to 70 percent.) The pion-neutrino flux is considered reliable to approximately 30 percent down to 300 MeV/c, but the flux below this momentum does not contribute to the results we wish to present.

The main shielding wall thickness, 13.5 m for most of the run, absorbs strongly interacting particles by nuclear interaction and muons up to 17 BeV by ionization loss. The absorption mean free path in iron for pions of 3, 6, and 9 BeV has been measured to be less than 0.24 m [COO n.d.]. Thus the shield provides an attenuation of the order of 10^{-24} for strongly interacting particles. This attenuation is more than sufficient to reduce these particles to a level compatible with this experiment. The background of strongly interacting particles within the detector shield probably enters through the concrete floor and roof of the 5.5-m-thick side wall. Indications of such leaks were, in fact, obtained during the early phases of the experiment and the shielding subsequently improved. The argument that our observations are not

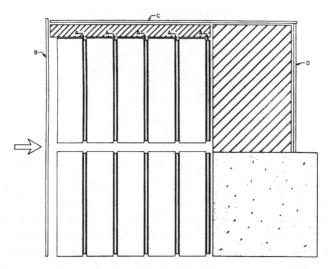

Fig. 3 Spark chamber and counter arrangement. A are the triggering slabs; B, C, and D are anticoincidence slabs. This is the front view seen by the four-camera stereo system.

induced by strongly interacting particles will also be made on the basis of the detailed structure of the data.

The spark chamber detector consists of an array of 10 one-ton modules. Each unit has nine aluminum plates 44 in. × 44 in. × 1 in. thick, separated by $\frac{3}{8}$-in. Lucite spacers. Each module is driven by a specially designed high-pressure spark gap and the entire assembly triggered as described below. The chamber will be more fully described elsewhere. Figure 3 illustrates the arrangement of coincidence and anticoincidence counters. Top, back, and front anticoincidence sheets (a total of 50 counters, each 48 in. × 11 in. × $\frac{1}{2}$ in.) are provided to reduce the effect of cosmic rays and AGS-produced muons which penetrate the shield. The top slab is shielded against neutrino events by 6 in. of steel and the back slab by 3 ft. of steel and lead.

Triggering counters were inserted between adjacent chambers and at the end (see Fig. 3). These consist of pairs of counters, 48 in. × 11 in. × $\frac{1}{2}$ in., separated by $\frac{3}{4}$ in. of aluminum, and in fast coincidence. Four such pairs cover a chamber; 40 are employed in all.

The AGS at 15 BeV operates with a repetition period of 1.2 sec. A rapid beam deflector drives the protons onto the 3-in.-thick Be target over a period of 20–30 μsec. The radiation during this interval has rf structure, the individual bursts being 20 nsec wide, the separation 220 nsec. This structure is employed to reduce the total "on" time and thus minimize cosmic-ray background. A Čerenkov counter exposed to the pions in the neutrino "beam" provides a train of 30-nsec gates, which is placed in coincidence with the triggering events. The correct phasing is verified by raising the machine energy to 25 BeV and counting the high-energy

Fig. 4   Land print of cosmic-ray muons integrated over many incoming tracks.

muons which now penetrate the shield. The tight timing also serves the useful function of reducing sensitivity to low-energy neutrons which diffuse into the detector room. The trigger consists of a fast twofold coincidence in any of the 40 coincidence pairs in anticoincidence with the anticoincidence shield. Typical operation yields about 10 triggers per hour. Half the photographs are blank; the remainder consist of AGS muons entering unprotected faces of the chamber, cosmic rays, and "events". In order to verify the operation of circuits and the gap efficiency of the chamber, cosmic-ray test runs are conducted every four hours. These consist of triggering on almost horizontal cosmic-ray muons and recording the results both on film and on Land prints for rapid inspection (see Fig. 4).

A convenient monitor for this experiment is the number of circulating protons in the AGS machine. Typically, the AGS operates at a level of 2–4×10^{11} protons per pulse, and 3000 pulses per hour. In an exposure of 3.48×10^{17} protons, we have counted 113 events satisfying the following geometric criteria: The event originates within a fiducial volume whose boundaries lie 4 in. from the front and back walls of the chamber and 2 in. from the top and bottom walls. The first two gaps must not fire, in order to exclude events whose origins lie outside the chambers. In addition, in the case of events consisting of a single track, an extrapolation of the track backwards (towards the neutrino source) for two gaps must also remain within the fiducial volume. The production angle of these single tracks relative to the neutrino line of flight must be less than $60°$.

These 113 events may be classified further as follows:

a Forty-nine short single tracks. These are single tracks whose *visible* momentum, if interpreted as muons, is less than $300 \, \text{MeV}/c$. These presumably include some energetic muons which leave the chamber. They also include low-energy neutrino events and the bulk of the neutron produced background. Of these, 19 have 4 sparks or less. The second half of the run (1.7×10^{17} protons) with improved shielding yielded only three tracks in this category. We will not consider these as acceptable "events".

b Thirty-four "single muons" of more than $300 \, \text{MeV}/c$. These include tracks which, if interpreted as muons, have a visible range in the chambers such that their momentum is at least $300 \, \text{MeV}/c$. The origin of these events must not be accompanied by more than two extraneous sparks. The latter requirement means that we include among "single tracks" events showing a small recoil. The

Table 1. *Classification of "events"*

Single tracks			
$p_\mu < 300\,\mathrm{MeV}/c^a$	49	$p_\mu > 500$	8
$p_\mu > 300$	34	$p_\mu > 600$	3
$p_\mu > 400$	19	$p_\mu > 700$	2

Total "events" 34

Vertex events

Visible energy released $< 1\,\mathrm{BeV}$ 15

Visible energy released $> 1\,\mathrm{BeV}$ 7

[a]These are not included in the "event" count (see text).

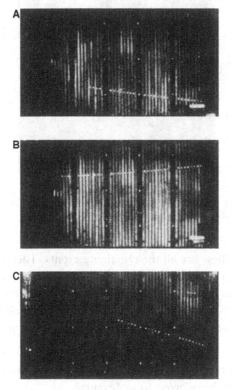

Fig. 5 Single muon events. (A) $p_\mu > 540\,\mathrm{MeV}/c$ and δ ray indicating direction of motion (neutrino beam incident from left); (B) $p_\mu > 700\,\mathrm{MeV}/c$; (C) $p_\mu > 440\,\mathrm{MeV}/c$ with δ ray.

34 events are tabulated as a function of momentum in Table 1. Figure 5 illustrates three "single muon" events.

c Twenty-two "vertex" events. A vertex event is one whose origin is characterized by more than one track. All of these events show a substantial energy release. Figure 6 illustrates some of these.

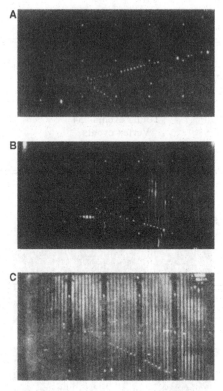

Fig. 6 Vertex events. (A) Single muon of $p_\mu > 500\,\mathrm{MeV}/c$ and electron-type track; (B) possible example of two muons, both leave chamber; (C) four-prong star with one long track of $p_\mu > 600\,\mathrm{MeV}/c$.

d Eight "showers". These are all the remaining events. They are in general single tracks, too irregular in structure to be typical of μ mesons, and more typical of electron or photon showers. From these eight "showers," for purposes of comparison with (b), we may select a group of six which are so located that their potential range within the chamber corresponds to μ mesons in excess of $300\,\mathrm{MeV}/c$.

In the following, only the 56 energetic events of type (b) (long μ's) and type (c) (vertex events) will be referred to as "events".

Arguments on the neutrino origin of the observed "events".

1. The "events" are not produced by cosmic rays.   Muons from cosmic rays which stop in the chamber can and do simulate neutrino events. This background is measured experimentally by running with the AGS machine off on the same triggering arrangement except for the Čerenkov gating requirement. The actual triggering rate then rises from 10 per hour to 80 per second (a dead-time circuit prevents jamming of the spark chamber). In 1,800 cosmic-ray photographs thus obtained, 21 would be accepted as neutrino events. Thus 1 in 90 cosmic-ray events is neutrinolike. Čerenkov gating and the short AGS pulse effect a reduction by a factor of $\sim 10^{-6}$ since the circuits are "on" for only 3.5 μsec per pulse. In fact, for the body of data represented by Table 1, a total of 1.6×10^6 pulses were counted. The equipment was therefore sensitive for a total time of 5.5 sec. This should lead to $5.5 \times 80 = 440$ cosmic-ray tracks which is consistent with observation. Among these, there should be 5 ± 1 cosmic-ray induced "events." These are almost evident in the small asymmetry seen in the angular distributions of Fig. 7. The remaining 51 events cannot be the result of cosmic rays.

2. The "events" are not neutron produced. Several observations contribute to this conclusion:

a The origins of all the observed events are uniformly distributed over the fiduciary volume, with the obvious bias against the last chamber induced by the $p_\mu > 300 \, \text{MeV}/c$ requirement. Thus there is no evidence for attenuation, although the mean free path for nuclear interaction in aluminum is 40 cm and for electromagnetic interaction 9 cm.

b The front iron shield is so thick that we can expect less than 10^{-4} neutron-induced reactions in the entire run from neutrons which have penetrated this shield. This was checked by removing 4 ft. of iron from the front of the thick shield. If our events were due to neutrons in line with the target, the event rate would have increased by a factor of one hundred. No such effect was observed (see Table 2). If neutrons penetrate the shield, it must be from other directions. The secondaries would reflect this directionality. The observed angular distribution of single-track events is shown in Fig. 7. Except for the small cosmic-ray contribution to the vertical plane projection, both projections are peaked about the line of flight to the target.

c If our 29 single-track events (excluding cosmic-ray background) were pions produced by neutrons, we would have expected, on the basis of known production cross sections, of the order of 15 single π^0's to have been produced. No cases of unaccompanied π^0's have been observed.

3. The single particles produced show little or no nuclear interaction and are therefore presumed to be muons. For the purpose of this argument, it is convenient to first discuss the second half of our data, obtained after some shielding improvements were effected. A total traversal of 820 cm of aluminum by single tracks was observed, but no "clear" case of nuclear interaction such as large angle or charge exchange scattering was seen. In a spark chamber calibration experiment at the Cosmotron, it was found that for 400-MeV pions the mean free path for "clear" nuclear interactions in the chamber (as distinguished from stoppings) is no more than 100 cm of aluminum. We should, therefore, have observed of the order of eight "clear" interactions; instead we observed none. The mean free path for the observed single tracks is then more than eight times the nuclear mean free path.

Table 2. *Event rates for normal and background conditions*

	Circulating protons × 10^16	No. of events	Calculated cosmic-ray[a] contribution	Net rate per 10^16
Normal run	34.8	56	5	1.46
Background I[b]	3.0	2	0.5	0.5
Background II[c]	8.6	4	1.5	0.3

[a] These should be subtracted from the "single muon" category.
[b] 4 ft. of Fe removed from main shielding wall.
[c] As above, but 4 ft. of Pb placed within 6 ft. of Be target and subtending a horizontal angular interval from 4° to 11° with respect to the internal proton beam.

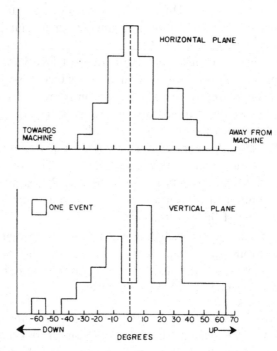

Fig. 7   Projected angular distributions of single track events. Zero degrees is defined as the neutrino direction.

Included in the count are five tracks which stop in the chamber. Certainly a fraction of the neutrino secondaries must be expected to be produced with such small momentum that they would stop in the chamber. Thus, none of these stoppings may, in fact, be nuclear interactions. But even if all stopping tracks are considered to represent nuclear interactions, the mean free path of the observed single tracks must be 4 nuclear mean free paths.

The situation in the case of the earlier data is more complicated. We suspect that a fair fraction of the short single tracks then observed are, in fact, protons produced in neutron collisions. However, similar arguments can be made also for these data which convince us that the energetic single-track events observed then are also noninteracting.[3]

It is concluded that the observed single-track events are muons, as expected from neutrino interactions.

4. The observed reactions are due to the decay products of pions and K mesons. In a second background run, 4 ft. of iron was removed from the main shield and replaced by a similar quantity of lead placed as close to the target as feasible. Thus, the detector views the target through the same number of mean free paths of shielding material. However, the path available for pions to decay is reduced by a factor of eight. This is the closest we could come to "turning off" the neutrinos. The results of this run are given in terms of the number of events per 10^{16} circulating protons in Table 2. The rate of "events" is reduced from 1.46 ± 0.2 to 0.3 ± 0.2 per 10^{16} incident protons. This reduction is consistent with that which is expected for neutrinos which are the decay products of pions and K mesons.

Are there two kinds of neutrinos? The earlier discussion leads us to ask if the reactions (2) and (3) occur with the same rate. This would be expected if ν_μ, the neutrino coupled to the muon and produced in pion decay, is the same as ν_e, the neutrino coupled to the electron and produced in nuclear beta decay. We discuss only the single-track events where the distinction between single muon tracks of $p_\mu < 300 \, \text{MeV}/c$ and showers produced by high-energy single electrons is clear. See Figs. 8 and 4, which illustrate this difference.

We have observed 34 single muon events of which 5 are considered to be cosmic-ray background. If $\nu_\mu = \nu_e$, there should be of the order of 29 electron showers with a mean energy greater than $400 \, \text{MeV}/c$. Instead, the only candidates which we have for such events are six "showers" of qualitatively different appearance from those of Fig. 8. To argue more precisely, we have exposed two of our one-ton spark chamber modules to electron beams at the Cosmotron. Runs were taken at various electron energies. From these we establish that the triggering efficiency for 400-MeV electrons is 67 percent. As a quantity characteristic of the calibration showers, we have taken the total number of observed sparks. The mean number is roughly linear with electron energy up to $400 \, \text{MeV}/c$. Larger showers saturate the two chambers which were available. The spark distribution for $400 \, \text{MeV}/c$ showers is plotted in Fig. 9, normalized to the $\frac{2}{3} \times 29$ expected showers. The six "shower" events are also plotted. It is evident that these are not consistent with the prediction based on a universal theory with $\nu_\mu = \nu_e$. It can perhaps be argued that the absence of electron events could be understood in terms of the coupling of a single neutrino to

---

[3] These will be published in a more complete report.

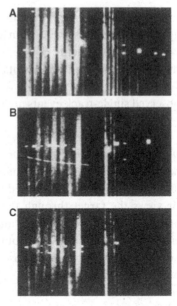

Fig. 8   400-MeV electrons from the cosmotron.

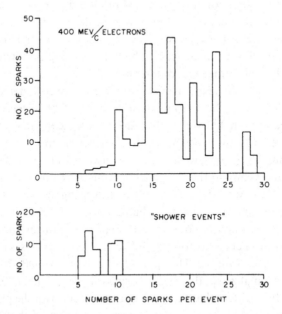

Fig. 9   Spark distribution for 400-MeV/c electrons normalized to expected number of showers. Also shown are the "shower" events.

the electron which is much weaker than that to the muon at higher momentum transfers, although at lower momentum transfers the results of β decay, μ capture, μ decay, and the ratio of $\pi \rightarrow \mu + \nu$ to $\pi \rightarrow e + \nu$ decay show that these couplings are equal [AND 60; CUL 61; HIL 62; BLE 62]. However, the most plausible explanation for the absence of the electron showers, and the only one which preserves universality, is then that $\nu_\mu \neq \nu_e$; that is, that there are at least two types of neutrinos. This also resolves the problem raised by the forbiddenness of the $\mu^+ \rightarrow e^+ + \gamma$ decay.

It remains to understand the nature of the six "shower" events. All of these events were obtained in the first part of the run during conditions in which there was certainly some neutron background. It is not unlikely that some of the events are small neutron produced stars. One or two could, in fact, be μ mesons. It should also be remarked that of the order of one or two electron events are expected from the neutrinos produced in the decays $K^+ \rightarrow e^+ + \nu_e + \pi^0$ and $K_2^0 \rightarrow e^\pm + \nu_e + \pi^\mp$.

3 The intermediate boson

It has been pointed out [LEE 60b] that high-energy neutrinos should serve as a reasonable method of investigating the existence of an intermediate boson in the weak interactions. In recent years many of the objections to such a particle have been removed by the advent of V-A theory [FEY 58a; MAR 58] and the remeasurement of the ρ value in μ decay [PLA 60]. The remaining difficulty pointed out by Feinberg [FEI 58], namely the absence of the decay $\mu \rightarrow e + \gamma$, is removed by the results of this experiment. Consequently it is of interest to explore the extent to which our experiment has been sensitive to the production of these bosons.

Our neutrino intensity, in particular that part contributed by the K-meson decays, is sufficient to have produced intermediate bosons if the boson had a mass m_w less than that of the mass of the proton (m_p). In particular, if the boson has a mass equal to $0.6 m_p$, we should have produced ~ 20 bosons by the process $\nu + p \rightarrow w^+ + \mu^- + p$. If $m_w = m_p$, then we should have observed two such events [LEE 61].

Indeed, of our vertex events, five are consistent with the production of a boson. Two events, with two outgoing prongs, one of which is shown in Fig. 6(B), are consistent with both prongs being muons. This could correspond to the decay mode $w^+ \rightarrow \mu^+ + \nu$. One event shows four outgoing tracks, each of which leaves the chamber after traveling through 9 in. of aluminum. This might in principle be an example of $w^+ \rightarrow \pi^+ + \pi^- + \pi^+$. Another event, by far our most spectacular one, can be interpreted as having a muon, a charged pion, and two gamma rays presumably from a neutral pion. Over 2 BeV of energy release is seen in the chamber. This could in principle be an example of $w^+ \rightarrow \pi^+ + \pi^0$. Finally, we have one event, Fig. 6(A), in which both a muon and an electron appear to leave the same vertex. If this were a boson production, it would correspond to the boson decay mode $w^+ \rightarrow e^+ + \nu$. The alternative explanation for this event would require (1) that a neutral pion be produced with the muon; and (2) that one of its gamma rays convert in the plate of the interaction while the other not convert visibly in the chamber.

The difficulty of demonstrating the existence of a boson is inherent in the poor resolution of the chamber. Future experiments should shed more light on this interesting question.

4 Neutrino cross sections

We have attempted to compare our observations with the predicted cross sections for reactions (2) using the theory [LEE 60b; YAM 60; CAB 60]. To include the fact that the nucleons in (2) are, in fact, part of an aluminum nucleus, a Monte Carlo calculation was performed using a simple Fermi model for the nucleus in order to evaluate the effect of the Pauli principle and nucleon motion. This was then used to predict the number of "elastic" neutrino events to be expected under our conditions. The results agree with simpler calculations based on Fig. 2 to give, in terms of number of circulating protons,

$$\text{from } \pi \rightarrow \mu + \nu, \quad 0.60 \text{ events}/10^{16} \text{ protons,}$$
$$\text{from } K \rightarrow \mu + \nu, \quad 0.15 \text{ events}/10^{16} \text{ protons,}$$
$$\text{Total} \quad 0.75 \text{ events}/10^{16} \pm \sim 30\%.$$

The observed rates, assuming all single muons are "elastic" and all vertex events "inelastic" (i.e., produced with pions) are

$$\text{"Elastic": } 0.84 \pm 0.16 \text{ events}/10^{16} (29 \text{ events})$$
$$\text{"Inelastic": } 0.63 \pm 0.14 \text{ events}/10^{16} (22 \text{ events}).$$

The agreement of our elastic yield with theory indicates that no large modification to the Fermi interaction is required at our mean momentum transfer of 350 MeV/c. The inelastic cross section in this region is of the same order as the elastic cross section.

5 Neutrino flip hypothesis

Feinberg, Gursey, and Pais [FEI 61] have pointed out that if there were two different types of neutrinos, their assignment to muon and electron, respectively, could in principle be interchanged for strangeness-violating weak interactions. Thus it might be possible that

$$\pi^+ \rightarrow \mu^+ + \nu_1 \qquad \qquad K^+ \rightarrow \mu^+ + \nu_2$$
$$\text{while}$$
$$\pi^+ \rightarrow e^+ + \nu_2 \qquad \qquad K^+ \rightarrow e^+ + \nu_1.$$

This hypothesis is subject to experimental check by observing whether neutrinos from $K_{\mu 2}$ decay produce muons or electrons in our chamber. Our calculation of the neutrino flux from $K_{\mu 2}$ decay indicates that we should have observed five events from these neutrinos. They would have an average energy of 1.5 BeV. An electron of this energy would have been clearly recognizable. None have been seen. It seems unlikely therefore that the neutrino flip hypothesis is correct.

The authors are indebted to Professor G. Feinberg, Professor T. D. Lee, and Professor C. N. Yang for many fruitful discussions. In particular, we note here that the emphasis by Lee and Yang on the importance of the high-energy behavior of weak interactions and the likelihood of the existence of two neutrinos played an important part in stimulating this research.

We would like to thank Mr. Warner Hayes for technical assistance throughout the experiment. In the construction of the spark chamber, R. Hodor and R. Lundgren of BNL, and Joseph Shill and Yin Au of Nevis did the engineering. The construction of the electronics was largely the work of the Instrumentation Division of BNL under W. Higinbotham. Other technical assistance was rendered by M. Katz and D. Balzarini. Robert Erlich was responsible for the machine calculations of neutrino rates, M. Tannenbaum assisted in the Cosmotron runs.

The experiment could not have succeeded without the tremendous efforts of the Brookhaven Accelerator Division. We owe much to the cooperation of Dr. K. Green, Dr. E. Courant, Dr. J. Blewett, Dr. M. H. Blewett, and the AGS staff including J. Spiro, W. Walker, D. Sisson, and L. Chimienti. The Cosmotron Department is acknowledged for its help in the initial assembly and later calibration runs.

The work was generously supported by the U.S. Atomic Energy Commission. The work at Nevis was considerably facilitated by Dr. W. F. Goodell, Jr., and the Nevis Cyclotron staff under Office of Naval Research support.

References

[AND 60]   H. L. Anderson, T. Fujui, R. H. Miller, and L. Tsu, *Phys. Rev.* 119 (1960) 2050.

[BAK 61]   W. F. Baker *et al.*, *Phys. Rev. Lett.* 7 (1961) 101.

[BAR 62]   D. Bartlett, S. Devons, and A. Sacha. *Phys. Rev. Lett.* 8 (1962) 120.

[BLE 62]   E. Bleser, L. Lederman, J. Rosen, J. Rothberg, and E. Zavattini, *Phys. Rev. Lett.* 8 (1962) 288.

[BLU 59]   S. A. Bludman, *Bull. Am. Phys. Soc.* 4 (1959) 80.

[CAB 60]   N. Cabibbo and R. Gatto, *Nuovo Cimento* 15 (1960) 304.

[COO n.d.]   R. L. Cool, L. Lederman, L. Marshall, A. C. Melissinos, M. Tannenbaum, J. H. Tinlot, and T. Yamanouchi, Brookhaven National Laboratory Internal Report UP-18 (unpublished).

[CUL 61]   G. Culligan, J. F. Lathrop, V. L. Telegdi, R. Winston, and R. A. Lundy, *Phys. Rev. Lett.* 7 (1961), 458

[FEI 58]   G. Feinberg, *Phys. Rev.* 110 (1958) 1482.

[FEI 61]   G. Feinberg, F. Gürsey, and A. Pais, *Phys. Rev. Lett.* 7 (1961) 208.

[FEY 58]   R. Feynman and M. Gell-Mason, *Phys. Rev.* 109 (1958) 193.

[FRA 62]   S. Frankel, J. Halpern, L. Holloway, W. Wales, M. Yearian, O. Chamberlain, A. Lemonick, and F. M. Pipkin, *Phys. Rev. Lett.* 8 (1962) 123.

[HIL 62]   R. Hildebrand, *Phys. Rev. Lett.* 8 (1962) 34.

[KAW 57]   I. Kawakami, *Progr. Theoret. Phys.* (Kyoto) 19 (1957) 459.

[KON 53]   E. Konopinski and H. Mahmoud, *Phys. Rev.* 92 (1953) 1045.

[KON 58]   M. Konuma, *Nuclear Phys.* 5 (1958) 504.

[LEE 60a]   T. D. Lee and C. N. Yang, *Proc. Int'l. Conf. on High-Energy Physics.* Rochester, Interscience (1960) 567.

[LEE 60b]   T. D. Lee and C. N. Yang, *Phys. Rev. Lett.* 4 (1960) 307.
[LEE 61]   T. D. Lee, P. Markstein, and C. N. Yang, *Phys. Rev. Lett.* 7 (1961) 429.
[MAR 58]   R. Marshak and E. Sudershan, *Phys. Rev.* 109 (1958) 1860.
[NIS 57]   K. Nishijima, *Phys. Rev.* 108 (1957) 907.
[ONE 59]   S. Oneda and J. C. Pati, *Phys. Rev. Lett.* 2 (1959) 125.
[PLA 60]   R. J. Plano, *Phys. Rev.* 119 (1960) 1400.
[PON 59]   B. Pontecorvo, *JETP* (USSR) 37 (1959) 1751.
[SCH 57]   J. Schwinger, *Ann. Phys.* (1957) 407.
[SCH 60]   M. Schwartz, *Phys. Rev. Lett.* 4 (1960) 306.
[YAM 60]   Y. Yamaguchi, *Progr. Theoret. Phys.* 23 (1960) 1117.

1.8 Neutrino interactions in the CERN heavy liquid bubble chamber

M. M. BLOCK, H. BURMEISTER, D. C. CUNDY, B. EIBEN, C. FRANZINETTI,

J. KEREN, R. MØLLERUD, G. MYATT, M. NIKOLIC, A. ORKIN-LECOURTOIS,

M. PATY, D. H. PERKINS, C.A. RAMM, K. SCHULTZE, H. SLETTEN,

K. SOOP, R. STUMP, W. VENUS AND H. YOSHIKI, 1964*

Preliminary results of the analysis of neutrino interactions in the CERN 500-liter freon chamber have been reported at the Sienna Conference [BIN 63; BEL 63]. This paper presents results of a more detailed analysis of a total of 459 events. The new data were obtained in experimental conditions similar to those described previously, apart from modifications to the inner conductor of the magnetic horn, which increased the high-energy flux.

Of the 459 events, 454 contain a negative muon candidate, and 5 a negative electron of energy exceeding 400 MeV. We expect 3.3 electron events from the ν_e flux resulting from Ke_3 decay; thus our data confirm the earlier findings of the Brookhaven–Columbia group regarding the two-neutrino hypothesis [DAN 62: This paper is reprinted in this volume]. Of the muon events, 236 contain no pions (nonpionic), 209 contain pions, and 9 contain pions and strange particles. Figure 1 shows the visible energy distributions of the different classes of events. It must be emphasized that all events occur in complex nuclei, and that the characteristics of the elementary neutrino–nucleon interaction are modified both by Fermi motion and by secondary nuclear processes.

1 Elastic process

From the 236 nonpionic events containing a muon and one or more protons, we have attempted to extract those due to the elastic process:

$$\nu + n \rightarrow \mu^- + p. \tag{1}$$

The events observed will contain background. By selecting those of visible energy $E_{vis} > 1$ GeV, contamination from interactions due to neutrons or incoming

* M. M. Block, H. Burmeister, D. C. Cundy, B. Eiben, C. Franzinetti, J. Keren, R. Møllerud, G. Myatt, M. Nikolic, A. Orkin-Lecourtois, M. Paty, D. H. Perkins, C. A. Ramm, K. Schultze, H. Sletten, K. Soop, R. Stump, W. Venus, and H. Yoshiki, CERN, Geneva, Switzerland, Reprinted from *Phys. Lett.*, October 1, 1964, Vol. 12, No. 3, pages 281–5. (Received 21 September 1964.)

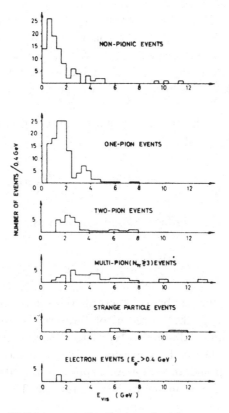

Fig. 1   Visible energy distribution of various event types.

charged particles can be shown to be negligible. The most serious remaining source of contamination is from neutrino events in which a pion is produced in the primary collision, and subsequently reabsorbed in its passage through the parent nucleus. For example, if N^* $(\frac{3}{2}, \frac{3}{2})$ production is assumed to dominate in the observed one-pion events, the absorption probability can be estimated. For $E_{vis} > 1$ GeV, the expected contamination of elastic events would be ~ 25 percent. The corresponding loss of elastic events, in the case where the outgoing proton creates a pion in a secondary collision, is known to be negligible.

The nucleon form factors for the weak interactions in elastic processes can be computed from the distribution of the squared four-momentum transfer $q^2 = (P_\nu - P_\mu)^2$. P_ν and P_μ are the four-momenta of the incoming neutrino and the outgoing muon. For each event, q^2 can be determined experimentally from the momentum and direction of the muon, assuming the event to be elastic and the target nucleon at rest. The background from inelastic events was subtracted on the basis of a comparison of the q^2 distribution of both pionic and nonpionic events. For $E_{vis} > 1$ GeV, 120 selected nonpionic events were analyzed.

Assuming the CVC hypothesis [FEY 58], G-symmetry, and time reversal invariance, and neglecting pseudoscalar terms and possible effects due to the intermediate boson, $d\sigma(E_\nu)/dq^2$ can be expressed in terms of the electromagnetic isovector form-factors F_1 and F_2, and the axial form-factor F_A. Electron scattering experiments are consistent with $F_1 = F_2 = F_V = (1 + q^2/M_V^2)^{-2}$ with $M_V = 0.84$ GeV. Therefore $d\sigma(E_\nu)/dq^2$ is determined except for F_A. If the parametric form $F_A = (1 + q^2/M_2^A)^{-2}$ is assumed, M_A can be determined from the experimental data.

To avoid errors due to the uncertainty in the neutrino spectrum, the expected q^2 distribution summed over all energies:

$$\frac{dN}{dq^2} dq^2 = dq^2 \int \varphi(E_\nu) \frac{d\sigma(E_\nu | M_A)}{dq^2} dE_\nu \qquad (2)$$

was then calculated by taking the neutrino spectrum as:

$$\varphi(E_\nu) dE_\nu \approx \frac{1}{\sigma(E_\nu | M_A)} \frac{dN}{dE_\nu} \Delta E_\nu,$$

where $(dN/dE_\nu)\Delta E_\nu$ is the number of observed events in the energy interval E_ν to $E_\nu + \Delta E_\nu$, and $\sigma(E_\nu | M_A)$ is the theoretical total cross section for the elastic process. The relation (2) was compared with the observed q^2 distribution and M_A determined by likelihood methods. The analysis then does not depend on assumptions about the absolute ν flux. Appropriate corrections were applied to the calculated distributions for the effects of Fermi motion and exclusion principle. This analysis yields

$$M_A = 1.0^{+0.35}_{-0.20} \text{ GeV}$$

where the errors are purely statistical, the observed and expected q^2 distributions for this value of M_A are shown in Fig. 2. If the background subtraction is doubled, the best value of M_A is 1.4 ± 0.4 GeV; it is clear that the systematic error due to uncertainty in the background contribution may be comparable with the statistical errors. We conclude that

$$M_A = 1.0^{+0.5}_{-0.3} \text{ GeV}$$

showing that $F_A \approx F_V$.

It is also possible to extract F_A directly from the data. The absolute neutrino flux can be estimated from the events in the range $0 < q^2 < 0.2$ $(\text{GeV}/c)^2$, where the approximation $F_A = F_V$ can be used. From this flux and the overall q^2 distribution the relationship between F_A/F_V and q^2 can be deduced. The result is shown in Fig. 3.

2 Single pion production

Single pion production has been predicted [BER n.d.] to take place mainly through excitation of the $(\frac{3}{2}, \frac{3}{2})$ nucleon isobar:

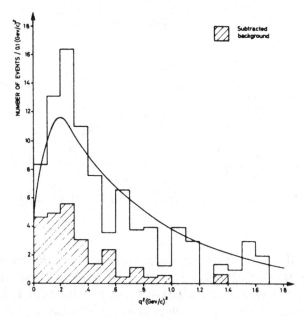

Fig. 2   Corrected four-momentum distribution for elastic events with $E_{\text{vis}} > 1$ GeV.

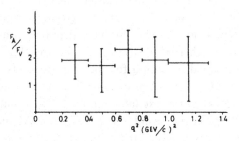

Fig. 3 F_A/F_V as a function of q^2.

$$\nu + N \rightarrow \mu^- + N^*$$
$$\hookrightarrow \pi + N. \tag{3}$$

Other processes such as peripheral π or ω exchange have cross sections which, at the neutrino energies considered, are calculated to be one or two orders of magnitude smaller [HEN n.d.].

$N^*(\frac{3}{2}, \frac{3}{2})$ production, together with the $\Delta I = 1$ rule, implies a ratio of final states $p\pi^+ : p\pi^0 : n\pi^+ = 9 : 2 : 1$, or an overall ratio $\pi^+/\pi^0 = 5/1$. For the observed one-pion events, the ratio is 1.9 ± 0.4. The observed ratio could be compatible with a large contribution of N^* production since the charge distribution will be severely distorted by charge-exchange interactions of the N^* decay products in the parent nucleus.

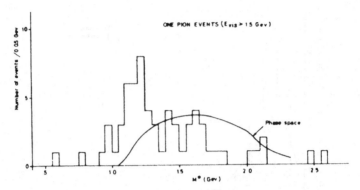

Fig. 4 M^* distribution for single pion events with $E_{vis} > 1.5\,\text{GeV}$.

In the two-body reaction (3) the mass M^* of the isobar can be deduced from the relation $M^{*2} = M^2 - q^2 + 2M(E_\nu - E_\mu)$, where M is the nucleon mass, and the neutrino energy E_ν is taken equal to the visible energy in the event. In practice, even if M^* were unique, the observed distribution would be broadened by ~ 15 percent by measurement errors and Fermi motion and distorted to slightly lower values by energy losses of the pion and nucleon in the parent nucleus. At low energy, a "phase space" distribution of M^* of the final products is grouped around the value of the isobar mass, for kinematical reasons. We have therefore considered only those events with $E_{vis} > 1.5\,\text{GeV}$; their M^* distribution is shown in Fig. 4. The curve shows the estimated phase space distribution for the final products (μ, π, p). The peak between $M^* = 1.0$ and $M^* = 1.4\,\text{GeV}$ is consistent with the assumption that single pion production proceeds through excitation of the $(\frac{3}{2}, \frac{3}{2})$ isobar in more than half the events. However, ~ 30 percent of events have $M^* > 1.4\,\text{GeV}$, and most of these are associated with high-energy protons. It is difficult therefore to attribute them to peripheral processes.

A cutoff at $M^* = 1.4\,\text{GeV}$ will contain most of the N^* events. Figure 5 shows the one-pion event rate and cross section for $M^* < 1.4\,\text{GeV}$; the data have been corrected for pion absorption in both the one- and two-pion events. As can be seen, the theoretical cross section calculated using the form-factors $F_A = F_V = (1 + (q/0.9)^2)^{-2}$ is too high by a factor of two. The cross section for one-pion events in the range $0 < q^2 < 0.2\ (\text{GeV}/c)^2$ and $1.0 < E_{vis} < 3.0\,\text{GeV}$, after correction for absorbed pions, is observed to be

$$\frac{d\sigma}{dq^2} = (0.5 \pm 0.2) \times 10^{-38} \frac{\text{cm}^2}{(\text{GeV}/c)^2} \text{ per nucleon}$$

in agreement with the predicted value [BER 64] of $\sim 0.7 \times 10^{-38}\,\text{cm}^2\,(\text{GeV}/c)^{-2}$ per nucleon. The experimental cross sections are evaluated from the calculated neutrino spectrum; the effect of the exclusion principle (estimated < 30 percent) has been

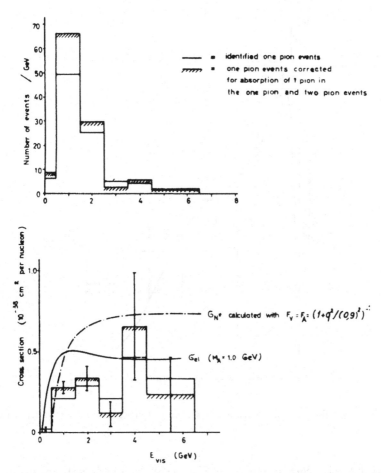

Fig. 5 Energy distribution and cross section for single pion events with $M^* < 1.4\,\text{GeV}$. Event rate corrected for pion absorption.

neglected. A choice of a smaller value of M_A could improve the agreement between experiment and theory in the whole q^2 range.

3 Neutrino flux and total inelastic cross section

Figure 6 shows the energy distribution of the neutrino flux up to 4 GeV, derived from the elastic event rate, the cross section computed for $M_A = 1.0\,\text{GeV}$.[1] Except at low energy, it is consistent with the flux calculated by Van der Meer on the basis of measured pion and kaon production spectra [DEK n.d.]. Assumptions of target efficiencies in this calculation have been cross-checked by measuring the muon range spectrum in the shielding. At high energies the neutrino spectrum cannot be considered to be known to better than a factor of two. The calculated variation of

---

[1] See also the data obtained in the spark chamber experiment that extend up to 8 GeV [BER 64].

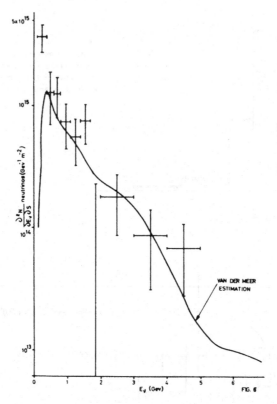

Fig. 6   Neutrino flux calculated from event rate and cross section, compared with predicted flux.

neutrino flux with energy may be used to estimate the inelastic neutrino cross sections, σ_{inel} at high energy (Fig. 7). It indicates a marked increase of the inelastic cross section with neutrino energy.

4 Intermediate vector boson

The intermediate vector boson [SCH 57; PON 59; LEE 60] is predicted to have a lifetime of $\lesssim 10^{-17}$ sec and to decay in the modes

$$W^+ \rightarrow e^+ + \nu$$
$$\mu^+ + \nu$$
$$\pi\pi\text{'s, etc.}$$

The two leptonic modes are assumed to have equal decay probability; the e^+ decay mode should be easily observable in a bubble chamber and the μ^+ decay in the spark chamber [BER 64]. To increase the weight of the data, half of the 450 events originating outside the fiducial volume of the chamber have been taken together

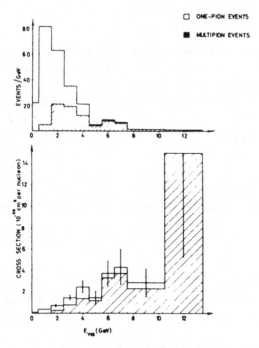

Fig. 7 Energy distribution and cross section of all inelastic events. Event rate corrected for pion absorption.

with all the 459 events inside. From ∼ 700 events only one possible "candidate" for e^+ decay has been observed. We expect about 1 event from $\bar{\nu}_e$ background. If the boson mass $M_W = 1.8$ GeV, and the branching ratio for e^+ decay is 50 percent, we expect to observe 2.5 such events [NEU n.d.], thus $M_W > 1.8$ GeV unless the pionic decay mode is predominant.

To discuss evidence for the pionic mode of decay of W^+, we restrict ourselves to events of $E_\nu > 6$ GeV, where W production is more likely to dominate over other processes. Of the 23 events observed, 14 have mesonic charge $+1$, as required for "elastic" W production. There is no clear peak in the spectrum of invariant mass of these pions; 8 events occur in the interval 1 to 2 GeV, to be compared with 11 expected if $M_W = 1.5$ GeV and the branching ratio for decay into points is > 90 percent.

To summarize, we have no evidence for the existence of the W boson in agreement with our previous conclusion [BIN 63]. If the boson does exist, $M_W > 1.5$ GeV irrespective of branching ratios; and $M_W > 1.8$ GeV unless leptonic decay is rare.

5 Strange particle production

As shown in Fig. 1, the events with strange particles occur mostly at high neutrino energy. A pair of strange particles is seen in four cases, which is consistent with associated production in the total of nine events. In the elastic and single pion events with $E_{vis} < 4\,GeV$, where associated production is unlikely, only one hyperon is found. Since single hyperon production would violate the $\Delta Q/\Delta S$ rule, this observation can be used to set a limit to the violation of < 20 percent.

Secondary production of these hyperons by collision of pions in the parent nucleus seems unlikely considering the observed pion spectrum. We conclude that the strange particles could not be produced by associated production in secondary processes. Thus the strange particles may come from the primary neutrino interaction.

6 Other conclusions

A number of other fundamental questions were discussed in our previous report [BIN 63]; their present status is

Violation of muon number conservation
$(\nu_e \neq \nu_\mu) < 1$ percent;
Violation of leptonic conservation < 6 percent;
$(\Delta S = 0$, neutral current coupling$/\Delta S = 0$,
charged current coupling$) < 3$ percent;
Neutrino flip intensity/no neutrino flip intensity
$$= \frac{K \to \mu + \nu_e}{K \to \mu + \nu_\mu} < 10 \text{ percent.}$$

A boson has been postulated which produces a resonance of the type $\nu_\mu + n \to W'_\mu \to \mu^- + p$. If its properties are as predicted [TAN 59; KIN 60] its mass is $> 5.5\,GeV$.

We are especially indebted to Professors V. W. Weisskopf and G. Bernardini for their continual support. We are grateful for the collaboration of J. S. Bell, H. Bingham, J. Løvseth, and M. Veltman.

These experiments have been made possible by technological development in many groups: the enhanced neutrino beam by M. Giesch, B. Kuiper, B. Langeseth, S. Van der Meer, S. Pichler, G. Plass, G. Pluym, K. Vahlbruch, H. Wachsmuth, and colleagues; and the operation and development of the heavy liquid bubble chamber by P. C. Innocenti and colleagues. The continued efforts of the members of the Proton Synchrotron Division to obtain the highest possible beam is of fundamental importance.

Finally our best thanks to "our scanning girls."

References

[BEL 63]   J. S. Bell. J. Løvseth, and M. Veltman, *Proc. Int'l. Conf. on Elementary Particles*, Sienna (1963) 587.

[BER n.d.] S. Berman and M. Veltman, to be published.

[BER 63]   G. Bernardini, G. Von Dardel, P. Egli, H. Faissner, F. Ferrero, C. Franzinetti, S. Fukui, J. M. Gaillard, H. J. Gerber, B. Hahn, R. R. Hillier, V. Kaftanov, F. Krienen, M. Reinharz, and R. A. Salmeron, *Proc. Int'l. Conf. on Elementary Particles*, Sienna, vol. 1 (1963) 571.

[BER 64]   G. Bernardini, J. K. Bienlein, G. Von Dardel, H. Faissner, F. Ferrero, J.-M. Gaillard, H. J. Gerber, B. Hahn, V. Kaftanov, F. Krienen, C. Manfredotti, M. Reinharz, and R. A. Salmeron, *Phys. Lett.* 13 (1964).

[BIN 63]   H. H. Bingham, H. Burmeister, D. Cundy, P. G. Innocenti, A. Lecourtois, R. Mollerud, G. Myatt, M. Paty, D. Perkins, C. A. Ramm, K. Schultze, H. Sletten, K. Soop, R. G. P. Voss, and H. Yoshiki, *Proc. Int'l. Conf. on Elementary Particles*, Sienna, vol. 1 (1963) 555.

[DAN 62]   G. Danby, J. M. Gaillard, K. Goulianos, L. M. Lederman, N. Mistry, M. Schwartz, and J. Steinberger, *Phys. Rev. Lett.* 9 (1962) 36.

[DEK n.d.] D. Dekkers, J. A. Geibel, R. Mermod, G. Webber, T. R. Willitts, K. Winter, B. Jordan, M. Vivargent, N. M. King, and E. J. N. Wilson, NP/Int. 64-5. CERN Internal Report, to be published.

[FEY 58]   R. Feynman and M. Gell-Mann, *Phys. Rev.* 109 (1958) 193.

[HEN n.d.] G. R. Henry, J. Løvseth, and J. D. Walecka, to be published.

[KON 60]   T. Konishita, *Phys. Rev. Lett.* 4 (1960) 378.

[LEE 60]   T. D. Lee and C. N. Yang, *Phys. Rev. Lett.* 4 (1960) 307.

[NEU n.d.] Neutrino Bubble Chamber Group, CERN, Dubna Conference (1964), to be published.

[PON 59]   B. Pontecorvo and R. Ryndin, *Proc. Ninth Int. Conf. on High-Energy Physics*, Kiev 2 (1959) 233.

[SCH 57]   J. Schwinger, *Ann. Phys.* (New York) 1 (1957) 407.

[TAN 59]   Y. Tanikawa and S. Watanabe, *Phys. Rev.* 113 (1959) 1344.

1.9 Spark chamber study of high-energy neutrino interactions

J. K. BIENLEIN, A. BÖHM, G. VON DARDEL, H. FAISSNER, F. FERRERO,

J.-M. GAILLARD, H. J. GERBER, B. HAHN, V. KAFTANOV, F. KRIENEN,

M. REINHARZ, R. A. SALMERON, P. G. SEILER, A. STAUDE,

J. STEIN, AND H. J. STEINER, 1964*

High-energy neutrino experiments were suggested by Pontecorvo and Ryndin [PON 59a,b], M. Schwartz [SCH 60] and others [NIS 57; FAK 58; REI 60; MAR 63]. Their importance was emphasized in particular by Lee and Yang [LEE 60; see also CAB 60, YAM 60]. The first experiment by Danby *et al.* [DAN 62] demonstrated that neutrinos from $\pi-\mu$ decay ν_μ are different from β-decay neutrinos ν_e. The present experiment confirms this result with considerably increased statistics. Elastic

* J. K. Bienlein, A. Böhm, G. Von Dardel, H. Faissner, F. Ferrero, J.-M. Gaillard, H. J. Gerber, B. Hahn, V. Kaftanov, F. Krienen, M. Reinharz, R. A. Salmeron, P. G. Seiler, A. Staude, J. Stein, and H. J. Steiner, CERN, Geneva, Switzerland. Reprinted from *Phys. Lett.*, November 1, 1964, Vol. 13, No. 1, pages 80–6. (Received 23 September 1964.)

electron production has been observed at a rate compatible with the amount of neutrinos from K-*e* decays in the beam. The measured rate excludes the possibility that neutrinos from $K_{\mu 2}$ decay are identical to ν_e (ν flip [BLU 61; FEI 61]). The measured ratio of μ^+ to μ^- production provides a test on lepton conservation. From the angular distribution of elastically produced muons we infer a cutoff in the axial vector form factor close to the one in the vector part. The electron events show structure effects similar to those in the muon case. The search for leptonic decays of an intermediate boson is described elsewhere [BER 64].

1 Experimental arrangements

A multiton spark chamber setup was exposed, behind the CERN heavy liquid bubble chamber [BIN 63; BLO 64] to the CERN neutrino beam [BER 63; GIE 63]. There were two runs, of about 30 days each, one in 1963, the other in 1964. In both runs the magnetic horn [GIE 63] focused positive particles emerging from a target hit by the extracted 24–25 GeV/*c* proton beam [BER 63]. Most of the neutrinos are due to $\pi^+ - \mu^+$ decays, and have energies below 2 GeV; higher energy neutrinos come mainly from K^+ decays. Due to a different horn shape in the 1964 run, the neutrino spectrum was harder than during the 1963 run; and the antineutrino contamination was estimated to be only 2.2 percent in 1964 as compared to 3 percent in 1963.

In both runs the spark chamber consisted of three sections: a thin-plate region, a magnet, and a thick-plate range region. In the 1963 setup (Fig. 1) the thin-plate region was first composed of a mixture of aluminum and brass three-plate spark chamber modules [FAI 63a], followed by an assembly of brass spark chambers. Later on the mixture was replaced by pure aluminum. The magnet was a pair of Helmholtz-type coils, with seven Al spark chambers sampling the path of the particles. The range region was made of lead and iron walls with spark chamber modules in between. The setup was triggered by counters, and was photographed from the side by pairs of stereocameras. The details have been described previously [FAI 63b,c,d].

In 1964, the magnet coils were replaced by an array of spark chambers and 5-cm-thick iron walls which were magnetized to 18 kgauss. Two slabs of magnetized iron were placed at the end of the range region [BER 64, fig. 1]. Some technical data about the two setups are given in Table 1.

At a PS intensity of 7×10^{11} circulating protons per pulse, the trigger rate was 40 per hour in 1963 and 60 per hour in 1964. The rate of events originating in the setup was 18 in 1963, and almost 40 in 1964. The remaining triggers were due to stray muons and interacting neutrons, coming mainly from neutrino interactions in the shielding. Cosmic rays contribute < 1 percent of the triggers. From an analysis of the arrival times of the initiating particles, measured relative to the phase of the proton bunches in the proton synchrotron, it was concluded that about 95 percent of

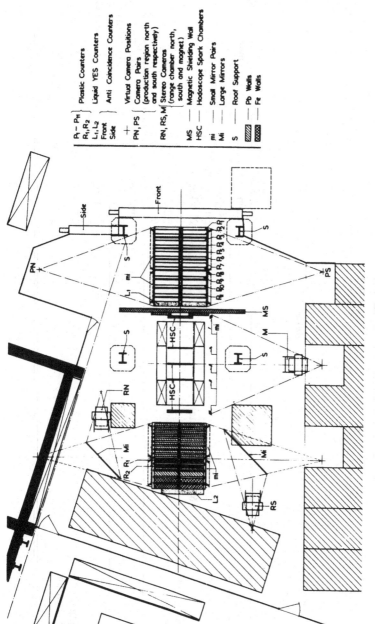

Fig. 1 Plan view of the spark chamber setup used in 1963.

P₁ – Pₙ Plastic Counters
R₁, R₂ Liquid YES Counters
L₁, L₂ Anti Coincidence Counters
Front
Side } Virtual Camera Positions
PN, PS Camera Pairs (production region north and south respectively)
RN, RS, M Stereo Cameras (range chamber north, south and magnet)
MS --- Magnetic Shielding Wall
HSC --- Hodoscope Spark Chambers
mi --- Small Mirror Pairs
Mi --- Large Mirrors
S --- Roof Support
        Pb Walls
        Fe Walls

Table 1. *Technical data about the setups*

Thin-plate region

Thin-plate region	Low density (Initially)	Low density (Later)	High density	1964
Average plate thickness 0.75 cm				
Total mass	8.3	5.1	12	4.6
Material	$\frac{2}{5}$ Al + $\frac{3}{5}$ brass	Al	brass	Al
Average density (g/cm^3)	1.55	0.95	3.05	0.9
Length (interaction lengths)	2.7	1.8	3.8	1.6
Number of counters	8	8	4	4

Magnet region

Magnet region	Low density	magnetized part	unmagnetized part
Fiducial mass (t)	—	25 (Fe)	20 (Fe)
Average density (g/cm^3)	—	4.1	4.1
Bending strength	6.4 kgauss × meter	21.6 kgauss × meter	—
Maximum detectable momentum	10 GeV/c	15 GeV/c	—
Length (interaction lengths)	—	9.2	8.4
Number of counters	—	6	5

Thick-plate region

Thick-plate region	Low density	unmagnetized part	magnetized plates at the end
Fiducial mass (t)	45 (Pb + Fe)	30 (Pb)	(2 × 15 cm Fe)
Average density	6.5	7.1	—
Length (interaction lengths)	12	6.7	—
Maximum detectable momentum	—	—	8 GeV/c
Number of counters	3	2	0

the events are neutrino-induced [FAI 63b,c,d; BRU 63; LØV 63a,b].[1] This is confirmed by the fact that less than 1% of the events show interactions on all tracks.

2 Rate of elastic muon production

Candidates for elastic muon production:

$$\nu_\mu + n \rightarrow \mu^- + p \tag{1}$$

were selected in a reduced fiducial volume consisting of half the pure aluminium setup. The selection criteria were: (a) The event must show at most two tracks and no shower; (b) the long track must not interact, and must have been able to trigger; (c) the short track (if any) must end in the chamber. Range and multiple scattering must be compatible with a recoil proton from reaction (1).

In total, 418 events satisfied these criteria. One-third of them had two tracks. An example is given in Fig. 2. With our criteria we have rejected elastic events where the recoil proton was energetic enough to produce a visible meson; but this effect was estimated to be at most 3 percent [BLO 64]. About 2 percent of the events are due to antineutrinos. A serious contamination comes from events where a pion was produced in the elementary interaction, but was either absorbed inside the parent nucleus, or was of too low energy to qualify as a track or as a shower. Estimates indicate that this contamination may be as large as 30 percent.

After correction for detection efficiency (77 percent), the selected sample corresponds to a rate of 5.8 events per ton and 10^{16} protons incident on the target. From the theoretical cross sections [LEE 60b; LØV 63a,b],[2] where the axial vector form-factor was taken equal to the vector form factor, and from the computed neutrino spectrum [GIE 63], one would have expected 3.2 events, in the same units. Considering the contamination with inelastic events and the uncertainty in the neutrino spectrum, the agreement is reasonable.

3 Elastic electron production: two-neutrino question and neutrino flip

We have observed elastic electron production:

$$\nu_e + n \rightarrow e^- + p. \tag{2}$$

The events were selected in a fiducial volume containing 80 percent of the thin-plate region. The event must show a single shower of energy above 500 MeV. It could be accompanied by one track, provided it was compatible with a proton (Fig. 3). The shower energy is inferred from the total number of sparks with an average accuracy of ±30 percent. The shower axis can be determined on the average to ±3°.

[1] Also, Løvseth, personal communication.
[2] Also, Løvseth, personal communication.

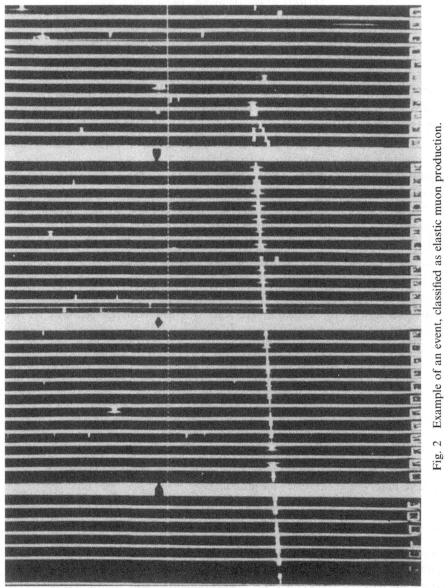

Fig. 2   Example of an event, classified as elastic muon production.

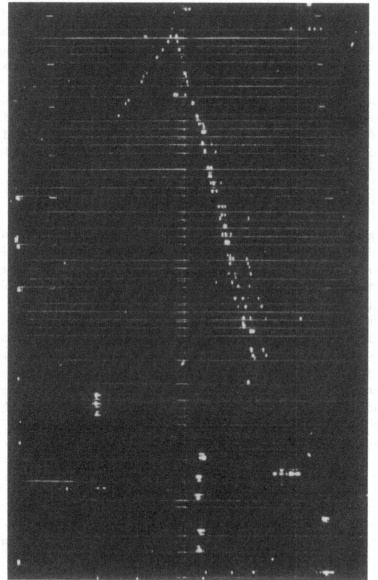

Fig. 3   Example of an event, calssified as elastic electron production.

Out of 4,400 events, 39 passed the selection criteria. The contamination with inelastic events should be comparable to the one in the elastic muon sample. After correction for electron detection efficiency we obtain a ratio of elastic electron to muon production of (1.7 ± 0.5) percent. The error includes an estimated systematic uncertainty.

The ratio expected from the cross sections [LEE 60] and the computed spectrum [GIE 63] is ≈ 1 for the one-neutrino hypothesis: $\nu_e = \nu_\mu$, 8 percent for the neutrino-flip hypothesis: $\nu_e \neq \nu_\mu$ but $K_{\mu2} \rightarrow \nu_e$, and 0.6 percent if there are two neutrinos ν_μ and ν_e and all ν_e's come from electronic K decays. Our result confirms the conclusions drawn from the first Columbia experiment [DAN 62]. It also rules out the neutrino flip hypothesis [BLU 61; FEI 61], at least in its extreme form. The maximum allowed admixture of electron neutrinos is about 20 percent in $K_{\mu2}$ decay and 1 percent in π-μ decay. The disagreement between observed and expected rate may be due to an underestimate of the flux of K-decay neutrinos, as indicated by a study of ν_μ reactions above 4 GeV [BER 64].

4 Test on lepton conservation

The determination of the sign of particles penetrating the magnets, together with the purity of the ν_μ beam, allowed a test on lepton conservation, in the sense that in all reactions neutrinos transform into negative muons but not into positive ones.

We measured the sign of the charge of 924 particles of momentum above 600 MeV/c, which penetrated the magnetized iron slabs at the end of the setup. The fraction of positive tracks R_+ was (0.027 ± 0.006). The fraction expected from antineutrino reactions is about 0.02. Allowing for uncertainties in this number, we conclude that R_+ does not differ from the value predicted with lepton conservation by more than 0.02. If there was a small violation of lepton conservation, with a relative amplitude α, it would contribute twice in our experiment: in the π decay and in the reaction itself. Our result therefore sets a limit to α^2 of 0.01. Previously, an upper limit of 0.027 could be derived [FRI 62], within the framework of the (V, A) theory, from the ρ value in μ decay [PLA 60].

5 Form-factors due to strong interactions

We can get some information about the form-factors in the elastic ν_μ reaction (1) from the observed angular distribution of the emitted muon. Løvseth [LØV 63a,b] has computed the expected distributions using the shape of the neutrino spectrum [GIE 63], and form-factors of the type:

$$F_i(q^2) = (1 + q^2/M_i^2)^{-2} \tag{3}$$

where q is the four-momentum transfer. The cutoff M_i in the vector form-factors equals 0.84 GeV according to the conserved vector current theory. Several values were used for the cutoff M_A in the axial vector form-factor. The intermediate boson

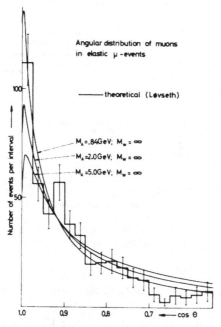

Fig. 4 Angular distribution of elastically produced muons. The histogram gives the experimental data, corrected for angle cutoff; the curves are predictions for several values of the axial vector cutoff parameter M_A.

was assumed to be infinitely heavy. Nuclear effects were included in terms of a Fermi gas model. The curves were normalized to the area of the experimental histogram.

Three of these curves are shown in Fig. 4, together with the experimental histogram, which was corrected for the angular cutoff imposed by the triggering counters. Acceptable fits are obtained with $M_A \leq 1.5\,\mathrm{GeV}$. Values of $M_A \leq 0.5\,\mathrm{GeV}$ are unlikely, since they would correspond to a ratio of expected to observed rate ≤ 0.4. Both limits are subject to uncertainties arising from the incomplete knowledge of the neutrino spectrum, and from the inelastic contamination. They are compatible with the value

$$M_A = \left(1.0^{+0.5}_{-0.3}\right)\mathrm{GeV}$$

derived by the bubble chamber group on the basis of a more reliable analysis.

Our measurements permit, for the first time, a study of structure effects in the elastic electron reaction (2). Since the electron energy and its emission angle can be measured in most of the events, it was possible to obtain a distribution of q^2. Figure 5 shows the experimental histogram together with theoretical curves, computed as before, and normalized to the number of events. The extreme case: $M_A = \infty$ is excluded; a value of M_A close to the muon value is consistent with the measurement. Although the statistics are poor, and systematic errors are relatively large, the data

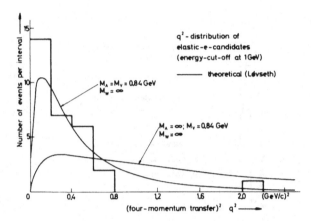

Fig. 5 Tentative q^2 distribution for elastic electron events with energies above 1 GeV (histogram). The curves give the distributions expected for two values of M_A.

suggest that muon-electron universality holds up to momentum transfers of the order of 1 GeV/c.

We thank G. Bernardini, on whose initiative the CERN neutrino experiment was started, for many suggestions and criticisms. We gratefully acknowledge the contributions of L. Cucančić, P. Egli, C. Franzinetti, S. Fukui, A. Ghani, E. Heer, R. R. Hillier, and T. B. Novey in the early stages of the experiment and the engineering effort of G. Muratori, M. Morpurgo, W. Albrecht, F. Blythe, T. Vuong Kha, E. H. Alleyn, F. Doughty, and G. Gendre. We are greatly indebted to our scanning and measuring staff, in particular to Miss M.-A. Huber. J. Løvseth made important contributions to the theoretical interpretation of the experiment.

We are very grateful to P. Germain, P. Standley, and their staff for the reliable operation of the CERN proton synchrotron and to C. Ramm, B. Kuiper, G. Plass, and S. Van der Meer, responsible for the extracted proton beam and the neutrino horn.

We thank V. F. Weisskopf, whose initiative has been essential throughout the experiment. We are grateful to B. Gregory, P. Preiswerk, and G. Puppi for their continuous interest and support.

References

[BER 63]   R. Bertolotti, H. Van Breugel, I. Caris, E. Consigny, H. Dijkhuizen, J. Goni, J. J. Hirsbrunner, B. Kuiper, B. Langeseth, S. Milner, S. Pichler, G. Plass, G. Pluym, H. Wachsmuth, and J. P. Zanasco, *Proc. Int'l. Conf. on Elementary Particles*, Sienna, vol. 1 (1963) 523.

[BER 64]   G. Bernardini, J. K. Bienlein, G. Von Dardel, H. Faissner, F. Ferrero, J. M. Gaillard, H. J. Gerber, B. Hahn, V. Kaftanov, F. Krienen, C. Manfredotti, M. Reinharz, and R. A. Salmeron, *Phys. Lett.* 13 (1964).

[BIN 63]   H. H. Bingham, H. Burmeister, D. Cundy, P. G. Innocenti, A. Lecourtois, R. Mollerud, G. Myatt, M. Paty, D. Perkins, C. A. Ramm, K. Schultze, H. Sletten, K. Soop, R. G. P.

Voss, and H. Yoshiki, *Proc. Int'l. Conf. on Elementary Particles.* Sienna, vol. 1 (1963) 555.

[BLO 64] M. M. Block, H. Burmeister, D. C. Cundy, B. Eiben, C. Franzinetti, J. Keren, R. Mollerud, G. Myatt, M. Nikolic, A. Orkin-Leourtois, M. Paty, D. Perkins, C. A. Ramm, K. Schultze, H. Sletten, K. Soop, R. Stump, W. Venus, and H. Yoshiki, *Phys. Lett.* 12 (1964) 281.

[BLU 61] S. A. Bludman, UCRL Report 9667.

[CAB 60] N. Cabibbo and R. Gatto, *Nuovo Cimento* 15 (1960) 304.

[DAN 62] G. Danby, J. M. Gaillard, K. Goulianos, L. M. Lederman, N. Mistry, M. Schwartz, and J. Steinberger, *Phys. Rev. Lett.* 9 (1962) 36.

[FAI 63a] H. Faissner, F. Ferreo, A. Ghani, E. Heer, F. Krienen, G. Muratori, T. B. Novey, M. Reinharz, and R. A. Salmeron, *Nucl. Instr. Methods* 20 (1963) 213.

[FAI 63b] H. Faissner, *Proc Int'l. Conf. on the Fundamental Aspects of Weak Interactions*, Brookhaven (1963) 137.

[FAI 63c] H. Faissner, *Physikertagung Hamburg*, ed. E. Brüche, Physik-Verlag, Mosbach/Baden (1963) 125.

[FAI 63d] H. Faissner, F. Ferrero, S. Fukui, J.-M. Gaillard, H. J. Gerber, B. Hahn, F. Krienen, G. Muratori, M. Reinharz, and R. A. Salmeron, *Proc. Int'l. Conf. on Elementary Particles*, Sienna, vol. 1 (1963) 571.

[FAK 58] D. Fakirov, Faculté des Sciences de Sofia, vol. 53 (1958) livre 2.

[FEI 61] G. Feinberg, F. Gürsey, and A. Pais, *Phys. Rev. Lett.* 7 (1961) 208.

[FRI 62] R. Frienberg, *Phys. Rev.* 129 (1962) 2298.

[GIE 63] M. Giesch, S. Van der Meer, G. Pluym, and K. M. Vahlbruch, *Proc. Int. Conf. on Elementary Particles*, Sienna, vol. 1 (1963) 536.

[LEE 60] T. D. Lee and C. N. Yang, *Phys. Rev. Lett.* 4 (1960) 307.

[LØV 63a] J. Løvseth, *Physics Letters* 5 (1963) 199.

[LØV 63b] J. Løvseth, CERN Report 63-37 (1963) 203.

[MAR 63] M. A. Markov, *The Neutrino*, Dubna Report D 1269 (1963) 62.

[NIS 57] K. Nishijima, *Phys. Rev.* 108 (1957) 907.

[PLA 60] R. J. Plano, *Phys. Rev.* 119 (1960) 1400.

[PON 59a] B. Pontecorvo, *JETP* (USSR) 37 (1959) 1751.

[PON 59b] B. Pontecorvo and R. Ryndin, *Proc. Ninth Int. Conf. on High-Energy Physics*, Kiev 2 (1959) 233.

[REI 60] F. Reines, *Ann. Rev. Nucl. Sci.*, 10 (1960) 16.

[SCH 60] M. Schwartz, *Phys. Rev. Lett.* 4 (1960) 306.

[YAM 60] Y. Yamaguchi, *Progr. Theoret. Phys.* 23 (1960) 1117.

1.10 Search for intermediate boson production in high-energy neutrino interactions

G. BERNARDINI, J. K. BIENLEIN, G. VON DARDEL, H. FAISSNER,

F. FERRERO, J.-M. GAILLARD, H. J. GERBER, B. HAHN, V. KAFTANOV,

F. KRIENEN, C. MANFREDOTTI, M. REINHARZ AND R. A. SALMERON, 1964*

If weak interactions are mediated by a boson W, it should be produced in high-energy neutrino experiments, provided its mass is not too high. The reaction would

* G. Bernardini, J. K. Bienlein, G. Von Dardel, H. Faissner, F. Ferrero, J. M. Gaillard, H. J. Gerber, B. Hahn, V. Kaftanov, F. Krienen, C. Manfredotti, M. Reinharz, and R. A. Salmeron, CERN, Geneva, Switzerland. Reprinted from *Phys. Lett.*, November 1, 1964, Vol. 13, no. 1, pages 86–91. (Received 23 September 1964.)

be [SCH 57; PON 59; LEE 60; BEL 63; VEL 61]

$$\nu_\mu + Z \rightarrow W^+ + \mu^- + Z \tag{1}$$

where Z is a nucleus or a proton.

The boson would then decay within 10^{-18} sec into a neutrino and a charged lepton or into a system of pions and kaons

$$W^+ \rightarrow \mu^+ + \nu_\mu \tag{2}$$

$$W^+ \rightarrow e^+ + \nu_e \tag{3}$$

$$W^+ \rightarrow \text{pions and/or kaons.} \tag{4}$$

The rates of (2) and (3) are expected to be almost equal, whereas the rate for (4) is unknown. Previously these reactions have been searched for by Danby *et al.* [DAN 62] in the Columbia neutrino spark chamber.

In the CERN neutrino spark chamber, we have made a systematic search for lepton pairs, $\mu^- \mu^+$ and $\mu^- e^+$ which would show up as a result of reactions (1) plus (2) or (3).

Figure 1 shows the spark chamber setup as it has been during the 1964 experiment. For details of the 1963 experiment setup see the previous letter [BIE 64: This paper is reproduced in the present volume].

1 Search for muon pairs: interaction analysis

In the thin-plate region, out of a total of 5200 events, we have selected all events satisfying the following conditions:

a There are only two tracks which have a visible range $> 0.5 \Lambda_0$ ($\Lambda_0 =$ geometrical interaction length).
b The longer track must have a range $> 1.5 \Lambda_0$, the shorter one a range $> 0.8 \Lambda_0$.
c The projected angles of the two tracks with the neutrino direction are smaller than $45°$ in both stereoviews, in order to avoid possible biases in the interaction detection.

This sample contains about 350 events. An examples is shown in Fig. 2. On the basis of the angular and energy distribution in W-production and decay [BEL 63], these events would include 70 percent of the muonic decays of elastically produced intermediate bosons. We investigate if the number of visible interactions (scattering $> 10°$ and stars) is compatible with the assumption that there are no muon pairs in the sample, that is, that every event contains at least one strongly interacting particle.

In each event we assume as a strongly interacting particle that track which gives the shortest length, where, with certain criteria, interactions could be seen with full efficiency. We sum up these track lengths for events in each material "i" separately.

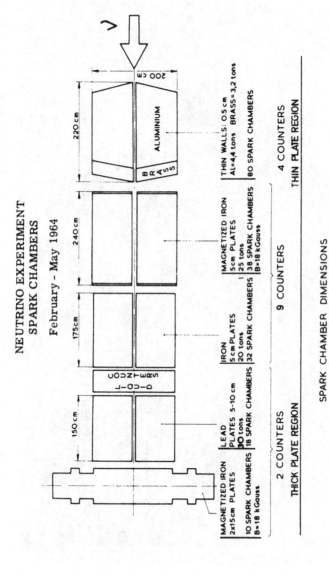

NEUTRINO EXPERIMENT
SPARK CHAMBERS

February - May 1964

Fig. 1 Top view of the apparatus as used in 1964.

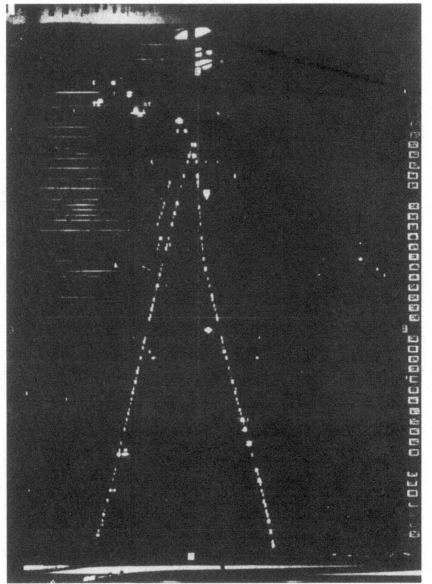

Fig. 2   Event satisfying the selection criteria used in the muon-pair search. The lower track shows an interaction.

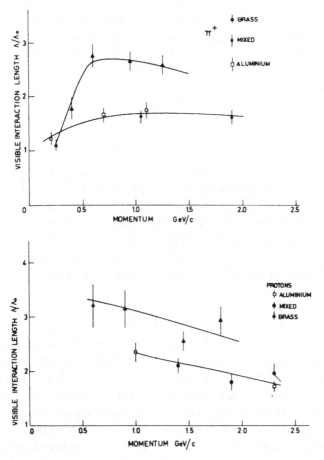

Fig. 3 Apparent interaction length of pions and protons as a function of momentum.

From the resulting sums L_i the expected number of interactions is given by

$$I_{\text{expected}} = \sum_i \frac{L_i}{\Lambda_i}$$

where Λ_i is the apparent interaction length in material "i".

The apparent interaction length of pions, protons, and kaons has been determined in the various thin-plate setups (brass, aluminium, and aluminium-brass mixture). The pion and proton calibration curves are shown in Fig. 3. Kaons gave a curve similar to that for pions. From the bubble chamber results we know that for the events of our sample at least 50 percent of the track length is not due to protons. We assume in what follows that 50 percent of the track length is due to protons and 50 percent to pions. The momentum of the particles is not known, therefore one has to use for Λ_i the peak value of the curves for each material.

Table 1. *Expected and observed number of interactions*

Experiment	I_expected	I_observed
1963	63	56
1964	33	36
Total	96	92

Under the assumption that there is at least one strongly interacting particle in every event, the observed number of interactions should be greater or equal to I_expected. The results are given in Table 1. They supersede preliminary numbers given earlier [BER 63; GAI 64] which were based on incomplete calibrations. Clearly this result is no evidence for muon pairs. Since we have no means for determining how much the expected number of interactions exceeds the minimum values of Table 1, we cannot with this method give a lower limit to the W-mass.

2 Sign analysis

The purity of the neutrino beam [MEE 63] and the possibility to determine signs of particles in the magnet allows us to use a more powerful method to reject the dominant background from $(\mu^- p)$ and $(\mu^- \pi)$ events. In muon pairs from W^+ decay the μ^+ is typically more energetic than the μ^- [BEL 63], whereas in the background energetic μ^+ are due only to the small antineutrino contamination.

Among the events produced during the 1964 experiment in the first part of the magnetized iron (15 tons) and in the first part of the thick-plate chamber (15 tons), we looked for events with two noninteracting tracks, compatible with a positive muon with range $> 7\ \Lambda_0$ and a negative muon with range $> 2.4\ \Lambda_0$. These ranges correspond to 1.2 and $0.47\ \text{GeV}/c$.

The background due to neutrino or antineutrino events producing noninteracting protons and pions is extremely small for such large ranges. In fact, we find no event which fulfils our conditions. The number one should expect due to boson production depends on four factors: the production cross section [BEL 63], which varies strongly with the mass of the boson, the detection efficiency, the neutrino spectrum at high energies, and the branching ratio R between leptonic and nonleptonic decays.

The detection efficiency for muonic decays of W events has been determined for our geometry. Using the kinematics computed by Bell and Veltman [BEL 63] we find that the efficiency is 11 percent for $M_W = 1.3\ GeV$; *it should not change much for slightly higher masses.*

Van der Meer has computed the neutrino spectrum at the spark chamber position [MEE 63]. There is considerable uncertainty in this spectrum, mainly from our lack of knowledge of the K^+ spectrum and angular distribution. We have tried, therefore

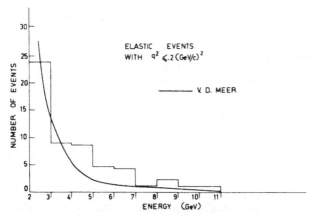

Fig. 4   Energy distribution of elastic events with $q^2 \leq 0.2$ (GeV/c^2). The curve follows from Van der Meer's spectrum [MEE 63] using Block's cross section [BLO n.d.].

to evaluate the high-energy part of the neutrino spectrum from the energy spectrum of "elastic" events with low four-momentum transfer q, as described by Block [BLO n.d.]. The method makes use of the fact that for low-momentum transfers the cross section does not depend much on the form-factors. We have used all the "elastic" events, produced inside the fiducial volume of the thin-plate chamber, which contain a muon going through the entire magnetic chamber. The criteria for the selection of "elastic" events have been discussed in the previous paper [BIE 64: This paper is reproduced in the present volume]. The neutrino energy and q^2 have been computed in each case from the muon angle and momentum. This momentum was deduced from sagitta and range measurements. The energy distribution of the events is shown by the histogram of Fig. 4. The distribution has been corrected for escape probability; the average correction amounts to 20 percent. The curve drawn in Fig. 4 gives the expected rate computed using the Van der Meer spectrum and the cross section for low q^2 [BLO n.d.].

Among the "elastic" events there is a contamination due to inelastic events as discussed in the preceding letter [BIE 64]. In our case the contamination could be as high as 50 percent, especially for the highest energy part of the spectrum. From our measurement we conclude that the spectrum computed by Van der Meer is very likely a lower limit in the energy range where intermediate bosons could be produced.

For the branching ratio R between leptonic and nonleptonic decays of the W, we assume $R = 1$.

Under these conditions, Table 2 gives the expected number of events which should fulfil our criteria as a function of the W mass. To deduce the spectrum from the low q^2 measurement an inelastic contamination of 30 percent has been subtracted from the histogram of Fig. 4.

Table 2. *Expected and observed number of $\mu\mu$ pairs as a function of the mass M_W of the intermediate boson*

M_W (GeV)	Expected		Observed
	Van der Meer spectrum	Low q^2 spectrum	
1.3	21	51	
1.5	11	26	0
1.8	4	9	

3 Search for μe events

We use for this analysis only the events produced in the aluminium thin-plate region. In the other sections of the setup the background of showers due to π^0's and single γs is large. The energy of showers was determined from calibration pictures with electrons.

From a total of 1500 events produced in aluminium we have found six events satisfying the following conditions:

a There is only one shower and it must start from the apex. Its energy is greater than 500 MeV.

b There is one track with a range greater than 0.8 Λ_0 and no other track with a range greater than 0.5 Λ_0. The projected angles of the longer track with the neutrino direction are smaller than 45° in both views.

This sample should include about 70 percent of the electronic decays of the boson. Figure 5 shows one of the six events.

In one of the events the track shows an interaction and is clearly not due to a (μe) pair, but probably to a ν_e interaction giving (ep) or $(e\pi)$. To get the amount of track length corresponding to this interaction, we expect at least one more event to be of the same type without interaction.

Events of the type we consider may also be simulated by inelastic ν_μ reactions with π^0 production in the following cases:

a One gamma ray converts internally or close to the vertex.

b The other gamma ray is missed because it overlaps with the first gamma ray, escapes without materializing, or has too low energy.

The contribution of such events in our sample can be computed from the π^0 spectrum in events where both photons are seen. These sources of background together with ν_μ events with internal bremsstrahlung [LØV 64] are estimated to account for one event.

Another kind of background consists of cases in which one gamma converts at a distance from the apex but a track makes a bridge between the shower and the apex and is afterwards buried inside the shower. This background is estimated to contribute one or two events.

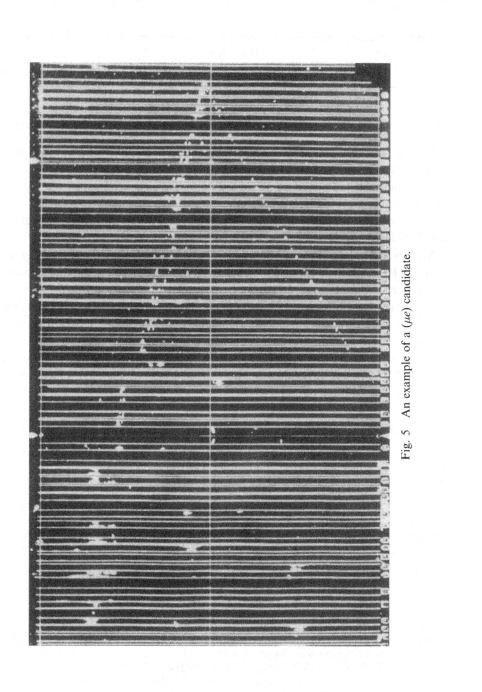

Fig. 5   An example of a (μe) candidate.

Table 3.   *Expected and observed lepton pairs for an immediate boson mass of*
1.8 GeV

	Expected		
Events	Van der Meer	Low q^2	Observed
μe	6	16	≤ 3
$\mu\mu$	4	9	0

With the uncertainties in all those subtractions, we can only set an upper limit of
three for the number of (μe) events in our sample. An analysis of the events produced
in the aluminium-brass mixture would lead to a similar maximum (μe) rate, but the
background is more difficult to evaluate.

Therefore, for $M_W = 1.8$ GeV and under the assumption that $R = 1$, we get the
numbers listed in Table 3.

We concluded that if $R \geq 1$, the mass of the intermediate boson is greater than
1.8 GeV. If instead there are no leptonic decays, $R = 0$, the result of the CERN heavy
liquid bubble chamber group [BLO 64] gives a lower limit of 1.5 GeV.

In addition to the acknowledgments of our previous report [BIE 64], we wish to
express our deep gratitude to Drs. R. Meunier, M. Spighel, and J. P. Stroot for their
collaboration in calibration runs. For discussions clarifying theoretical aspects of
this work we are indebted to Drs. J. S. Bell, M. Veltman, and J. D. Walecka. Finally
we thank A. Böhm, P. G. Seiler, J. Stein, H. J. Steine and M. Holder for their help in
the evaluation of the data.

References

[BEL 63]	J. S. Bell and M. Veltman, *Phys. Lett.* 5 (1963) 94, 151.
[BER 63a]	G. Bernardini, G. Von Dardel, P. Egli, H. Faissner, F. Ferrero, C. Franzinetti, S. Fukui, J. M. Gaillard, H. J. Gerber, B. Hahn, R. R. Hillier, V. Kaftanov, F. Krienen, M. Reinharz, and R. A. Salmeron, *Proc. Int'l. Conf. on Elementary Particles*, Sienna, vol. 1 (1963) 571.
[BIE 64]	J. K. Bienlein, A. Böhm, G. Von Dardel, H. Faissner, F. Ferrero, J. M. Gaillard, H. J. Gerber, B. Hahn, V. Kaftanov, F. Krienen, M. Reinharz, R. A. Salmeron, P. G. Seiler, A. Staude, J. Stein, and H. J. Steiner, *Phys. Lett.* 13 (1964) 80.
[BLO 64]	M. M. Block, H. Burmeister, D. C. Cundy, B. Eiben. C. Franzinetti, J. Keren, R. Mollerud, G. Myatt, M. Nikolic, A. Orkin-Leourtois, M. Paty, D. Perkins, C. A. Ramm, K. Schultze, H. Sletten, K. Soop, R. Stump, W. Venus, and H. Yoshiki, *Phys. Lett.* 12 (1964) 281.
[BLO n.d.]	M. M. Block, *Phys. Rev. Lett.* to be published.
[DAN 62]	G. Danby, J.-M. Gaillard, K. Goulianos, L. M. Lederman, N. Mistry, M. Schwartz, and J. Steinberger, *Phys. Res. Lett.* 9 (1962) 36.
[GAI 64]	J.-M. Gaillard, *Bull. Am. Phys. Soc.* 9 (1964) 40.
[GIE 63]	M. Giesch, S. Van der Meer, G. Pluym, and K. M. Vahlbruch, *Proc. Int. Conf. on Elementary Particles*, Sienna, vol. 1 (1963) 536.
[LEE 60]	T. D. Lee and C. N. Yang, *Phys. Rev. Lett.* 4 (1960) 307.
[LØV 64]	J. Løvseth and J. D. Walecka, preprint CERN (1964).

[LØV 63a]   J. Løvseth, *Physics Letters* 5 (1963) 199.
[LØV 63b]   J. Løvseth, CERN Report 63-37 (1963) 203.
[MEE 63]   S. Van der Meer, *Proc. Intern. Conf. on Elementary Particles*, Sienna, vol. 1 (1963) 536.
[PLA 60]   R. J. Plano, *Phys. Rev.* 119 (1960) 1400.
[PON 59]   B. Pontecorvo, *JETP* 36 (1959) 1615.
[SCH 57]   J. Schwinger, *Ann. Phys.* (New York) 1 (1957) 407.
[VEL 61]   M. Veltman, *Physica* 29 (1963) 161.

1.11 Search for elastic muon–neutrino electron scattering

F. J. HASERT *et al.*, 1973*

Recently many theoretical models have been postulated in an attempt to resolve the divergency of the classical current-current theory by unifying the weak and electromagnetic interactions. All these theories require neutral currents, heavy leptons, or both. One of these theories, that of Salam and Ward [SAL 64] and Weinberg [WEI 67], gives specific predictions about the amplitudes of the neutral currents which are susceptible to experimental tests.

In particular, using this model, t'Hooft [t'HO 71] has calculated the differential cross sections for the purely leptonic processes

$$\nu_\mu + e^- \rightarrow \nu_\mu + e^- \tag{1}$$

$$\bar{\nu}_\mu + e^- \rightarrow \bar{\nu}_\mu + e^- \tag{2}$$

which are forbidden to first order in the conventional Feynman–Gell-Mann theory. The predicted cross sections are of the order of $10^{-41}\,\mathrm{cm}^2$/electron at $1\,\mathrm{GeV}$, depending on the Weinberg angle θ_W, which is the only free parameter of the theory.

A search for these processes has been carried out in the large heavy liquid bubble chamber Gargamelle, useful volume $6.2\,\mathrm{m}^3$, filled with freon CF_3Br, exposed to both the neutrino and antineutrino beams at the CERN PS. The large length of the chamber, $4.8\,\mathrm{m}$, compared to the radiation length of freon, $11\,\mathrm{cm}$, ensured that electrons were unambiguously identified.

These interactions are characterized by a single electron (e^-) originating in the liquid, unaccompanied by nuclear fragments, hadrons, or γ-rays correlated to the

* F. J. Hasert, H. Faissner, W. Krenz, J. Von Krogh, D. Lanske, J. Morfin, K. Schultze, and H. Weerts, III. Physikalisches Institut der technischen Hochschule, Aachen, Germany; G. H. Bertrand-Coremans, J. Lemonne, J. Sacton, W. Van Doninck, and P. Vilain, Interuniversity Institute for High Energies, U.L.B., V.U.B. Brussels, Belgium; C. Baltay, D. C. Cundy, D. Haidt, M. Jaffre, P. Musset, A. Pullia, S. Natali, J. B. M. Pattison, D. H. Perkins, A. Rousset, W. Venus, and H. W. Wachsmuth, CERN, Geneva, Switzerland; V. Brisson, B. Degrange, M. Haguenauer, L. Kluberg, U. Nguyen-Khac, and P. Petiau, Laboratoire de Physique des Hautes Energies, Ecole Polytechnique, Paris, France; E. Bellotti, S. Bonetti, D. Cavalli, C. Conta, E. Fiorini, and M. Rollier, Istituto di Fisica dell 'Università, Milano and I.N.F.N. Milano, Italy; B. Aubert, L. M. Chounet, P. Heusse, A. Lagarrigue, A. M. Lutz, and J. P. Vialle, Laboratoire de l'Accélérateur Linéaire, Orsay, France; and F. W. Bullock, M. J. Esten, T. Jones, J. McKenzie, A. G. Michette, G. Myatt, J. Pinfold, and W. G. Scott, University College, University of London, England. *Phys. Lett.*, September 3, 1973, Vol. 46B, No. 1, pages 12–4. (Received 2 July 1973.)

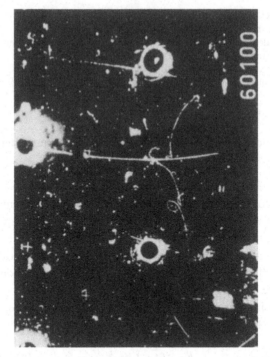

Fig. 1   Possible event of the $\bar{\nu}_\mu + e^- \to \bar{\nu}_\mu + e^-$.

vertex. The kinematics of the reactions are such that the electron is emitted at small angle, θ_e, with respect to the neutrino beam; the electron is expected to carry typically one-third of the energy of the incident neutrino, which is peaked between 1 and 2 GeV. As the neutrino interactions in the surrounding magnet and shielding produce a low-energy background of photons and electrons, a lower limit on the electron energy was set at 300 MeV. This energy cut ensures that all electrons from reactions (1) and (2) will have $\theta_e < 5°$.

A total of 375 000 ν and 360 000 $\bar{\nu}$ pictures were scanned twice and one single electron event satisfying the selection criteria was found in the $\bar{\nu}$ film. This event is shown in Fig. 1. The curvature of the initial part of the track shows the negative charge, and the spiralization and bremsstrahlung prove unambiguously that the track is due to an electron. The electron energy is 385 ± 100 MeV, and the angle to the beam axis is $1.4°^{+1.6°}_{-1.4}$. The electron vertex is 60 cm from the beginning of the visible volume of the chamber and 16 cm from the chamber axis.

The scanning efficiency for single electrons with an energy > 300 MeV was determined to be 86 percent using the isolated electron–positron pairs found in the chamber.

The main source of background is from the process

$$\nu_e + n \to e^-(\theta_e < 5°) + p \tag{3}$$

where the proton is either of too low an energy to be observed or is captured in the nucleus and no visible evaporation products are formed. This is due to the small ($< 1\%$) ν_e flux present in the predominantly ν_μ or $\bar{\nu}_\mu$ beam.

This background has been determined empirically using the observed events of the type

$$\nu_\mu + n \rightarrow \mu^- (\theta < 5°) + p \tag{4}$$

where the proton is not observed, and the ν_e flux calculated from the observed electron–neutrino events.

This is a good estimate as the two processes are kinematically similar at these energies and the ν_μ and ν_e spectra have nearly the same shape. In a partial sample of the film we have observed 450 events, occurring in a fiducial volume of $3\,\mathrm{m}^3$, of the type:

$$\mu^- + m \text{ protons } (m \geq 0)$$

where the visible energy is $> 1\,\mathrm{GeV}$, and the momentum in the beam direction is $> 0.6\,\mathrm{GeV}/c$. These cuts eliminate the background due to incoming charged particles.

In these events, only 3 have no protons and a μ^- angle $< 5°$. The scanning efficiency for single μ^- has been assumed to be the same as that for the single μ^+ found in the antineutrino film. This was determined to be 50 percent using the sample of 200 single μ^+.

Hence we obtain that

$$\frac{\mu^- (\theta_\mu < 5°) + 0_p}{\mu^- + mp} = 1.3 \pm 7\%.$$

This ratio is an overestimate as the inclusion of events of energies $< 1\,\mathrm{GeV}$ would be expected on kinematical grounds to lower it.

In the neutrino film 15 ν_e events of the type $e^- + m$ protons $(m > 0)$ have been observed in the fiducial volume ($3\,\mathrm{m}^3$). This number is in agreement with the one expected from the estimated ν_e/ν_μ flux ratio (0.7 percent). Hence one deduces a background from this source 0.3 ± 0.2 events.

Another estimate using the calculated ν_e and ν_μ fluxes and expected cross sections gives 0.4 ± 0.2.

In the $\bar{\nu}$ film zero $e^- + m$ proton events have been observed and a background estimate is obtained as above using the calculated ν_e and ν_μ fluxes. The ν_e flux in the antineutrino film is an order of magnitude less than in the neutrino film. Hence the background from the above sources in the $\bar{\nu}$ film is 0.03 ± 0.02 events.

The other sources of background could be due to Compton electrons or asymmetric electron pairs. Only two isolated electron–positron pairs having an energy greater than $300\,\mathrm{MeV}$ and making an angle of less than $5°$ with the beam direction were observed in the visible volume of the chamber in the ν film, and none in the $\bar{\nu}$ film.

Given these events and using the ratio of Compton to pair production cross sections as well as the differential cross section for pair production for the energy repartition among the electron and positron, this source of background is estimated to be 0.04 ± 0.02 events in ν and negligible in $\bar{\nu}$.

As the ν_e flux is less than 1 percent of the ν_μ flux, the background from the $V\text{-}A$ reactions

$$\begin{pmatrix} \nu_e \\ \bar{\nu}_e \end{pmatrix} + e^- \rightarrow \begin{pmatrix} \nu_e \\ \bar{\nu}_e \end{pmatrix} + e^-$$

of which the cross sections are of the same order as processes (1) and (2), are negligible. Similarly the lack of high-energy neutrons ($> 16\,\text{GeV}$) eliminates the background contribution from the electromagnetic interaction $n + e^- \rightarrow n + e^-$.

To calculate the detection efficiency, i.e., the fraction of reaction (1) and (2) that would survive the selection criteria, the electron laboratory energy and angular distributions have to be known. These spectra are not uniquely predictable but depend on the model assumed to introduce the neutral currents into the weak interactions. However, the detection efficiency in the present experiment is not very sensitive to these uncertainties since the electron minimum energy accepted is small compared to the incident neutrino energy.

In the case of isotropy in the center of mass, the detection efficiency is 87 percent.

In this case, the 90 percent confidence upper limits for the cross sections for the processes (1) and (2) are

$$0.26E_\nu \times 10^{-41}\ \text{cm}^2/\text{electron}$$

and

$$0.88E_\nu \times 10^{-41}\ \text{cm}^2/\text{electron}$$

respectively.

Table 1 shows the upper and lower event rates expected from the Weinberg model, taking into account the detection efficiencies, and using the measured ν_μ and $\bar{\nu}_\mu$ fluxes. The estimated backgrounds are also shown. These are to be compared with the one event found in the $\bar{\nu}$ film.

Figure 2 shows the number of expected ν and $\bar{\nu}$ events as a function of the Weinberg parameter $\sin^2\theta_W$.

In order to combine the neutrino and antineutrino results a maximum likelihood method has been used, taking into account the fluxes and backgrounds. The 90 percent confidence limit gives

$$0.1 < \sin^2\theta_W < 0.6.$$

It may be remarked that, in the context of the Weinberg theory, the proportion of electrons with $E_e > 1\,\text{GeV}$ is much lower in neutral current events than in the ν_e

Table 1. *Number of single e^- events of $E_e > 300 \, MeV$, $\theta_e < 5°$*

Flux neutrinos/m^2		Weinberg predictions		Background	Observed
		Minimum	Maximum		
ν	1.8×10^{15}	0.6	6.0	0.3 ± 0.2	0
$\bar{\nu}$	1.2×10^{15}	0.4	8.0	0.03 ± 0.02	1

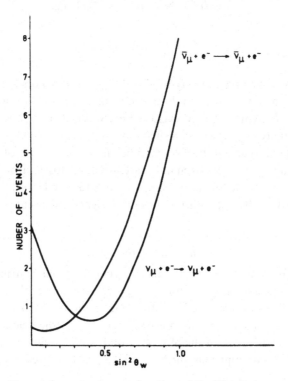

Fig. 2   Expected event rate as a function of the Weinberg parameter.

background, and hence our quoted background is overestimated. We conclude that the probability that the single event observed in the $\bar{\nu}$ film is due to nonneutral current background is less than 3 percent.

It is a pleasure to express our thanks to the members of the CERN TC-L group who have carried the technical responsibility for the experiment. We also thank the CERN PS operational staff, and the scanning and programming personnel in the various laboratories.

References

[SAL 64]   A. Salam and J. G. Ward, *Phys. Lett.* 13 (1964) 168.
[t'HO 71]   G. t'Hooft, *Phys. Lett.* 37B (1971) 195.
[WEI 67]   S. Weinberg, *Phys. Rev. Lett.* 19 (1967) 1264.

1.12 Observation of neutrino-like interactions without muon or electron in the Gargamelle neutrino experiment

F. J. HASERT et al., 1973*

We have searched for the neutral current (NC) and charged current (CC) reactions:

$$\text{NC } \nu_\mu/\bar{\nu}_\mu + \text{N} \rightarrow \nu_\mu/\bar{\nu}_\mu + \text{hadrons} \tag{1}$$

$$\text{CC } \nu_\mu/\bar{\nu}_\mu + \text{N} \rightarrow \mu^-/\mu^+ \text{hadrons} \tag{2}$$

which are distinguished respectively by the absence of any possible muon, or the presence of one, and only one, possible muon. A small contamination of $\nu_e/\bar{\nu}_e$ exists in the $\nu_\mu/\bar{\nu}_\mu$ beams giving some CC events which are easily recognized by the e^-/e^+ signature. The analysis is based on 83 000 ν pictures and 207 000 $\bar{\nu}$ pictures taken at CERN in the Gargamelle bubble chamber filled with freon of density 1.5 × 10^3 kg/m^3.[1] The dimensions of this chamber are such that most hadrons are unambiguously identified by interaction or by range–momentum and ionization. Any track which could possibly be due to a muon has consigned the event to reaction (2).

1 Analysis of the signal

To estimate the background of neutral hadrons coming from neutrino interactions in the shielding and simulating reaction (1), events where a visible charged current interaction produces an identified neutron star in the chamber (associated, AS, events) were also studied. To obtain a good estimate of the true neutral hadron direction from the direction of the observed total momentum a cut in visible total energy of > 1 GeV was applied to the NC and AS events, as well as to the hadronic part of the CC events.

We have observed, in a fiducial volume of 3 m^3, 102 NC, 428 CC, and 15 AS in the ν run and 64 NC, 148 CC, and 12 AS in the $\bar{\nu}$ run. Using these numbers without

* F. J. Hasert, S. Kabe, W. Krenz, J. Von Krogh, D. Lanske, J. Morfin, K. Schultze, and H. Weerts, III. Physikalisches Institut der Technischen Hochschule, Aachen, Germany; G. H. Bertrand-Coremans, J. Sacton, W. Van Doninck, and P. Vilain, Interuniversity Institute for High Energies. U.L.B., V.U.B. Brussels, Belgium; U. Camerini, D. C. Cundy, R. Baldi, I. Danlichenko, W. F. Fry, D. Haidt, S. Natali, P. Musset, B. Osculati, R. Palmer, J. B. M. Pattison, D. H. Perkins, A. Pullia, A. Rousset, W. Venus, and H. Wachsmuth, CERN, Geneva. Switzerland; V. Brisson, B. Degrange, M. Haguenauer, L. Kluberg, U. Nguyen-Khac, and P. Petiau, Laboratoire de Physique Nucléaire des Hautes Energies, Ecole Polytechnique, Paris, France; E. Belotti, S. Bonetti, D. Cavalli, C. Conta, E. Fiorini, and M. Rollier, Istituto di Fisica dell'Università, Milano and I.N.F.N. Milano, Italy; B. Aubert, D. Blum, L. M. Chounet, P. Heusse, A. Lagarrigue, A. M. Lutz, A. Orkin-Lecourtois, and J. P. Vialle, Laboratoire de l'Accélérateur Linćaire, Orsay, France; F. W. Bullock, M. J. Esten, T. W. Jones, J. McKenzie, A. G. Michette, G. Myatt, and W. G. Scott, University College, London, England. *Phys. Lett.*, September 3, 1973, Vol. 46B, No. 1, pages 138–40. (Received 23 July 1973.)
[1] A more detailed account of the analysis of this experiment appears in a paper to be submitted to *Nuclear Physics*.

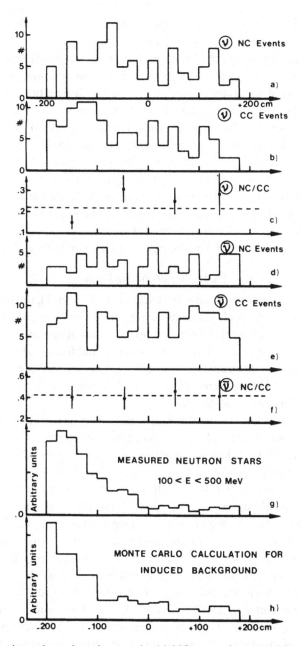

Fig. 1 Distributions along the ν-beam axis. (a) NC events in ν. (b) CC events in ν (this distribution is based on a reference sample of $\sim\frac{1}{4}$ of the total ν film). (c) Ratio NC/CC in ν (normalized). (d) NC events in $\bar{\nu}$. (e) CC events in $\bar{\nu}$. (f) Ratio NC/CC in $\bar{\nu}$. (g) Measured neutron stars with $100 < E < 500$ MeV having protons only. (h) Computed distribution of the background events from the Monte Carlo.

background subtraction the ratios NC/CC are then 0.24 for ν and 0.42 for $\bar{\nu}$, whilst the NC/AS ratios are 6.8 and 5.3, respectively.

The spatial distributions of the NC events have been compared to those of the CC events and found to be similar. In particular, the distribution along the beam direction of NC (Fig. 1) has the same shape as the CC distribution. In contrast, the observed distribution of low-energy neutral stars shows a typical exponential attenuation as expected for neutron background. The distributions of radial position, hadron total energy, and angle between measured hadron total momentum and beam direction are also indistinguishable for NC and CC.

Using the direction of measured total momentum of the hadrons in NC and CC events, a Bartlett method has been used to evaluate the apparent interaction mean free paths, λ_a, for NC and CC, which are found to be compatible with infinity. For the NC events we find $\lambda_a > 2.6$ m at 90 percent CL; this corresponds to 3.5 times the neutron interaction length for high-energy (> 1 GeV) inelastic collisions in freon.

2 Evaluation of the background

Since the outgoing neutrinos cannot be detected in reaction (1), the NC events may be simulated by neutral hadrons coming from the ν beam or elsewhere.

As a check for cosmic-ray origin, the up-down asymmetries of NC events in vertical position and momenta have been measured and found to be (3 ± 8) percent and (-8 ± 8) percent, respectively. In addition, a cosmic-ray exposure of 15 000 pictures shows no NC type event satisfying the selection criteria. We conclude that the cosmic background is negligible.

The low-energy muons (< 100 MeV/c) captured at rest in the ν run could be mistaken as protons. A study of the observed muon spectrum in CC events, as well as a theoretical estimate of the low end of this spectrum, shows that the correction to be applied is 0 ± 5 events.

Interactions of neutral hadrons produced by the primary protons up to and including the target should produce events at an equal rate in ν and $\bar{\nu}$ runs. On the contrary, we observe an absolute rate 4 times larger in the ν run than in the $\bar{\nu}$ run. If the neutral hadrons are due to defocused secondary pions and kaons, the disagreement is larger since we expect 1–2 times more events in $\bar{\nu}$ than in ν. Since the whole installation is shielded from below by earth we should again expect up-down asymmetries in the NC events. This is not observed.

The most important source of background is the interaction of neutral hadrons produced by the undetected neutrino interactions in the shielding. The high elasticity (0.7) of the neutrons causes a cascade effect in propagation through the shielding. The neutron energy spectrum at production can, in principle, be obtained from the AS events together with available nucleon-nucleus data. Due to the limited statistics in the AS events we make the extreme assumption that all the NC events are neutron produced and use their observed energy spectrum to calculate the neutron spectrum from neutrino interactions. This gives an energy dependence described by E^{-2}. The effective interaction length λ_e of neutrons in the shielding is then found to be 2.5 times the inelastic interaction length, λ_i. A smaller effective interaction length is found for K_L^0 although the background from this source must be negligible since we find no examples of Λ^0 hyperon production among the NC events.

From the absolute value of the number of AS events, we can calculate the number of background events. This has been done by Monte Carlo generation of events in the shielding surrounding the fiducial volume according to the radial intensity distribution of the beam. The ratio of background events (B) to AS events is found to be B/AS = 0.7 for $\lambda_e = 2.5\lambda_i$. If the NC sample has to be explained as being entirely due to neutral hadrons, the Monte Carlo requires $\lambda_e/\lambda_i > 10$, instead of the best estimate of 2.5. Both ratios would predict distributions along the beam direction in the chamber in strong disagreement with those observed.

Another evaluation of this type of background has been made using the simple assumption that an equilibrium of neutral hadrons with neutrinos exists throughout the entire chamber/shielding assembly. For a radially uniform ν flux it gives B/AS < 1.0, which confirms the Monte Carlo prediction.

3 Conclusion

We have observed events without secondary muon or electron, induced by neutral penetrating particles. We are not able to explain the bulk of the signal by any known source of background, unless the effective interaction length of neutrons and K_L^0 is at least 10 times the inelastic interaction length. These events behave similarly to the hadronic part of the charged current events. They could be attributed to neutral-current-induced reactions, other penetrating particles than ν_μ and ν_e, heavy leptons decaying mainly into hadrons, or by penetrating particles produced by neutrinos and in equilibrium with the ν beam.

On subtraction of the best estimate of the neutral hadron background, and taking into account the $\nu - \bar{\nu}$ contamination in the $\bar{\nu}(\nu)$ beam, our best estimates of the NC/CC ratios are

$$(NC/CC)_\nu = 0.21 \pm 0.03$$
$$(NC/CC)_{\bar{\nu}} = 0.45 \pm 0.09$$

where the stated errors are statistical only. If the events are due to neutral currents, these two results are compatible with the same value of Weinberg parameter, $\sin^2\theta_W$ [WEI 72; PAI 72; PAS 73] in the range 0.3 to 0.4.

References

[PAI 72]   A. Pais and S. B. Treiman, *Phys. Rev.* D6 (1972) 2700.
[PAS 73]   E. A. Paschos and L. Wolfenstein, *Phys. Rev.* D7 (1973) 91.
[WEI 72]   S. Weinberg, *Phys. Rev.* D5 (1972) 1412.

2

Intrinsic properties of neutrinos

2.1 Particle-antiparticle properties of neutrinos*

2.1.1 Motivation for considering the possibility that $\bar{\nu} = \nu$

Electrons and protons are obviously not their own antiparticles, since they are electrically charged. Similarly, neutrons are clearly not their own antiparticles, since they carry baryon number. By contrast, it is possible that neutrinos *are* their own antiparticles, since they carry neither electric charge, nor, as far as we know, any other chargelike attribute. It might be objected that neutrinos carry "lepton number," the quantum number that distinguishes an antilepton from a lepton. However, as we shall see, there is in reality no evidence that any such quantum number exists. Thus, it is indeed possible that neutrinos, unlike all the other known fermions, are their own antiparticles.

From the theoretical standpoint, this possibility is a very attractive one. To see why, let us first note that, in general, grand unified theories lead us to expect that neutrinos are massive. In any grand unified theory, the neutrino of a given generation is placed in a multiplet together with the charged lepton and the quarks of the same generation (and sometimes together with additional particles as well). Now, the charged lepton and quarks of any generation are all known to be massive. Thus, being in a multiplet with them, the neutrino would have to be exceptional to be massless. Nevertheless, we know that the neutrino in each generation is, at the heaviest, much lighter than the corresponding charged lepton and quarks. Assuming that the neutrino is indeed massive, we have to understand why its mass is so much smaller than the masses of these other particles. The most popular explanation of this fact is the "see-saw mechanism" [GEL 79; YAN 79; MOH 80, 81]. This predicts that each neutrino mass M_ν obeys a "see-saw relation" of the form $M_\nu M \approx$ [Typical quark or charged lepton mass]^2, where M is a very large mass scale. Very importantly, *the see-saw mechanism also predicts that neutrinos are their own antiparticles*. For a discussion of neutrino mass terms in gauge field theories, and a detailed explanation of the see-saw mechanism, see [KAY 88a].

2.1.2 The precise meaning of $\bar{\nu} = \nu$: Dirac and Majorana neutrinos

What, precisely, do we mean when we say that a neutrino ν is its own antiparticle? We do *not* mean that $C|\nu\rangle = \tilde{\eta}_c|\nu\rangle$, where C denotes charge-conjugation and $\tilde{\eta}_c$ is the

---

* B. Kayser, Division of Physics, National Science Foundation, Washington, D.C. 20550.

C parity of ν. After all, the weak interactions that dress the state $|\nu\rangle$ are maximally *C*-non-conserving. Hence, if $|\nu\rangle$ has some definite *C* parity at one instant, it will not have this same *C* parity at a later instant. Thus, a neutrino that is its own antiparticle must be defined by its transformation properties under *CPT*, which presumably is completely conserved. Under *CPT*, *any neutrino* $|\nu(\mathbf{p}, h)\rangle$ of momentum $\mathbf{p}$ and helicity h transforms according to

$$CPT|\nu(\mathbf{p}, h)\rangle = \tilde{\eta}_{CPT}^{h}|\bar{\nu}(\mathbf{p}, -h)\rangle. \tag{2.1.1}$$

Here the helicity reversal is due to the *P* operation, and the phase factor $\tilde{\eta}_{CPT}^{h}$ depends on the helicity, as we shall see. If the neutrino is *not* its own antiparticle, then the particle $\nu(\vec{p}, h)$ and the particle $\bar{\nu}(\vec{p}, h)$ of the *same* helicity are different objects. By this we mean that they interact differently with matter. When this is the case, ν is referred to as a Dirac neutrino ν^{D}. Obviously, in its rest frame, such a neutrino consists of four states: two spin states for the neutrino and an additional two for the antineutrino. By contrast, when the neutrino ν *is* its own antiparticle, the particles $\nu(\vec{p}, h)$ and $\bar{\nu}(\vec{p}, h)$ are identical. That is, *for given momentum and helicity*, the particles we call the "neutrino" and the "antineutrino" have identical interactions with matter. When this is the case, ν is called a Majorana neutrino ν^{M}. In its rest frame, such a neutrino consists of only two states: one with spin up along some reference direction, and one with spin down.

2.1.3 Why we do not know if $\bar{\nu} = \nu$

Why is it that we do not know whether neutrinos are their own antiparticles? The reason is that the experimentally available neutrinos are always polarized, and, in particular, the "neutrinos" are polarized oppositely from the "antineutrinos." The particles we call "neutrinos" are always left-handed, whereas those we refer to as "antineutrinos" are always right-handed. As a result, we have not been able to compare the interactions with matter of neutrinos and antineutrinos of the *same* helicity. To be sure, we know very well that the left-handed neutrinos interact very differently from the right-handed antineutrinos. However, there is no way of knowing whether this difference is due simply to the difference in polarization in the two cases, or to a real distinction between neutrinos and antineutrinos that goes beyond mere polarization.

A good illustration of this state of affairs is provided by the neutrinos from pion decay. The neutral lepton emitted in the decay $\pi^{+} \rightarrow \mu^{+} + \nu_{\mu}$, which by convention we call a neutrino rather than an antineutrino, is always of left-handed (i.e., negative) helicity. Let us indicate this fact by labeling it $\nu_{\mu}(-)$. By contrast, the neutral lepton emitted in the decay $\pi^{-} \rightarrow \mu^{-} + \bar{\nu}_{\mu}$, which by convention we call an antineutrino, is always of right-handed (positive) helicity. We shall indicate this fact by labeling it $\bar{\nu}_{\mu}(+)$. Now, it is observed that when a $\nu_{\mu}(-)$ strikes a nucleon *N* which is at rest, the reaction $\nu_{\mu}(-) + N \rightarrow \mu^{-} + X$ may occur, but the reaction

$\nu_\mu(-) + N \to \mu^+ + X$ will not. By contrast, when a $\bar{\nu}_\mu(+)$ strikes a nucleon which is at rest, the reaction $\bar{\nu}_\mu(+) + N \to \mu^+ + X$ may occur, but the reaction $\bar{\nu}_\mu(+) + N \to \mu^- + X$ will not. Unfortunately, this difference in interaction patterns has two possible explanations: (1) The difference may be due simply to the fact that $\nu_\mu(-)$ and $\bar{\nu}_\mu(+)$ have different polarizations. (2) It may be that there exists a conserved lepton number L, with $L(\nu_\mu) = L(\mu^-) = +1$ but $L(\bar{\nu}_\mu) = L(\mu^+) = -1$, so that the unobserved reactions are forbidden, and ν_μ and $\bar{\nu}_\mu$ are genuinely different.

To settle the issue of whether ν_μ and $\bar{\nu}_\mu$ differ, we must find out how the interactions of a ν_μ and a $\bar{\nu}_\mu$ of the *same* helicity compare. Suppose, for example, that we could somehow reverse the helicity of a $\bar{\nu}_\mu(+)$ created in π^- decay. We could then ask whether the resultant left-handed particle, $\bar{\nu}_\mu(-)$, interacts with nucleons in the same way as the left-handed $\nu_\mu(-)$ born in π^+ decay. If the answer is yes, then $\bar{\nu}_\mu(+)$ and $\nu_\mu(-)$ differ only in helicity; that is, ν_μ is a Majorana neutrino. If the answer is no, then $\nu_\mu(-)$ and $\bar{\nu}_\mu(+)$ evidently differ in a way that goes beyond helicity; that is, ν_μ is a Dirac neutrino. Regrettably, the reversal of neutrino helicity is very difficult, and has not been done.

Indeed, when a neutrino is massless, the reversal of its helicity is completely impossible, assuming there are no right-handed currents. For a massless neutrino, the helicity cannot be reversed by viewing the neutrino from a frame in which the direction of its momentum is reversed, since for a massless particle there is no such frame. It is not hard to show that, in addition, if all weak currents are left-handed, the helicity of a massless neutrino cannot be reversed by interactions between the neutrino and matter. Thus, in the massless case there is no way to produce a particle such as $\bar{\nu}_\mu(-)$, so it becomes meaningless to ask how this particle behaves. Consequently, the distinction between a Majorana neutrino and a Dirac one disappears. Furthermore, the approach to the massless limit is a smooth one, so that even if, as we suspect, neutrinos have non-zero masses, it is nevertheless very difficult to tell whether they are Majorana or Dirac particles because their masses are so tiny compared to their energies and other mass scales. This difficulty, which is illustrated in [KAY 88b], has been referred to as the "practical Dirac–Majorana confusion theorem" [KAY 82].

2.1.4 CP and CPT properties of Majorana neutrinos

We have defined a Majorana neutrino ν^M as one that is its own mirror image under *CPT*:

$$CPT|\nu^M(\mathbf{p}, h)\rangle = \tilde{\eta}^h_{CPT}|\nu^M(\mathbf{p}, -h)\rangle. \tag{2.1.2}$$

To the extent that *CP* is conserved, such a neutrino is also an eigenstate of *CP*:

$$CP|\nu^M(\mathbf{p}, h)\rangle = \tilde{\eta}_{CP}|\nu^M(-\mathbf{p}, -h)\rangle. \tag{2.1.3}$$

Here the momentum and helicity reversals are due to the P operation, and the phase factor $\tilde{\eta}_{CP}$ is the intrinsic *CP* parity of the neutrino ν^M. Different neutrinos can have different values of $\tilde{\eta}_{CP}$, but the permissible values of this quantum number are $\pm i$, rather than ± 1. An easy way to see this is to consider the decay of the neutral weak boson into a pair of identical Majorana neutrinos: $Z^0 \rightarrow \nu\nu$. In the Standard Model, this decay conserves CP. To find the consequences of this conservation, it suffices to suppose that the outgoing neutrinos are non-relativistic. Then, since they are identical fermions, they must be in a $^3P_1$ state, since this is the only antisymmetric non-relativistic state with total angular momentum equal to the spin of the Z^0. Now, from Eq. (2.1.3) it follows that if the intrinsic CP parity of ν is $\tilde{\eta}_{CP}(\nu)$, then our $\nu\nu$ final state, with orbital angular momentum $L = 1$, obeys

$$CP|\nu\nu; {}^3P_1\rangle = \tilde{\eta}_{CP}^2(\nu)(-1)^L|\nu\nu; {}^3P_1\rangle$$

$$= -\tilde{\eta}_{CP}^2(\nu)|\nu\nu; {}^3P_1\rangle. \qquad (2.1.4)$$

Since the Z^0 has $CP = +1$, conservation of CP in $Z^0 \rightarrow \nu\nu$ then implies that $-\tilde{\eta}_{CP}^2(\nu) = +1$. Hence, the allowed values of the intrinsic CP parity of a Majorana neutrino are [KAY 84]

$$\tilde{\eta}_{CP}(\nu) = \pm i. \qquad (2.1.5)$$

To illustrate the consequences of $\tilde{\eta}_{CP}$, and of the fact that it is imaginary, let us consider the process $e^- e^+ \rightarrow N_1 N_2$, where N_1 and N_2 are two distinct heavy Majorana neutral leptons [PET 86]. Assuming that the process is engendered by W boson exchange, the only incoming helicity configuration that couples is $e^-(-)e^+(+)$. In the $e^- e^+$ c.m. frame, this state is a CP eigenstate, and it is not hard to show that it has $CP = +1$. Now, consider the process just above $N_1 N_2$ production threshold, and suppose that the final particles are in a state with definite orbital angular momentum L. Then the final state is also a CP eigenstate, and from Eq. (2.1.3) its CP is $\tilde{\eta}_{CP}(\nu_1)\tilde{\eta}_{CP}(\nu_2)(-1)^L$. Thus, if CP is conserved in our reaction,

$$+1 = \tilde{\eta}_{CP}(\nu_1)\tilde{\eta}_{CP}(\nu_2)(-1)^L. \qquad (2.1.6)$$

Bearing in mind that the possible values of $\tilde{\eta}_{CP}$ are imaginary, we see that if $\tilde{\eta}_{CP}(\nu_1) = \tilde{\eta}_{CP}(\nu_2)$, the allowed partial wave near $N_1 N_2$ threshold is the p wave, while if $\tilde{\eta}_{CP}(\nu_1) = -\tilde{\eta}_{CP}(\nu_2)$, it is the s wave. Had the values of $\tilde{\eta}_{CP}$ been real, it would have been the other way around.

Now what can be said about the CPT phase factor $\tilde{\eta}_{CPT}^h$ in Eq. (2.1.2)? With $\zeta \equiv CPT$, in the rest frame of ν^M this equation reads

$$\zeta|\nu^M(s)\rangle = \tilde{\eta}_\zeta^s|\nu^M(-s)\rangle \qquad (2.1.7)$$

where $s = \pm\frac{1}{2}$ is the projection of the spin of ν^M along some reference direction. This equation implies that, as long as we act only on the states $|\nu^M(s)\rangle$, $\zeta \mathbf{J} = -\mathbf{J}\zeta$, where $\mathbf{J}$ is the angular momentum operator. It follows that $\zeta J_+ = -J_-\zeta$, where $J_\pm = J_x \pm iJ_y$ are the raising and lowering operators, and we have used the fact that

ζ is antiunitary. If we apply this anticommutation relation to the state $|\nu^M(-\tfrac{1}{2})\rangle$, we obtain [KAY 84]

$$\zeta J_+ |\nu^M(-\tfrac{1}{2})\rangle = \tilde{\eta}_\zeta^{+\frac{1}{2}} |\nu^M(-\tfrac{1}{2})\rangle$$

$$= -J_-\zeta |\nu^M(-\tfrac{1}{2})\rangle = -\tilde{\eta}_\zeta^{-\frac{1}{2}} |\nu^M(-\tfrac{1}{2})\rangle. \tag{2.1.8}$$

Thus, $\tilde{\eta}_\zeta^s$ does indeed depend on the direction of the spin:

$$\tilde{\eta}_\zeta^{+\frac{1}{2}} = -\tilde{\eta}_\zeta^{-\frac{1}{2}}. \tag{2.1.9}$$

However, apart from this constraint, $\tilde{\eta}_\zeta^s$ is arbitrary, because the states $|\nu^M(s)\rangle$ and $|\nu^M(-s)\rangle$ appearing in Eq. (2.1.7) can always be redefined through multiplication by arbitrary phase factors.

2.1.5 Electromagnetic properties of Majorana neutrinos

How do the electromagnetic properties of neutrinos depend on whether they are Dirac or Majorana particles? From Lorentz invariance and current conservation, it follows that for any spin-$\tfrac{1}{2}$ fermion f, the matrix element of the electromagnetic current J_μ^{EM} has the form

$$\langle f(p_f, h_f) | J_\mu^{EM} | f(p_i, h_i)\rangle = i\bar{u}_f [F\gamma_\mu + G(q^2\gamma_\mu - 2m_f iq_\mu)\gamma_5$$

$$+ M\sigma_{\mu\nu}q_\nu + Ei\sigma_{\mu\nu}q_\nu\gamma_5]u_i. \tag{2.1.10}$$

Here p_i, h_i are the initial momentum and helicity of f, p_f, h_f are the final ones, $q = p_i - p_f$, m_f is the mass of f, and F, G, M, and E are form-factors that depend on q^2. If f is a Majorana neutrino ν^M, then the electromagnetic matrix element obeys the CPT constraint

$$\langle \nu^M(p_f, h_f)| J_\mu^{EM} |\nu^M(p_i, h_i)\rangle = -\langle \zeta\nu^M(p_i, h_i)| J_\mu^{EM} |\zeta\nu^M(p_f, h_f)\rangle$$

$$= -\tilde{\eta}_\zeta^{h_i*}\tilde{\eta}_\zeta^{h_f} \langle \nu^M(p_i, -h_i) \cdot |J_\mu^{EM}|\nu^M(p_f, -h_f)\rangle. \tag{2.1.11}$$

The minus sign in this relation arises from the fact that J_μ^{EM} is CPT-odd, and the interchange of the initial and final states from the fact that $\zeta \equiv CPT$ is antiunitary. Using the relation $\tilde{\eta}_\zeta^{h_i*}\tilde{\eta}_\zeta^{h_f} = (-1)^{h_i-h_f}$ which follows from Eq. (2.1.9), and writing both the first and third "sides" of the constraint (2.1.11) in the form (2.1.10), one can show that this constraint implies that $F = M = E = 0$ [NIE 82; KAY 83; MCK 82]. That is, for a Majorana neutrino, the most general form of the electromagnetic matrix element is [KAY 82; NIE 82; SCH 81]

$$\langle \nu^M(p_f, h_f)| J_\mu^{EM} |\nu^M(p_i, h_i)\rangle = i\bar{u}_f G(q^2)(q^2\gamma_\mu - 2m_\nu iq_\mu)\gamma_5 u_i \tag{2.1.12}$$

involving only a G-type form-factor. By contrast, for a Dirac neutrino there is no analog of the constraint (2.1.11), and the electromagnetic matrix element can have the full structure of Eq. (2.1.10), with all four form-factors.

The magnetic and electric dipole moments of any fermion are, respectively, the value of its M and E form-factors at $q^2 = 0$. Thus, a Majorana neutrino has no dipole moments. The electric charge radius of any fermion is, apart from a numerical factor, the derivative of its F form-factor at $q^2 = 0$. Thus, a Majorana neutrino has no charge radius either.

For a Majorana neutrino with mass, the absence of dipole moments is easy to understand. Suppose that such a neutrino has a magnetic dipole moment $\mu_{Mag}\mathbf{s}$ and an electric dipole moment $\mu_{El}\mathbf{s}$, where $\mathbf{s}$ is the neutrino spin. Then, when this neutrino is at rest in static, uniform magnetic and electric fields $\mathbf{B}$ and $\mathbf{E}$, it has a dipole interaction energy $-\mu_{Mag}\mathbf{s}\cdot\mathbf{B}-\mu_{El}\mathbf{s}\cdot\mathbf{E}$. Now, in the CPT-reflected state, the spin $\mathbf{s}$ is reversed, but (as one may easily show) $\mathbf{B}$ and $\mathbf{E}$ are unchanged. Thus, the dipole interaction energy is reversed. Hence, if the world is to be invariant under CPT reflection, μ_{Mag} and μ_{El} must vanish.

The absence of a charge radius is also easy to understand. Suppose, for example, that some Majorana neutrino has a charge radius arising from the presence, in the (neutral) neutrino, of a positively charged core surrounded by a compensating negatively charged shell. Under CPT, this charge distribution transforms into a negative core surrounded by a positive shell, something quite different from its original self. However, a Majorana neutrino must transform into itself under CPT, apart from a spin reversal. Thus, a Majorana neutrino actually cannot contain a positive core and negative shell. This illustrates why, more generally, such a neutrino cannot have a charge radius.

Despite the absence of dipole moments and a charge radius, a Majorana neutrino *can* couple to a photon. It does this through its G-type form-factor. The electromagnetic structure to which this form-factor corresponds [RAD 85] may be pictured as a torus formed by bending a flexible straight solenoid into the shape of a circle and joining the ends. The $\mathbf{B}$ field formerly present inside the solenoid will now circulate around the interior of the torus.

Unfortunately, it is extremely unlikely that we will be able to determine whether neutrinos are Dirac or Majorana particles by studying their electromagnetic properties. Indeed, the insensitivity of electromagnetic studies to the Dirac–Majorana distinction is an example of the practical Dirac–Majorana confusion theorem referred to earlier. It is true that, while a Majorana neutrino can never have a magnetic dipole moment, the Standard Model (with neutrino masses added) predicts that a Dirac neutrino of mass M_ν will have a dipole moment $\mu_{Mag} = 6 \times 10^{-19}(M_\nu/1\text{ eV})\mu_B$, where μ_B is the Bohr magneton [LEE 77a]. However, for M_ν below the existing upper bounds, this moment is far too small to be detected experimentally [SHR 82]. It is also true that, while a Majorana neutrino can never have an F-type form-factor, the Standard Model predicts that a Dirac neutrino will have one (Section 2.6 of this volume and [DEG 88]). However, this model also predicts that both a Dirac and a Majorana neutrino will have a G-type form-factor. Now, the only experimentally available neutrinos are highly relativistic and left-handed. For such neutrinos, the F and G form-factors lead to electromagnetic matrix elements which are helicity-preserving and of identical structure. Furthermore, the Standard Model (or any model with no right-handed currents) predicts that for any highly relativistic left-handed neutrino, the matrix element arising from the G form-factor if the neutrino is of Majorana character is *identical*, not only in structure but also in size, with that arising from the F and G form-factors together if the neutrino is of Dirac character [KAY 82, 88b]. See also [BAR 88d].

For further discussion of the electromagnetic structure of neutrinos, see Section 2.6 of this volume.

2.1.6 CP violation when $\bar{\nu} = \nu$

In the Standard Model, CP violation in the weak interactions of quarks arises from complex phase factors in the quark mixing matrix. However, unless there are at least three generations, all phase factors in this matrix can be rotated away, and so have no physical significance. Thus, in the Standard Model, the quark interactions could not violate CP at all if there were fewer than three generations [KOB 73].

In analogy with the quark interactions, the leptonic interactions can violate CP (in the Standard Model) as a result of complex phase factors in the leptonic mixing matrix. However, if neutrinos are their own antiparticles, then, for a given number of generations, fewer of the phases in the leptonic mixing matrix than of those in the quark matrix can be rotated away [BIL 80; KOB 80; SCH 80; DOI 81]. In particular, one phase already survives when there are only two generations. As a result, even if only two of the three known lepton generations mix appreciably, so that in effect there *are* only two generations, there can still be sizable CP-violating effects in the leptonic sector.

One can understand why more lepton phases than quark phases have physical significance when neutrinos are their own antiparticles by noting that when this is the case, certain leptonic processes have more Feynman diagrams than do the analogous quark processes [KAY 88c]. Now, complex phase factors in the lepton or quark mixing matrix can lead to physical CP-violating effects only when Feynman diagrams, to which these phase factors have imparted complex overall phases, interfere with one another. If some leptonic processes involve more Feynman diagrams than the corresponding quark processes, there can be additional inter-ferences between the diagrams in the leptonic case. These additional interferences can allow phase factors, which have no consequences when they occur in the quark mixing matrix, to lead to physical CP-violating effects when they occur in the lepton matrix.

As an illustration, let us compare the radiative decay $\nu_2 \rightarrow \nu_1 + \gamma$ of a heavy Majorana neutrino into a lighter one with the analogous decay $c \rightarrow u + \gamma$ of the charmed quark into the up quark. We shall suppose for simplicity that only the first two generations exist. Then the quark decay is engendered by diagrams in which the c quark turns either into a virtual dW^+ pair, or into a virtual sW^+ pair, and the photon is radiated by one of the particles in the pair. The pair then coalesces into the daughter u quark. It is very easy to show that the interferences between the various diagrams are completely insensitive to any complex phase factors in the (two-by-two) quark mixing matrix. The related neutrino decay arises from diagrams in which the ν_2 turns either into a virtual e^-W^+ pair, or into a virtual μ^-W^+ pair, and the photon is radiated by one of the charged particles in the pair. The pair then coalesces into the daughter ν_1. So far, everything is in complete analogy with the quark decay. However, if the ν_2 is its own antiparticle, then, "confused" about whether it is a lepton or an antilepton, it can turn not only into the virtual pairs already mentioned, but also into e^+W^- and μ^+W^-. Thus, there are additional diagrams in which one of these new pairs replaces e^-W^+ or μ^-W^+. These additional diagrams, which have no analog in the quark case, interfere with the diagrams containing e^-W^+ or μ^-W^+. It is not difficult to show that these new interferences *are* sensitive to a complex phase factor in the lepton mixing matrix. Through these added interferences, this phase factor, if present, can lead to a physical *CP*-violating effect [KAY 88c].

2.1.7 Neutrinoless double beta decay

In spite of the difficulty of telling whether neutrinos are Majorana or Dirac particles, there is one reaction which could provide evidence that they are Majorana particles even if their masses are well below 1 eV. This reaction is the nuclear decay $(A, Z) \rightarrow (A, Z+2) + 2e^-$, known as neutrinoless double beta decay ($\beta\beta_{0\nu}$). This decay can arise from a diagram in which the parent nucleus emits a pair of virtual W bosons, and then these W bosons exchange a neutrino ν_m, of mass M_m, to produce the outgoing electrons. The amplitude is a sum over the contributions of all the ν_m that may exist.

At the vertex where it is emitted, the exchanged ν_m is created together with an e^-. Thus, should there be a difference between leptons and antileptons and lepton number be conserved, this "ν_m" would have to be a $\bar{\nu}_m$. However, at the vertex where it is absorbed, this same particle creates a second e^-, so it must be a ν_m. Thus, the diagram vanishes unless $\bar{\nu}_m = \nu_m$. Even then, it is suppressed by a helicity mismatch at the two vertices touched by the virtual ν_m. Where this particle is emitted, it is behaving like an antineutrino. Hence, assuming the leptonic weak current is left-handed, the ν_m will be emitted in a predominantly right-handed state. On the other hand, where it is absorbed, it is behaving like a neutrino, so the current prefers to absorb it from a left-handed state.

Now, there is an amplitude of order $M_m/[\text{Energy of } \nu_m]$ for the ν_m to be emitted *left-handed*. If it is a Majorana particle, it can then be reabsorbed without further suppression. Thus, in effect, $\beta\beta_{0\nu}$ is a realization of the type of gedanken experiment we described when discussing neutrinos from pion decay. In $\beta\beta_{0\nu}$, we produce a particle – the exchanged ν_m – which is identified as an antineutrino by the fact that it is emitted together with an e^-. However, at least some of the time, this "antineutrino" is produced left-handed. We can then see whether this left-handed "antineutrino" interacts as would a left-handed neutrino at the vertex where it is absorbed.

If the leptonic weak current contains a small right-handed piece, then this piece will lead to emission of a virtual $\bar{\nu}_m(-)$ in $\beta\beta_{0\nu}$, just as does the ν_m mass. As before, if $\bar{\nu}_m(-) = \nu_m(-)$, this particle can then be reabsorbed without suppression.

As already mentioned, $\beta\beta_{0\nu}$ can provide evidence that neutrinos are Majorana particles even if their masses are small compared to 1 eV. The primary reason for this special sensitivity is that the decays which can in principle compete with $\beta\beta_{0\nu}$ are highly suppressed. So long as one chooses a parent nucleus that is stable against single beta decay, this competing mode is totally absent. Of course, competition with $\beta\beta_{0\nu}$ can always come from decay by emission of two electrons *and* two antineutrinos, a mode which can occur whether or not neutrinos are Majorana particles. However, this mode is phase-space suppressed, typically by six orders of magnitude, relative to $\beta\beta_{0\nu}$.

The amplitude $A[\beta\beta_{0\nu}]$ for $\beta\beta_{0\nu}$ can be written in the form

$$A[\beta\beta_{0\nu}] = M_{\mathit{eff}} N \tag{2.1.13}$$

where N is a very non-trivial nuclear matrix element [HAX 84a], and M_{eff}, the effective neutrino mass for neutrinoless double beta decay, contains the particle physics of the process. Assuming that there are no right-handed currents, and that all neutrino masses are small compared to the typical momentum transfer in $\beta\beta_{0\nu}(\sim 10\,\mathrm{meV})$, M_{eff} given by [DOI 81; WOL 81; KAY 83, 84; BIL 84]

$$M_{\mathit{eff}} = \sum_m \omega_{em}|U_{em}|^2 M_m. \tag{2.1.14}$$

In this sum over neutrino exchange contributions, the contribution of ν_m is proportional to its mass M_m because of the helicity considerations we have discussed. The quantity U_{em} is an element of a unitary mixing matrix describing the coupling of neutrinos to charged leptons, and ω_{em} is a phase factor.

Suppose that $\beta\beta_{0\nu}$ were actually to be observed. From the observed decay rate, and a calculated value for the nuclear matrix element N, one could then obtain an experimental value for M_{eff}. Since $\sum_m |U_{em}|^2 = 1$, we see from Eq. (2.1.14) that this experimental value could not exceed the largest of the actual neutrino masses M_m. That is, the observation of $\beta\beta_{0\nu}$ would imply a lower bound on neutrino mass: At least one neutrino would have to have a mass no smaller than the measured M_{eff}. By contrast, the observed absence of $\beta\beta_{0\nu}$ at some level does not imply an upper bound on the masses of any neutrinos. This absence only limits M_{eff}, and M_{eff} can be much smaller than the actual neutrino masses M_m, owing to the possible cancellations in Eq. (2.1.14).

If right-handed currents, and/or W bosons beyond the known one, do exist, then M_{eff} can become much more complicated than the expression in Eq. (2.1.14). In particular, the contribution to M_{eff} of a given ν_m exchange need no longer vanish with M_m. Nevertheless, a simple argument shows that it is still true that the observation of $\beta\beta_{0\nu}$ would imply *non-zero* neutrino mass, even if the origin of this decay is not neutrino exchange but some more exotic mechanism [SCH 82; TAK 84]. To be sure, the non-zero mass which would be implied according to this argument is of very high order in the weak interaction, and consequently could be extremely infinitesimal. However, if one does assume that $\beta\beta_{0\nu}$ is caused (at least primarily) by neutrino exchange, then, for a broad class of gauge theories, the observation of this reaction would imply a rather interesting lower bound on neutrino mass [KAY 87, 89]. Namely, even if right-handed currents and numerous W bosons exist, the bound discussed previously assuming their absence would still hold. That is, at least one neutrino would have to have a mass no smaller than the experimentally measured M_{eff} defined by Eq. (2.1.13) and determined from the observed decay amplitude and a calculated nuclear matrix element.

To date, $\beta\beta_{0\nu}$ has not been seen, so we have only upper bounds on M_{eff}. The most stringent of these is 0.46 eV, coming from an upper limit on $\beta\beta_{0\nu}$ decay of ^{76}Ge [KLA 98a]. It is hoped that future $\beta\beta_{0\nu}$ searches can be made sensitive to values of M_{eff} as small as 0.01 eV, or even 0.001 eV [KLA 98a].

Recently, strong evidence that neutrinos have non-zero masses has been reported [FUK 98]. This evidence, coming from the behavior of atmospheric neutrinos, suggests that one neutrino mass eigenstate has a mass of 0.02 eV or more. Other hints of neutrino mass suggest masses of 10^{-5} eV to a few eV.

If neutrinos do have mass, as now appears likely, then, as we have discussed, there is a distinction between Majorana and Dirac neutrinos even if all weak currents are left-handed. Thus, it will be very interesting to find out whether the neutrinos in nature are of Majorana or of Dirac type. If the neutrino masses are well below 1 eV, this will be quite challenging.

2.2 Searching for double beta decay*

2.2.1 Motivation for the search

Double beta decay yields – besides proton decay – the most promising possibilities for probing beyond Standard Model physics, beyond accelerator energy scales. Propagator physics has to replace direct observations. That this method is very effective is obvious from important earlier research work and has been stressed, for example, by [RUB 96]. Examples are the properties of W and Z bosons derived from neutral weak currents and β-decay, and the top mass deduced from LEP electroweak radiative corrections. Double beta decay has been with us now for more than

* H.V. Klapdor-Kleingrothaus, Max-Planck-Institut für Kernphysik, P.O. Box 10 39 80, D-69029 Heidelberg, Germany.

60 years, since the first calculation of corresponding matrix elements by [GOE 35]. The interest at that time was in the stability of even-even nuclei compared with the second-order weak interaction. In 1939, Furry [FUR 39] observed that a Majorana neutrino could induce a process in which a neutrino emitted as a virtual particle by one neutron could be absorbed by another neutron, leading to neutrinoless double beta decay. Only in the early 1980s was it understood that double beta decay allows information about the Majorana mass of the exchanged neutrino [SCH 81; KAY 88] to be obtained, and only in the last 10 years or so has the much more far-reaching potential of double beta decay for probing physics beyond Standard Model physics been discovered (see, for example, [MOH 86; MOH 96]).

Today, the potential of double beta decay includes information about the neutrino and sneutrino mass, SUSY models, compositeness, leptoquarks, right-handed W bosons, and others. It has been found that the $0\nu\beta\beta$ decay already probes the TeV scale on which new physics should manifest itself according to present theoretical expectations.

To give just one example, inverse double beta decay $e^-e^- \rightarrow W^-W^-$ requires an energy of at least 4 TeV for observability, according to present constraints from double beta decay [BEL 95b]. Similar energies are required to study leptoquarks [H1 95; HIR 96a; BAV 95; LEU 94].

2.2.2 Double beta decay and particle physics

We present an introductory outline of the potential of $\beta\beta$ decay for some representative examples, including some brief comments on the status of the required nuclear matrix elements.

Double beta decay can occur in several decay modes (Figs. 2.2.1 and 2.2.2)

$$^A_ZX \rightarrow\, ^A_{Z+2}X + 2e^- + 2\bar{\nu}_e \tag{2.2.1}$$

$$^A_ZX \rightarrow\, ^A_{Z+2}X + 2e^- \tag{2.2.2}$$

$$^A_ZX \rightarrow\, ^A_{Z+2}X + 2e^- + \phi \tag{2.2.3}$$

$$^A_ZX \rightarrow\, ^A_{Z+2}X + 2e^- + 2\phi, \tag{2.2.4}$$

the last three of them violate lepton number conservation by $\Delta L = 2$. Figure 2.2.1 shows the corresponding spectra, for the neutrinoless mode (2.2.2) a sharp line at $E = Q_{\beta\beta}$, for the two-neutrino mode, and the various Majoron-accompanied modes classified by their spectral index, and continuous spectra. Important for particle physics are the decay modes (2)–(4).

The neutrinoless mode (2) need not necessarily be connected with the exchange of a virtual neutrino. *Any* process violating lepton number can, in principle, lead to a process with the same signature as the usual $0\nu\beta\beta$ decay (see below). There is, however, a generic relation between the amplitude of $0\nu\beta\beta$ decay and the $(B-L)$ violating Majorana mass of the neutrino. It was recognized about 15 years ago [SCH 81] that if either of these two quantities vanish, the other one vanishes too, and

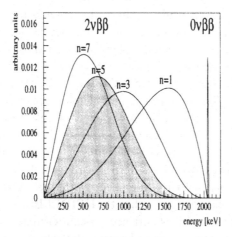

Fig. 2.2.1 Spectral shapes of the different modes of double beta decay, n denotes the spectral index, $n = 5$ for $2\nu\beta\beta$ decay.

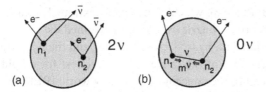

Fig. 2.2.2 Schematic representation of 2ν and 0ν double beta decay.

vice versa, if one of them is non-zero, the other one also differs from zero. This Schechter–Valle theorem is valid for any gauge model with spontaneously broken symmetry at the weak scale, independent of the mechanism of $0\nu\beta\beta$ decay. A generalization of this theorem to supersymmetry has recently been given [HIR 97b]. This Hirsch–Klapdor-Kleingrothaus–Kovalenko theorem claims for the neutrino Majorana mass, the $B-L$ violating mass of the sneutrino and neutrinoless double beta decay amplitude: if one of them is non-zero, the others are also non-zero and vice versa, independent of the mechanisms of $0\nu\beta\beta$ decay and (s-)neutrino mass generation. It connects double beta research with new processes potentially observable with future colliders such as the NLC (next linear collider) [HIR 97b, 98].

2.2.2.1 Mass of the electron neutrino

The neutrino plays, by its nature (Majorana or Dirac particle), and its mass, a key role for the structure of modern particle physics theories (GUTs, SUSYs, SUGRAs, etc.) [KLA 95; KLA 97a; GRO 90; LAN 88; MOH 91]. At the same time, it is a candidate for non-baryonic dark matter in the universe, and the neutrino mass is connected – by the sphaleron effect – to the matter-antimatter asymmetry of the early universe [KUZ 90]. Neutrino physics has entered an era of new actuality in connection with several possible indications of physics beyond the Standard Model (SM) of particle physics: The lack of solar (^{7}Be) neutrinos (see Chapter 6.1), the atmospheric ν_μ deficit and mixed dark matter models could all be explained by non-vanishing neutrino masses. Recent GUT models, for example an extended $SO(10)$ scenario with S_4 horizontal symmetry, could explain these observations by requiring degenerate neutrino masses of the order of 1 eV [LEE 94; MOH 94; PET 94; IOA 94; FRI 95; MOH 95]. Such degenerate scenarios are the more general solution of the well-known see-saw mechanism, of which the often discussed strongly hierarchical neutrino mass pattern is just a special solution (see [MOH 96]).

This brings double beta decay experiments into a key position, as with some second-generation $\beta\beta$ experiments such as the HEIDELBERG–MOSCOW experiment which uses large amounts of enriched $\beta\beta$-emitter material, the predictions or assumptions in such scenarios can now be tested. If the first of the above scenarios of neutrino mass textures is ruled out by tightening the double beta limit on m_{ν_e}, then the only way to understand *all* neutrino results may require an additional sterile neutrino [CAL 93; PEL 93], which couples only extremely weakly to the Z boson. Then the solar neutrino puzzle would be explained by ν_e–ν_S oscillation, and atmospheric neutrino data by ν_μ–ν_τ oscillations, and the $\nu_{\mu,\tau}$ would constitute the hot dark matter (HDM) of the universe. The request for a light sterile neutrino would naturally lead to the concept of a shadow world [BER 95]. Such a scenario could explain all *four* of the present indications for the non-vanishing neutrino mass [MOH 96]. The expectation for the effective neutrino mass (see below) to be seen in double beta decay would then be $\langle m_{\nu_e} \rangle \simeq 0.002$ eV [MOH 97]. It could therefore be checked by the new Genius project (see Sections 2.2.3.2 and 2.2.4).

At present, $\beta\beta$ decay is the most sensitive of the various existing methods for determining the mass of the electron neutrino. It also provides a unique possibility of deciding between a Dirac and a Majorana nature of the neutrino (see Section 2.1). Neutrinoless double beta decay can be triggered by the exchange of a light or heavy left-handed Majorana neutrino, as shown in Figs. 2.2.2 and 2.2.3. The propagators in the first and second case show a different m_ν dependence: Fermion propagator $\sim m/q^2 - m^2 \Rightarrow$

$$(a) \quad m \ll q \to \sim m \quad \text{`light' neutrino} \tag{2.2.5}$$

$$(b) \quad m \gg q \to \sim \frac{1}{m} \quad \text{`heavy' neutrino.} \tag{2.2.6}$$

The half-life for $0\nu\beta\beta$ decay induced by the exchange of a light neutrino is given by [MUT 88]

$$[T_{1/2}^{0\nu}(0_i^+ \to 0_f^+)]^{-1} = C_{mm} \frac{\langle m_\nu \rangle^2}{m_e^2} + C_{\eta\eta} \langle \eta \rangle^2 + C_{\lambda\lambda} \langle \lambda \rangle^2 + C_{m\eta} \frac{m_\nu}{m_e}$$

$$+ C_{m\lambda} \langle \lambda \rangle \frac{\langle m_\nu \rangle}{m_e} + C_{\eta\lambda} \langle \eta \rangle \langle \lambda \rangle, \tag{2.2.7}$$

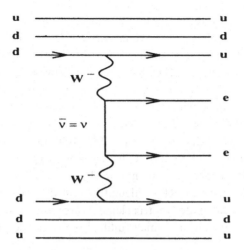

Fig. 2.2.3 Feynman graph for neutrinoless double beta decay triggered by exchange of a left-handed light or heavy neutrino.

or, when neglecting the effect of right-handed weak currents, by

$$[T_{1/2}^{0\nu}(0_i^+ \to 0_f^+)]^{-1} = C_{mm}\frac{\langle m_\nu \rangle^2}{m_e^2} = (M_{GT}^{0\nu} - M_F^{0\nu})^2 G_1 \frac{\langle m_\nu \rangle^2}{m_e^2}, \qquad (2.2.8)$$

where G_1 denotes the phase space integral, and $\langle m_\nu \rangle$ denotes an effective neutrino mass

$$\langle m_\nu \rangle = \sum_i m_i U_{ei}^2, \qquad (2.2.9)$$

thus respecting the possibility of the electron neutrino being a mixed state (mass matrix not diagonal in the flavor space)

$$|\nu_e\rangle = \sum_i U_{ei}|\nu_i\rangle. \qquad (2.2.10)$$

The effective mass $\langle m_\nu \rangle$ could be smaller than m_i for all i and for appropriate CP phases of the mixing coefficients U_{ei} [WOL 81]. In general, not too pathological GUT models yield $m_{\nu_e} = \langle m_{\nu_e} \rangle$ (see [LAN 88]).

η, λ describe an admixture of right-handed weak currents, and $M^{0\nu} \equiv M_{GT}^{0\nu} - M_F^{0\nu}$ denote nuclear matrix elements.

Nuclear matrix elements A detailed discussion of $\beta\beta$ matrix elements for neutrino-induced transitions, including the substantial (well understood) differences in the precision with which 2ν and $0\nu\beta\beta$ rates can be calculated, can be found in [GRO 90; MUT 88, 89; STA 90]. After the major step of recognizing the importance of nuclear ground state correlations for calculating $\beta\beta$ matrix elements

[KLA 84; GRO 86], in recent years the main groups have used the QRPA (quasiparticle random phase approximation) model for calculating $M^{0\nu}$. The different groups obtained very similar results for $M^{0\nu}$ when using a realistic nucleon–nucleon interaction [MUT 89; STA 90; TOM 87], consistent with shell model approaches [MUT 91; HAX 84], where the latter are possible. Some deviations are found only when a non-realistic nucleon–nucleon interaction is used (e.g. δ force, see [VOG 86]). Furthermore, use of a by far too small configuration space, as in recent shell model Monte Carlo (SMMC) calculations, can hardly lead to reliable results. On the other hand, refinements of the QRPA approach by going to higher-order QRPA lead only to minor changes for the $0\nu\beta\beta$ ground state transitions. The most recent QRPA calculations [SIM 97] do not fulfil the Ikeda sum rule by 30%. The calculated matrix elements are (correspondingly) about 40% smaller than earlier calculations which fulfilled the sum rule property [MUT 89; STA 90].

Since the usual QRPA approach ignores deformations, some larger uncertainties in these approaches may occur in deformed nuclei. This shows up, for example, in different results obtained for ^{150}Nd by QRPA and by a pseudo SU(3) model as used by [HIR 95c]. Calculations of matrix elements of all double beta emitters have been published by [GRO 85; STA 90]. Typical uncertainties of calculated $0\nu\beta\beta$ rates originating from the limited knowledge of the particle–particle force, which is the main source of the uncertainty in those nuclei where this QRPA approach is applicable, are shown in [STA 90]. They are of the order of a factor of 2.

2.2.2.2 *Exchange of other particles*

The exchange of other particles can contribute to double beta decay in Feynman graphs of the type shown in Fig. 2.2.3. Such contributions have been worked out in detail, see e.g. [HIR 95a,d, 96c; MOH 91, 86a] for the case of supersymmetry, [HIR 96d; DOI 93; MOH 86b] for the case of heavy neutrinos and right-handed W boson, and [MOH 92; SOU 92] for composite quarks and leptons. The limited space available for this review does not allow a detailed discussion.

2.2.2.3 *Majorons*

In many theories of physics beyond the Standard Model neutrinoless double beta decay could occur with the emission of Majorons

$$2n \rightarrow 2p + 2e^- + \phi \qquad (2.2.11)$$

$$2n \rightarrow 2p + 2e^- + 2\phi. \qquad (2.2.12)$$

In the classical Majoron model invented by Gelmini and Roncadelli in 1981 [GEL 81], the Majoron is the Nambu–Goldstone boson associated with the spontaneous breaking of the $B-L$-symmetry, and generates Majorana masses of neutrinos. This was expected [GEO 81] to give a sizeable contribution to double beta decay. It was, however, ruled out, as was the doublet Majoron [AUL 82] by LEP [STE 91], as it would have contributed the equivalent of two neutrino species to the width of the Z^0. On the other hand, Majoron models in which the Majoron is an electroweak isospin singlet [CHI 81, BER 92] are still viable. The drawback of the singlet Majoron is that it requires severe fine-tuning in order to preserve existing bounds on neutrino masses and, at the same time, to get an observable rate for Majoron-accompanied $0\nu\beta\beta$ decay.

To avoid such unnatural fine-tuning, in recent years several new Majoron models have been proposed [BUR 93; BAM 95b; CAR 93], where the term Majoron denotes, in a more general sense, light or massless bosons with couplings to neutrinos.

The half-lives of Majoron-accompanied $0\nu\beta\beta$ decay are according to [MOH 88; DOI 85] in an approximation given by

$$[T_{1/2}]^{-1} = |\langle g_\alpha \rangle|^2 \cdot |M_\alpha|^2 \cdot G_{BB_\alpha} \qquad (2.2.13)$$

for $\beta\beta\phi$-decays, or

$$[T_{1/2}]^{-1} = |\langle g_\alpha \rangle|^4 \cdot |M_\alpha|^2 \cdot G_{BB_\alpha} \qquad (2.2.14)$$

for $\beta\beta\phi\phi$-decays. The index α indicates that the effective neutrino–Majoron coupling constants g, the matrix elements M, and the phase spaces G differ for different models.

2.2.2.4 Sterile neutrinos

It has been claimed that the introduction of sterile neutrinos will solve simultaneously the conflict between dark matter neutrinos, LSND and supernova nucleosynthesis (see S. Petcov in [KLA 96b]), and that light sterile neutrinos are part of the popular neutrino mass textures for understanding the various hints for neutrino oscillations (see Section 2.2.2.1).

Assuming that we have a light neutrino with a mass $\ll 1$ eV, mixing it with a much heavier ($m \geq 1$ GeV) sterile neutrino can yield, under certain conditions, a detectable signal in current $\beta\beta$ experiments [BAM 95a].

2.2.2.5 Leptoquarks

The interest in leptoquarks (LQ) has been renewed during the last few years as ongoing collider experiments have good prospects for searching for these particles [BUC 87]. LQs are vector or scalar particles carrying both lepton and baryon numbers and therefore have a well distinguished experimental signature. Direct searches for LQs in deep inelastic ep-scattering at HERA [H1 96] placed lower limits on their mass $M_{LQ} \geq 225$–275 GeV, depending on the LQ type and couplings.

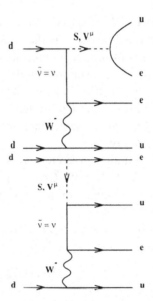

Fig. 2.2.4 Examples of Feynman graphs for $0\nu\beta\beta$ decay within LQ models. S and V^μ stand symbolically for scalar and vector LQs, respectively (from [HIR 96a]).

In addition to the direct searches for LQs, there are many constraints which can be derived from the study of low-energy processes [DAV 94].

In models with LQ–Higgs interaction [HIR 96a] contributions to $0\nu\beta\beta$ decay appear via the Feynman graphs of Fig. 2.2.4. Here, S and V^μ stand symbolically for scalar and vector LQs, respectively. The half-life for $0\nu\beta\beta$ decay arising from leptoquark exchange is given by [HIR 96a]

$$T_{1/2}^{0\nu} = |M_{GT}|^2 \frac{2}{G_F^2} [\tilde{C}_1 a^2 + C_4 b_R^2 + 2 C_5 b_L^2], \qquad (2.2.15)$$

with

$$a = \frac{\epsilon_S}{M_S^2} + \frac{\epsilon_V}{M_V^2}, \quad b_{L,R} = \frac{\alpha_S^{(L,R)}}{M_S^2} + \frac{\alpha_V^{(L,R)}}{M_V^2}, \quad \tilde{C}_1 = C_1 \left(\frac{M_1^{(\nu)}/(m_e R)}{M_{GT} - \alpha_2 M_F} \right)^2.$$

For the definition of the C_n see [DOI 85], and for the calculation of the matrix element $M_1^{(\nu)}$ see [HIR 96a]. This allows information on leptoquark masses and leptoquark–Higgs couplings to be deduced (see Section 2.2.3.1).

2.2.3 Double beta decay experiments: status and perspectives

We can differentiate between two classes of direct (non-geochemical) $\beta\beta$ decay experiments:

(a) active source experiments (source = detectors)
(b) passive source experiments.

In the first class of experiments, the $\beta\beta$ process is usually identified only on the basis of the distribution of the total energy of the emitted electrons. The second class of experiments yields, in principle, more complete information on the $\beta\beta$ events by measuring time coincidence, tracks and vertices of the emitted electrons, and their energy distribution. Time projection chambers (TPCs) such as the Gotthard ^{136}Xe experiment using $\beta\beta$ active counting gas belong to the first class.

Figure 2.2.5 shows an overview of measured $0\nu\beta\beta$ half-life limits and deduced mass limits. The largest sensitivity for $0\nu\beta\beta$ decay is at present obtained by active source experiments, in particular ^{76}Ge [HMC 95; HMC 97; KLA 94; KLA 97a] and ^{136}Xe [GER 96]. The main reason is that large source strengths can be used (simultaneously with high-energy resolution), in particular when enriched $\beta\beta$ emitter materials are used.

Other criteria to ensure the 'quality' of a $\beta\beta$ emitter are:

- a small product $T_{1/2}^{0\nu} \cdot \langle m_\nu \rangle^2$, i.e. a large matrix element $M^{0\nu}$ or phase space;
- a $Q_{\beta\beta}$ value beyond the limit of natural radioactivity (2.614 MeV).

The future of $\beta\beta$ experiments will be dominated by the use of enriched detectors, ^{76}Ge at present playing a particular favorable role here, and enriched source material such as ^{136}Xe, ^{100}Mo, and ^{116}Cd. Some of these experiments may probe the neutrino mass in the next years down to 0.1 eV (see Fig. 2.2.5b). A detailed discussion of the various experimental possibilities can be found in [KLA 95; KLA 96b]. A useful listing of existing data from the various $\beta\beta$ emitters is given in [TRE 95].

2.2.3.1 Present limits on parameters

The sharpest limits for $0\nu\beta\beta$ decay presently come from the Heidelberg–Moscow experiment [HMC 95; KLA 94; HMC 97; KLA 97a]. With five enriched (86% of ^{76}Ge) detectors with a total mass of 11.5 kg taking data in the Gran Sasso underground laboratory, the experiment has reached its final set-up and is now exploring the sub-eV range for the mass of the electron neutrino. Fig. 2.2.6 shows the spectrum taken in a measuring time of 35 kg y.

Half-life of neutrinoless double beta decay

The deduced half-life limit for $0\nu\beta\beta$ decay is

$$T_{1/2}^{0\nu} > 1.2 \times 10^{25} \text{y} \quad (90\% \text{ CL}) \tag{2.2.16}$$

$$> 2.0 \times 10^{25} \text{y} \quad (68\% \text{ CL}). \tag{2.2.17}$$

Neutrino mass

Light neutrinos: The deduced upper limit of an (effective) electron neutrino Majorana mass is, with the matrix element from [STA 90]

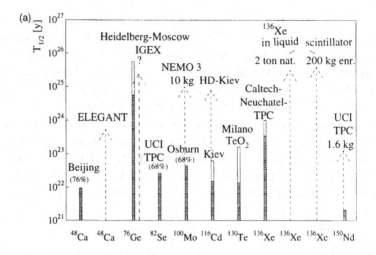

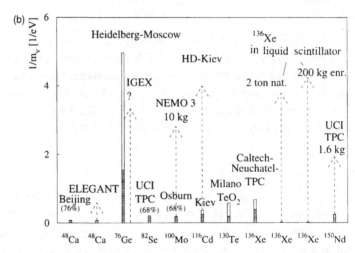

Fig. 2.2.5 Present situation (1998) and expectations for the near future (until the year 2002 and beyond) of the most promising $\beta\beta$-experiments with respect to accessible half-life (a) and neutrino mass limits (b). The filled-in parts of the columns correspond to the present achievements, the 'empty' parts of the columns correspond to the future achievements for about the year 2000, and the dashed lines correspond to long-term planned or hypothetical experiments (from [KLA 96a]).

$$\langle m_\nu \rangle < 0.44\,\text{eV} \quad (90\%\ \text{CL}) \tag{2.2.18}$$

$$< 0.34\,\text{eV} \quad (68\%\ \text{CL}). \tag{2.2.19}$$

This is the sharpest limit for a Majorana mass of the electron neutrino so far.

Superheavy neutrinos: For a superheavy *left*-handed neutrino, exploiting the mass dependence of the matrix element (for the latter see [MUT 89]), a lower limit of

$$\langle m_H \rangle \geq 100\,\text{TeV} \tag{2.2.20}$$

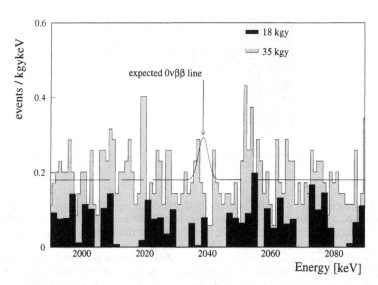

Fig. 2.2.6 Integral spectrum in the region of interest after subtraction of the first 200 days of measurement of each detector, leaving 35 kg y of measuring time. The dashed histogram corresponds to the signal excluded with 90% CL. It corresponds to $T_{1/2}^{0\nu} > 1.2 \times 10^{25}\,y$. The darkened histogram corresponds to data accumulated in the mean time using a new pulse shape analysis method [HEL 96] in a measuring time of 18 kg y.

can be deduced [HMC 95; BEL 95b]. For a heavy *right*-handed neutrino the relation obtained for the mass of the right-handed W boson is given in [HIR 96d].

Right-handed W boson

For the right-handed W boson a lower limit of

$$m_{W_R} \geq 1.2\,\text{TeV} \tag{2.2.21}$$

is obtained [HIR 96d].

SUSY parameters

The constraints on the parameters of the minimal supersymmetric Standard Model with explicit R-parity violation deduced [HIR 95a, 96c, 96e] from the $0\nu\beta\beta$ half-life limit are more stringent than those from other low-energy processes and from the largest high-energy accelerators (Fig. 2.2.7). The limits are

$$\lambda'_{111} \leq 3.8 \times 10^{-4} \left(\frac{m_{\tilde{q}}}{100\,\text{GeV}}\right)^2 \left(\frac{m_{\tilde{g}}}{100\,\text{GeV}}\right)^{\frac{1}{2}}, \tag{2.2.22}$$

with $m_{\tilde{q}}$ and $m_{\tilde{g}}$ denoting squark and gluino masses, respectively, and with the assumption $m_{\tilde{d}_R} \simeq m_{\tilde{u}_L}$. This result is important for discussing new physics in connection with the high Q^2 events seen at HERA. It excludes the possibility of

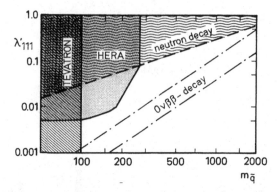

Fig. 2.2.7 Comparison of limits on the R-parity violating MSSM parameters from different experiments in the $\lambda'_{111} - m_{\tilde{q}}$ plane. The dashed line is the limit from charged current universality according to [BAR 89]. The vertical line is the limit from the Tevatron data [ROY 92]. The solid line is the region which might be explored by HERA [BUT 93]. The two dash-dotted lines to the right are the limits obtained from the half-life limit for $0\nu\beta\beta$ decay of ^{76}Ge, for gluino masses of (from left to right) $m_{\tilde{g}} = 1$ TeV and 100 GeV, respectively. The regions to the upper left of the lines are forbidden (from [HIR 95a]).

first-generation squarks (of R-parity violating SUSY) being produced in high Q^2 events [ALT 97a; HIR 97a].

Double beta decay (the Heidelberg–Moscow experiment) yields the limits

$$\lambda'_{113}\lambda'_{131} \leq 1.1 \times 10^{-7} \tag{2.2.23}$$

$$\lambda'_{112}\lambda'_{121} \leq 3.2 \times 10^{-6}. \tag{2.2.24}$$

For the $(B-L)$ violating sneutrino mass $\tilde{m}_M$ the following limits are obtained [HIR 98]

$$\tilde{m}_M \leq 2\left(\frac{m_{\text{SUSY}}}{100\,\text{GeV}}\right)^{\frac{3}{2}} \text{GeV}, \quad \chi \simeq \tilde{B} \tag{2.2.25}$$

$$\tilde{m}_M \leq 11\left(\frac{m_{\text{SUSY}}}{100\,\text{GeV}}\right)^{\frac{7}{2}} \text{GeV}, \quad \chi \simeq \tilde{H} \tag{2.2.26}$$

for the limiting cases when the lightest neutralino is a pure Bino $\tilde{B}$, as suggested by the SUSY solution of the dark matter problem [JUN 96], or a pure Higgsino. Actual values for $\tilde{m}_M$ for other choices of the neutralino composition should lie in between these two values.

Compositeness

Evaluation of the $0\nu\beta\beta$ half-life limit for the exchange of excited Majorana neutrinos ν^* yields, under some assumptions, bounds on the compositeness scale roughly of the same order of magnitude as those coming from high-energy experiments (see Panella and Takasugi in [KLA 98]).

Leptoquarks

Assuming that either scalar or vector leptoquarks contribute to $0\nu\beta\beta$ decay, some constraints on the effective LQ parameters can be derived [HIR 96a].

Since the LQ mass matrices appearing in $0\nu\beta\beta$ decay are (4×4) matrices [HIR 96a], it is difficult to solve their diagonalization in full generality algebraically. However, if one assumes that only one LQ–Higgs coupling is present at a time, the (mathematical) problem is simplified greatly and one can deduce that either the LQ–Higgs coupling must be smaller than $\sim 10^{-(4-5)}$ or that there cannot be any LQ with, for example, couplings of electromagnetic strength with masses below $\sim 250\,\mathrm{GeV}$. These bounds from $\beta\beta$ decay are of interest in connection with recently discussed evidence for new physics from HERA [BAB 97; HIR 97A].

Half-life of $2\nu\beta\beta$ decay

For the first time an experiment produced a high statistics $2\nu\beta\beta$ spectrum ($\sim 20\,000$ counts) compared with the 40 counts on which the first detector observation of $2\nu\beta\beta$ decay by [ELL 87] (for the decay of ^{82}Se) had to rely. The deduced half-life is [HMC 97]

$$T_{1/2}^{2\nu} = \left[1.77^{+0.01}_{-0.01}(\text{stat.})^{+0.13}_{-0.11}(\text{syst.})\right] \times 10^{21}\,\mathrm{y}. \qquad (2.2.27)$$

For the first time this result brings $\beta\beta$ research into the region of 'normal' nuclear spectroscopy and allows a statistically reliable investigation of Majoron-accompanied decay modes.

Majoron-accompanied decay

From simultaneous fits of the 2ν spectrum and one selected Majoron mode, experimental limits for the half-lives of the decay modes of the newly introduced Majoron models (see C.P. Burgess in [KLA 96b]) are given [HMC 96].

The small matrix elements and phase spaces for these modes [HIR 96b] have already determined that these modes cannot be seen in experiments with the present sensitivity if we assume typical values for the neutrino–Majoron coupling constants around $\langle g \rangle = 10^{-4}$.

2.2.3.2 Perspectives

In addition to the present status and future perspectives of the main $\beta\beta$ decay experiments, Figs. 2.2.5a and b also include ideas for the next decade. The Heidelberg–Moscow experiment will probe the neutrino mass within the next five years down to the order of $0.1\,\mathrm{eV}$. The best presently existing limits besides the Heidelberg–Moscow experiment (filled in parts of the columns in Fig. 2.2.5), with half-life limits above 10^{21} a, were obtained with the isotopes: ^{48}Ca [YOU 95], ^{82}Se [ELL 92], ^{100}Mo [ALS 89], ^{116}Cd [DAN 95], ^{130}Te [ALE 94], ^{136}Xe [VUI 93], and ^{150}Nd [MOE 94]. These and other double beta decay set-ups, presently under construction or partly in operation such as NEMO [NEM 94], the Gotthard ^{136}Xe TPC experiment [JÖR 94], the ^{130}Te cryogenic experiment [ALE 94], a new ELEGANT ^{48}Ca experiment using 64 g of ^{48}Ca (see Kumein in [KLA 96b]), a hypothetical experiment with an improved UCI TPC [MOE 94] assumed to use 1.6 kg of ^{136}Xe, etc., will reach the 'empty' parts of the columns in Figs. 2.2.5a,b but will not exceed the ^{76}Ge limits. The goal of 0.3 eV for the year 2004 by the NEMO experiment (see Piquemal in [KLA 96b] and Fig. 2.2.5) may be very optimistic if claims about the effect of proton–neutron pairing on the $0\nu\beta\beta$ nuclear matrix elements by [SIM 96] turn out to be true, and also if the energy resolution is not improved considerably (for the latter problem see [TRE 95]). As pointed out by Raghavan [RAG 94], even using about 200 kg of enriched ^{136}Xe or 2 t of natural Xe added to the scintillator of the KAMIOKANDE detector or similar amounts added to BOREXINO (both primarily devoted to solar neutrino investigation) would hardly lead to a sensitivity larger than the present ^{76}Ge experiment (see also the new KAMLAND proposal [SUZ 97]). An interesting future candidate was for some time a ^{150}Nd bolometer exploiting the relatively large phase-space of this nucleus (see [MOE 94]). The approach outlined by [MOE 91] proposing a TPC filled with 1 t of liquid enriched ^{136}Xe and identification of the daughter nucleus by laser fluorescence may not be feasible in a straightforward way.

It is obvious that the Heidelberg–Moscow experiment will give the best limit for the electron neutrino mass for the next few years. For further improvements beyond the region of < 0.1 eV one has to think of *very large* experiments with a *much bigger* source strength (see [KLA 98]).

A corresponding proposal – the GENIUS project – which could cover the neutrino mass region down to 0.02 eV, and which at the same time could provide high sensitivity for dark matter detection probing almost the entire parameter space of SUSY predictions for neutralinos as dark matter, has recently been published [KLA 97a].

2.2.4 Conclusion

Double beta decay has a broad potential for providing important information about modern particle physics beyond present and future high-energy accelerator energy scales. This includes SUSY models, compositeness, left–right symmetric models, leptoquarks, and the neutrino and sneutrino mass. For the neutrino mass, double beta decay has now been pushed into a key position by the recent possible indications of beyond the Standard Model physics from solar and atmospheric neutrinos, and dark matter COBE results. The Heidelberg–Moscow experiment has presently a leading position among these new $\beta\beta$ experiments as the first of them now yields results in the sub-eV range.

2.3 Direct measurement of the neutrino masses*

2.3.1 Introduction

The neutrino masses pose one of the important problems of today's particle physics. The problem is also of surprising complexity. It would not be sensible to investigate it by one type of experiment. In this section we discuss direct measurements of the masses of the three known neutrinos. By "direct" we mean the analysis of the kinematics of suitable decays without assuming unknown neutrino properties. So far no indication for a non-zero neutrino mass has been found. The present upper limits are

$$m_{\nu_e} < 10\,\text{eV} \simeq 2.0 \times 10^{-5} m_e$$
$$m_{\nu_\mu} < 170\,\text{keV} \simeq 1.6 \times 10^{-3} m_\mu$$
$$m_{\nu_\tau} < 24\,\text{MeV} \simeq 1.4 \times 10^{-2} m_\tau.$$

The given limit for the electron neutrino is just a round number. The reason for this will become clear later on. We have also compared the limits with the masses of the related charged leptons. This is only for convenience and does not imply that a physical relation of this sort should be expected.

Limits for the mass of the electron neutrino are determined from measurements of the tritium β-spectrum. Many such experiments have been done. With this method by far the sharpest limit, absolute and relative to the charged lepton mass, can be obtained. Further motivation for these efforts is that the limit is of interest for cosmology. It has become possible during the last ten years to push the limit for the electron neutrino mass well below the cosmological limit for the neutrino masses (see Section 2.7). It appears hopeless that this could also be achieved with direct methods for the other neutrinos. This might be one reason why comparatively little effort has been made to measure the masses of the muon and the tau neutrinos. Experiments to measure the mass of the muon neutrino, from pion decay, have been done at only one laboratory during the last twenty years. The mass of the tau neutrino can be obtained from various decay modes of the tau lepton. It would not be improper to call the present results by-products of large accelerator experiments which were primarily designed for other purposes.

There are other types of experiments which are possibly much more sensitive to effects of non-zero neutrino masses. Double beta decay requires for a signal that the neutrinos are Majorana particles (or that there are right-handed currents). This has already been discussed in Section 2.2. Oscillation experiments can probe extremely small neutrino masses, or more precisely differences of squared masses. However, in order to deduce anything about the neutrino masses, it is necessary that there is

* E. Holzschuh, Physics Institute, University of Zürich, 8001 Zürich, Switzerland.

neutrino mixing of sufficient magnitude. This is by itself a very interesting subject which is treated in Section 2.4.

The detection of a neutrino burst from the supernova SN 1987A provided a unique opportunity to determine neutrino properties, in particular the mass of the electron neutrino. In principle, only the time of flight as a function of energy needs to be considered and this is certainly a "direct" method. In reality it is much more involved and the subject is therefore discussed in Section 2.7.

The subject of this section has been treated in various recent review articles. A more general discussion may be found in [GEL 95] which contains also many references. Specific reviews about the tritium experiments are [HOL 92a; OTT 95].

2.3.2 The electron neutrino mass

The best direct limit for the mass of the electron neutrino has traditionally been obtained from studies of the beta decay of tritium.

$$^3\text{H} \longrightarrow {}^3\text{He}^+ \; e^- \; \bar{\nu}_e.$$

The energy distribution of the decay electrons is basically given by the available phase-space. For a bare nucleus, it can be written in the form

$$\frac{dN}{dE} = N(E) = CF(Z, W)pW\epsilon^2 \sqrt{1 - \frac{m_{\nu_e}^2}{\epsilon^2}} \Theta(\epsilon - m_{\nu_e}), \qquad (2.3.1)$$

where

$$\epsilon = E_0 - E \qquad (2.3.2)$$

is the energy carried away by the neutrino and m_{ν_e} is its mass. p, E, and $W = E + m_e$ are momentum, kinetic and total energy of the electron. The step function Θ indicates that the spectrum is zero above the true endpoint $E_0 - m_{\nu_e}$. Occasionally, the parameter E_0 ($E_0 \simeq 18.6\,\text{keV}$) is also called endpoint, although this is strictly correct only for zero neutrino mass. The Fermi function $F(Z, W)$ is a phase-space correction to account for the deceleration in the Coulomb field of the daughter nucleus (He^{++}, $Z = 2$). It will be discussed below together with further corrections to the spectrum. The tritium β-spectrum is shown in Fig. 2.3.1. The spectrum is sensitive to a non-zero neutrino mass only close to the endpoint, i.e. for small neutrino energies. This is shown in the inset. A measurement of the electron neutrino is thus basically a careful study of the shape of the spectrum around the endpoint.

2.3.2.1 Early tritium experiments

The beta decay of tritium was discovered around 1940 [ALV 40] and it was soon realized that it would be a very favorable case to measure the mass of the electron

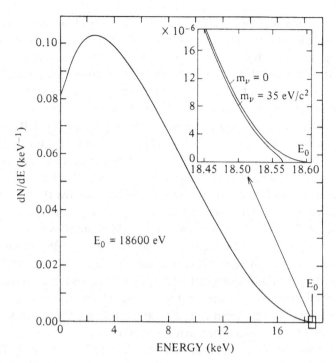

Fig. 2.3.1   The tritium β-spectrum. The inset shows a region around the endpoint on an expanded scale for two assumed neutrino masses.

neutrino. Tritium has a convenient half-life ($T_{1/2} = 12.3$ y) and the smallest endpoint energy of all allowed beta decays. The signature of a non-zero neutrino mass can be made more apparent by transforming Eq. (2.3.1) to

$$K(E) = \sqrt{N(E)/(FpW)} \sim \epsilon(1 - m_{\nu_e}^2/\epsilon^2)^{1/4}. \qquad (2.3.3)$$

A spectrum represented in this way is called a Kurie plot. One expects a straight line for $\epsilon \gg m_{\nu_e}$ and a steep decrease with a vertical slope at $\epsilon = m_{\nu_e}$ if $m_{\nu_e} > 0$. In practice, this distinctive feature of a non-zero neutrino mass would of course only be visible if the energy resolution of the measurement is much narrower than the value of m_{ν_e}.

First measurements of the tritium β-spectrum were performed by Curran et al. and by Hanna and Pontecorvo using a proportional counter [CUR 49; HAN 49]. A small amount of tritium was mixed into the counting gas and the pulse height distribution was recorded. No indication for a non-zero neutrino mass was observed and an upper limit of about 1 keV was derived.

Measurements with better resolution were subsequently performed by Langer and Moffat [LAN 52] and by Hamilton et al. [HAM 53], both reporting an upper limit of 250 eV. Langer and Moffat used a magnetic spectrometer with 0.9% momentum resolution. A nice straight Kurie plot was obtained which was later reproduced in many textbooks. Hamilton *et al.* employed an electrostatic device which recorded all electrons above a certain threshold energy, i.e. the integral of the spectrum was measured. Because of the high voltages involved, such methods work well in practice only for the very low energy electrons from tritium decay.

In the following years not much happened for more than a decade. Several new tritium experiments were initiated in the 1960s. From these the work of Bergkvist [BER 72] has become what is now considered a landmark in the field. Bergkvist used a magnetic spectrometer of the $\pi\sqrt{2}$-type. He found a solution for the seemingly contradictory requirements of a strong but very thin and therefore extended source on the one hand and high instrumental resolution on the other hand. The trick was to divide the source backing into many narrow strips (total area $10 \times 20\,\mathrm{cm}^2$) and to correct the large source extension by appropriate electric potentials of the strips. A resolution of 40 eV FWHM (Full Width at Half Maximum) was achieved.

Bergkvist recognized that the decay product He^+ may be left in some excited electronic state and that by energy conservation the decay electron must have lost the corresponding excitation energy [BER 71]. This effect has the same magnitude as Bergkvist's resolution and was thus not negligible. He made a correction for this effect and obtained an upper limit of 55 eV (90% CL) for the neutrino mass.

In the history of the tritium experiments, one recognizes that major progress was always related to new ideas in instrumentation design. In the 1970s Tret'yakov invented a new iron-free β-spectrometer [TRE 75]. It has cylindrical symmetry like the famous orange spectrometer, but it has only straight current conductors. This makes it relatively easy to build large instruments with high mechanical precision. Tret'yakov's spectrometer was used by a group at ITEP (Institute of Theoretical and Experimental Physics in Moscow). The design, with some modifications, was later employed in three more tritium experiments (see below).

In 1980, a first claim for a non-zero neutrino mass was reported by the ITEP group [LUB 80, 81] causing quite some excitement. The result, $m_{\nu_e} \simeq 34\,\mathrm{eV}$, was statistically highly significant, but what made the measurement really exciting was that it seemed possible to establish a model-independent non-zero lower limit for m_{ν_e}. This can be explained as follows. The ITEP source consisted of tritium labeled valine, an amino-acid. At the time, little was known about the excitation energies and probabilities [BER 71] of a molecule as complex as valine. It can however be shown that the fitted value of m_{ν_e} (actually $m_{\nu_e}^2$, see Eq. (2.3.1)) follows a simple rule when the data are analysed with a distribution of excitation energies which is different from the actual one. If the distribution is taken to be too narrow, the fitted value comes out too small. The most extreme assumption is to ignore the distribution altogether (zero width, no excitations, 'nucleus model'). For the ITEP data from 1980, this gave a positive result, $m_{\nu_e} \geq 14\,\mathrm{eV}$. Any other, more physical distribution would give only a *larger* value. The argument is of course only valid if there were no other systematic errors. This was not the case, as was quickly found out by others. The resolution function of the spectrometer and probably other things also were flawed.

The further course of the events is somewhat complicated and has been described in reference [HOL 92a]. The ITEP group made improvements, corrected mistakes, and produced more and more data. The basic conclusion however, $m_{\nu_e} \sim 30\,\mathrm{eV}$, never changed significantly. The last publication [BOR 87] dates from 1987.

We known now that the ITEP result is wrong, but it motivated some 20 groups to start tritium projects during the 1980s. The first result of these new experiments was reported in 1986 by the Zürich group [FRI 86]. The spectrometer built was of the Tret'yakov type with some modifications. Sources were prepared by implanting tritium ions into a thin layer of carbon. The measurements were made with 17 eV (FWHM) resolution [FRI 91]. The problem of the excitation of electronic final states in complex sources turned out not to be as severe as once assumed, mainly due to the extensive work of Kaplan et al. [KAP 82,83,85,88]. Using their results, the analysis of the data gave no indication for a non-zero neutrino mass and a cautiously estimated upper limit, $m_{\nu_e} < 18\,\mathrm{eV}$, could be set.

While the 1980s were dominated by discussions and controversies about the ITEP result, the 1990s brought new problems. These will be discussed below.

2.3.2.2 The tritium β-spectrum

In tritium experiments, the neutrino mass is determined from the shape of the β-spectrum. Measurements are made over an energy range of typically 1 keV or less below the endpoint. Also the energy E_0 is treated as a free parameter in the data analyses. With this in mind, we will here discuss various corrections to the shape given by Eq. (2.3.1).

Theoretical considerations

The Fermi function $F(Z, W)$ describes the influence of the Coulomb potential of the nucleus on the wavefunction of the emitted electron. It is usually written as a product [BEH 82], (p. 105)

$$F = F_0 L_0, \tag{2.3.4}$$

where the factor F_0 is obtained from an exact solution of the Dirac equation with a point charge nucleus. Neglecting terms of order $(\alpha Z)^2$, with α the fine structure constant, it is given by

$$F_0 \simeq 2\pi Z\eta/(1 - \exp(-2\pi Z\eta)). \tag{2.3.5}$$

The parameter η is defined by

$$\eta = \alpha/\beta \tag{2.3.6}$$

with $\beta = v/c$ the speed of the emitted electron with respect to the speed of light.

The factor L_0 is a correction for the finite charge distribution in the nucleus with charge radius R. For tritium it is close to 1 and changes slowly with energy. A convenient formula is given in [BEH 82, p. 141], which to order αZ reads

$$L_0 = 1 - \alpha Z \left(\frac{W}{m_e} + \frac{m_e}{2W}\right) \frac{Rm_e}{\hbar} \tag{2.3.7}$$

with m_e the mass of the electron.

Radiative corrections (QED) for the spectrum shape arise from the emission of real photons (internal bremsstrahlung) and from virtual photon exchange. They have been calculated for zero neutrino mass by several authors (cf [BEH 82, p. 432]). The result to order α and in the non-relativistic limit is given by

$$S = 1 + \frac{\alpha}{2\pi}\left\{3\ln(m_p/m_e) - \frac{27}{4} + \frac{4}{3}\left(\ln(2\epsilon/m_e) - \frac{13}{6}\right)\beta^2\right\}, \tag{2.3.8}$$

where m_p is the proton mass and where neglected terms inside braces are of order β^4. The logarithmic singularity at the endpoint ($\epsilon = 0$) is extremely weak. Even for $\epsilon = 1$ eV the log term contributes only -1.4×10^{-3} to S. In fact the singularity is only present in the correction but not in the spectrum, because $\epsilon^2 \ln(2\epsilon) \to 0$ as $\epsilon \to 0$.

The formulas discussed so far are applicable to a bare nucleus. In a real source, the tritium is part of a molecule RT (or solid) and the molecular electrons will dynamically respond to the decay. The remaining ion RHe^+ may be left in some excited state. As already noted, the outgoing electron must lose the corresponding excitation energy, which causes a broadening of the measured spectrum. Hence we must consider a multi-channel process

$$RT \longrightarrow (RHe^+)_n \; e^- \; \bar{\nu}_e,$$

where n denotes all quantum numbers of the product ion (which may be unbound).

It is easy to see that the decay electron leaves the molecule in a time which is two orders of magnitude smaller than typical orbital periods. Therefore the sudden approximation should be applicable. If recoil effects are ignored, it simply means that the molecular electronic wave function does not change during the decay, i.e. $\psi_i = \psi_f$. Assuming the molecule is initially in its groundstate $\varphi_0 = \psi_i$, the transition probability to a state n with wavefunction ψ_n of RHe^+ is given by

$$W_{0n} = |\langle\varphi_0|\psi_n\rangle|^2. \tag{2.3.9}$$

The energy gained by the decay electron is the energy difference of the initial state and the state n.

$$\Delta E_{0n} = E_0(RT) - E_n(RHe^+). \tag{2.3.10}$$

Table 2.3.1. *Calculated results of the electronic final states distribution for various molecules*

Molecule	W_{00}	ΔE_{00} (eV)	$\overline{\Delta E^*}$ (eV)	Reference
T	0.702	40.82	13.61	—
T_2	0.574	49.08	—	[KOL 85]
HT	0.577	48.65	18.62	[KAP 86]
CH_3T	0.606	49.42	18.98	[KAP 86]
CH_3-CH_2T	0.611	49.15	18.92	[KAP 88]
$CH_3-CHT-CH_3$	0.603	49.59	19.24	[KAP 88]
CH_3T	0.579	50.03	19.10	[SCH 91]
CH_3-CH_2T	0.568	50.57	19.73	[SCH 91]
$CH_3-CHT-CH_3$	0.571	50.03	19.02	[SCH 91]

Some useful information about the distribution can be obtained from two simple sum rules [KAP 88], which can be evaluated by using wavefunctions of groundstates only. They are easily derived from the completeness of the states n. For the average energy gain, one obtains

$$\overline{\Delta E} = \sum_n W_{0n} \Delta E_{0n} = \langle \varphi_0 | \Delta H | \varphi_0 \rangle, \tag{2.3.11}$$

where ΔH is the difference of the initial and final Hamiltonian, which is simply a Coulomb energy term. It is convenient to measure the energy with respect to the groundstate of RHe^+, i.e.

$$\overline{\Delta E^*} = \Delta E_{00} - \overline{\Delta E}. \tag{2.3.12}$$

In a similar way, the variance

$$\sigma^2 = \overline{\Delta E^2} - \overline{\Delta E}^2 = \langle \varphi_0 | (\Delta H)^2 | \varphi_0 \rangle - \overline{\Delta E}^2 \tag{2.3.13}$$

can be obtained. Some typical results for various molecules are shown in Table 2.3.1. The differences of the results for the same molecule are due to limitations in the basis sets used, i.e. due to incomplete convergence of the computations. Besides this, the results are remarkably similar for quite different molecules. Particularly striking are the hydrocarbons. Here, the differences between various molecules are so small that they are probably not real, as they are smaller than estimated errors of the computation [SCH 91]. The most detailed calculations have been performed for the T_2 molecule. Here also the excitation of molecular vibrations and rotation have been included [FAC 85; KOL 85]. This leads to a broadening of otherwise sharp electronic lines. The result is shown in Fig. 2.3.2.

The accuracy of the sudden approximation for tritium decay has never been tested experimentally. It is thus important to explore the size of possible corrections theoretically. The transition amplitude to a state n can systematically be expanded in

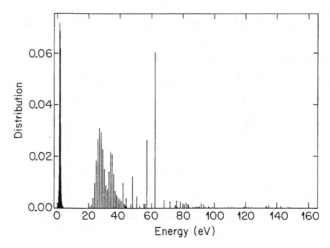

Fig. 2.3.2   Distribution of the electronic final states, including nuclear motion, for the T_2 molecule [FAC 85]. The equally spaced lines represent a continuum.

terms of the final state interaction, i.e.

$$T_n = T_n^{(0)} + T_n^{(1)} + T_n^{(2)} + \cdots \qquad (2.3.14)$$

where the small parameter is η, defined in Eq. (2.3.6). Numerically, we have $\eta \simeq 0.027$ near the endpoint. The probability is proportional to

$$|T_n|^2 = |T_n^{(0)}|^2 + 2\mathrm{Re}\,T_n^{(0)} T_n^{(1)} + |T_n^{(1)}|^2 + 2\mathrm{Re}\,T_n^{(0)} T_n^{(2)} + \cdots. \qquad (2.3.15)$$

The zero order amplitude is given by the sudden approximation, i.e. $T_n^{(0)} = g\langle \varphi_0 | \psi_n \rangle$ where g, which includes the weak coupling constant, is independent of n.

Results for the tritium atom have been reported by several authors [WIL 83; ARA 86b; DRU 87]. Initially there was some confusion, which now however is resolved. The expression in Eq. (2.3.15) must be normalized by summing over all final states n. This is because in neutrino mass experiments, only the shape of the β-spectrum is analyzed and a change in the total decay rate or half-life of tritium due to the final states interaction is insignificant. The amplitude $T_n^{(1)}$ is to first order in η purely imaginary, whereas the real part of $T_n^{(1)}$ is of order η^2. Because $T_n^{(0)}$ is real, this implies, as can be seen from Eq. (2.3.15), that the leading order correction is of order η^2 and all terms written out must be included in a correct calculation. The normalization is greatly simplified by a remarkable cancelation of terms

$$\sum_n \left\{ |T_n^{(1)}|^2 + 2\mathrm{Re}\,T_n^{(0)} T_n^{(2)} \right\} = 0, \qquad (2.3.16)$$

first found in [DRU 87] and later generalized in [LOP 88]. The cancelation for individual terms is not complete. Nevertheless, the corrections for the atom turn out to be very small. For example, the probability for the groundstate ($n = 1s$) is reduced by $\delta W_{00} = -2 \times 10^{-4}$ and the changes for excited states are smaller still.

One might suspect that the smallness of the correction for the atom is accidental and that for molecules, without spherical symmetry, there might be corrections of order η. This has turned out to be not the case. Recently, the part $2\mathrm{Re}\,T_n^{(0)}T_n^{(1)}$ was computed for the T_2 molecule and found to lead to corrections of order 0.1 $\eta^2 \simeq 10^{-4}$ for the normalized probabilities [FRO 96; SAE 97]. This is negligible for the present experiments. Provided that perturbation theory is applicable at all, it is probably safe to conclude that the sudden approximation is sufficiently accurate.

In our discussion of the β-spectrum, we have so far neglected the possibility of neutrino mixing. However, the generalization is straight forward. Let the energy spectrum for a definite neutrino mass m_ν be denoted by $N(E, E_0, m_\nu)$. Then the spectrum with mixing is given by

$$N_{\mathrm{mix}}(E) = \sum_i |U_{ei}|^2 N(E, E_0, m_i), \qquad (2.3.17)$$

where the sum includes neutrinos with masses smaller than E_0 and the $|U_{ei}|^2$ are mixing probabilities. A characteristic kink at an energy $E_0 - m_i$ is expected which may be used to search for heavy neutrinos. A summary of limits for $|U_{ei}|^2$ derived from various β-decays was recently given in [DEU 90]. For a review of the unfortunate 17 keV neutrino, we refer to [FRA 96].

Experimental considerations

The finite resolution of a spectrometer causes a broadening of the measured spectrum. We define the spectrometer resolution function (R) as the normalized distribution a spectrometer records when a source is measured which emits monoenergetic electrons with energy E'. Thus, in principle, R is a function of two variables, the measured energy E, and E'. If we assume for simplicity that R depends only on the difference $E - E'$, then the measured spectrum of a source, which emits a spectrum $N(E)$, is given by the convolution integral

$$N_{\mathrm{exp}}(E) = R \otimes N(E) = \int R(E - E')N(E')dE'. \qquad (2.3.18)$$

For a differential spectrometer, R is a peak-like function with a certain width. For an integrating spectrometer, R has a step-like shape and the width is defined by the energy range over which R rises from 0 to 1. In either case, the finite width modifies a measured β-spectrum significantly only close to the endpoint, i.e. in the energy range where most of the information about the neutrino mass is to be found. As the neutrino mass is deduced from the spectrum shape, it is important that the shape of R be known accurately, not just the width.

Table 2.3.2. *Results for the electron neutrino mass from recent experiments. The column $m_{\nu_e}^2$ (all data) gives the results when all measured data are analyzed. The upper limits for m_{ν_e} (UL) are at 95% confidence level*

Experiment	Source	$m_{\nu_e}^2$ (eV^2)	$m_{\nu_e}^2$ (all data)	UL (eV)
Los Alamos [ROB 91]	T$_2$ gas	$-147 \pm 68 \pm 41$	-230	9.3
Zürich [HOL 92b]	CHT	$-24 \pm 48 \pm 61$	same	11
Mainz [WEI 93]	Solid T$_2$	$-39 \pm 34 \pm 15$	-120	7.2
Livermore [STO 95]	T$_2$ gas	$-130 \pm 20 \pm 15$	same	—
Troitsk [BEL 95a]	T$_2$ gas	-22 ± 5	-60	4.35
Mainz [BAK 96]	Solid T$_2$	$-22 \pm 17 \pm 14$	?	5.6
Troitsk [LOB 96]	T$_2$ gas	$3.8 \pm 7.4 \pm 2.9$	-4.4	4.4

It is clearly desirable to use a strong source, not only for a high counting rate, but also for a high signal to background ratio in the mass sensitive endpoint region. Hence the issue of energy loss in the source must be addressed. A β-particle loses energy by discrete, inelastic interactions with the source material. There is thus a finite probability that the β-particle leaves the source without energy loss. This no-loss fraction must be close to 1, i.e. tritium sources are very thin. Assuming a homogeneous medium, the probability for exactly n interactions in a thin layer of thickness x is given by the Poisson distribution

$$P_n(x) = \frac{1}{n!}\left(\frac{x}{\lambda}\right)^n \exp(-x/\lambda), \qquad (2.3.19)$$

where λ is the mean free path. Let the energy loss distribution for exactly one interaction be $\omega(E)$, then the energy loss distribution for an initial energy E_i can in general be computed by

$$L(E) = \sum_{n=0}^{\infty} \overline{P_n} \omega^{\otimes n}, \qquad (2.3.20)$$

where

$$\omega^{\otimes n} = \omega \otimes \cdots \otimes \omega, \qquad \omega^{\otimes 0} = \delta(E - E_i) \qquad (2.3.21)$$

is the n-fold convolution of $\omega(E)$ with itself and where $\overline{P_n}$ is the average of Eq. (2.3.19) over all possible path lengths x. We have written Eq. (2.3.20) as an infinite sum although for the thin tritium sources only a few terms are necessary.

It is convenient to define an effective resolution by the convolution of the spectrometer resolution and the energy loss distribution, i.e.

$$R_{\text{eff}} = R \otimes L. \qquad (2.3.22)$$

This is the quantity which determines the quality of a measurement.

2.3.2.3 Recent tritium experiments

A reviewer of the recent tritium experiments does not have an easy task. The reason can be seen in Table 2.3.2. There is a strong tendency for negative values of the fitted $m_{\nu_e}^2$ which is certainly not just a statistical fluctuation. Moreover, when only part of the measured spectra above a certain energy E_{cut} are analyzed, the fitted $m_{\nu_e}^2$ depends in a non-statistical and unphysical way on E_{cut} in most experiments. Such a dependence indicates that data and fitted model are not compatible. As systematic errors become smaller when a smaller energy range of the spectrum (E_{cut} closer to the endpoint) is analyzed, some authors have chosen to use only part of the measured data for their final results. This can be seen by comparing the two columns labeled $m_{\nu_e}^2$ in Table 2.3.2.

The cause of this problem is presently not clear. The particle data group in its latest evaluation [PDG 96] therefore recommends an upper limit $m_{\nu_e} < 15\,\text{eV}$. This is certainly overly conservative, but what number precisely should be recommended is a difficult question. It is not an important question, however. It is more important to solve the problems. For that reason we have given just a round number in the introduction.

In the following, we briefly discuss the various recent tritium experiments. For details we refer to the original publications as cited in Table 2.3.2.

Zürich

The spectrometer used in the Zürich experiment is shown in Fig. 2.3.3 during installation. It was of the Tret'yakov type. The toroidal magnetic field was produced by 36 rectangular current loops. The tritium source was located in the lower part and the detector in the upper part of the spectrometer. Electrons from the source are focused onto the detector in four 180° bends separated by annular baffles. The focal distance was 2.65 m.

The spectrometer was operated in a somewhat unusual way. The magnetic field was set to analyze electrons with a low energy E_{mag} (2.2 keV for the tritium runs) and spectra were recorded by stepping a positive high voltage applied to the source. Thus the electrons from the source were decelerated by a large factor before being analyzed.

The detector was a position sensitive proportional counter with 5 cm diameter and 10 cm length of the entrance window. While scanning a spectrum, the electrons which arrived at the detector had a constant energy, determined by E_{mag}. In that way, it was guaranteed that the measured spectrum shape was not distorted by an energy dependent efficiency of the detector.

The source activity was on the surface of a cylinder with the same dimensions as the detector. The cylinder consisted of ten isolated discs with appropriate electrostatic potentials to compensate the axial source extension. The spectrometer resolution function was determined by Monte Carlo simulation and

Fig. 2.3.3   The spectrometer of the Zürich group while being mounted.

by measurements with a conversion line source, the width being 17 eV (FWHM) with both methods.

A model of the tritium source is shown in Fig. 2.3.4. It was produced by chemically growing a monolayer of hydrocarbon chains on a suitable surface. There were six tritium atoms per molecule at the indicated positions.

The method is known as the spontaneous formation of a self-assembling monolayer. First, tritiated molecules, consisting of a simple hydrocarbon chain with 18 C atoms and a reactive $SiCl_3$ end group, are prepared (OTS for octadecyltrichlorosilane). The surface of the substrate consists of SiO_2, treated such that it is densely covered by OH groups. A highly diluted solution of OTS is

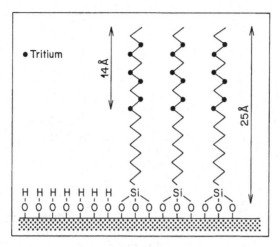

Fig. 2.3.4   The monolayer tritium source used by the Zürich group.

prepared into which the substrate is simply dipped for a few minutes. The $SiCl_3$ groups of OTS react with the OH groups on the surface, forming strong Si–O bonds and HCl is released into the solution. The reactions come to an end when the surface is densely covered by a monolayer.

The tritium sources produced in this way were very thin and had a well determined structure. Only 2% of the detected electrons had lost energy by inelastic interactions in the source layer. The distribution of electronic final states were taken from [SCH 91].

Data were recorded from 920 eV below to 180 eV above the endpoint. The measured spectrum is shown in Fig. 2.3.5. As a test of the consistency of data and fitted model, some data below a certain energy E_{cut} were excluded from the fit. This is shown in Fig. 2.3.6. Within a narrow band of statistical fluctuations, the physical parameters are independent of E_{cut}, as it should be. The final result given in Table 2.3.2 was obtained from all data.

Los Alamos and Livermore

The first experiment using a gaseous tritium source was performed in Los Alamos. The experiment in Livermore was similar in principle and we can discuss both under the same heading. A difficulty with a gaseous source is that no windows can be used and that contamination of the spectrometer must be avoided to a *very* high level. This was realized in both experiments, quite an achievement.

A schematic diagram of the Los Alamos set-up is shown in Fig. 2.3.7. The source consisted of a long tube. Tritium gas entered the tube in the middle and streamed to the ends, where it was pumped away by large mercury diffusion pumps. The tritium gas was recycled through a palladium foil. Decay electrons were transported in a strong longitudinal magnetic field from the source tube through a collimator into a

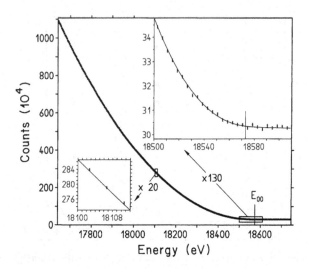

Fig. 2.3.5   Measured tritium spectrum and best fit (Zürich).

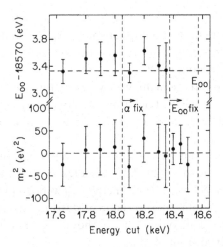

Fig. 2.3.6   Fitted neutrino mass squared and endpoint energy when data points below the indicated energy are excluded from the analysis (Zürich).

Tret'yakov type spectrometer with a special entrance section. The diameter of the collimator was chosen such that decay electrons originating from the tube walls could not enter the spectrometer. This was important since it was necessary to assume that the large amount of tritium adsorbed by the walls was in an unknown chemical form.

In both experiments, a position sensitive semiconductor detector was used. The spectrometers were operated with constant current to analyze electrons of a fixed energy several keV above the tritium endpoint energy. In that way, contamination induced background could be reduced efficiently. Spectra were recorded by

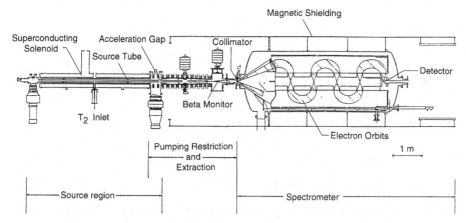

Fig. 2.3.7 Overview of the Los Alamos tritium experiment. The overall length of the apparatus is 16 m. By courtesy of T. J. Bowles.

changing a negative voltage applied to the source tube, i.e. the decay electrons were accelerated. The energy resolution was about 22 eV for the Los Alamos and 18 eV FWHM for the Livermore experiment. The fraction of electrons making an inelastic interaction before leaving the source tube was also similar, being 8.5% (Los Alamos) and 12% (Livermore).

Because of the acceleration, the effective solid angle of the source accepted by the spectrometer became a function of energy. Both groups reported difficulties with this effect. The problem was caused by the complicated transition region between the strong longitudinal guiding field of the source (several kG) and the relatively weak toroidal field of the spectrometer. Apparently the effect could not be computed reliably. The Los Alamos group took the effect into account by fitting a phenomenological shape parameter to the spectrum. The Livermore group made extensive measurements of the tritium spectrum over large energy ranges to find a setting of the guiding field where the effective solid angle depended linearly on the acceleration.

The result for $m_{\nu_e}^2$ from Los Alamos was below zero by 1.9σ. The group believed that this could be explained by a statistical fluctuation (3.2% probability). The Livermore group found an anomalous bump in their spectrum close to the endpoint, which seemingly could not be accounted for by experimental effects. As a consequence, they did not claim an upper limit for m_{ν_e} but only indicated that it would be 7 eV if $m_{\nu_e}^2$ were assumed to be zero.

Mainz and Troitsk

In the history of tritium experiments quite a few attempts have been made with integrating spectrometers. To obtain a good resolution (narrow width of the transmission step), the electrons must penetrate the electrostatic potential barrier on trajectories (nearly) perpendicular to the barrier. This can be achieved in various

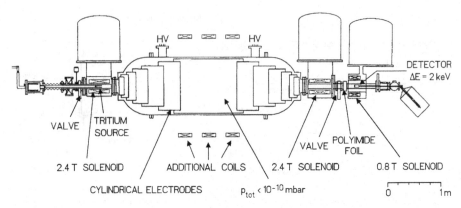

Fig. 2.3.8   Schematic view of the Mainz spectrometer. By courtesy of Ch. Weinheimer.

ways, for instance by forming a parallel beam with collimators or by a spherical arrangement of grids. So far only limited success can be claimed for these methods.

The groups in Mainz and Troitsk (Moscow) use an ingenious new method which they call adiabatic magnetic collimation. The set-up from Mainz is shown in Fig. 2.3.8. The source is located in a strong magnetic field B_i. The electrons emitted in the forward direction spiral along the field lines into a large vacuum tank where the magnetic field gradually drops to a small value B_f, typically $B_f/B_i = 1/3000$. The adiabaticity theorem [JAC 75, p. 509] shows that most of the transverse energy $E_{i\perp}$ at the source is converted into longitudinal energy

$$E_{f\parallel} = E_i - \frac{B_f}{B_i} E_{i\perp}. \qquad (2.3.23)$$

No grids or collimators are necessary. The isotropically diverging electrons from the small source are converted into a very wide but nearly parallel beam by the field only. At the centre of the tank, an electrostatic potential barrier is generated by a set of cylindrical high voltage electrodes. Electrons with sufficient energy pass the barrier and are reaccelerated and focused onto a detector, all other electrons returning to the source. The energy resolution, defined as the energy range over which the transmission curve drops from one to zero, was 6 eV at Mainz and 3.7 eV at Troitsk.

The source at Mainz was frozen T_2 on a small backing cooled with liquid helium. Initially, the backing was made of aluminum, later of graphite. A gaseous source was used at Troitsk with a set-up similar to Los Alamos. The guiding field of the source and the spectrometer field fit together nicely. There is no complicated transition region as was necessary with the Tret'yakov type spectrometers and gaseous sources. The fields can be chosen such that the motion of the electrons is adiabatic throughout the whole apparatus.

From the information given in the publications, the fraction of electrons interacting in the source can be estimated. We find 13% for Mainz and 16% to 26% for Troitsk (rather thick).

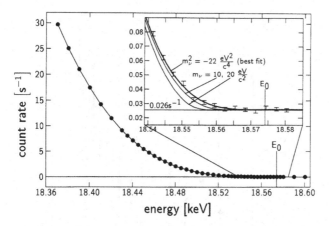

Fig. 2.3.9 Data from the Mainz group showing the last 200 eV of the measured spectrum and the best fit. The inset shows a small part around the endpoint with fits assuming some fixed values of the neutrino mass parameter [BAK 96].

Both groups have collected data with very high statistical precision. Part of a measured spectrum from Mainz is shown in Fig. 2.3.9. One of the problems mentioned before can be seen Fig. 2.3.10, which shows the fitted parameter $m^2_{\nu_e}$ when the data below the indicated energy are excluded from the analysis. The cause of this problem is presently not known and only the data above the energy E_0-140 eV were used for the final results. The Mainz group recently indicated that they cannot completely rule out the possibility that the tritium layer is not flat and homogeneous as assumed but rough. This would cause a larger energy loss than assumed in the analysis and could, at least partly, explain the effect.

A similar effect was seen in the first measurement of Troitsk. However, the group recently reported that they have found the cause. Decay electrons, produced with a large angle with respect to the source axis, are trapped in the guiding field, which acts as a magnetic bottle. These electrons can escape into the spectrometer only by scattering, i.e. with energy loss. Taking this into account, the dependence of $m^2_{\nu_e}$ on the data range disappeared within errors. However, the Troitsk data show a further anomaly. The group claims that there is a small step-like feature just below the endpoint. It would correspond to a peak in a differential spectrum. This was taken into account in the analysis by two phenomenological parameters, an amplitude of the step and an energy. It seems that the step energy varies between runs, being 7.6 eV in the first and 12.3 eV below the endpoint in the second run. An anomaly of this kind has not been observed in any other experiment.

Tritium is dangerous stuff. Not only is it radioactive and contaminates just about everything, it is much worse. A colleague recently said that at the time he was contemplating the start of a new tritium experiment, he was strongly warned that 'tritium experiments have already ruined many scientific reputations'. Be that as it may, it should be recognized that these are not simple experiments.

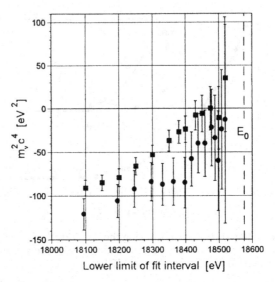

Fig. 2.3.10   Fitted $m^2_{\nu_e}$ as a function of the data range for two runs (dots and squares) from the Mainz experiment [OTT 95].

2.3.2.4 Other electron neutrino mass experiments

The mass of the electron neutrino can also be inferred from studies of electron capture decays. Various methods have been proposed [RUJ 81; BEN 81]. A favourable isotope is ^{163}Ho, decaying to ^{163}Dy with a Q-value of 2.7 keV. A recent measurement of the partial capture rates from various atomic subshells gave an upper limit of $m_{\nu_e} < 460$ eV [YAS 94]. This is far from what can at present be achieved with tritium. It should be noted, however, that with ^{163}Ho the mass of the electron *neutrino* is measured in contrast to the *antineutrino* mass with tritium. These masses are of course required to be identical by *CPT* and the experiments can also be seen as tests of this theorem.

There is an isotope which makes a β-decay with a smaller endpoint energy than tritium. This is ^{187}Re which occurs naturally with 62% abundance. The endpoint is $E_0 \simeq 2.6$ keV. The decay is forbidden however, and the half-life is very large, being $T_{1/2} \simeq 5 \times 10^{10}$ y. This implies that in an experiment with any significant counting rate, source and detector must be identical. Otherwise the problem of energy loss would be insurmountable. It has recently been shown that it is indeed possible to make a detector from rhenium [SWI 96]. The device consisted of a small piece of rhenium (0.003 mm^3) thermally coupled with a germanium thermistor. The device was cooled to 0.1 K and operated as a calorimeter. A decay caused a small rise of the temperature which was measured with the thermistor. An impressive energy resolution $\sigma = 13$ eV (corresponding to 31 eV FWHM) was achieved. The group believes that the method can be developed to be an alternative to tritium.

2.3.3 The muon neutrino mass

The mass of the muon neutrino can been determined from a study of the pion decay.

$$\pi^+ \longrightarrow \mu^+ \nu_\mu.$$

All recent experiments have been performed with the pion being at rest [ASS 94, 96]. Taking for this case the square of the relation for four-momentum conservation, $p_\nu = p_\pi - p_\mu$, immediately gives a formula for the mass of the muon neutrino.

$$m_{\nu_\mu}^2 = m_\pi^2 + m_\mu^2 - 2m_\pi \sqrt{m_\mu^2 + p_\mu^2}. \tag{2.3.24}$$

Assuming that the masses of the pion (m_π) and muon (m_μ) are known, the three-momentum (p_μ) of the muon must be determined. As the formula above involves the difference of two large terms, a measurement with high accuracy is required.

Such measurements have been performed at the proton accelerator of the Paul Scherrer Institute (PSI, formally SIN). The pions are produced by the proton beam (590 MeV) in a graphite target. Some of the pions are stopped in the target close to its surface, where they decay. The muons are transported through a quadrupole beam line to a spectrometer for momentum measurement. As the muons have to pass through some material before they leave the target, they lose some energy and are not monoenergetic. The energy or momentum distribution has a step-like shape. Only muons originating from decays right at the surface have the momentum required for Eq. (2.3.24). As there are large amounts of pions available, this is not a disadvantage.

The spectrometer is shown in Fig. 2.3.11. It is an iron magnet with a homogeneous field. The muons enter the spectrometer through a hole in the yoke. Those passing a collimator slit (0.12 mm width) are focused after a 180° bend onto a silicon microstrip detector with 50 μm pitch. The momentum of a muon is determined by the magnetic field and the distance between the entrance slit and the detector strip being hit. The field was measured with NMR probes. The distance (approximately 72 cm) was measured with a laser interferometer with an accuracy better than 1 μm.

The pions cannot be truly at rest, of course. Assuming the pions to be non-relativistic, it is easy to derive the shift of the muon momentum

$$\Delta p_\mu = \frac{E_\mu}{m_\pi} p_{\pi\|}. \tag{2.3.25}$$

Here, $E_\mu \simeq 109.9$ MeV is the total muon energy and $p_{\pi\|}$ is the component of the pion momentum parallel to the muon momentum. If the distribution of the pion momentum vector is isotropic, the variance of the muon momentum can be written

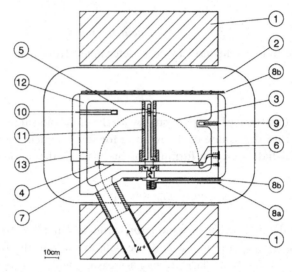

Fig. 2.3.11   Magnetic spectrometer for measuring the muon momentum. Magnet yoke (1) and coils (2), central muon trajectory (3), collimators (4–6), support (7), cooling water pipes (8), NMR probes (9,10), lead shielding (11), vacuum chamber (12), port for vacuum pump (13). From [ASS 96].

in the form

$$\sigma_p^2 = \frac{2}{3}\bar{T}_\pi \frac{E_\mu^2}{m_\pi}, \qquad (2.3.26)$$

where $\bar{T}_\pi$ is the mean kinetic energy of the pions.

The measured distribution of the muon momentum is mainly determined by three effects: spectrometer resolution, energy loss in the target, and pion motion. The data showed that the third effect is quite significant. An assumed Maxwell–Boltzmann distribution for the pions would give a pion kinetic energy of 0.13 eV for a typical target temperature of 1000 K (resulting from heating by the proton beam). Using Eq. (2.3.26), the data implied a much larger value, being $\bar{T}_\pi = 0.425 \pm 0.016$ eV. This can be understood as follows. The pion can be considered to be a light isotope of hydrogen. As such it can form a strong chemical bond with carbon. Assuming that this bond can be represented by an isotropic harmonic oscillator potential, the oscillator energy is given by $\hbar\omega = 4\bar{T}_\pi/3$. Scaling with the appropriate masses, a comparison with what is known about hydrogen in graphite gave agreement within errors. Thus $\bar{T}_\pi$ is due to zero-point motion, a remarkable result. As with tritium, measurements of the muon neutrino mass have reached a precision where chemical effects are non-negligible.

The result for the muon momentum reported is

$$p_\mu = 29\,792\,000 \pm 110\,\text{eV}, \qquad (2.3.27)$$

a 3.7 ppm measurement, which was achieved with a spectrometer resolution of 170 ppm. Using the muon and the pion masses, the mass of the muon neutrino can be computed from Eq. (2.3.24). Unfortunately, there is a long-standing, twofold ambiguity of the pion mass [JEC 94]. One value of the pion mass gives a $m_{\nu_\mu}^2$ which is negative by 6σ and, so it is concluded, can be ruled out. With the other pion mass, $m_{\nu_\mu}^2$ comes out to be compatible with zero.

$$m_{\nu_\mu}^2 = -0.016 \pm 0.023\,\text{MeV}^2. \qquad (2.3.28)$$

The upper limit computed from this result is

$$m_{\nu_\mu} < 170\,\text{keV} \quad (90\%\ \text{CL}). \qquad (2.3.29)$$

The limit could be significantly improved with a more accurate value of the pion mass. Such an experiment is presently in progress at PSI.

2.3.4 The tau neutrino mass

Tau leptons are produced in pairs at e^+e^- storage rings. Decay modes which have been used to constrain the mass of the tau neutrino are of the type

$$\tau^- \longrightarrow nh^{(+,-,0)}\nu_\tau,$$

where nh stands for n hadrons, being either pions or kaons. It should be noted that the particle (the tau neutrino, ν_τ), the mass of which is to be measured, has never been observed directly in the sense that a ν_τ was absorbed in some material, producing a charged tau lepton. Its existence is however not in doubt.

The measurable quantities of the decay are the beam energy (equal to the energy of one tau), the mass of the tau (m_τ), and energies E_i and momenta $\vec{p}_i$ of the n hadrons. Using four-momentum conservation,

$$p_\tau = \sum_{i=1}^{n} p_i + p_\nu, \qquad (2.3.30)$$

the invariant mass of the hadrons can be related to the mass of the tau neutrino.

$$m_{nh}^2 = \left(\sum_i p_i\right)^2 = \left(\sum_i E_i\right)^2 - \left(\sum_i \vec{p}_i\right)^2$$
$$= m_\tau^2 + m_{\nu_\tau}^2 - 2m_\tau E_\nu^0.$$

Here, E_ν^0 is the energy of the tau neutrino in the rest frame of the decaying tau. With $E_\nu^0 \geq m_{\nu_\tau}$, a kinematic limit follows.

$$m_{nh} \leq m_\tau - m_{\nu_\tau}. \qquad (2.3.31)$$

Table 2.3.3. *Recent results for the tau neutrino mass from 5(6) pion mode*

Events observed	Upper limit (MeV), 95%CL	Collaboration
20	31	ARGUS [ALB 92]
113	32.6	CLEO [CIN 93]
25	24	ALEPH [BUS 95]

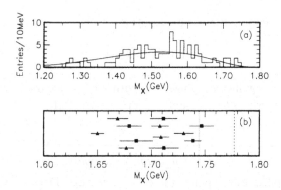

Fig. 2.3.12 Measured data from CLEO [CIN 93]. The invariant mass is denoted by $M_x = m_{n\pi}$. (a) Histogram of all events. (b) Single events close to the kinematic limit.

Recent investigations have concentrated on the very rare decays

$$\tau^- \longrightarrow 5\pi^{(+,-)} \left(\pi^0\right) \nu_\tau,$$

$$\tau^- \longrightarrow 3\pi^{(+,-)} 2\pi^0 \nu_\tau$$

with five or possibly six pions. The idea is to convert as much decay energy into restmass as possible. In that way, the sensitivity (per event) to the restmass of the tau neutrino is maximized. The largest event sample for the five pion modes has been obtained by CLEO [CIN 93]. The data are plotted in Fig. 2.3.12.

The distribution of the invariant mass $m_{n\pi}$ is not known exactly. However, it can be argued that the distribution should be sensitive to m_{ν_τ} only close to the kinematic limit, and there the shape of the distribution is dominated by the steep drop of the phase-space factor.

There is a strong dependence of the results on single events. This can be seen in Table 2.3.3 by comparing the limits of ARGUS and CLEO, which were obtained by essentially the same analysis. There is nothing wrong with this, in principle. It reflects the fact that there can be large statistical fluctuations with small sample sizes. The distribution of $m_{n\pi}$ is broad and most events provide no information about m_{ν_τ}. Only the very few events (possibly only one) close to the kinematic limit dominate the final results.

The ALEPH collaboration performed a different analysis. In addition to $m_{n\pi}$ they used the sum of the pion energies ΣE_i as a second variable in the analysis.

The use of this additional information improved the upper limit quite significantly. They report that their upper limit would rise to 40 MeV if only $m_{n\pi}$ were used in the analysis.

Occasionally some uneasiness has been expressed about the strong dependence on single events. It is thus interesting to investigate also decay modes for which much larger event samples can be obtained. Recently the OPAL collaboration [ALE 96] analyzed a sample of 2514 events of the type $\tau \rightarrow 3h\nu$ and obtained an upper limit $m_{\nu_\tau} < 35.3$ MeV with 95% confidence. Although this limit is slightly larger than the limits from the 5(6) pion modes, it depends much less on single events.

2.4 Neutrino mixing*

2.4.1 Introduction

It is well known that quark fields enter into the charged weak current in a mixed form (the Cabibbo–Kobayashi–Maskawa mixing). Does mixing take place in the lepton sector?

The hypothesis of neutrino mixing was put forward by B. Pontecorvo [PON 57, 58] in the late 1950s, immediately after the $V-A$ theory was formulated. Only one type of neutrino was known at that time. Afterward, the hypothesis was extended to the case of two and more types of neutrinos [PON 67,71; MAK 62; BIL 78]. After the formulation of unified gauge theories interest in lepton mixing began to increase. In fact, mixing of fields is a natural consequence of these theories and it is related to the mechanism of spontaneous symmetry breaking.

However, between quark and neutrino mixing there may be an essential difference since the electrical charges of quarks differ from zero, whereas those of neutrinos are zero. Consequently, quarks are Dirac particles, whereas neutrinos may be either Dirac particles (when there is a conserved lepton charge) or truly neutral Majorana particles (when there are no conserved lepton charges). In Sections 2.4.2–2.4.4, we present the general phenomenological theory of neutrino mixing; in Section 2.4.5, the case of two neutrino fields is examined in detail; Section 2.4.6 deals with neutrino oscillations; and in Section 2.4.7 we briefly report data of the latest experiments in the search for neutrino oscillations.

2.4.2 The Dirac mass term

The neutrino Dirac mass term is of the form[1]

$$\mathscr{L}^{\mathscr{D}} = -\sum_{l',l} \bar{\nu}_{l'R} M_{l'l} \nu_{lL} + \text{h.c.} = -\bar{\nu}'_R M \nu'_L + \text{h.c.}, \qquad (2.4.1)$$

---

* S. Bilenky, JNR, Dubna, U.S.S.R.

[1] In gauge models. the mass term $\mathscr{L}^{\mathscr{D}}$, as well as other mass terms to be considered below, arise from the Lagrangian of interaction of leptons and Higgs bosons when the symmetry is spontaneously broken.

where M is a complex-valued $n \times n$ matrix, and

$$\nu'_L = \begin{pmatrix} \nu_{eL} \\ \nu_{\mu L} \\ \vdots \end{pmatrix}, \quad \nu'_R = \begin{pmatrix} \nu_{eR} \\ \nu_{\mu R} \\ \vdots \end{pmatrix}.$$

The matrix M is in general nondiagonal. To write the mass term $\mathscr{L}^{\mathscr{D}}$ in a standard (i.e., diagonal in the fields) form, it is necessary to diagonalize the matrix M. An arbitrary complex-valued M can be made diagonal by a biunitary transformation. Thus, we have

$$M = V m U^+ \tag{2.4.2}$$

where V and U are unitary $n \times n$ matrices, and m is a diagonal matrix with positive elements. Inserting (2.4.2) into (2.4.1) we obtain

$$\mathscr{L}^{\mathscr{D}} = -\bar{\nu}_R m \nu_L - \bar{\nu}_L m \nu_R = -\bar{\nu} m \nu = -\sum_{k=1}^{n} m_k \bar{\nu}_k \nu_k. \tag{2.4.3}$$

Here

$$\nu_L = U^+ \nu'_L, \ \nu_R = V^+ \nu'_R, \tag{2.4.4}$$

$$\nu = \nu_L + \nu_R = \begin{pmatrix} \nu_1 \\ \nu_2 \\ \vdots \end{pmatrix}. \tag{2.4.5}$$

So, ν_k is the neutrino field with mass m_k. As is seen from (2.4.4), the neutrino current fields and left-handed components of the neutrino fields with definite masses are connected by a unitary transformation

$$\nu'_L = U \nu_L \tag{2.4.6}$$

or, in terms of components,

$$\nu_{lL} = \sum_{k=1}^{n} U_{lk} \nu_{kL}. \tag{2.4.7}$$

Thus, if the neutrino mass term is given by (2.4.1), the left-handed current fields ν_{lL} are connected with the left-handed components of the neutrino fields with definite masses ν_{kL}, by (2.4.7).

The Lagrangian $\mathscr{L}^{\mathscr{D}}$ does not separately conserve the electron L_e, muon L_μ, and other lepton charges. It is obvious that in the theory under consideration, the total lepton charge

$$L = L_e + L_\mu + \cdots$$

is conserved. Indeed, it follows that the total Lagrangian of the system is invariant under the global gauge transformation

$$\nu'_{lL}(x) = e^{i\lambda}\nu_{lL}(x), \nu'_{lR}(x) = e^{i\lambda}\nu_{lR}(x), l'(x) = e^{i\lambda}l(x) \tag{2.4.8}$$

where λ is an arbitrary constant. Invariance with respect to (2.4.8) means that the lepton charge L, the same for all types of the neutrinos and leptons, is conserved. From (2.4.7) and (2.4.8) it is also seen that neutrinos with definite masses possess a conserved lepton charge and are consequently Dirac particles.

The conservation of the lepton number L forbids the processes like the neutrinoless double beta decay, the decay $K^+ \to \pi^- + e^+ + \mu^+$, the process $\mu^- + A \to e^+ + \cdots$, and others. Oscillations of the neutrinos $\nu_e \rightleftarrows \nu_\mu$, $\nu_\mu \rightleftarrows \nu_\tau, \ldots$ are possible in the theory with Dirac mass term.

2.4.3 The Majorana mass term

The Dirac mass term considered in the previous subsection is constructed by using both the left-handed ν_{lL} and the right-handed ν_{lR} fields, whereas the right-handed fields do not enter into the conventional Lagrangian of weak interaction. It is clear that the most economical scheme of neutrino mixing would be the one whose mass term would contain only the current fields ν_{lL}. A scheme of that sort was first proposed in [GRI 69] for the case of two types of neutrinos.

Let us consider the general case of n types of neutrinos.

The mass term of the Lagrangian is a Lorentz scalar made by a product of the left- and right-handed components of fields. Obviously $(\nu_{lL})^c = C\bar{\nu}_{lL}^T$ is the right-handed field (where C is the matrix of charge conjugation obeying the conditions $C^T\gamma_\alpha C^{-1} = -\gamma_\alpha$, $C^T = -C$, $C^+C = 1$). Using ν_{lL} and $(\nu_{lL})^c$ we may construct the following neutrino mass term of the Lagrangian

$$\mathscr{L}^M = -\frac{1}{2}\sum_{l',l}(\bar{\nu}_{l'L})^c M_{l'l}\nu_{lL} + \text{h.c.} = -\frac{1}{2}(\bar{\nu}'_L)^c M\nu'_L + \text{h.c.} \tag{2.4.9}$$

where M is the complex-valued $n \times n$ matrix. It can be verified that the matrix M is symmetric.[2]

Now let us bring the Lagrangian $\mathscr{L}^M$ into the diagonal form. To this end, we diagonalize the complex-valued symmetric matric M using the transformation

$$M = (U^+)^T m U^+ \tag{2.4.10}$$

---

[2] In fact,

$$(\bar{\nu}'_L)^c M\nu'_L = -\nu'^T_L C^{-1} M\nu'_L = \nu'^T_L (C^{-1})^T M^T \nu'_L = (\bar{\nu}'_L)^c M^T \nu'_L$$

from which it follows that $M^T = M$.

(with the unitary matrix U and $m_{ik} = m_k \delta_{ik}$, $m_k > 0$). Substituting (2.4.10) into (2.4.9) we get

$$\mathscr{L}^M = \frac{1}{2} \nu_L'^T C^{-1} (U^+)^I m U^+ \nu_L' + \text{h.c.} = -\frac{1}{2} \bar{\chi} m \chi$$

$$= -\frac{1}{2} \sum_{k=1}^{n} m_k \bar{\chi}_k \chi_k \tag{2.4.11}$$

where

$$\chi = U^+ \nu_L + (U^+ \nu_L')^c = \begin{pmatrix} \chi_1 \\ \chi_2 \\ \vdots \end{pmatrix}. \tag{2.4.12}$$

As a result,

$$\chi_k^c = \chi_k \tag{2.4.13}$$

that is, χ_k is the Majorana field. This way, we arrive at the conclusion that (2.4.9) is a Majorana mass term, and χ_k is the field of a particle with a Majorana mass m_k.

The meaning of this result is the following: From (2.4.9) it is clear that there are no gauge transformations under which the Lagrangian $\mathscr{L}^M$ would be invariant. This means that the theory does not contain any conserved lepton number that would allow us to distinguish between neutrino and antineutrino. Particles with definite masses should consequently be Majorana particles.

Let us now derive the relations between the current fields and the neutrino fields with Majorana masses. To this end, we multiply (2.4.12) by $(1 + \gamma_5)/2$ from the left, which gives

$$\chi_L = U^+ \nu_L'. \tag{2.4.14}$$

From this relation and unitarity of the matrix U we obtain

$$\nu_L' = U \chi_L \tag{2.4.15}$$

or

$$\nu_{lL} = \sum_{k=1}^{n} U_{lk} \chi_{kL}. \tag{2.4.16}$$

Thus, if the neutrino mass term is given by (2.4.9), then n left-handed current fields ν_{lL} ($l = e, \mu, \tau, \ldots$) are superpositions of the left-handed components of n neutrino fields with Majorana masses m_k ($k = 1, 2, 3, \ldots, n$).

2.4.4 The Dirac–Majorana mass term

The most general mass term that can be constructed both with the current left-handed fields ν_{lL} and with the right-handed fields ν_{lR} is of the form [BIL 76]

$$\mathcal{L}^{\mathcal{D}-M} = -\tfrac{1}{2}g[(\bar{\nu}'_L)^c M_L \nu'_L + \bar{\nu}'_R M_R (\nu'_R)^c + \bar{\nu}'_R M_\mathcal{D} \nu'_L$$
$$+ (\bar{\nu}'_L)^c M_\mathcal{D}^T (\nu'_R)^c g] + \text{h.c.}$$
$$= -\tfrac{1}{2}(\bar{n}_L)^c M n_L + \text{h.c.} \tag{2.4.17}$$

Here M_L, M_R, and $M_\mathcal{D}$ are complex-valued $n \times n$ matrices,

$$n_L = \begin{pmatrix} \nu'_L \\ (\nu'_R)^c \end{pmatrix} \quad \text{and} \quad M = \begin{pmatrix} M_L & M_\mathcal{D} \\ M_\mathcal{D}^T & M_R \end{pmatrix}$$

is a symmetric $2n \times 2n$ matrix. The first term in (2.4.17) is the left-handed Majorana mass term considered in subsection 2.4.3; the second term is constructed only with ν'_R and is the right-handed Majorana mass term; finally, the sum of the third and fourth terms is the Dirac mass term.[3]

The mass term $\mathcal{L}^{\mathcal{D}-M}$ can be brought into diagonal form as was done for the Majorana mass term in 2.4.3. Considering that

$$M = (U^+)^T m U^+ \tag{2.4.18}$$

where U is a unitary $2n \times 2n$ matrix, and $m_{ik} = m_k \delta_{ik}$, $m_K > 0$ we get

$$\mathcal{L}^{\mathcal{D}-M} = -\frac{1}{2} \bar{\chi} m \chi = -\frac{1}{2} \sum_{k=1}^{2n} m_k \bar{\chi}_k \chi_k, \tag{2.4.19}$$

where

$$\chi = U^+ n_L + (U^+ n_L)^c = \begin{pmatrix} \chi_1 \\ \chi_2 \\ \vdots \end{pmatrix} \tag{2.4.20}$$

is the Majorana field.

So, the particles with definite masses are $2n$ Majorana neutrinos. From (2.4.20) we obtain

$$n_L = U \chi_L \tag{2.4.21}$$

which for the n upper current components of the column gives

$$\nu_{lL} = \sum_{k=1}^{2n} U_{lk} \chi_{kL}. \tag{2.4.22}$$

Thus, in the case of the Dirac–Majorana mass term, n current left-handed fields ν_{lL} are superpositions of $2n$ left-handed components of the neutrino fields with Majorana masses. The essential feature of the Dirac–Majorana scheme is that

---

[3] We have

$$(\bar{\nu}'_L)^c M_\mathcal{D}^T (\nu'_R)^c = -\nu'^T_L C^{-1} M_\mathcal{D}^T C \bar{\nu}'^T_R = \bar{\nu}'_R M_\mathcal{D} \nu'_L$$

The mass term $\mathcal{L}^{\mathcal{D}-M}$ is called the Dirac–Majorana mass term.

the n fields $(\nu_{lR})^c$ are superpositions of the same $2n$ components X_{kL}. Indeed, for n lower components of the column n_L from (2.4.21) we have

$$(\nu_{lR})^c = \sum_{k=1}^{2n} U_{\bar{l}k} X_{kL} \qquad (2.4.23)$$

(the index l runs over the n lower rows of the matrix U). The quanta of the fields ν_{lR} are the right-handed neutrinos $\nu_{eR}, \nu_{\mu R}, \dots$ and left-handed antineutrinos $\bar{\nu}_{eL}, \bar{\nu}_{\mu L}$, ... which are sterile particles in the sense that they do not participate in the conventional weak interaction. Owing to (2.4.22) and (2.4.23), in the considered scheme of mixing, along with the transitions $\nu_l \to \nu_{l'}$ the transitions of current neutrinos into sterile states $\nu_l \to \bar{\nu}_{l'L}$ are also possible.

2.4.5 Neutrino oscillations in vacuum

If there is neutrino mixing

$$\nu_{lL} = \sum_i U_{li} \nu_{iL} \qquad (2.4.24)$$

(ν_i is a neutrino field (Dirac or Majorana) with mass m_i) for the state vector of flavor neutrino ν_l ($l = e, \mu, \tau$) with momentum p, we have [BIL 78, 87]

$$|\nu_l\rangle = \sum_i U_{li}^* |i\rangle. \qquad (2.4.25)$$

Here $|i\rangle$ is the state vector of a neutrino with momentum p, energy $E_i = \sqrt{m_i^2 + p^2} \simeq p + (m_i^2/2p)(p \gg m_i)$ and negative helicity.

Thus if neutrino mixing takes place, state vectors of flavor neutrinos are *coherent superpositions* of the state vectors of neutrinos with different masses (like vectors $|K^0\rangle$ and $|\bar{K}^0\rangle$ are superpositions of $|K_S\rangle$ and $|K_L\rangle$). Let us notice that this is valid in the case of small neutrino mass differences.

Flavor neutrinos are produced in weak processes. If at $t = 0$ the vector of state of neutrinos is $|\nu_l\rangle$, at a time t we have

$$|\nu_l\rangle_t = \sum |i\rangle e^{-iE_i t} U_{li}^*. \qquad (2.4.26)$$

The beam of neutrinos is analyzed by the observation of weak processes. Expanding the vector (2.4.26) over a complete set of states of flavor neutrinos we have

$$|\nu_l\rangle = \sum_{l'} |\nu_{l'}\rangle A_{\nu_{l'};\nu_l}(t), \qquad (2.4.27)$$

where

$$A_{\nu_{l'};\nu_l}(t) = \sum_i U_{l'i} e^{-iE_i t} U_{li}^*. \qquad (2.4.28)$$

Thus at time t the vector of the beam of neutrinos that was produced (at $t = 0$) in a state with definite flavor is a *coherent superposition* of the vectors of states of all possible flavor neutrinos. The amplitude $A_{\nu_{l'};\nu_l}(t)$ is the amplitude of the transition $\nu_l \to \nu_{l'}$ for the time t. The amplitude of the transition of antineutrinos $\bar{\nu}_l \to \bar{\nu}_{l'}$ is given by

$$A_{\bar{\nu}_{l'};\bar{\nu}_l}(t) = \sum U_{l'i}^* e^{-iE_i t} U_{li}. \qquad (2.4.29)$$

Let us assume that $m_1 \leq m_2 \leq m_3$. Using the unitarity of the mixing matrix from (2.4.28) we find for the transition probability

$$P_{\nu_{\alpha'};\nu_\alpha} = \left| \sum_i U_{\alpha'i}(e^{-i\Delta m_{i1}^2 (L/2p)} - 1)U_{\alpha i}^* - \delta_{\alpha'\alpha} \right|^2, \qquad (2.4.30)$$

where $\Delta m_{i1}^2 = m_i^2 - m_1^2$ and $L \cong t$ is the distance between neutrino source and neutrino detector. Let us notice that this expression is valid not only for the case of transitions between flavor neutrinos $\nu_l \to \nu_{l'}(l, l' = e, \mu, \tau)$ but also for the case of transitions between flavor and sterile neutrinos. In order to obtain the probability of a transition between antineutrinos $\bar{\nu}_\alpha \to \bar{\nu}_{\alpha'}$ it is necessary to make in the expression (2.4.30) the following change: $U_{\alpha'i} U_{\alpha i}^*$.

The transition probability (2.4.30) depends periodically on the quantity L/p and describes *neutrino oscillations*. If $\Delta m_{i1}^2 (L/2p) \ll 1$ at all i, in this case there are no neutrino oscillations: $P_{\nu_{\alpha'};\nu_\alpha} = \delta_{\alpha'\alpha}$. In order to observe neutrino oscillations it is necessary that at least for one neutrino mass squared difference the following inequality is satisfied:

$$\Delta m^2 \frac{L}{p} \gtrsim 1. \qquad (2.4.31)$$

In (2.4.31) Δm^2 is in eV^2, L is in m and p is in MeV. The inequality (2.4.31) allows us to estimate the sensitivity of neutrino oscillation experiments to the neutrino mass squared difference Δm^2. Thus, for example, for long-baseline reactor experiments ($L \simeq 1000$ m, $p \simeq 1$ MeV), for accelerator experiments with neutrinos from decays of muons at rest ($L \simeq 30$ m, $p \simeq 30$ MeV), for atmospheric neutrino experiments ($L \simeq 1000$ km, $p \simeq 1$ GeV) and for solar neutrino experiments ($L \simeq 10^{11}$ m, $p \simeq 1$ MeV) we have: $\Delta m^2 \geq 10^{-3}$ eV^2, 1 eV^2, 10^{-3} eV^2, 10^{-11} eV^2, respectively.

The data of neutrino oscillation experiments are usually analyzed under the simplest assumption of the mixing of two massive neutrinos. In this case from (2.4.30) we have

$$P_{\nu_{\alpha'};\nu_\alpha} = |U_{\alpha'2} U_{\alpha i}^* (e^{-i\Delta m^2 (L/2p)} - 1) + \delta_{\alpha'\alpha}|^2, \qquad (2.4.32)$$

where $\Delta m^2 = m_2^2 - m_1^2$. The mixing matrix in the case of two massive Dirac neutrinos is a 2×2 orthogonal matrix

$$U = \begin{pmatrix} \cos \theta & \sin \theta \\ -\sin \theta & \cos \theta \end{pmatrix}, \tag{2.4.33}$$

where θ is the mixing angle. From (2.4.32) and (2.4.33) we easily obtain the standard expressions for the probabilities of two-neutrino transitions

$$P_{\nu_{\alpha'};\nu_\alpha} = \frac{1}{2} \sin^2 2\theta \left(1 - \cos \Delta m^2 \frac{L}{2p} \right), \alpha' \neq \alpha$$

$$P_{\nu_\alpha;\nu_\alpha} = 1 - \frac{1}{2} \sin^2 2\theta \left(1 - \cos \Delta m^2 \frac{L}{2p} \right). \tag{2.4.34}$$

Thus the transition probabilities in the simplest two-neutrino case are characterized by two oscillation parameters: Δm^2 and $\sin^2 2\theta$.

The probability $P_{\nu_{\alpha'};\nu_\alpha}(\alpha' \neq \alpha)$ can be written in the form

$$P_{\nu_{\alpha'};\nu_\alpha} = \frac{1}{2} \sin^2 2\theta \left(1 - \cos 2\pi \frac{L}{L_0} \right), \tag{2.4.35}$$

where

$$L_0 = 4\pi \frac{p}{\Delta m^2} \tag{2.4.36}$$

is the oscillation length. We have

$$L_0 \simeq 2.5m \frac{p/\text{MeV}}{\Delta m^2/\text{eV}^2}. \tag{2.4.37}$$

Neutrino oscillations can be observed if the oscillation length is less than or comparable with the source–detector distance L. It is obvious that this condition is equivalent to (2.4.31). If the oscillation length is much larger than the distance L, neutrino oscillations will not be observed.

In conclusion let us notice that from (2.4.28) and (2.4.29) we can easily obtain the following general relation between the probabilities of neutrino and antineutrino transitions:

$$P_{\nu_{\alpha'};\nu_\alpha} = P_{\bar\nu_\alpha;\bar\nu_{\alpha'}}. \tag{2.4.38}$$

This relation is a consequence of the *CPT* theorem. If *CP* invariance in the lepton sector holds,

$$P_{\nu_{\alpha'};\nu_\alpha} = P_{\bar\nu_{\alpha'};\bar\nu_\alpha}. \tag{2.4.39}$$

Let us notice that in the case of oscillations between two neutrino types possible phases of the mixing matrix do not enter into the probability (see (2.4.32)) and relation (2.4.39) is always satisfied.

2.4.6 Experimental status of neutrino oscillations

Investigation of the problem of neutrino masses and mixing is the major goal of today's neutrino physics. After many years of searching for neutrino oscillations in different experiments, strong evidence in favor of oscillations of atmospheric neutrinos was found in the Super-Kamiokande experiment [KAJ 98; FUK 98]. The data of Super-Kamiokande can be explained by neutrino oscillations with a neutrino mass-squared difference $\Delta m^2 \simeq$ a few $10^{-3}\,\mathrm{eV}^2$.

Indications in favour of neutrino masses and mixing were obtained also in solar neutrino experiments (see Section 6.1). From the analysis of the data of these experiments it follows that $\Delta m^2 \simeq 10^{-5}\,\mathrm{eV}^2$ (in the case of matter MSW transitions) or $\Delta m^2 \simeq 10^{-10}\,\mathrm{eV}^2$ (in the case of vacuum oscillations).

Finally, indications in favor of $\bar{\nu}_\mu\,\bar{\nu}_e$ oscillations were obtained in the accelerator experiment LSND [ATH 96]. From the analysis of the data of this experiment it follows that there exists the third scale $\Delta m^2 \simeq 1\,\mathrm{eV}^2$.

The atmospheric neutrino range of Δm^2 will be investigated in details in long-baseline (LBL) neutrino oscillation experiments. Accelerator LBL experiments K2K [NIS 97], MINOS [AYR 95] and ICARUS [CEN 94] are now under preparation. Recently the results of the first reactor LBL experiment CHOOZ [APO 98] were published. The other reactor LBL experiment Palo Verde [BOE 96] is taking data.

The investigation of the problem of solar neutrinos will be continued in the next generation experiments SNO, BOREXINO, GNO and others (see Section 6.1). The future reactor neutrino experiment KAM-LAND [SUZ 98] will also be able to reach the solar range of Δm^2.

The LSND indications in favor of neutrino oscillations are checked now in the KARMEN [ZEI 98] experiment. The LSND range of Δm^2 will also be investigated in detail in the future BooNE [WOJ 98] experiment.

Finally, in the short-baseline (SBL) neutrino oscillation experiments CHORUS [SAT 98] and NOMAD [GOM 98] $\nu_\mu \rightleftarrows \nu_\tau$ oscillations are searched for in a beam of high energy neutrinos from the CERN SPS.

We will start with the discussion of the atmospheric neutrino experiments. The main source of atmospheric neutrinos are decays

$$\pi \rightarrow \mu\nu_\mu, \quad \mu \rightarrow e\nu_e\nu_\mu, \tag{2.4.40}$$

pions being produced in the interaction of cosmic protons and nuclei in the earth's atmosphere. At relatively low neutrino energy ($\lesssim 3\,\mathrm{Gev}$) the ratio of the flux of ν_μ and $\bar{\nu}_\mu$ to the flux of ν_e and $\bar{\nu}_e$ is equal to 2. At higher energy this ratio is larger than 2 (not all muons decay in the atmosphere).

The fluxes of muon and electron neutrinos are predicted with an accuracy of about 20%. However, in the ratio of the fluxes of muon and electron neutrinos uncertainties connected with the absolute fluxes are canceled and the ratio is predicted with an accuracy better than 5% [GAI 96].

The results of atmospheric neutrino experiments are usually presented in the form of a double ratio R of the ratio of observed muon and electron events to the Monte Carlo predicted ratio

$$R = \left(\frac{N_\mu}{N_e}\right)_{MES} / \left(\frac{N_\mu}{N_e}\right)_{MC}. \tag{2.4.41}$$

The Kamiokande and Super-Kamiokande collaborations divide their events into two categories: sub-GeV events with $E_{vis} \leq 1.33\,\text{GeV}, p_e \geq 100\,\text{MeV}, p_\mu \geq 200\,\text{MeV}$, and multi-GeV events with $E_{vis} > 1.33\,\text{GeV}$. For the double ratio R in the Kamiokande experiment [HIR 92] in the sub-GeV region a value significantly less than one was found:

$$R = 0.60^{+0.06}_{-0.05} \pm 0.05. \tag{2.4.42}$$

This result was confirmed by the Super-Kamiokande experiment in which during 535 days of data taking about six times more events were observed than in the Kamiokande experiment. In the Super-Kamiokande experiment in the sub-GeV region it was found that [KAJ 98]

$$R = 0.63 \pm 0.03 \pm 0.05. \tag{2.4.43}$$

In the multi-GeV region it was found that:

$$\begin{aligned} R &= 0.57^{+0.08}_{-0.07} \pm 0.08 \quad \text{(Super-Kamiokande)} \\ R &= 0.65 \pm 0.05 \pm 0.08 \quad \text{(Kamiokande)}. \end{aligned} \tag{2.4.44}$$

Kamiokande and Super-Kamiokande are water Cherenkov detectors (1 kton and 22.5 kton fiducial mass of water, respectively). The results of the IMB experiment [BEC 92] (water Cherenkov detector) and the Soudan-2 experiment [PET 98] (iron calorimeter) are in agreement with the Kamiokande results:

$$\begin{aligned} R &= 0.54 \pm 0.05 \pm 0.11 \quad \text{(IMB)} \\ R &= 0.61 \pm 0.15 \pm 0.05 \quad \text{(Soudan-2)}. \end{aligned} \tag{2.4.45}$$

Notice that in the earlier iron calorimeter experiments Frejus [DAU 95] and NUSEX [AGI 89] values compatible with one were found for the ratio R:

$$\begin{aligned} R &= 1.00 \pm 0.15 \pm 0.08 \quad \text{(Frejus)} \\ R &= 0.96^{+0.32}_{-0.28} \quad \text{(Nusex)}. \end{aligned} \tag{2.4.46}$$

The fact that the ratio R is significantly less than one could mean disappearance of ν_μ or appearance of ν_e (or both). The Super-Kamiokande collaboration found compelling evidence in favor of disappearance of ν_μ due to neutrino oscillations.

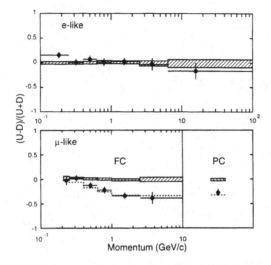

Fig. 2.4.1   Up-down asymmetry; Super-Kamiokande experiment.

Relatively large statistics allow the Super-Kamiokande collaboration to investigate the zenith angle θ dependence of the numbers of electron and muon events.
Vertically down-going neutrinos ($\cos\theta = 1$) pass the distance $L \simeq 20\,\mathrm{km}$ and
vertically up-going neutrinos ($\cos\theta = -1$) pass the distance $L \simeq 13\,000\,\mathrm{km}$. The
electron (muon) direction and energy are correlated with the momentum of the
neutrino. In the multi-GeV region the angle between the lepton and neutrino
momenta is $< 20°$.

The Super-Kamiokande collaboration found a significant up-down asymmetry A
of muon events in the multi-GeV region. The up-down asymmetry is determined as
follows:

$$A = \frac{U - D}{U + D}, \tag{2.4.47}$$

where U is the number of up-going events ($-1 \leq \cos\theta \leq -0.2$) and D is the number
of down-going events ($0.2 \leq \cos\theta \leq 1$). The asymmetry of electron and muon events
due to the magnetic field of the earth is less than 0.02 in sub-GeV region and less than
0.01 in the multi-GeV region. In Fig. 2.4.1 the Super-Kamiokande-measured
asymmetry as a function of lepton momentum is presented. The results of MC
calculations (with statistical and systematical errors) under the assumption that
there are no neutrino oscillations are shown by the hatched region. As is seen from
Fig. 2.4.1, the measured asymmetry of electron events is close to zero and in a good
agreement with MC prediction. The significant muon asymmetry is observed at
$p \geq 1\,\mathrm{GeV}$. The integral asymmetry of muon events in the multi-GeV region is

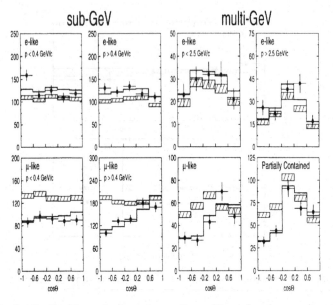

Fig. 2.4.2   Zenith angle dependence; Super-Kamiokande experiment.

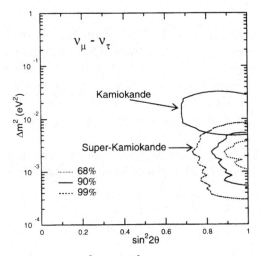

Fig. 2.4.3   Allowed regions of Δm^2 and $\sin^2 2\theta$ from the Kamiokande and the Super-Kamiokande experiments.

different from zero by about six standard deviations:

$$A = -0.296 \pm 0.048 \pm 0.01. \qquad (2.4.48)$$

In Fig. 2.4.2 the zenith angle dependence of the numbers of electron and muon events in sub-GeV and multi-GeV regions are shown. The hatched region is the result of MC calculations (with statistical error) under the assumption that there are

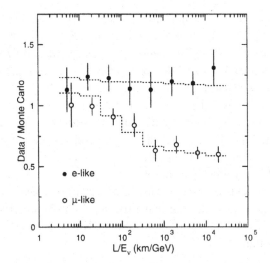

Fig. 2.4.4 Dependence of the ratio data/MC on L/E in the Super-Kamiokande experiment.

no neutrino oscillations. The significant deviation of the numbers of up-going muon events from MC predicted numbers is observed at $p \geq 0.4\,\text{GeV}$.

The zenith angle dependence of the number of events is well described if we assume that two-flavor $\nu_\mu \rightleftarrows \nu_\tau$ oscillations take place. The minimum value of χ^2 corresponds to the following values of oscillation parameters:

$$\Delta m^2 \simeq 2.2 \cdot 10^{-3}\,\text{eV}^2, \quad \sin^2 2\theta = 1, \tag{2.4.49}$$

($\chi^2/n_{dof} = 65.2/67$). In Fig. 2.4.3 the allowed regions of the parameters are shown (correspondingly 68% CL, 90% CL and 99% CL). The allowed region of the parameters obtained from the analysis of earlier Kamiokande data (90% CL) is also presented. The description of Super-Kamiokande data under the assumption that there are no neutrino oscillations is statistically unacceptable: $\chi^2/n_{dof} = 135/69$. If we assume $\nu_\mu \rightleftarrows \nu_\tau$ oscillations with parameter values given by (2.4.49) the expected number of CC tau-events is less than 20. The authors neglected the effect of production and decay of tau in their fit of the data. Thus they can not distinguish $\nu_\mu \rightleftarrows \nu_\tau$ oscillation from $\nu_\mu \rightleftarrows \nu_{ster}$ oscillations.

In the simplest case of two neutrino flavors the survival probability is given by Eq. (2.4.34). This expression can be written in the form

$$P_{\nu_\mu;\nu_\mu} = 1 - \frac{1}{2}\sin^2 2\theta \left(1 - \cos 2.54 \Delta m^2 \frac{L}{E}\right), \tag{2.4.50}$$

where Δm^2 is in eV^2, L is in km and E is neutrino energy in GeV.

In Fig. 2.4.4 the dependence on the parameter L/E of the ratios of the numbers of muon (electron) events to the numbers that were predicted by MC under the assumption that there are no oscillations is presented ($p > 400\,\text{MeV}$). As is seen from Fig. 2.4.4, the ratio practically does not depend on L/E for electron events and there is a deficit of muon events at large values of L/E (at large values of L/E the argument of the cosine in (2.4.50) is large and the cosine term disappears due to averaging over energies and distances; as a result we have in this region $\bar{P}_{\nu_\mu;\nu_\mu} \simeq 1 - \frac{1}{2}\sin^2 2\theta \simeq \frac{1}{2}$).

The dashed lines in Figs. 2.4.1 and 2.4.4 and bold lines in Fig. 2.4.2 were obtained under the assumption of $\nu_\mu \rightleftarrows \nu_\tau$ oscillations with $\Delta m^2 \simeq 2.2 \cdot 10^{-3}\,\text{eV}^2$ and $\sin^2 2\theta = 1$. In Fig. 2.4.4 flux normalizations are considered as free parameters.

The most important new result of Super-Kamiokande is the statistically significant up-down asymmetry of muon events. This result is model-independent evidence in favor of neutrino oscillations. If we take into account other indications in favor of oscillations of atmospheric neutrinos (Kamiokande [HIR 92] up-going muons: MACRO [RON 98], Kamiokande, Super-Kamiokande [KAJ 98]) we come to the conclusion that

$$10^{-3} \le \Delta m^2 \le 10^{-2}, \quad \sin^2 2\theta \ge 0.8. \tag{2.4.51}$$

In the next years this region of Δm^2 will be investigated in detail in the long-baseline neutrino experiment. The first results of the reactor LBL experiment CHOOZ were published recently [APO 98]. In this experiment the detector (5 tons of liquid scintillator) is at a distance of about 1 km from the reactor. Electron antineutrinos from the reactor are detected by the observation of the reaction

$$\bar{\nu}_e + p \to e^+ + n. \tag{2.4.52}$$

The $\bar{\nu}_e$ signature is a coincidence between the e^+ signal and the delayed signal from neutron capture by Gd.

This experiment is of the disappearance type; it allows us to obtain information about transitions of $\bar{\nu}_e$s into all possible neutrino states.

No indications in favor of neutrino oscillations were found in the CHOOZ experiment. For the ratio R_0 of the numbers of measured and expected events it was found that

$$R_0 = 0.98 \pm 0.04 \pm 0.04. \tag{2.4.53}$$

The data obtained in the CHOOZ experiment allow us to exclude $\Delta m^2 \ge 0.9 \cdot 10^{-3}\,\text{eV}^2$ at $\sin^2 2\theta = 1$. At $\Delta m^2 \ge 10^{-2}\,\text{eV}^2$ the experiment allows us to exclude $\sin^2 2\theta \ge 0.18$ (90% CL). The results of the CHOOZ experiment are not compatible with the possible interpretation of the atmospheric neutrino anomaly as $\nu_\mu \rightleftarrows \nu_e$ oscillations with large mixing angle. This conclusion is in agreement with the results of the analysis of the Super-Kamiokande data [KAJ 98; FUK 98].

The first accelerator LBL experiment K2K was started in January 1999 [NIS 97]. In this experiment neutrinos from the KEK 12 GeV accelerator (average neutrino energies $\simeq 1$–$2\,\text{GeV}$) are detected by the Super-Kamiokande detector (the distance is about 250 km). Two front detectors in the KEK site monitor the beam and allow measurement of the contamination of ν_es in the beam. The appearance of ν_es in the Super-Kamiokande detector due to possible $\nu_\mu \rightarrow \nu_e$ oscillations and disappearance of ν_μ are being searched for. The investigation of the distortion of the ν_μ energy spectrum will allow us to obtain information about $\nu_\mu \rightarrow \nu_x$ oscillations. The region $\Delta m^2 \geq 1 \cdot 10^{-3}\,\text{eV}^2$, $\sin^2 2\theta \geq 0.1$ for the $\nu_\mu \rightarrow \nu_e$ channel and $\Delta m^2 \geq 3 \cdot 10^{-3}\,\text{eV}^2$, $\sin^2 2\theta \geq 0.4$ for the $\nu_\mu \rightarrow \nu_\tau$ channel will be studied in the K2K experiment.

The LBL experiment MINOS [AYR 95] is planned to start in 2002. In this experiment neutrinos from the Fermilab Main Injector will be detected by an 8 kton detector in the Soudan Underground Laboratory (distance about 730 km; neutrino energy about 10 GeV). The goal of the experiment is to reach a sensitivity of $\Delta m^2 \simeq 10^{-3}\,\text{eV}^2$ in the $\nu_\mu \rightarrow \nu_\tau$ and $\nu_\mu \rightarrow \nu_e$ channels.

It is planned that neutrinos from the CERN 450 GeV SPS will be detected by ICARUS [CEN 94] in the Gran Sasso Underground Laboratory (about 730 km). Other LBL CERN-Gran Sasso projects NOE, RICH and OPERA are developing [PIE 98].

In several short-baseline experiments neutrino oscillations in different channels are under investigation. In the accelerator LSND experiment [ATH 96a] (Los Alamos linear accelerator) and KARMEN experiment [ZEI 98] (ISIS neutron spallation source, Rutherford Laboratory) $\bar{\nu}_\mu \, \bar{\nu}_e$ oscillations are searched for. Neutrinos in these experiments are produced in decays of π^+ and μ^+ at rest in beam stop targets. In decays of π^+ and μ^+ particles ν_μ, $\bar{\nu}_\mu$ and e^+ are produced. In the LSND detector (167 tons of liquid scintillator) at a distance of about 30 m from the source, $\bar{\nu}_e$ are searched for by the observation of the process

$$\bar{\nu}_e p \rightarrow e^+ n$$
$$np \rightarrow d\gamma \tag{2.4.54}$$

(both e^+ and delayed γ are detected). In the 36–60 MeV interval of neutrino energies $20.8 \pm 5.4 \bar{\nu}_e$ events were found with an expected background of 3.0 ± 0.6 events. The LSND collaboration interprets the observed excess of $\bar{\nu}_e$ events as $\bar{\nu}_\mu \, \bar{\nu}_e$ oscillations. For the transition probability it was found that $P = 0.31 \pm 0.12 \pm 0.05\%$.

In the KARMEN experiment [ZEI 98] (56 t scintillator calorimeter, detector–source distance about 18 m) no $\bar{\nu}_e$ events were found. For the transition probability in this experiment it was found that $P \leq 2.6 \cdot 10^{-3}$.

In Fig. 2.4.5 the LSND allowed region of oscillation parameters is shown (shadowed region). Exclusion regions that were obtained from the data of the reactor BUGEY [ACH 95] experiment and of the accelerator CCFR [ROM 97], BNL E776 [BOR 92] and KARMEN [ZEI 98] experiments are shown. As is seen from Fig. 2.4.5 the region

$$0.3\,\text{eV}^2 \leq \Delta m^2 \leq 1\,\text{eV}^2 \tag{2.4.55}$$

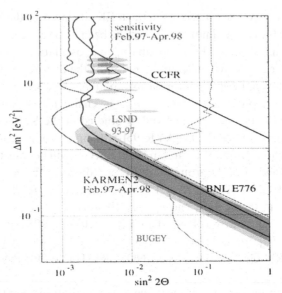

Fig. 2.4.5 Regions of oscillation parameters allowed by LSND and excluded by KARMEN and other experiments.

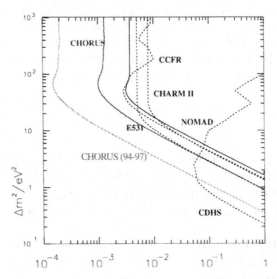

Fig. 2.4.6   Exclusion plot (90% CL) $\nu_\mu \to \nu_\tau$ oscillation in the mass region relevant for hot dark matter.

cannot be excluded by other experiments. In 1–2 years the KARMEN collaboration plans to reach LSND sensitivity. The experiment BooNE [WOJ 98] that is under preparation at Fermilab (according to plan it will start in 2001) will have a sensitivity to $\nu_\mu \rightleftarrows \nu_e$ oscillations much higher than LSND and will solve the problem of the LSND anomaly.

Two SBL experiments searching for $\nu_\mu \rightleftarrows \nu_\tau$ oscillations are going on at CERN on the neutrino beam from SPS (average neutrino energy about 30 GeV). The experiment CHORUS [SAT 98] is an emulsion experiment (800 kg of emulsion). The Production and decay of τ in the emulsion is searched for. In the experiment NOMAD [GOM 98] a magnetic detector is used. The production of τ is identified by kinematical criteria. No indications for $\nu_\mu \rightleftarrows \nu_\tau$ oscillations were found in either experiment. For large Δm^2 in CHORUS and NOMAD it was found that, correspondingly: $\sin^2 2\theta \leq 1.3 \cdot 10^{-3}$ and $\sin^2 2\theta \leq 2.2 \cdot 10^{-3}$. In Fig. 2.4.6 the exclusion plots obtained from the data of the CHORUS and NOMAD experiments are presented. The search for $\nu_\mu \rightarrow \nu_\tau$ oscillations will be continued in the experiment TOSCA [CAM 98].

There is no doubt that after the Super-Kamiokande experiment a new era in the investigation of neutrino oscillations started. The idea of neutrino mixing, that was put forward by B. Pontecorvo [PON 57, 58], which during many years was only a courageous hypothesis, now became reality. Transitions between neutrino flavors due to neutrino masses and mixing are new phenomena in physics and there is general belief that these are phenomena of physics beyond the Standard Model. We need new experimental data to understand what is the physical origin of neutrino masses and mixing, but the first decisive step has been done.

This contribution is dedicated to the memory of BRUNO PONTECORVO whose pioneer contribution to the problem of neutrino oscillations is difficult to overestimate.

2.5 How many generations of fermions?*

2.5.1 Introduction

The success of the Standard Model as the theory of electroweak and strong interactions, up to the highest energies investigated so far, cannot hide the many questions that remain unanswered. In particular, the fact that all observed elementary fermions organise themselves in three families that differ only by their mass spectrum is a purely experimental fact. Although the Standard Model could certainly accommodate any number of families of the same type as those already observed, these are subject to a number of constraints coming both from experiment and from theory. The object of this section is to review these constraints, with emphasis on the most stringent one, coming from the counting of light neutrino species.

* Alain Blondel, L.P.N.H.E. Ecole Polytechnique, 91128 Palaiseau, France, and CERN, Geneva, Switzerland and Daniel Denegri D.A.P.N.I.A.–S.P.P., Centre d'Etudes de Saclay, 91191 Gif-sur-Yvette, France, and CERN, Geneva, Switzerland.

2.5.2 The known generations

The observed quarks and leptons are organised in three families (or generations).

$$\begin{pmatrix} u \\ d' \end{pmatrix} \begin{pmatrix} c \\ s' \end{pmatrix} \begin{pmatrix} t \\ b' \end{pmatrix} \qquad \text{doublets of left-handed quarks}$$

$$\begin{matrix} (u) & (c) & (t) \\ (d) & (s) & (b) \end{matrix} \qquad \text{singlets of right-handed quarks}$$

$$\begin{pmatrix} \nu_e \\ e \end{pmatrix} \begin{pmatrix} \nu_\mu \\ \mu \end{pmatrix} \begin{pmatrix} \nu_\tau \\ \tau \end{pmatrix} \qquad \text{doublets of left-handed leptons}$$

$$\begin{matrix} (?\nu_e) & (?\nu_\mu) & (?\nu_\tau) \\ (e) & (\mu) & (\tau) \end{matrix} \qquad \text{singlets of right-handed leptons.}$$

The symbols d', s', b' stand as usual for the eigenstates of the weak interaction, as mass eigenstates are mixed through the Cabbibo–Kobayashi–Maskawa matrix. The above contains all elementary fermions that have yet been discovered, the latest one being the top quark [CDF 95; D0 95]. It should be said that, to date, direct observation of neutrinos has been achieved only for the neutrinos ν_e and ν_μ. The electron neutrino, ν_e, is defined as the particle, postulated by Pauli in 1931 to account for energy–momentum and spin conservation, that is produced together with the electron in β-decays. The first observation of ν_e-induced interactions, in 1953–1959, used the inverse neutron β-decay reaction $\bar{\nu}_e + p \rightarrow e^+ + n$. The detection became possible when nuclear reactor power provided sufficiently intense flux of $\bar{\nu}_e$s with energies in the MeV range [REI 53]. Observation of high energy neutrino interactions producing a muon – and not an electron – in neutrino beams generated at accelerators by pion and kaon decays [DAN 62] established the existence of another type of neutrino, the muon neutrino ν_μ, defined as the particle that is produced in association with the muon, in pion or kaon decay.

The tau neutrino has not been directly detected in that sense, no interaction of a neutrino producing a tau lepton in the final state having yet been observed, by lack of intense tau neutrino beams.[1] Its existence and its quantum numbers are, however, well established from τ-lepton decays, from which one can set a limit on its mass [ALEPH 98a; PDG 98], verify with per mil precision, by combining the lifetime measurement with the measurement of the leptonic branching ratios, that it has the same charged-current couplings as the other neutrinos [ALEPH 97; PDG 98], and measure its left-handed helicity [ALEPH 98; PDG 95]. Tau neutrinos are also produced directly in $W \rightarrow \tau\nu_\tau$ decays, where it has been verified that universality holds (to within a few percent) also at high energies [PDG 98; LEP 98a].

In searching for further families of neutrinos, it will be assumed that their couplings are the same as those of ν_e, ν_μ and ν_τ. Universality is deeply embedded in

---

[1] A possible candidate has recently been reported by the Fermilab E872 (DONUT) collaboration [DONUT 98].

the Standard Model, identical multiplets being obliged to have the same coupling constant. It is very well verified for charged current interactions of $e-\mu-\tau$ leptons, including the neutrinos [VID 96], and for neutral current interactions of charged leptons [LEP 98a] and of neutrinos [VIL 93].

All experiments are consistent with the fact that only left-handed neutrinos are produced and interact, and actually right-handed neutrinos need not exist if neutrinos are massless. This is well accounted for in the Standard Model, where right-handed neutrinos, having no charge, weak isospin 0 and no colour, will not be produced and will not interact.

The fermions listed above are classified in order of increasing mass. This scheme, where for instance the muon and the strange quark belong to the second family, is rather arbitrary, since nothing relates them except for the fact that they are each the second-heaviest particle of their type. The cancellation of axial anomalies in the Standard Model [ADL 72], however, requires that the quarks and leptons obey the relation $\sum_{\text{fermions}} Q = 0$, which is respected by each generation independently. It thus appears that fermions must come in units of one family at a time. The number of such generations remains, however, unspecified, and constitutes a major unknown still today.

Experimental investigation for further charged fermions beyond the existing ones is standard practice at, e.g., e^+e^- colliders. Given that the mass spectrum of charged fermions extends over many orders of magnitude it is only possible in this way to state that no new quark or lepton with mass in excess of about 91 GeV has been observed. From the Tevatron it is likely that other quarks with masses below that of the top quark would have been observed, too.

The strongest constraint on further families of fermions comes from the so-called "neutrino-counting" experiments. As we will see, these experiments rule out further families of fermions with *light* neutrinos, with couplings, or equivalently, isospin multiplet assignment, identical to those of the three known families. Although this fact was already well supported by experiment before 1989, the question was settled in a definitive way by e^+e^- experiments on the Z pole at SLC and LEP. Given that the first three neutrinos have very small masses, it is natural to conclude that there are probably only three generations of quarks and leptons. This conclusion can however be evaded by families with heavy neutrinos, for which the limits are far less stringent, or for "sterile" neutrinos, i.e. neutrinos for which both left- and right-handed states are isospin singlets.

This chapter will be organised as follows. Although the number of "light neutrinos" can nowadays be considered a closed subject, it is interesting to review the phenomena that are sensitive to it, as often the question can be turned around. This will be done in Section 2.5.3, where a summary of the situation before the advent of LEP and SLC will be given. Then we will turn to the determination of the number of light neutrino species by SLC and LEP from total cross section measurements (Section 2.5.4). Emphasis will be placed on the experimental techniques and on the theoretical assumptions imbedded in the analysis of these

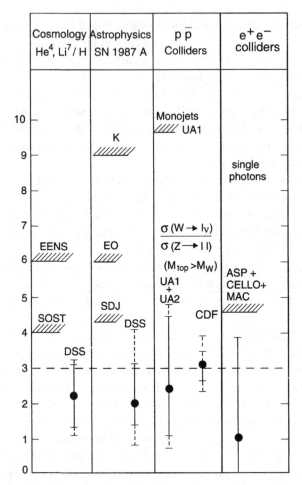

Fig. 2.5.1   Before LEP: Compilation of central values and 90% CL limits on the number of neutrino species N_ν from cosmology, astrophysics, and particle physics. From [DEN 89].

results. Finally limits on further families from electroweak radiative corrections will be described (Section 2.5.6).

2.5.3 Before LEP started

In a review article by Denegri *et al.* [DEN 89] published in summer 1989, just before SLC and LEP started, limits on the number of light neutrinos were given from the following sources:

- e^+e^- experiments looking for the signal in $e^+e^- \to \nu\bar{\nu}\gamma$ (single-photon experiments);

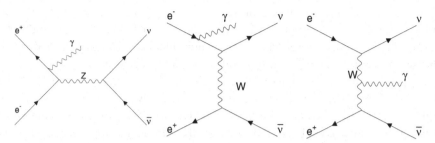

Fig. 2.5.2 The three diagrams contributing to single photon production in e^+e^- annihilation.

- The CERN and FERMILAB $p\bar{p}$ experiments, from monojets ($Z \to \nu\bar{\nu}$ + gluon events) or from the ratio of $Z \to \ell^+\ell^-$ to $W \to \ell\nu$ events;
- Astrophysical limits based on supernova SN 1987 A;
- Cosmological constraints from, e.g. the observed ^{4}He/H ratio;

The review concluded with $N_\nu = 2.1^{+0.6}_{-0.4}$, Fig. 2.5.1, and the following statement: "$N_\nu = 3$ is perfectly compatible with all data. Although the consistency is significantly worse, four families still provide a reasonable fit".

These methods will be briefly summarised below. For full details the reader should consult [DEN 89].

Number of light neutrinos from single photon experiments

The determination of the number of neutrino species N_ν through the study of single photon production in e^+e^- annihilation was first suggested by Dolgov *et al.* [DOL 72] and Ma and Okada [MA 78]. It is based on the process $e^+e^- \to \nu\bar{\nu}\gamma$, which can occur via the diagrams sketched in Fig. 2.5.2. The signature for such events is an isolated photon, with large missing energy and missing mass.

The contribution from the first diagram is proportional to the number of neutrinos, while the other two only produce $\nu_e\bar{\nu}_e$ pairs. The contribution of the last one is very small and was neglected in most analyses. It is relevant as a test of W–γ couplings. The differential cross section is given by

$$\frac{d\sigma}{dE_{\perp\gamma}d\cos\theta_\gamma} = \frac{G_F^2\alpha}{6\pi^2}\frac{s'}{E_{\perp\gamma}\sin\theta_\gamma^2}\left(\frac{s'}{s} + \frac{x_\gamma^2}{4}(1+\cos\theta_\gamma^2)\right)$$
$$\left[\frac{2M_Z^4 N_\nu(g_{Le}^2 + g_{Re}^2)(g_{L\nu}^2) + g_{Le}g_{L\nu}M_Z^2(M_Z^2 - s')}{(s' - M_Z^2)^2 + M_Z^2 G_Z^2} + 2\right]. \quad (2.5.1)$$

The terms in the square bracket are easily understood as originating from the Z diagram, $(g_{Le}^2 + g_{Re}^2)$ (Fig. 2.5.2), from the W exchange, and from the interference term $g_{Le}g_{L\nu}$ for which only left-handed electrons contribute. In this formula, s is the centre-of-mass energy squared, $s' = s(1 - x_\gamma)$ the invariant mass squared of the

$\nu\bar{\nu}$ system, $x_\gamma = E_\gamma / E_{\text{beam}}$ is the fraction of beam energy carried away by the photon and θ_γ is the angle of the photon w.r.t. the beam direction. The photons tend to be produced at low angles. Their energy spectrum differs considerably depending on whether the centre-of-mass energy is below, around or above the Z mass.

Energies below M_Z were the only ones available before the start-up of LEP and SLC in 1989. At these energies the reaction is peaked towards low energies and low angles. This renders the detection of the isolated photons difficult, this region of phase space being close to beam-induced as well as physics backgrounds, the largest ones of these being caused by the reaction $e^+e^- \to e^+e^- + \gamma$ where the final state electrons escape detection at low angles. The cross section within acceptance at PEP and PETRA energies is small. For $\sqrt{s} = 29\,\text{GeV}$, $E_\gamma \geq 1\,\text{GeV}$, $\theta_\gamma \geq 20°$, it amounts to $0.04\,\text{pb}$, giving a few events for the typical exposures of $\leq 100\,\text{pb}^{-1}$.

Several experiments attempted the identification of a single photon signal, including a dedicated experiment, ASP. A good experiment requires a detector with good coverage down to small angles, to ensure a veto against the $e^+e^- \to e^+e^- + \gamma$ background. Fine granularity and longitudinal segmentation are important as they allow reconstruction of the photon direction, to verify that the photon points to the e^+e^- interaction region and thus eliminate cosmic ray backgrounds.

The analysis explicitly requires that there be no other activity in the detector than the photon itself. Monitoring of the selection efficiency in presence of such a veto condition is always difficult. It could be done with $e^+e^- \to e^+e^- + \gamma$ events where both electrons are detected. By kinematic reconstruction, it is possible to predict the energy and angle of the additional photon, and verify its detection.

Selection efficiencies were typically of 60% and rather low purities, requiring a statistical extraction of the signal. Results were reported by MAC [MAC 85] and ASP [ASP 87] at PEP ($\sqrt{s} = 29\,\text{GeV}$) obtaining 1 event (1.1 expected for $N_\nu = 3$) in MAC and 1.6 events (2.7 expected for $N_\nu = 3$) in ASP. At PETRA, ($\sqrt{s} = 34 - 45\,\text{GeV}$) CELLO [CELLO 88] reported a signal of 1.3 events (1.9 expected for $N_\nu = 3$). The total of 3.9 events (6.1 expected for $N_\nu = 3$) leads to

$$N_\nu = 1.0^{+2.9}_{-1.0}$$

as available before 1989.

The single photon search was continued at and above the Z peak by the LEP experiments. An example of an event is shown in Fig. 2.5.3. At and above the Z peak the photon spectrum exhibits an enhancement at an energy of $E_\gamma = (s - M_Z^2)/2\sqrt{s}$, with a width commensurate to the Z width. The cross section becomes sizeable, several tens of pb at large angles, with a sharp rise when centre-of-mass energy crosses the Z peak, as shown in Fig. 2.5.4.

The four LEP experiments have performed this analysis, unfortunately not all on the full LEP1 data sample. The results are [OPAL 95; DELPHI 97;

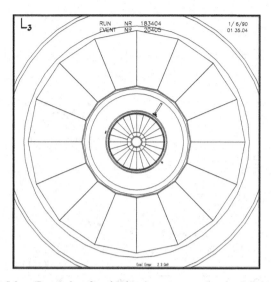

Fig. 2.5.3   Example of a single photon event in the L3 detector.

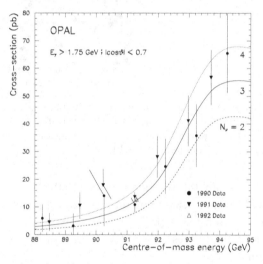

Fig. 2.5.4 Cross-section for single photon production as measured by OPAL, for $E_\gamma \geq 1.75\,\mathrm{GeV}$ and $\cos\theta_\gamma < 0.7$.

ALEPH 93; L3 98].

$$
\begin{array}{rll}
\text{ALEPH} & N_\nu = 2.68 \pm 0.20 \pm 0.20 & 1990\text{--}1991 \text{ runs} \\
\text{DELPHI} & N_\nu = 2.89 \pm 0.32 \pm 0.19 & 1993\text{--}1994 \text{ runs} \\
\text{L3} & N_\nu = 2.98 \pm 0.07 \pm 0.07 & 1990\text{--}1994 \text{ runs} \\
\text{OPAL} & N_\nu = 3.23 \pm 0.16 \pm 0.10 & 1990\text{--}1992 \text{ runs} \\
\text{average} & N_\nu = 3.00 \pm 0.09.
\end{array}
$$

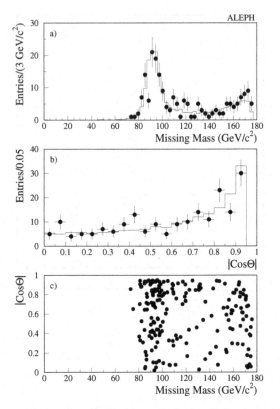

Fig. 2.5.5 Energy and angular distribution of single photon events in ALEPH at $\sqrt{s} = 183\,\text{GeV}$. The quantity that is displayed is the measured recoil mass $\sqrt{s'}$ against the photon, which is clearly peaked at $\sqrt{s'} = M_Z$. The histogram shows the Standard Model prediction for $N_\nu = 3$; from [ALEPH 98b].

Before LEP/SLC start-up, this reaction was seen as the best way to measure the number of light neutrinos, as being more direct. It was quickly realised, however, that the measurement from the Z peak cross sections was far more precise and efficient. The physics meaning is essentially the same, with small caveats that will be discussed in Section 2.5.5.

Since 1995 the LEP experiments have been running above the Z peak, (LEP2) and the single photon spectrum exhibits a clear peak (radiative Z returns) (see Fig. 2.5.5). This process is very useful [ALEPH 98b, LEP 98b] even at these high energies to search for new physics signals, such as gravitino production $e^+e^- \rightarrow \tilde{G}\tilde{G}\gamma$ and other gauge-mediated-susy-breaking supersymmetric scenarios.

Monojets in p$\bar{\text{p}}$ *collisions*

The analysis, performed by the UA1 Collaboration [UA1 87a], is based on $p\bar{p} \rightarrow$ jet $+ (Z \rightarrow \nu\bar{\nu})$ production, which is the QCD equivalent of the single photon experiment in e^+e^- annihilation. The simplest $Z +$ jet gluon bremsstrahlung process

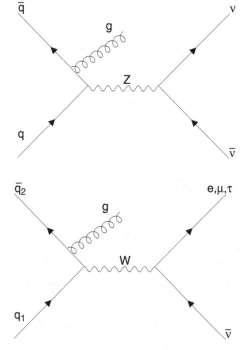

Fig. 2.5.6   Monojet production in $p\bar{p}$ collisions.

leading to monojets is sketched in Fig. 2.5.6, together with the main background source, where a W is produced and one of the decay leptons is partially observed or not at all, in particular in case of a tau decay. The presence of a gluon allows detection of the event at the trigger level, and provides a tag for the analysis. The signal consists of events with large missing transverse energy and momentum. The production of jets in association with vector bosons is controlled by well identified Z + jet events where the Z decays in e^+e^-, which are however rare, or better, by W + jet events where the W decays in $e\nu$.

UA1 detected 24 events with at least one jet with transverse energy in excess of 12 GeV, and a missing energy at a level of significance of more than 4σ, and special rejection of the $W \rightarrow \tau\nu$ background. The known sources would contribute 21 ± 5 events, including $Z \rightarrow \nu\bar{\nu}$ events with $N_\nu = 3$. Each additional neutrino would contribute two more events. This allowed the following limit to be placed:

$$N_\nu < 10 \text{ at } 90\% \text{ C.L.}$$

Limit from the measurement of $R = \sigma(W \rightarrow \ell\nu)/\sigma(Z \rightarrow \ell\ell)$ in $p\bar{p}$ collisions

This method provided the most sensitive measurement of the number of light neutrinos before LEP/SLC. It is based on the observation that, although the production cross sections for W and Z bosons are not precisely predicted in $p\bar{p}$

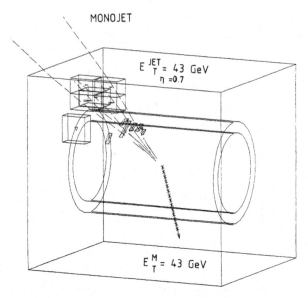

Fig. 2.5.7   A typical monojet event in UA1.

collisions, taking the ratio reduces those uncertainties related to parton structure functions, and experimentally eliminates those related to luminosity measurement and lepton detection efficiency.

In 1989 the ratio R had been measured by the experiments UA1 and UA2 at CERN at a centre-of-mass energy of 630 GeV [UA1 87b; UA2 87], and by CDF at FERMILAB's TEVATRON at a centre-of-mass of 1800 GeV [CDF 91].

This ratio can be written as:

$$R = \frac{\sigma(W \to \ell\nu)}{\sigma(Z \to \ell\ell)} = \frac{\sigma_W}{\sigma_Z} \times \frac{\mathrm{Br}(W \to \ell\nu)}{\mathrm{Br}(Z \to \ell\ell)}$$

where the branching ratios depend not only on the number of neutrinos but also on the hadronic decay widths of the W and Z:

$$\mathrm{Br}(W \to \ell\nu) = \frac{\Gamma^W_{\ell\nu}}{3\Gamma^W_{\ell\nu} + \Gamma^W_{u\bar{d}'} + \Gamma^W_{c\bar{s}'} + \Gamma^W_{t\bar{b}'}}$$

$$\mathrm{Br}(Z \to \ell\ell) = \frac{\Gamma^Z_{\ell\ell}}{N_\nu \Gamma^Z_{\nu\bar{\nu}} + 3\Gamma^Z_{\ell\ell} + \Gamma^Z_{u\bar{u}} + \Gamma^Z_{d\bar{d}} + \Gamma^Z_{s\bar{s}} + \Gamma^Z_{c\bar{c}} + \Gamma^Z_{b\bar{b}} + \Gamma^Z_{t\bar{t}}}. \quad (2.5.2)$$

The method assumes the Charged Current and Neutral Current Standard Model couplings of quarks and leptons. Beyond this assumption, there remain two uncertainties in relating the measurement of R to that of N_ν.

First the ratio σ_W/σ_Z has to be determined. Producing a W requires a u quark annihilating with a $\bar{d}$ or vice versa, producing a Z arises from $u\bar{u}$ or $d\bar{d}$ annihilation. Since the initial particles are proton and antiproton, the relative probability depends

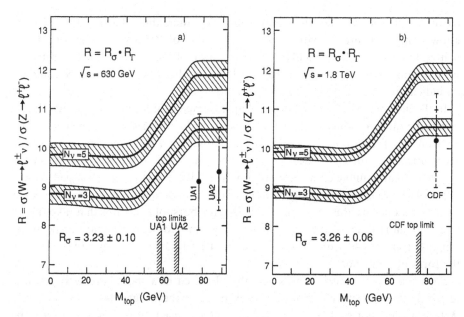

Fig. 2.5.8   Comparison of the measured values of R with the prediction as a function of (a) M_t and (b) N_ν. The grey bands correspond to the uncertainty in the production cross section ratio, σ_W/σ_Z.

on the quark density ratio $d(x)/u(x)$ around $x = m_W/\sqrt{s}$ and $x = m_Z/\sqrt{s}$. This ratio of quark densities can be extracted from the ratio of quark structure functions on neutron and protons, as obtained from deep-inelastic muon scattering experiments. From this one obtains at centre-of-mass energy of 630 GeV a value of $\sigma_W/\sigma_Z = 3.25 \pm 0.10$.

The second uncertainty stems from the unknown top quark mass, on which the partial widths $\Gamma_{t\bar{b}}^W$ and $\Gamma_{t\bar{t}}^Z$ depend. This led to predictions of R as function of both M_t and N_ν, as in Fig. 2.5.8.

Taking the top quark to be heavier than 80 GeV, (nowadays an evidence), one could derive

$$N_\nu = 1.05 \ ^{+2.5}_{-1.7}(\text{exp.}) \quad \pm 0.40(\text{sys.}) \quad (\text{UA1})$$
$$N_\nu = 1.45 \ ^{+1.3}_{-1.1}(\text{exp.}) \quad \pm 0.40(\text{sys.}) \quad (\text{UA2})$$
$$N_\nu = 2.50 \ \pm 1.3(\text{exp.}) \quad \pm 0.30(\text{sys.}) \quad (\text{CDF}).$$

It is clear that the method was limited by systematic errors at the level of $\Delta N_\nu = 0.3$.

Supernova 1987b and the number of light neutrinos

Besides the frightening possibility of washing out the Z pole, a large number of neutrinos would have deep consequences for the phenomena that occur in the universe at very high temperatures, for kinetic energies in excess of a few MeV per particle. This involves neutrinos with a mass less than about 1 MeV, so we do not know if the tau neutrino, with a present mass limit of 19 MeV, is to be considered light in this context. In particular a determination of the number of light neutrinos from astrophysics or cosmology that would be significantly less than three could be an indication of a relatively heavy tau neutrino.

In the final stages of a supernova collapse, temperatures reach a very high level. When the range of kinetic energies around 3–6 MeV is reached, in the process of neutron star formation, e^+e^- pairs are in equilibrium with neutrino pairs of all kinematically available species. The neutrinos, however, still interact weakly with the medium and unlike electrons and photons, will escape the star. A large fraction of the collapse energy is actually released in the form of neutrinos. According to the preferred scenarios, this results in a phenomenal burst of neutrinos that takes place in a very short time, of the order of 10 s. The emitted neutrinos have a spectrum ranging from 3 to 40 MeV. It is estimated that 1.5 to 3.5×10^{53} ergs are released in the form of neutrinos.

Such a burst of neutrinos was actually observed in a spectacular way when a relatively near supernova explosion took place in the Magellanic cloud in 1987. Within a time slot of 12 s, 11 candidates for the reaction $\bar{\nu}_e + p \rightarrow e^+ + n$ were observed in the Kamiokande [KAM 87] experiment in Japan and 8 in the IMB [IMB 87] experiment in Ohio (USA). Both are very large water Cherenkov detectors, designed to detect proton decays. They are nevertheless sensitive to energy deposits as low as 8.5 MeV for Kamiokande and 20 MeV for IMB. This allows detection of neutrinos from the upper end of the spectrum. One can hardly imagine the colossal energy release that leads to the observation of 20 neutrino interactions in 20 ktons on earth from a star that is situated 154 000 light years away.

Of course, only the $\bar{\nu}_e$ component of the neutrino burst is able to interact. This would be decreased if $\bar{\nu}_e$ production were in competition with further species of neutrinos: the more neutrino species there are, the fewer $\bar{\nu}_e$s.

The calculation of the energy release in the supernova takes into account the measurements of light spectra and intensity that were recorded starting several hours after the initial neutrino burst. It is estimated to be precise to $\pm 30\%$.

The authors of [DEN 89] conclude that the observations were consistent with the model of stellar collapse, provided:

$$N_\nu = 2.0^{+1.1+1.0}_{-0.4-0.8}(\text{stat, syst.}).$$

In retrospect this constitutes a superb success of the theory of supernovas. Clearly the next supernova is awaited with impatience!

Primordial nucleosynthesis and the number of light neutrinos

This subject is discussed in more detail in Section 2.7. It is now widely accepted that our universe originates from a Big Bang. The number of neutrinos is an important parameter in determining [STE 77; OLI 95] the evolution of the universe's properties from the time when the temperature was around 2 MeV to the time when it reached 0.75 MeV, which was incidently about 1 second. The interactions involving neutrinos become very small when the temperature decreases below 2 MeV, so that the number of neutrino species has an influence on the fraction of kinetic energy that remains available at that time. Among the reactions that are affected are those which govern the ratio of protons to neutrons:

$$
\begin{aligned}
n + \nu_e &\leftrightarrow p + e^- \\
p + \bar{\nu}_e &\leftrightarrow n + e^+ \\
n &\leftrightarrow p + e^- + \bar{\nu}_e,
\end{aligned}
\tag{2.5.3}
$$

a larger number of light neutrinos would lead to a larger ratio of neutrons to protons generated in that early phase [WEI 72]. This in turn modifies the relative abundances of various nuclei during the subsequent cooling.

The predictions for the abundances of H, D, He, ^{7}Li require modelling of the subsequent reactions, down to temperatures of 100 KeV. These predictions depend strongly on the n/p ratio and on N_ν. Of course the measurement of primordial abundances is quite problematic, and is not devoid of ambiguities.

In 1989, it was possible [DEN 89] to fit the observed abundances provided the number of neutrinos was in the range from 1 to 4.2, the best fit giving $N_\nu = 2.3 \pm 0.8$.

The issue has been revisited since, see e.g. [OLI 95; COP 96]. Although the number of light neutrinos is now determined at LEP with high precision, Big Bang nucleosynthesis could be sensitive to degrees of freedom (particles) which do not couple to the Z, but are produced otherwise in some hypothetical other scheme of new physics. The more recent analyses, although admitting the many hypotheses involved in extracting the limits, conclude that the equivalent number of light neutrinos is strictly less than four, $N_\nu < 4$.

2.5.4 Determination of the number of light neutrino species around the Z peak

Historical background

The most precise determination of the number of light neutrino species is obtained from measurements of the visible cross sections of e^+e^- annihilation at and around the Z pole, as is made explicit in Fig. 2.5.9. The realization that the visible cross sections might be sensitive to the number of light neutrinos is rather ancient, one finds the question asked in John Ellis 'Zedology' [ELL 76], where one can find this anguished question: *The Z peak is large and dramatic, as long as there are not too*

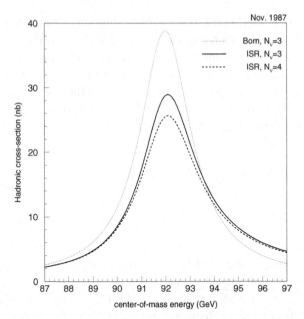

Fig. 2.5.9   The $e^+e^- \rightarrow$ hadrons cross section as a function of centre-of-mass energy. This curve was drawn before LEP start-up in 1987. At that time the Z mass was measured to be around 92 GeV with an error larger than 1.5 GeV. The dotted line represents the Born approximation prediction for three species of light neutrinos. The full line includes the effect of initial state radiation. The dashed line represents the effect of adding one more type of light neutrino with the same couplings as the first three. It is clear from this picture that the cross section at the peak of the resonance contains most of the information on the number of light neutrino species.

many generations of fermions. Is it conceivable that there might be so many generations as to wash out the Z peak? Since at that time the bound on the number of light neutrinos was very weak (about 6000), this certainly was a frightening possibility for those planning to build LEP. Dramatic also were the few first weeks of SLC and LEP operation where it was quickly realised that the Z peak was there indeed, large and dramatic, and that, alas, the number of light neutrinos was three.

The determination of the number of light neutrinos was, indeed, the object of a intense competition between the SLC at SLAC (California, USA) and LEP at CERN (Switzerland). The two projects were built with rather different aims in mind. LEP was built as presumably the largest possible conventional e^+e^- storage ring, with a circumference of 27 km. The standard technique was to lead to few surprises and assure reliable high luminosity, and schedule. SLC, on the other hand, was the prototype of a new concept of accelerator, the linear collider. SLC was re-using the old Stanford linac, with improvements in the acceleration technique (SLED) and addition of arcs to bring e^+ and e^- in collisions, damping rings, etc.

The commissioning of SLC started in early 1987, and led to a number of technical difficulties, not surprising in retrospect for such a new project. The first Z hadronic

Table 2.5.1. *First results from LEP and SLC on the Z mass and the number of light neutrino species, as published around 12 October 1989*

Experiment	Hadronic Zs	Z mass (GeV)	N_ν
L3	2538	91.13 ± 0.06	3.42 ± 0.48
ALEPH	3112	91.17 ± 0.05	3.27 ± 0.30
MARKII	450	91.14 ± 0.12	2.8 ± 0.60
OPAL	4350	91.01 ± 0.05	3.10 ± 0.40
DELPHI	1066	91.06 ± 0.05	2.4 ± 0.64
Average		91.10 ± 0.05	3.12 ± 0.19

decay was produced on 11 April 1989, and recorded in the MarkII detector. Luminosity was very low, a few $10^{+27}/\text{cm}^2/\text{s}$, leading to a few Z hadronic decays per day. Advertised to start up on 14 July, the time where SLC would hold the lead was going to be short, and intense. Nevertheless the SLC collaboration was able to collect a total of 106 Z decays by 24 July and submit a publication [MARKII 89], where the Z mass was determined to be $M_Z = 91.11 \pm 0.23$ GeV, and the number of light neutrinos species $N_\nu = 3.8 \pm 1.4$.

LEP did not start collisions on 14 July but on 13 August, for a week. The small β^* optics were not yet commissioned and events came at a rate of one a day for the four experiments. Running resumed on 20 September with superconducting quadrupoles working and thus the small β^* optics, and in just three weeks, until 9 October, typically 3000 Zs were collected. By October 13, a seminar was organised at CERN where the four collaborations presented their first results [ALEPH 89; DELPHI 89; L3 89; OPAL 89; MARKII 89b], shown in Table 2.5.1. The day before, SLC had organised a public conference where updated results had been presented [MARKII 89b], based on 480 events. The 'online average' of these results is also shown in Table 2.5.1. Clearly, the average being $N_\nu = 3.12 \pm 0.19$, the number of light neutrinos was then determined to be three.

Following this important contribution, SLC was shaken by an earthquake on 24 October 1989, from which it took more than a year to recover, and then concentrated on polarised beam physics. LEP went on, to the end of 1989 and for six more years (1989 to 1995), each experiment collecting four million hadronic Z decays. Essentially final results are now available, with the number of light neutrinos determined to be $N_\nu = 2.994 \pm 0.011$.

The early results were somewhat unexpectedly precise, and the final one is amazingly precise, if one compares them with the expectations that could be found in e.g. the 1986 studies [LEP 86], where ± 0.2, was stated as asymptotic reachable precision on N_ν from the Z width measurement – which was assumed to be performed with muon pairs. Once the method is explained in more detail, it will become clear that the unexpected capacity of the experiments to perform precise measurement of hadronic cross sections is the reason for this success.

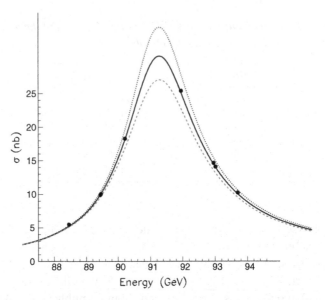

Fig. 2.5.10   The $e^+e^- \to$ hadrons cross section as a function of centre-of-mass energy, as measured by one of the LEP experiments, ALEPH. The curves represent the Standard Model predictions for two, three and four species of light neutrinos. It is clear from this picture that there is no further light neutrino species with couplings identical to the first three.

The method

Around the Z pole, the photon exchange is only a correction to the Z-channel, which dominates the cross section and can be written as:

$$\sigma_f = \frac{12\pi(\hbar c)^2}{M_Z} \cdot \frac{s\Gamma_e\Gamma_f}{(s - M_Z)^2 + s^2\Gamma_Z^2/M_Z}. \tag{2.5.4}$$

This formula is obtained in fact for the e^+e^- cross section into a visible channel f produced through a spin-one resonance under very general assumptions and is not particularly sensitive to Standard Model inputs, such as the assignment of fermions to multiplets, etc. The Standard Model predicts the values of the partial widths, but the formula for cross sections written above is very general. Photon radiation from the initial state electrons ISR leads to a sizeable but well calculated correction to this line shape, as can be seen in Fig. 2.5.9.

Measurements of cross section for a given final state f around the Z pole allows us to extract three parameters: the position of the peak, the width of the resonance and an overall normalisation, that is best obtained from the peak cross section,

$$\sigma_f^0 = \frac{12\pi(\hbar c)^2}{M_Z} \frac{\Gamma_e\Gamma_f}{\Gamma_Z^2} = \frac{12\pi(\hbar c)^2}{M_Z} B_e.B_f, \tag{2.5.5}$$

where B_f is the branching fraction of the Z into final state f. See Figs. 2.5.7, 2.5.10 and Table 2.5.2. The principle of the analysis is then as follows: all visible channels

Table 2.5.2. *Numerical values of quantum numbers, Neutral Current couplings, and Z decay partial width, for the four types of fermions. The value of* $\sin^2 \theta_W^{\text{eff}}$ *is 0.2315*

f	I_{3f}	Q_f	g_{Af}	g_{Vf}	Γ_f (MeV)
ν	1/2	0	1/2	1/2	167.1
e	−1/2	−1	−1/2	−0.04	83.92
u	1/2	2/3	1/2	0.19	299.8
d	−1/2	−1/3	−1/2	−0.35	382.7
b	−1/2	−1/3	−1/2	−0.35	375.5

Table 2.5.3. *Synopsis of parameters of the Z line shape.* R_ℓ *is defined as* $R_\ell \equiv \Gamma_{had}/\Gamma_{\ell\ell}$, *where* $\Gamma_{\ell\ell}$ *refers to the partial width into a pair of massless charged leptons.*

Quantity	Main technologies	Physics outputs	Relative precision
line shape			
M_Z	**Absolute energy scale** relative cross sections line shape fit (QED rad. corr.)	input	2×10^{-5}
Γ_Z	**Relative energy scale** relative cross sections line shape fit (QED rad. corr.)	$\Delta\rho$	1.1×10^{-3}
$\sigma_{\text{had}}^{\text{peak},0}$ $R_\ell \equiv \Gamma_{\text{had}}/\Gamma_{\ell\ell}$	**Absolute cross sections** lepton, hadron event selection	$N_\nu \cdot \Gamma_{\text{inv}}/\Gamma_{\ell\ell}$ universality universality $f(\alpha_s, \sin^2 \theta_W^{\text{eff}}, \delta_{vb})$	1.2×10^{-3} 1.0×10^{-3}

are detected by large acceptance detectors and classified according to four categories: (i) hadrons, (ii) electron pairs, (iii) muon pairs, (iv) tau pairs. By measuring these cross sections, one can obtain six numbers: the mass, the total width, and four other parameters which could be four branching ratios, or four partial widths, but are chosen to be the peak cross section for hadrons $\sigma_{\text{had}}^{\text{peak},0}$, and the ratios of hadrons to the various leptonic partial widths, $R_\ell \equiv \Gamma_{\text{had}}/\Gamma_\ell$. The Standard Model implies lepton universality, and if this is assumed, the number of parameters can be reduced to four, $m_z, G_z, \sigma_{\text{had}}^{\text{peak},0}, R_\ell$. The choice of these observables to fit the line-shape measurements is dictated by the fact that they are experimentally uncorrelated, both from the point of view of statistical and systematic errors. Table 2.5.3 lists the main experimental technology and the present accuracy in these parameters.

The sensitivity to the number of neutrino species appears in the Z width, which is the sum of all partial widths:

$$\Gamma_Z = \Gamma_{\text{had}} + 3 \cdot \Gamma_\ell + N_\nu \cdot \Gamma_\nu. \tag{2.5.6}$$

Now comes an important remark. By reporting the expression for Γ_Z of Eq. 2.5.6 into the peak cross section for hadrons, Eq. 2.5.5, one sees that the number of neutrinos can be extracted from quantities that are measured at the peak only:

$$N_\nu = \frac{\Gamma_\ell}{\Gamma_\nu} \cdot \left(\sqrt{\frac{12\pi R_\ell}{M_Z^2 \sigma_{had}^{peak,0}}} - R_\ell - 3 \right). \tag{2.5.7}$$

When studying this equation it is apparent that the sensitivity of N_ν to R_ℓ is small, as there is a cancellation between the two terms containing this quantity. As a result the experimental measurement that enters most in the determination of N_ν is the peak cross section, as already guessed intuitively from Fig. 2.5.9. The only quantity that is required from the Standard Model is the ratio Γ_ℓ/Γ_ν.

The Standard Model predicts the numerical values for the Z partial widths,

$$\Gamma_f = \frac{\alpha}{6 \sin^2 \theta_W \cos^2 \theta_W} M_Z (g_{Lf}^2 + g_{Rf}^2) \cdot N_c \cdot \left(1 + \frac{3 Q f^2}{4} \frac{\alpha}{\pi} \right) (1 + \alpha_s/\pi + \cdots). \tag{2.5.8}$$

where the couplings are given in a universal way by: $g_{L,Rf} = I_{L,Rf}^3 - Q_f \sin^2 \theta_W$, with $I_{Rf}^3 = 0$ for all known fermions. Equivalently one can define the coupling of the Z to the vector or axial vector fermion current:

$$g_{Vf} = (g_{Lf} + g_{Rf}) = I_{Lf}^3 - 2Q_f \sin^2 \theta_W$$

$$g_{Af} = (g_{Lf} - g_{Rf}) = I_{Lf}^3. \tag{2.5.9}$$

Electroweak corrections to these formulae are largely accounted for by using universal effective couplings at the Z energy scale, both for $\alpha \to \alpha(M_Z^2)$, $\alpha_s \to \alpha_s(M_Z^2)$, and for the weak mixing angle $\sin^2 \theta_W \to \sin^2 \theta_W^{eff}$. These amount to 6% for $\alpha(M_Z^2)$, and to 1–2% for $\sin^2 \theta_W^{eff}$ due to the large mass of the top quark. Small additional corrections are non-universal (vertex corrections) but they are small, a few 10^{-3}, and insensitive to such effects as the top quark or Higgs boson masses. A well known exception is the b partial width where the vertex correction involving the top quark amounts to 2%.

In the determination of the number of neutrinos from the line-shape parameters, Eq. 2.5.7, the only Standard Model assumption concerns the ratio Γ_ν/Γ_ℓ. It can be seen that because of the smallness of g^{ν_e} for the electron, the large universal radiative corrections cancel in the ratio. Consequently the numerical value of this ratio is predicted very accurately to be 1.9909 with variations of a few 10^{-4} upon top mass and Higgs mass variations or by using directly the measured value of $\sin^2 \theta_W^{eff}$ obtained from asymmetries at LEP/SLC.

One could argue however that, although the above ratio is predicted very precisely by the Standard Model, it is not experimentally known. The only information we have on neutrino couplings comes from muon-neutrino electron scattering experiments. The ν_μ coupling with the Z is found to be [VIL 93]

$$g^{\nu_\mu} = 0.5002 \pm 0.0165,$$

in good agreement with the LEP result on the average value of neutrino Z coupling derived from the invisible decay width of the Z, Γ_{inv},

$$g^\nu = 0.4988 \pm 0.0014$$

for three flavours.

The value of the ν_e coupling with the Z has been derived from the ratio of neutral-current to charged-current scattering [CHARM 86] and from the $\nu_e e$ scattering experiment at LAMPF [ALL 93] with the combined result

$$g^{\nu_e} = 0.50085 \pm 0.071,$$

again in good agreement with g^ν and flavour universality.

The argument can be, in retrospect, turned around, since the determination of N_ν at LEP is in strong support of universality.

2.5.5 Discussion

The measurement from single photon counting is often referred to as "direct", in contrast with that from the Z peak cross sections. It is true that the single photon measurement, by vetoing any additional activity in the detectors, ascertains that the Z decay is truly invisible. The total cross section, however is proportional to $\Gamma_\ell(\Gamma_{\text{inv}}/\Gamma_Z)$ i.e. the coupling of the Z to the initial state electrons times the branching ratio into invisible final states. Clearly this rate is sensitive to assumptions on the total Z width, and has larger sensitivity (at the level of 1%) to α_s and radiative corrections, to which the measurement from the peak cross sections is immune. Also this rate would decrease if the Z were to decay into new but visible final states which would increase Γ_Z without increasing Γ_{inv}.

The measurement from the peak cross sections, on the other hand, although it is based on hadronic cross section measurements, is only dependent on the assumption on Γ_ν/Γ_ℓ, which is free of hadronic corrections. In particular this measurement does not need to assume that the neutrino is perfectly invisible. If neutrinos were unstable, for instance decaying into $\nu' + \gamma$, this decay mode would not be counted as a hadronic or leptonic event candidate and the number of neutrinos would remain correctly measured. The measurement from the peak cross sections would be sensitive, however, to the presence of new Z decay modes which would not be selected by either the leptonic or hadronic event selections, or at least not with the same efficiency, the additional inefficiency being in this context called 'invisible'. Extensive searches for such decay modes have been pursued by the LEP experiments [ALEPH 92], in the context of the search for super-symmetric partners of the quarks and leptons, excited leptons and quarks and other exotic topologies. Limits are in general more stringent than from the Z width measurements, although in some difficult cases the Z widths provide the best limits. It is concluded that, with a very

Table 2.5.4. *Summary of the present experimental errors, physics sensitivity and theoretical uncertainties in SM predictions for the line shape parameters. The SM values are given for ZFITTER, with $M_t = 175$ GeV, $M_H = 300$ GeV, $\alpha_s = 0.118$*

| Observable | Summer '97 value (error) | SM | Physics sensitivity of electroweak observables Sensitivity to: | | | | | | |
			M_t (GeV) 175 ± 6	M_H (GeV) 60	M_H (GeV) 1000	α_s ±0.003	$\alpha(M_Z^2)^{-1}$ 128.89 (9)	(m_b) 4.7 ± 0.3	Higher orders
Γ_Z (MeV)	2494.8(2.5)	2493.3	+1.5	+4.2	−5.3	1.7	0.7	0.2	0.6
Γ_ℓ (MeV)	83.91 (0.10)	83.93	+0.06	+0.11	−0.14	0.02	—	—	0.02
$\sigma_{had}^{peak,0}$ (pb)	41486 (53)	41481	+3	−4	+4	−16	2	2	5
$R_\ell \times 10^3$	20775 (27)	20732	−1.8	+15	−13	21	4	2	5

few and specific possibilities, the Z undergoes no other decay than those into known quarks and leptons at the level of less than a few 10^{-3}. See Table 2.5.4.

2.5.6 Constraints on further families of quarks and leptons from radiative corrections

The excellent agreement of all electroweak observables sensitive to electroweak radiative corrections certainly places constraints on the existence of particles that would have a strong effect on them. Such analyses have been performed for e.g. super-symmetry [ALT 97; ERL 98], or for more general cases of new physics [LAN 92]. This last paper gives the general framework under which the effect of an additional family of quarks and leptons can be estimated.

Constraints from electroweak radiative corrections are very powerful but ambiguous in nature. Although precisely measured observables are potentially affected by *many* effects, placing bounds in the absence of deviations from the minimum hypothesis (validity of the Standard Model and nothing else) requires the assumption that the envisaged scenario of new physics is not undergoing accidental cancellation with another one. Therefore it is necessary to consider a scenario under which no other new physics is present than one or several new families of quarks and leptons identical to the known three.

The effect of a new family of quarks and leptons is two-fold: (i) if there is isospin violation in the mass spectrum of this new family, it will have very large effects on the ρ parameter, in much the same way as the top–bottom mass splitting; (ii) in addition, even a degenerate family will affect the Z and W self-energies, in a way that affects the so called S or ϵ_3 parameter [ALT 97a] (see Section 4.4).

The good agreement between the top quark mass predicted from precision data and its direct measurement excludes a fourth family with large mass splittings. More precisely, the approximate limit on the splitting of the would-be new family is obtained:

$$\Delta m_q^2 + \frac{1}{3}\Delta m_l^2 + \frac{1}{3}\Delta m_s^2 < M_{t\text{EW.fit}}^2 - M_{t\text{measured}}^2,$$

where Δm_q^2 is the difference between the mass squared of the new top-like quark and the mass squared of the new bottom-like quark, and similarly Δm_l^2 for the leptons, or possibly Δm_s^2 for a doublet of non-degenerate scalars. Since the top quark mass from the electroweak fits ($161 \pm 9\,\text{GeV}$) is lower than the measured one ($175 \pm 5\,\text{GeV}$), the limit on the second term is $4300\,\text{GeV}^2$. A new generation, to be consistent with precision measurements, must be nearly degenerate.

A degenerate family would affect ϵ_3, changing the predicted value of $\sin^2\theta_W^{\text{eff}}$ by $+0.00076$, the W mass by $-60\,\text{MeV}$, but leaving the leptonic width of the Z unchanged. This could play against the Higgs mass, but the value of $\sin^2\theta_W^{\text{eff}}$ is already such that the Higgs mass has to be very close to its experimental lower bound. A mass splitting could well be invoked for cancellation since the splitting would modify $\sin^2 \theta_W^{\text{eff}}$ downwards and M_W and Γ_ℓ upwards. Nevertheless, the effect of a fourth family would be to worsen the χ^2 of the electroweak fit by at least 2.5 units, if the Higgs mass is to remain within a physical range.

One can conclude that a fourth family of almost mass-degenerate quark and leptons could be (barely) accommodated by present precision data, but a fifth is certainly already excluded.

2.5.7 Conclusions

The number of standard fermion families with light neutrinos has been experimentally determined to be three, $N_\nu = 2.994 \pm 0.011$. The search applies to (i) families with weak isospin assignment identical to those already known and (ii) neutrinos with a mass less than 45 GeV. The search assumes family universality, as implied by the Standard Model.

The most sensitive measurement comes from the Z peak observables $\sigma_{\text{had}}^{\text{peak},0}$, R_{had}. Direct search for invisible Z decays by the single photon method leads to the same conclusion: $N_\nu = 3.00 \pm 0.09$.

This number provides an important input for the calculation of astrophysical and cosmological processes, such as supernovas and Big Bang nucleosynthesis, from which consistent and complementary limits are obtained.

For families with heavier neutrinos, the limits come from the consistency of precision electroweak data. Assuming the validity of the Standard Model and nothing else than one or two additional families of nearly degenerate heavy fermions, a fourth family can be marginally accommodated, while a fifth one is certainly excluded by the data.

2.6 Electromagnetic properties of neutrinos*

2.6.1 Introduction

As neutrinos rank among the most fascinating particles of nature, it is not surprising that physicists have long been interested in their electromagnetic properties.

* W. J. Marciano, Department of Physics, Brookhaven National Laboratory, Upton, New York 11973, and A. Sirlin, Department of Physics, New York University, New York, NY 10003.
This research was supported in part by the U.S. Department of Energy under Contract No. DE-C02-76CH00016 and the National Science Foundation under Grant No. PHY-8715995. Accordingly, the U.S. government retains a nonexclusive, royalty-free license to publish or reproduce the published form of this contribution, or allow others to do so, for U.S. government purposes.
This paper is dedicated to the cherished memory of Ralph E. Behrends, a remarkable human being who, in the decade of 1955–65, carried out early and pioneering work in several areas of theoretical particle physics.

Assuming charge conservation in $n \to p + e^- + \bar{\nu}_e$, experimental measurement of the charges of $p + e^-$ and n leads [ZOR 63] to $|e_{\bar{\nu}_e}| \leq 4 \times 10^{-17} e$, where e is the positron charge. Astrophysical arguments involving the neutrino luminosity of the sun and white dwarfs give $|e_\nu| < 10^{-13} e$ for neutrino species with a mass smaller than 150 eV and 20 keV, respectively [BER 63; DOL 81]. Henceforth, we will assume that neutrinos have zero charge.

In order to set the stage for our discussion, it is convenient to write down the most general expression for the matrix element of the electromagnetic current between invariantly normalized neutrino states [KAY 82].

$$\langle \nu(p_2, \lambda_2) | J_\mu^\gamma(0) | \nu(p_1, {}_1) \rangle$$
$$= \bar{u}(p_2, \lambda_2)\{[F(q^2) + \gamma_5 G(q^2)]\gamma^\alpha(q^2 g_{\alpha\mu} - q_\alpha q_\mu)$$
$$+ [M(q^2) + iE(q^2)\gamma_5]i\sigma_{\mu\alpha}q^\alpha\}u(p_1, \lambda_1) \tag{2.6.1}$$

where p_1 and p_2 represent the four-momenta of the initial and final neutrinos, $q = p_2 - p_1$, the u's are Dirac spinors; the γ matrices and metric follow the conventions of [BJO 64], and $\lambda = \pm 1$ stand for the eigenvalues of the helicity operator $\sigma \cdot \mathbf{p}/|\mathbf{p}|$. In general, the neutrino states may be of the same or different species. In the latter case, they may have different masses. We refer to these two cases as diagonal and off-diagonal (or transition) amplitudes, respectively.

Equation (2.6.1) follows from Lorentz covariance and electromagnetic current conservation. In order to ensure electric charge neutrality and nonsingular behavior of the form-factors, we will assume that $F(q^2)$ and $G(q^2)$ are regular as $q^2 \to 0$. Thus, calling $f(q^2) \equiv q^2 F(q^2)$ and $g(q^2) \equiv q^2 G(q^2)$, we have $f(0) = g(0) = 0$. The functions $f(q^2)$, $g(q^2)$, $M(q^2)$, and $E(q^2)$ are sometimes referred to as the electromagnetic, anapole, magnetic, and electric dipole form-factors, respectively. In the case of diagonal amplitudes, the hermiticity of $J_\mu^\gamma(0)$ implies that they are real functions in the physical domain $q^2 \leq 0$.

We now consider special cases:

1. If CP invariance holds, $E(q^2) = 0$ in diagonal amplitudes.
2. For Weyl neutrinos, that is, massless two-component neutrinos satisfying $a_- u = u$ where $a_- = (1 - \gamma_5)/2$ is the negative helicity projector, the terms proportional to $M(q^2)$ and $E(q^2)$ in Eq. (2.6.1) vanish and we are left with a single form-factor $F(q^2) + G(q^2)$.
3. For Majorana neutrinos, that is, neutrinos that are their own antiparticles, *CPT* invariance and the hermiticity of J_μ^γ imply the vanishing of the F, M, and E form-factors in diagonal amplitudes, and we are again left with a single form-factor $G(q^2)$ [KAY 82; NIE 82; SCH 81; RAD 85]. In spite of the apparently different structure of Eq. (2.6.1) under the special cases (2) and (3), a general equivalence theorem states that massless Majorana and Weyl neutrinos are physically indistinguishable (see, e.g., [MAR 69]). In practice, the same

equivalence holds in the massive case if all available $\nu(\bar{\nu})$ are extremely relativistic left- (right-) handed particles, a conclusion referred to as the "practical
Majorana–Dirac confusion theorem" [KAY 82, 88].

4. Both Dirac (i.e., four-component) and Majorana neutrinos can have off-diagonal amplitudes with nonvanishing M and (or) E form-factors.

The static values of the form-factors, that is, $f(0)$, $g(0)$, $M(0)$, and $E(0)$, can in
principle be probed by real photons and are expected to be observable, gauge-
invariant quantities. On the other hand, in general this is not the case for $q^2 \neq 0$. For
example, in the minimal version of the Standard Model (SM) the neutrinos fall
under the special case (2) and it is known that the single form-factor $f(q^2) + g(q^2)$ is
generally divergent and depends on the choice of the non-Abelian gauge [BAR 72;
LEE 77; LUC 85; DEG 88]. What occurs in physical scattering processes such as
neutral current ν-lepton and ν-hadron scattering, is that contributions mediated by
virtual photons, involving Eq. (2.6.1), combine with other terms of the same order to
give rise to gauge-invariant and finite form-factors. This is explained in greater
detail in Section 2.6.2.1, which describes general features of neutral-current neutrino
scattering in the SM and the recently introduced concept of effective electromagnetic form-factor of the neutrino [DEG 88]. In Section 2.6.2.2 we briefly discuss
the observability of the SM form-factor [MAR 80; SAR 83] $\Delta^{(\nu_\mu;l)}(q^2)$ and a strategy
to search for neutrino structure beyond the SM. In Section 2.6.3 we turn our
attention to neutrino electromagnetic dipole moments.

2.6.2 Neutral current ν scattering processes

2.6.2.1 General features in the SM

A general and convenient analysis of electroweak corrections to neutral current ν
scattering processes for $|q^2| \ll m_W^2$, in the framework of the SM, was given in
[MAR 80]. (References to other papers are given, for example, in [BEG 82].) The
analysis in that work was carried out in the simple renormalization scheme
developed in [SIR 80], in which the basic renormalized parameters are taken to
be e and the physical masses m_w, m_z, while the weak interaction angle is defined by
$\sin^2 \theta_W \equiv 1 - m_W^2/m_Z^2$. We will refer to the other particles involved in the scattering
(except for photons) as the target system. It was shown in [MAR 80] that the
electroweak corrections to ν-lepton scattering fall into two classes, according to
whether they are proportional to $\langle f|J_Z^\mu|i \rangle$ or $\langle f|J_\gamma^\mu|i \rangle$. Here J_Z^μ and J_γ^μ stand for the Z
and γ currents coupled to the target system, and i and f represent the initial and final
states of the latter. In the case of ν-hadron scattering the situation is analogous
except that there are two additional induced currents, not present at the tree level.
The latter arise from $W-W$ and $Z-Z$ box diagrams and have convergent cofactors.
A nice feature of this classification is that the corrections proportional to $\langle f|J_Z^\mu|i \rangle$
and $\langle f|J_\gamma^\mu|i \rangle$ can be readily combined with the zeroth-order amplitude $M^{(0)}$: the

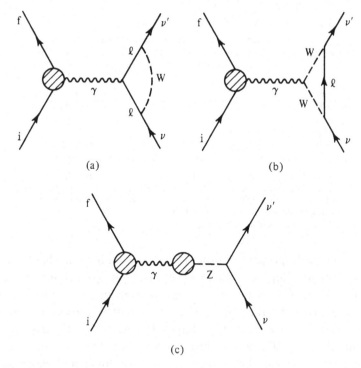

(a)   (b)

(c)

Fig. 2.6.1   Proper $\bar{\nu}\nu\gamma$ vertex and γZ mixing diagrams in ν scattering.

former amounts to an overall renormalization of $M^{(0)}$ while the latter simply renormalizes the cofactor of $\langle f|J^{\mu}_{\gamma}|i\rangle$ in $M^{(0)}$, namely $\sin^2\theta_W$. This leads to a radiatively corrected amplitude of the form

$$M = \frac{im_Z^2}{q^2 - m_Z^2} \frac{G_{\mu}}{\sqrt{2}} \rho_{NC}^{(\nu;t)} 4L^{\rho}$$
$$\langle f|\tfrac{1}{2}J^{(3)}_{\rho} - \kappa^{(\nu;t)}(q^2)J^{\gamma}_{\rho}|i\rangle + \cdots \qquad (2.6.2)$$

where $L^{\rho} \equiv \bar{u}_{\nu f}\gamma^{\rho}a_-uv_{\iota}$ is the neutrino factor, J^{γ}_{ρ} and $J^{(3)}_{\rho}$ stand for the electromagnetic current and the third component of the $SU(2)_L$ current in the target system, and the ellipses represent the induced currents mentioned previously. The effect of the electroweak corrections is contained in the factor $\rho_{NC}^{(\nu;t)}$ which, for $|q^2| \ll m_W^2$, is independent of q^2 and the form-factor $\kappa^{(\nu;t)}(q^2)$, which has a complicated dependence; they differ from unity by $0(\alpha)$ corrections. The superscript t specifies the nature of the target system: $t = l$ (lepton) or h (hadron).

Both $\rho_{NC}^{(\nu;t)}$ and $\kappa^{(\nu;t)}(q^2)$ are observable quantities and therefore finite and gauge-invariant. One readily discovers, however, that contributions from subsets of Feynman diagrams are in general divergent and dependent on the choice of the non-Abelian gauge [BAR 72; LUC 85; DEG 88]. This is true, for example, of the familiar

Table 2.6.1. $\Delta^{(\nu)}(0)$ for $\nu = \nu_\mu$ and $m_H = 100\,GeV$

m_t (GeV)	$10^2\Delta^{(\nu)}(0)$
45	0.33
60	0.39
90	−1.04
120	−2.16
150	−3.29
180	−4.53

Note: For $\nu = \nu_e$ and $\nu = \nu_\tau$, +1.79 and −0.95 should be added to all entries, respectively. To obtain $10^2\Delta^{\nu;\ell}(0)$ and $10^2\Delta^{(\nu;h)}(0)$, 0.42 and 0.13, respectively, should be added to all entries. The estimated error in the second column is ±0.20.
Source: Taken from [DEG 88].

diagrams involving the proper $\bar{\nu}\nu\gamma$ vertex (Fig. 2.6.1a,b). Indeed, it has been pointed out [DEG 88] that in order to obtain a finite and gauge-invariant answer in the calculation of $\kappa(q^2)$, one must include not only the contributions from Fig. 2.6.1a,b,c and appropriate counterterms, but also certain radiative corrections proportional to $\langle f|J_\mu^\gamma|i\rangle$ arising from Z^0 mediated amplitudes and $W\!-\!W$ box diagrams!

A recent analysis [DEG 88] has shown that the radiative correction $\Delta^{(\nu;t)}(q^2) \equiv 1 - \kappa^{(\nu;t)}(q^2)$ can be separated into two finite and gauge-invariant quantities according to

$$\Delta^{(\nu;t)}(q^2) = \Delta^{(\nu)}(q^2) + \Delta^{(t)} \qquad (2.6.3)$$

where $\Delta^{(l)} = 4.2 \times 10^{-3}$, $\Delta^{(h)} = 1.3 \times 10^{-3}$ are target-dependent contributions arising from the box diagrams. In contrast, the function $\Delta^{(\nu)}(q^2)$ involves a large subset of contributions including Fig. 2.6.1a,b,c and, significantly, it is independent of the nature of the target. Exploiting this fact, it has been argued [DEG 88] that the finite and gauge-invariant function

$$f(q^2) = -\frac{q^2}{2m_W^2}\Delta^{(\nu)}(q^2) \qquad (2.6.4)$$

can be interpreted as an effective electromagnetic form-factor of the neutrino in the framework of the low-energy theory derived from the SM at invariant mass scales $\ll m_W^2$. Our normalization convention in Eq. (2.6.4) corresponds to $\langle \nu'|j_\gamma^\mu|\nu\rangle = f(q^2)L^\mu$. The effective mean square charge radius derived from Eq. (2.6.4) is

$$\langle r^2 \rangle = -3\Delta^{(\nu)}(0)/m_W^2 = -17.8\Delta^{(\nu)}(0)\,(10^{-16}\,\mathrm{cm})^2$$

for $m_W = 81\,GeV$. Incorporating a recent update in the analysis of the hadronic contributions to κ this leads, for $m_t = 45\,GeV$ and $m_H = 100\,GeV$, to [DEG 88]

$$\langle r^2 \rangle \simeq \begin{cases} -(37.7 \pm 3.6) \times (10^{-17}\,\mathrm{cm})^2 & \text{for } \nu_e \\ -(5.9 \pm 3.6) \times (10^{-17}\,\mathrm{cm})^2 & \text{for } \nu_\mu \\ +(11.0 \pm 3.6) \times (10^{-17}\,\mathrm{cm})^2 & \text{for } \nu_r. \end{cases} \qquad (2.6.5)$$

The flavor dependence arises from Fig. 2.6.1a. The m_t dependence of $\Delta^{(\nu)}(0)$ originates in Fig. 2.6.1c and appropriate counterterms. Values of $\Delta^{(\nu)}(0)$ as a function of m_t are given in Table 2.6.1.

2.6.2.2 Detectability of $\Delta^{(\nu;t)}$ and search for new physics

Although, as we have seen, $\Delta^{(\nu)}(q^2)$ has some nice theoretical features, it is the overall correction $\Delta^{(\nu;t)}(q^2) \equiv 1 - \kappa^{(\nu;t)}(q^2)$ in the renormalization of $\sin^2\theta_W$ that is most directly related to experiment. For example, in $\nu_\mu - e$ scattering, the experimental physicist can eliminate $\rho_{NC}^{(\nu;l)}$ by considering the ratio $\sigma(\nu_\mu e)/\sigma(\bar\nu_\mu e)$ and determine the effective parameter

$$\sin^2\theta^{eff}(q^2) \equiv \kappa^{(\nu_\mu;l)}(q^2)\sin^2\theta_W = (1 - \Delta^{(\nu_\mu;l)}(q^2))\sin^2\theta_W.$$

If $m_H = 100\,\mathrm{GeV}$ and $m_t = 45\,\mathrm{GeV}$, we expect from Table 2.6.1 and the value of $\Delta^{(l)}$ that $\Delta^{(\nu_\mu;l)}(0) = 0.75 \times 10^{-2}$. We see that detection of the SM value $\Delta^{(\nu\nu_\mu;l)}(0)$ would require measurements of $\sin^2\theta_W$ to better than 1 percent both in $\nu_\mu - e$ scattering and in some other observable such as the vector boson masses, μ decay or $e^+e^- \to \mu^+\mu^-$. (At least two independent measurements are needed to extract both $\sin^2\theta_W$ and $\Delta^{(\nu_\mu;l)}(0)$). Needless to say, this is a very difficult task. On the other hand, for large values of m_t, $\Delta^{(\nu_\mu;l)}(0)$ increases significantly (see Table 2.6.1), and its detection becomes more feasible.

Let us now consider the possibility that, for some unknown reason associated with physics beyond the SM, the ν has an additional electromagnetic interaction described phenomenologically by a mean square charge radius $\langle r^2 \rangle n.ph.$ In this case the effective phenomenological parameter measured in $\nu_\mu e$ scattering becomes

$$\sin^2\theta^{eff}(q^2) = \kappa^{(\nu_\mu;\ell)}(q^2)\sin^2\theta_W[1 + \tfrac{1}{3}m_W^2\langle r^2\rangle n.ph.]. \qquad (2.6.6)$$

Thus, by measuring $\sin^2\theta^{eff}(q^2)$ in $\nu_\mu e$ scattering and $\sin^2\theta_W$ from processes other than neutral current ν scattering, and employing the SM calculation of $\kappa^{(\nu_\mu;l)}(q^2)$, the experimental physicist can attempt to determine $\langle r^2 \rangle\, n.ph.$ via Eq. (2.6.6) and, in this way, search for unknown ν structure! As an example, if $\sin^2\theta^{eff}(q^2)$ coincides with the SM model value with an error of ± 2 percent, then we would learn that $|\langle r^2\rangle n.ph.| \le (0.76 \times 10^{-16}\mathrm{cm})^2$ at 90 percent C.L. Recent experimental analyses [ABE 87; WIN 88] lead presently to $\langle r^2\rangle n.ph. < 6 \times 10^{-33}\,\mathrm{cm}^2$.

2.6.3 Electromagnetic moments

The existence of non-vanishing magnetic ($M(0)$) and/or electric ($E(0)$) dipole moments for Dirac neutrinos can lead to interesting consequences such as electromagnetic neutrino spin-flip scattering [BET 35; KYU 84; MAR 86], plasmon decay into $\nu\bar{\nu}$ [SUT 76; BEG 78; FUK 87a; NOT 88], and spin precession in magnetic fields [CIS 71; OKU 86]. Non-observation of such phenomena in terrestrial and astrophysical environments has been used to place bounds on such moments. For example, in the case of $\nu e \rightarrow \nu e$ scattering, the effect of a magnetic or electric dipole moment, which we parameterize generically by $\kappa e/2m_e$, is to increase the differential cross section by (neglecting terms of order m_e/E_ν) [KYU 84; MAR 86]

$$\Delta \frac{d\sigma(\nu e)}{dy} = |\kappa|^2 \frac{\pi\alpha^2}{m_e^2} \left(\frac{1}{y} - 1\right) \tag{2.6.7}$$

where $y = E_e'/E_\nu$. A larger than expected cross section could therefore be taken as evidence for a non-vanishing electromagnetic dipole moment, particularly if the excess exhibited the distinctive $1/y$ dependence in (2.6.7). Consistency of existing measurements of $(\nu)_e e$ and $(\nu)_\mu e$ scattering with the $SU(2)_L \times U(1)$ model's predictions leads to the bounds [REI 76; KYU 84; MAR 86; ABE 87; LIM 88]

$$\begin{aligned}
|\kappa_{\nu_e}| &< 4 \times 10^{-10} \\
|\kappa_{\nu_\mu}| &< 1 \times 10^{-9}.
\end{aligned} \tag{2.6.8}$$

Anticipated future measurements may lower these bounds by an order of magnitude. That turns out to be an interesting region to explore, since it has been pointed out that for $|\kappa_{\nu_e}| \gtrsim 10^{-11}$, spin precession $\nu_{e_L} \rightarrow \nu_{e_R}$ of solar neutrinos could be sizable. Indeed, Okun, Voloshin, and Vysotsky [OKU 86] have speculated that for a moment of that magnitude about $\frac{1}{2}$ of the expected ν_{e_L} solar flux could be converted into sterile ν_{e_R} as they propagate through the strong magnetic fields $10^3 \sim 10^4$G in the sun's convection zone. Such a depletion would explain the lower than expected [BAH 88] solar neutrino flux measured by Ray Davis [DAV 86] via the reaction $\nu_e + {}^{37}\text{Cl} \rightarrow e^- + {}^{37}\text{Ar}$ as well as apparent anticorrelations between solar ν_e flux variations and sunspot activity (which corresponds to magnetic field variations). To directly confirm or disprove that scenario will likely require many years of solar neutrino experiments; so, it is important to complement those measurements with laboratory constraints at the $\kappa \simeq 10^{-11}$ level.

Tighter indirect bounds can also be placed on neutrino electromagnetic dipole moments from astrophysical arguments. If neutrinos have such moments, then the decay plasmon $\rightarrow \nu\bar{\nu}$ can occur in stellar interiors. (A plasmon is an effectively massive photon that acquires its "mass," w_p the plasma frequency, and longitudinal degree of freedom from electron–hole excitations in the plasma.) The neutrinos produced in that process would easily escape from the star and carry away energy, thereby speeding up the stellar evolution. Since there is no evidence for such a speed-up, one can bound the electromagnetic moments if the neutrinos are light compared with the plasma frequency. Sutherland *et al.* [SUT 76] found from such an analysis

$$|\kappa_{\nu l}| \lesssim 8.5 \times 10^{-11} \quad (m_{\nu l} \gtrsim 10 \, \text{KeV}). \tag{2.6.9}$$

More recently, Fukugita and Yazaki [FUK 87a] and Raffelt and Dearborn [RAF 88] updated the red giant part of that analysis. Their results suggest an even better bound

$$|\kappa_{\nu l}| \lesssim 1 \sim 3 \times 10^{-11}. \tag{2.6.10}$$

An independent bound on neutrino electromagnetic moments can be obtained from the detection [BIO 87; HIR 87] of $\bar{\nu}_e$ neutrinos from Supernova 1987a. An electromagnetic dipole moment would allow neutrino electromagnetic spin-flip scattering off charged heavy nuclei in the very dense presupernova core. The right-handed neutrinos (left-handed antineutrinos) produced by that mechanism would be sterile with respect to ordinary weak interactions and thus more easily escape the dense supernova core. A significant neutrino flux loss would modify the collapse dynamics and supernova explosion. Observation of the $\bar{\nu}_e$ flux at the (approximately) anticipated level implies bounds in the range [LAT 88; BAR 88a; NOT 88]

$$|\kappa_{\nu_e}| < 10^{-12} \sim 10^{-13}. \tag{2.6.11}$$

That stringent constraint essentially eliminates spin-flip precession as a viable solution to the solar neutrino problem and sunspot anticorrelation hypothesis unless the solar magnetic fields are enormous $\gtrsim 10^5 \sim 10^6$ gauss. We note, however, that the bound in (2.6.11) does not apply to transition moments of Majorana neutrinos since in that case electromagnetic dipole scattering effectively changes the neutrino into an antineutrino of a different flavor (e.g., $\nu_e \to \bar{\nu}_\mu$ or $\bar{\nu}_\tau$) and vice versa. (We refer to $(\nu_L)^C$ as $\bar{\nu}$ in the case of Majorana neutrinos.) Both interact weakly with the medium and remain trapped in the dense core. So, spin-flavor precession [SCH 81] of solar neutrinos $\nu_e \to \bar{\nu}_\mu$ or $\bar{\nu}_\tau$ may still be relevant if neutrinos are Majorana and have transition electromagnetic moments $\approx 10^{-11} e/2m_e$. In fact, the changing solar density can lead to resonant neutrino spin-flavor conversion [AKH 88; LIM 88], a phenomenon that we subsequently describe.

What size neutrino magnetic, electric, or transition moment might one expect? If a Dirac four-component neutrino is given a small mass, m_ν, in the standard $SU(2)_L \times U(1)$ framework, then it acquires a magnetic dipole moment via weak loop corrections such that [MAR 77; LEE 77; PET 77; BEG 78]

$$\kappa_\nu = \frac{3}{4} \frac{G_\mu m_\nu m_e}{\sqrt{2}\pi^2} \simeq 3 \times 10^{-19} (m_\nu/1 \, \text{eV}). \tag{2.6.12}$$

That small a moment (for light neutrinos) would not give rise to observable physical consequences, except perhaps in a supernova where enormous magnetic fields $B \simeq 10^{12} \sim 10^{15}$ gauss may exist [FUJ 80]. To get a much larger magnetic moment would seem to require a new chiral changing interaction that can be most easily implemented by enlarging the Higgs sector [BAB 87; FUK 87b]. However, it is likely that such an interaction would also induce a relatively large neutrino mass [LIV 87]. So, it is generally felt that large neutrino magnetic moments ($\kappa_\nu \gg 10^{-19}$) are not naturally compatible with small neutrino masses ($m_\nu \lesssim 1\,\text{eV}$).

A neutrino electric dipole moment requires CP violation. Since a non-zero electric dipole moment has not been observed for any other particle, it would be surprising if neutrino electric dipole moments turned out to be anomalously large. To make that comment quantitative, let us assume that $|d_{\nu_e}| \leq |d_e|$ since their left-handed components are weak isodoublets under $SU(2)_L \times U(1)$. Present experimental bounds (PLA 70) on the electron's electric dipole moment, d_e, then imply

$$|d_{\nu_e}| < 5 \times 10^{-14}\, e/2m_e. \tag{2.6.13}$$

Of course, there can be ways of circumventing that constraint.

In the case of flavor transition moments, mixing is required and hence one might expect a suppression even beyond the value in (2.6.12). However, because transition moments are off-diagonal in flavor space, they could in principle be much larger than (2.6.12); that is, the mass argument does not directly apply. Examples of such models have been given in the literature [BAB 87; FUK 87b]. They are particularly plausible in the framework of Grand Unification. In the remainder of this section we keep an open mind about transition moments and examine their implications. For definiteness, we consider the Majorana case with a transition moment, $\kappa_{e\mu}e/2m_e$, connecting ν_e and $\bar{\nu}_\mu$ and assume mass eigenstates $m_{\nu_e} < m_{\nu_\mu}$. Generalization to three neutrinos with arbitrary mixing is straightforward but cumbersome.

A Majorana transition moment $\kappa_{e\mu}e/2m_e$ can lead to $\nu_\mu e \to \bar{\nu}_e e$ and $\bar{\nu}_e e \to \nu_\mu e$ scattering. The ν_e cross section would be increased by the formula in (2.6.7) with $\kappa_{e\mu}$ replacing κ. From the bound in (2.6.8) we therefore find

$$|\kappa_{e\mu}| < 4 \times 10^{-10}. \tag{2.6.14}$$

Such an interaction could also give rise to plasmon $\to \nu_e \bar{\nu}_\mu$ or $\nu_\mu \bar{\nu}_e$. The bounds in (2.6.9) and (2.6.10) imply

$$\begin{aligned}
|\kappa_{e\mu}| &\lesssim 8.5 \times 10^{-11} \\
|\kappa_{e\mu}| &\lesssim 1 \sim 3 \times 10^{-11}.
\end{aligned} \tag{2.6.15}$$

The latter is particularly stringent. It implies $|\kappa_{e\mu}|$ must be $\simeq 10^{-11}$, that is, near the red giant bound if it enters into the solar neutrino puzzle.

A neutrino transition moment would also lead to neutrino decay $\nu_\mu \to \bar{\nu}_e + \gamma$ at a rate [BEG 78]

$$\Gamma(\nu_\mu \to \bar{\nu}_e \gamma) = \frac{\alpha}{4m_e^2} |\kappa_{e\mu}|^2 \left(\frac{m_{\nu_\mu}^2 - m_{\nu_e}^2}{m_{\nu_\mu}}\right)^3. \qquad (2.6.16)$$

For $|\kappa_{e\mu}| \lesssim 10^{-11}$, that formula implies a radiative lifetime

$$\tau(\nu_\mu \to \bar{\nu}_e \gamma) > 4.5 \times 10^{17} \times \left(\frac{8\text{eV}}{m_{\nu_\mu}}\right)^3 \text{s} \qquad (2.6.17)$$

which for $m_{\nu_\mu} \lesssim 8\,\text{eV}$ is longer than the lifetime of the universe.

The final effect of a transition moment that we will consider is the phenomenon of spin-flavor precession ($\nu_e \to \bar{\nu}_\mu$) in large magnetic fields. In particular, the possibility of resonant spin-flavor precession in a medium of varying density [AKH 88; LIM 88]. The basic point is that a ν_e propagating in matter with a transverse magnetic field B will precess into $\bar{\nu}_\mu$ via its transition moment interaction. Neglecting mixing, the precession probability is given by (for a neutral medium) [AKH 88; LIM 88]

$$P(t)_{\nu_e \to \bar{\nu}_\mu} = \frac{(2\mu B)^2}{\Delta^2 + (2\mu B)^2} \sin^2\left(\sqrt{\Delta^2 + 4\mu^2 B^2}\, t/2\right)$$

$$\mu = \kappa_{e\mu} e/2m_e$$

$$\Delta = \sqrt{2} G_\mu (N_e - N_n) - \frac{m_{\nu_\mu}^2 - m_{\nu_e}^2}{2E_\nu} \qquad (2.6.18)$$

where N_e and N_n are electron and neutron number densities of the medium. Note that for $\Delta \neq 0$, precession is quenched. That can occur because of the neutrino mass difference or the different interactions of ν_e and $\bar{\nu}_\mu$ with the medium. For the special case $\Delta = 0$, the resonance condition [WOL 78; MIK 86], spin-flavor precession will proceed unimpeded. That condition is satisfied when

$$N_e - N_n = \frac{m_{\nu_\mu}^2 - m_{\nu_e}^2}{2\sqrt{2} E_\nu G_\mu}. \qquad (2.6.19)$$

In the case of a varying density profile such as the sun or a supernova, a neutrino can start out in a dense region where $\Delta > 0$, propagate into a resonance region $\Delta = 0$ where $\nu_e \to \bar{\nu}_\mu$ precession can occur, and then reach a $\Delta < 0$ domain where precession is again quenched. For the right conditions, nearly complete $\nu_e \to \bar{\nu}_\mu$ conversion can result. That mechanism has been suggested [AKH 88; LIM 88] as a solution to the solar neutrino puzzle and/or flux anticorrelation with sunspot activity.

For resonant spin-flavor precession to proceed, non-zero transition moments, large magnetic fields, and dense matter are required. In the sun, where $\langle B \rangle \simeq 10^3 \sim 10^4 \text{G}$ is expected, the transition moment $\kappa_{e\mu}$ must be relatively large

$\gtrsim 10^{-11}$ for appreciable spin-flavor precession of neutrinos with $E_\nu \simeq 10 \, \text{MeV}$ to occur, that is, near its present bound. Nevertheless, it will be interesting to see whether the hint of a time variation in the neutrino flux correlated with magnetic field fluctuations in the sun is confirmed. A more likely candidate for spin-flavor neutrino precession is a supernova. In that case, very large $B \simeq 10^{12} \sim 10^{15} \text{G}$ can occur. Hence, one is sensitive to $\kappa_{e\mu} \sim 10^{-19}$–$10^{-23}$, a realistic range. The signature of such a phenomenon will be the interchange of parts of the ν_e and $\bar{\nu}_\mu$ (or $\bar{\nu}_\tau$) supernova spectra. In addition, for $N_n > N_e$, one may have $\bar{\nu}_e \leftrightarrow \nu_\mu$ or ν_τ near the core. Observation of such an effect would be extraordinary. Hopefully, future supernova neutrino detectors will be sensitive to such a phenomenon.

In summary, even though the neutrino is electrically neutral, its electromagnetic properties are still very interesting in that they serve as a test of the Standard Model at the quantum loop level and as a valuable means of searching for new physics.

2.7 Astrophysical and cosmological constraints to neutrino properties*

Since the 1970s with the establishment of the big bang model, it has become clear that some of the most restrictive constraints on certain neutrino properties come from astrophysical and cosmological considerations. Furthermore, in 1987 the detection of neutrinos from the supernova in the Large Magellanic Cloud provided a new "neutrino laboratory" as well as confirming our basic understanding of gravitational collapse energetics. We review those constraints on neutrinos derived from cosmological and astrophysical considerations.

We first examine the freeze out of neutrinos in the early universe and derive the cosmological limits on masses for stable neutrinos. We then use the freeze out arguments coupled with observational limits to constrain decaying neutrinos as well. We also review the limits to neutrino properties that follow from SN 1987A. We then look at the constraint from big bang nucleosynthesis on the number of neutrino flavors. Before ending, we briefly look at astrophysical constraints on neutrino mixing as well as future astronomical observations of relevance to neutrino physics.

2.7.1 Cosmological mass and decay limits

Cosmological limits to neutrino mass and decay properties depend on their relic number density from the early universe. If a massive particle species remained in thermal equilibrium until the present, its abundance, $n/s \sim (m/T)^{\frac{3}{2}} \exp(-m/T)$, would be absolutely negligible because of the exponential factor ($s = $ entropy

* Edward W. Kolb, David N. Schramm, and Michael S. Turner, The University of Chicago and NASA/Fermilab Astrophysics Center.
This work was supported in part by NSF, NASA, and DOE at The University of Chicago and by NASA (NAGW-1340) at Fermilab.

density). If the interactions of the species freeze out (i.e., $\Gamma < H$ where Γ is the interaction rate and H is the cosmological expansion rate) at a temperature such that m/T is not much greater than 1, the species can have a significant relic abundance today. We will now calculate that relic abundance.

First, suppose that the species is stable (or very long-lived compared to the age of the universe when its interactions freeze out). Later we will consider the case where the species is unstable. Given that it is stable, only annihilation and inverse annihilation processes, for example,

$$\nu\bar{\nu} \ \leftrightarrow \ X\bar{X} \tag{2.7.1}$$

can change the number of ν's and $\bar{\nu}$'s in a comoving volume. Here X generically denotes all the species into which ν's can annihilate. In addition, we assume that there is no asymmetry between ν's and $\bar{\nu}$'s.

We also assume that all the species X, $\bar{X}$ into which ν, $\bar{\nu}$ annihilate have thermal distributions with zero chemical potential. Because these particles will usually have additional interactions that are "stronger" than their interactions with ν's, the assumption of equilibrium for the X's is almost always a good one. For example, let X, $\bar{X} = e^-$, e^+; while the neutrinos only have weak interactions, the $e^{\pm}$'s have weak and electromagnetic interactions.

The evolution of the number density n_ν can be expressed [KOL 90] in terms of the *total* annihilation cross section $\langle\sigma_A|\nu|\rangle$

$$\frac{dn_\nu}{dt} + 3Hn_\nu = -\langle\sigma_A|\nu|\rangle[n_\nu^2 - (n_\nu^{EQ})^2]. \tag{2.7.2}$$

This equation for the evolution of the abundance of a species is a particular form of the Riccati equation, for which there are no general, closed-form solutions. Before we solve the equation by approximate methods, let's consider the qualitative behavior of the solution. The annihilation rate Γ_A varies as n_{EQ} times the thermally averaged annihilation cross section $\langle\sigma_A|\nu|\rangle$. In the relativistic regime, $(m_\nu/T \ll 3)$ $n_{EQ} \sim T^3$, and like other rates, Γ_A will vary as some power of T. In the nonrelativistic regime, $(m_\nu/T \gg 3)$, $n_{EQ} \sim (mT)^{\frac{3}{2}}\exp(-m/T)$, so that Γ_A decreases exponentially. In either regime, Γ_A decreases as T decreases, and so eventually annihilations become impotent, roughly when $\Gamma_A \simeq H$, which we call freeze out.

Hot relics   First consider the case of a particle species such that $m/T \lesssim 3$ at freeze out. In this case, freeze out occurs when the species is still relativistic and the equilibrium number density per comoving volume $Y_{EQ} \equiv n_{EQ}/s$ is not changing with time. Since Y_{EQ} is constant, the final value of Y ($Y = n/s$) is very insensitive to the details of freeze out, and the asymptotic value of Y, $Y(m/T \rightarrow \infty) \equiv Y_\infty$, is just the equilibrium value at freeze out:

$$Y_\infty = Y_{EQ} = 0.278g_{eff}/g_{*_s} \tag{2.7.3}$$

where $g_{eff}=g$ (bosons), $0.75 g$ (fermions), and g counts the internal degrees of freedom. Thus the species freezes out with order unity abundance relative to entropy s (or the number density of photons). Assuming the expansion remains isentropic thereafter (constant entropy per comoving volume), the abundance of ν's today is (s_0 is the present entropy density)

$$n\nu = s_0 Y_\infty = 2970 Y_\infty \text{ cm}^{-3} \qquad (2.7.4)$$

$$= 825(g_{eff}/g_{*_s}) \text{ cm}^{-3}. \qquad (2.7.5)$$

If, after freeze out, the entropy per comoving volume of the universe should increase, say by a factor of γ, the present abundance of ν's in a comoving volume would be diminished by γ.

A species that decouples when it is relativistic is often called a *hot relic*. The present relic mass density contributed by a hot relic is simple to compute:

$$\rho_\nu = s_0 Y_\infty m = 2.97 \times 10^3 Y_\infty (m/\text{eV}) \text{ eV cm}^{-3} \qquad (2.7.6)$$

$$\Omega_\nu h^2 = 7.83 \times 10^{-2} [g_{eff}/g_{*_s}](m/\text{eV}). \qquad (2.7.7)$$

On the basis of the present age of the universe, we know that $\Omega_0 h^2 \lesssim 1$; applying this bound to the contribution of the species to $\Omega_0 h^2$ we obtain a cosmological bound to the mass of the species:

$$m \lesssim 12.8 \text{ eV}[g_{*_s}(x_f)/g_{eff}]. \qquad (2.7.8)$$

Light (mass $\lesssim$ MeV) neutrinos decouple when $T \sim$ few MeV, and $g_{*_s} = g_* = 10.75$. For a single, two-component neutrino species $g_{eff} = 2 \times (3/4) = 1.5$, so that $g_{eff}/g_{*_s} = 0.140$. This implies that

$$\Omega_\nu h^2 = \frac{m_\nu}{91.5 \text{ eV}} \qquad (2.7.9)$$

$$m_\nu \lesssim 91.5 h^2 \text{ eV} \lesssim 91.5 \text{ eV}. \qquad (2.7.10)$$

This cosmological bound to the mass of a stable, light neutrino is often referred to as the Cowsik–McClelland bound [COW 72]. In their original paper, Cowsik and McClelland considered a four-component neutrino ($g = 4$), and took $\Omega < 3.8$, $h = \frac{1}{2}$ and $T_\nu = T$, which resulted in the bound $m \lesssim 8 \text{ eV}$.

If there are more than one light ($\lesssim$ MeV) species, this bound applies to the sum of the masses of the light neutrinos.

Cold relics   Now consider the more difficult case where freeze out occurs when the species is nonrelativistic ($m/T \gtrsim 3$), and Y_{EQ} is decreasing exponentially with m/T. In this case, the precise details of freeze out are important.

First we will parameterize the temperature dependence of the annihilation cross section. On general theoretical grounds we expect the annihilation cross section to have the velocity dependence $\sigma_A|v| \propto v^p$, where $p=0$ corresponds to s-wave annihilation, $p=2$ to p-wave annihilation, and so on. This implies that $\langle \sigma_A|v| \rangle \propto T^n$, $n=0$ for s-wave annihilation, $n=1$ for p-wave annihilation, etc. Therefore we parameterize $\langle \sigma_A|v| \rangle$ as

$$\langle \sigma_A|v| \rangle \equiv \sigma_0 (T/m)^n. \tag{2.7.11}$$

With this parameterization, the Boltzmann equation for the abundance of v's becomes,

$$dY/dx = -\lambda x^{-n-2}(Y^2 - Y_{EQ}^2) \tag{2.7.12}$$

where

$$x = m/T \tag{2.7.13}$$

$$\lambda = 0.264(g_{*_s}/g_*^{\frac{1}{2}})m_{PL}m\sigma_0 \tag{2.7.14}$$

$$Y_{EQ} = 0.145(g/g_{*_s})x^{3/2}e^{-x}. \tag{2.7.15}$$

As shown in [KOL 90], Eq. (2.7.12) can be solved approximately to good accuracy, where it is found that

$$Y_\infty = \frac{3.79(n+1)x_f^{n+1}}{(g_{*_s}/g_*^{\frac{1}{2}})m_{PL}m\sigma_0} \tag{2.7.16}$$

where

$$x_f \simeq \ln[0.038(n+1)(g/g_*^{\frac{1}{2}})m_{PL}m\sigma_0]$$
$$- (n+1/2)\ln\ln[0.038(n+1)(g/g_*^{\frac{1}{2}})m_{PL}m\sigma_0]. \tag{2.7.17}$$

As with a hot relic, the present number density and mass density of a cold relic is easy to compute,

$$n_{v0} = s_0 Y_\infty = 2970 Y_\infty cm^{-3}$$

$$= 1.13 \times 10^4 \frac{(n+1)x_f^{n+1}}{(g_{*_s}/g_*^{\frac{1}{2}})m_{PL}m\sigma_0} \, cm^{-3} \tag{2.7.18}$$

$$\Omega_v h^2 = 1.07 \times 10^9 \frac{(n+1)x_f^{n+1}GeV^{-1}}{(g_{*_s}/g_*^{\frac{1}{2}})m_{PL}m\sigma_0} \tag{2.7.19}$$

(where the subscript f denotes the freeze out value). It is very interesting to note that the relic density is inversely proportional to the annihilation cross section and mass of the particle

$$Y_\infty = \frac{3.79(n+1)(g_*^{\frac{1}{2}}/g_{*_s})x_f}{mm_{PL}\langle \sigma_A|v| \rangle}. \tag{2.7.20}$$

The smaller its annihilation cross section, the greater its relic abundance – the weak prevail. Moreover, the present mass density only depends upon the annihilation cross section at freeze out, which for $n = 0$ (s-wave annihilation) is independent of temperature (and energy).

Let us now look at the specific application of this to massive neutrinos ($m \gg \text{MeV}$). Annihilation for such a species proceeds through Z^0 exchange to final states $i\bar{i}$; where $i = \nu_L, e, \mu, \tau, u, d, s, \ldots$ (ν_L denotes any lighter neutrino species). The annihilation cross section depends upon whether the heavy neutrino is a Dirac or Majorana species; for $T \lesssim m \lesssim M_z$, the annihilation cross section is

$$\langle \sigma_A | v \rangle_{\text{Dirac}} = \frac{G_F^2 m^2}{2\pi} \sum_i (1 - z_i^2)^{\frac{1}{2}} [(C_{V_i}^2 + C_{A_i}^2)$$

$$+ \frac{1}{2} z_i^2 (C_{V_i}^2 + C_{A_i}^2)]$$

$$\langle \sigma_A | v | \rangle_{\text{Majorana}} = \frac{G_F^2 m^2}{2\pi} \sum_i (1 - z_i^2)^{\frac{1}{2}} [(C_{V_i}^2 + C_{A_i}^2) 8\beta^2 / 3$$

$$+ C_{A_i}^2 2 z_i^2] \tag{2.7.21}$$

where $z_i = m_i/m$, β is the relative velocity, and C_V and C_A are given in terms of the weak isospin j_3, the electric charge q, and the Weinberg angle θ_W by $C_A = j_3$, $C_V = j_3 - 2q\sin^2\Theta_W$. (We have assumed that the neutrino is less massive than M_Z.)

In the Dirac case, annihilations proceed through the s-wave and $\langle \sigma_A | v | \rangle$ is velocity independent:

$$\sigma_0 \simeq c_2 G_F^2 m^2 / 2\pi \tag{2.7.22}$$

where $c_2 \sim 5$. Taking $g = 2$ and $g_* \simeq 60$, from our formulae we find

$$x_f \simeq 15 + 3\ln(m/\text{GeV}) + \ln(c_2/5)$$

$$Y_\infty \simeq 6 \times 10^{-9} \left(\frac{m}{\text{GeV}}\right)^{-3} \left[1 + \frac{3\ln(m/\text{GeV})}{15} + \frac{\ln(c_2/5)}{15}\right] \tag{2.7.23}$$

from which we compute that

$$\Omega_{\nu\bar{\nu}} h^2 = 3(m/\text{GeV})^{-2} \left[1 + \frac{3\ln(m/\text{GeV})}{15}\right] \tag{2.7.24}$$

where we have included the identical relic abundance of the antineutrino species ($\Omega_{\nu\bar{\nu}} = 2\Omega_\nu$). Note that freeze out takes place at $T_f \simeq m/15 \simeq 70\,\text{MeV}(m/\text{GeV})$ – before the interactions of light neutrinos freeze out. This is because as neutrinos annihilate and become rare, the annihilation process quenches. Requiring $\Omega_\nu h^2 \lesssim 1$ we obtain the so-called Lee–Weinberg bound:

$$m \gtrsim 2\,\text{GeV}. \tag{2.7.25}$$

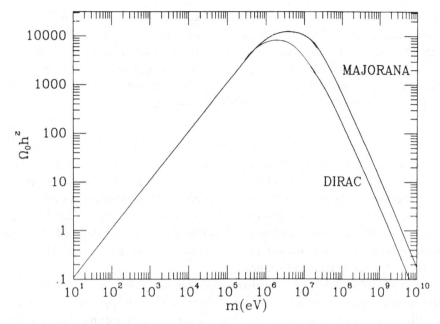

Fig. 2.7.1 The contribution to $\Omega_0 h^2$ for a stable neutrino species of mass m (from [KOL 90]).

Although it is often called the Lee–Weinberg bound, the basic argument [LEE 77] was noted a decade earlier by Zeldovich, Novikov, and Chiu.

For the Majorana case, annihilation proceeds through both the s- and p-waves; however the formulae that obtain for x_f, Y_∞, and $\Omega_\nu h^2$ are similar. In Fig. 2.7.1 we show the contribution to $\Omega_0 h^2$ for a stable, massive neutrino species. For $m \lesssim \mathrm{MeV}$, $\Omega_\nu h^2 \propto m$ as the relic abundance is constant. For $m \gtrsim \mathrm{MeV}$, $\Omega_\nu h^2 \propto m^{-2}$ as the relic abundance decreases as m^{-3}. The relic mass density achieves its maximum for $m \sim \mathrm{MeV}$.

Neutrino masses less than about $92 h^2 \,\mathrm{eV}$ or more than about $2\,\mathrm{GeV}$ (Dirac) or about $5\,\mathrm{GeV}$ (Majorana) are cosmologically acceptable.

These limits are quite impressive when compared with the laboratory limits, ν_μ at $250\,\mathrm{keV}$ and ν_τ at $35\,\mathrm{MeV}$, and imply that both must be below $92\,\mathrm{eV}$ if they are stable. Furthermore, recent searches for the products of neutrino annihilations in the sun and earth by Kamiokande and Irvine–Michigan–Brookhaven (IMB) probably constrain any stable massive neutrino to be $\lesssim 12\,\mathrm{GeV}$ or the high-energy neutrinos produced by annihilations would have been observed [SIL 85].

Before leaving stable massive neutrinos, it is worth noting that they can still be the dominant mass in the universe. Relic neutrinos of a few GeV mass provide closure density and behave as cold dark matter. Moreover, this possibility may soon be tested by more sensitive searches for their annihilation products, and/or use of cryogenic detectors. Relic neutrinos of mass $\sim 30\,\mathrm{eV}$ provide closure density and behave as hot dark matter. While laboratory experiments will eventually probe a ν_e mass as small as 10 eV, we will probably have to wait for the next nearby Supernova to prove ν_μ and ν_τ masses in the 30 eV range. While hot dark matter and adiabatic density perturbations (such as those produced by inflation) seem to be incompatible with observations, hot dark matter with cosmic strings (as the seed perturbations) is a very viable and interesting structure formation scenario.

2.7.2 Unstable ν's

Now consider the possibility of an unstable neutrino species whose decay products are relativistic, even at the present epoch. It is clear that the mass density bound for such a species must be less stringent: from the epoch at which they decay (say, $z = z_D$) until the present, the mass density of the relativistic neutrino decay products decreases as R^{-4}, as opposed to the R^{-3} had the neutrinos not decayed. Roughly speaking, then, the mass density today of the decay products is a factor of $(1 + z_D)^{-1}$ less than of a stable neutrino species.

The precise abundance of the neutrino decay products is very easy to compute. Denote the energy density of the relativistic decay products by ρ_D, and for simplicity we will assume that they do not thermalize. The equations governing the evolution of the daughter products are

$$\dot{\rho}_D + 4H_{\rho_D} = \rho_\nu/\tau,$$

$$\rho_\nu(R) = \rho_\nu(R_i) \left(\frac{R}{R_i}\right)^{-3} \exp(-t/\tau). \tag{2.7.26}$$

where R_i, t_i is some convenient epoch prior to decay, $t_i \ll \tau$. The relic density of the decay products is obtained by integrating (2.7.26):

$$\rho_D(t) = \rho_{\nu i}\tau^{-1} \left(\frac{R_i}{R}\right)^4 \int_{t_i}^t \frac{R(t')}{R_i} \exp(-t'/\tau)dt'. \tag{2.7.27}$$

Assuming that around the time the neutrinos decay $(t \sim \tau)$ the scale factor $R \propto t^n$ $(n = \frac{1}{2}$ radiation dominated; $n = \frac{2}{3}$ matter dominated) we can evaluate this integral directly, and find that the present density of relic, relativistic particles from neutrino decays is

$$\rho_D(t_0) = n!\rho_\nu(t_0)\frac{R(\tau)}{R_0} \tag{2.7.28}$$

where $\rho_\nu(t_0)$ is the present density that neutrinos and antineutrinos would have had they not decayed, and $R(\tau)$ is the value of the scale factor at the time $t = \tau$. As expected, the present energy density of the decay products is less than that of a stable neutrino species, by a factor of $n!R(\tau)/R_0 \sim (1 + z_D)^{-1}$.

To obtain the mass–lifetime constraint based upon the present mass density of the neutrino's relativistic decay products we assume that it saturates the bound:

$$\Omega_R h^2 \simeq \Omega_\nu h^2/(1 + z_D) \lesssim \Omega_0 h^2 \qquad (2.7.29)$$

which means that the universe has been radiation dominated since the decay epoch ($z = z_D$). This implies that the age of the universe at the decay epoch is related to z_D by:

$$\tau \simeq t(z = z_D) = 0.5 H_0^{-1} \Omega_0^{-1/2} (1 + z)^{-2}$$

$$\simeq 1.54 \times 10^{17} \sec (\Omega_0 h^2)^{-1/2} (1 + z_D)^{-2}. \qquad (2.7.30)$$

Using the results of our earlier calculations for $\Omega_\nu h^2$, we obtain the following constraint to the epoch of decay (for neutrino masses which fall in the previously disallowed range)

$$m \lesssim 3.6 \times 10^{10} \, \mathrm{eV} (\Omega_0 h^2)^{\frac{3}{4}} \tau_{sec}^{-\frac{1}{2}} \qquad \text{(light)}$$

$$m \gtrsim 8.7 \times 10^{-5} \, \mathrm{GeV} (\Omega_0 h^2)^{-\frac{3}{8}} \tau_{sec}^{\frac{1}{4}} \qquad \text{(heavy Dirac)}$$

$$m \gtrsim 2.0 \times 10^{-4} \, \mathrm{GeV} (\Omega_0 h^2)^{-\frac{3}{8}} \tau_{sec}^{\frac{1}{4}} \qquad \text{(heavy Majorana).} \qquad (2.7.31)$$

The excluded region of the neutrino mass–lifetime plane is shown in Fig. 2.7.2. (Consideration of the formation of structure in the universe leads to a significantly more stringent constraint to the mass density of the relativistic decay products; structure cannot grow in a radiation-dominated universe. For a discussion of these constraints see [FRE 83].)

The limits just discussed [VYS 77] apply irrespective of the nature of the decay products (so long as they are relativistic). If the decay products include "visible" particles such as photons, $e^\pm$ pairs, pions, and the like, much more stringent limits can be obtained [SAT 77]. We will now consider the additional constraints that apply when the decay products include a photon. (For the most part, these same limits also apply if the decay products include $e^\pm$ pairs.) The limits that obtain depend both qualitatively and quantitatively upon the decay epoch, and we will consider five distinct epochs.

Before discussing these limits, it is useful to calculate the time at which the energy density of the massive neutrino species would dominate the energy density in photons. The energy density in photons is $\rho_\gamma = (\pi^2/15) T^4$, and assuming the neutrinos are NR, their energy density is $\rho_\nu = Y_\infty ms$. Taking $g_{*S} \simeq 4$, the energy densities are equal when $T \simeq 3 Y_\infty m$. For heavy neutrinos, Y_∞ is given by Eq. (2.7.20), and for light neutrinos, $Y_\infty \simeq 0.04$. Thus we find that the relic neutrino energy density will exceed the photon energy density at $T/m \lesssim 0.1$ for light neutrinos, and $T/m \lesssim 2 \times 10^{-8} m_{\mathrm{GeV}}^{-3}$ for heavy neutrinos. Using $t \simeq 1 \sec / T_{\mathrm{MeV}}^2$ for

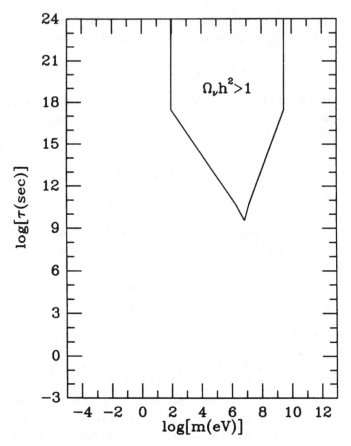

Fig. 2.7.2　The forbidden region of the neutrino mass–lifetime plane based upon the requirement that $\Omega_2 \leq 1$ (from [KOL 90]).

the age of the universe, the epoch of matter domination (by massive neutrinos) is given by

$$t(sec) \simeq \begin{cases} 10^{14}(m/1\text{eV})^{-2} & \text{light neutrinos} \\ 3 \times 10^{9} m_{\text{GeV}}^{4} & \text{heavy neutrinos.} \end{cases} \qquad (2.7.32)$$

(Here, and throughout the following discussion, "light" will refer to neutrinos of mass less than an MeV, and "heavy" will refer to neutrinos of mass greater that an MeV, but less than M_Z.).

$t_U \simeq 3 \times 10^{17} \text{ sec} \leq \tau$　If the neutrino lifetime is greater than the age of the universe, neutrinos will be decaying at the present and decay-produced photons will contribute to the diffuse photon background. Assuming that the neutrinos are unclustered (the most conservative assumption), the differential number flux of

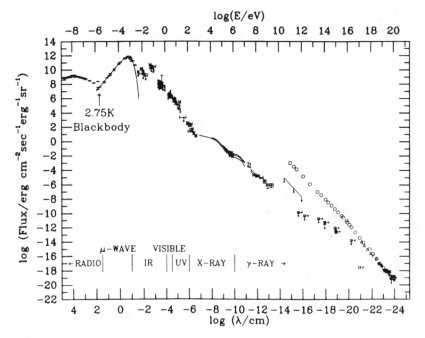

Fig. 2.7.3   The diffuse photon background. Vertical arrows indicate upper limits, and horizontal arrows indicate integrated fluxes ($> E$.) Circles and triangles indicate the total cosmic-ray flux (p's, nuclei, and photons), which provides an absolute upper limit to the photon flux at the highest energies (from [KOL 90]).

decay-produced photons (per cm^2 sr s erg) is

$$\frac{d\mathscr{F}_\gamma}{dEd\Omega} = \frac{n_\nu c}{4\pi\tau H_0} E^{-1} \left(\frac{E}{m/2}\right)^{\frac{3}{2}}, (E \leq m/2), \qquad (2.7.33)$$

where for simplicity we have assumed that each decay produces one photon of energy $m/2$ and that $\Omega_0 = 1$. Taking the number flux to be $d\mathscr{F}_\gamma/d\Omega \simeq Ed\mathscr{F}_\gamma/dE\,d\Omega$ and $H_0 = 50\,\mathrm{km\,s^{-1}\,Mpc^{-1}}$, we find

$$\frac{d\mathscr{F}_\gamma}{d\Omega} \simeq 10^{29}\tau_{\mathrm{sec}}^{-1}\,\mathrm{cm^{-2}sr^{-1}s^{-1}} \qquad \text{light neutrinos}$$

$$\simeq 3 \times 10^{22}\tau_{\mathrm{sec}}^{-1}m_{\mathrm{GeV}}^{-3}\,\mathrm{cm^{-2}sr^{-1}s^{-1}} \qquad \text{heavy neutrinos.} \qquad (2.7.34)$$

A summary of the observations of the diffuse photon background are shown in Fig. 2.7.3. The differential energy flux, $d\mathscr{F}/dE\,d\Omega$, is shown as a function of energy and wavelength. From this data, a very rough limit of

$$\frac{d\mathscr{F}_\gamma}{d\Omega} \lesssim \left(\frac{\mathrm{MeV}}{E}\right)\mathrm{cm^{-2}sr^{-1}s^{-1}} \qquad (2.7.35)$$

can be placed to the contribution of neutrino decay-produced photons to the photon background. Based upon this, the following lifetime limit results:

$$\tau_{sec} \geq \begin{cases} 10^{23} m_{\rm eV} & \text{light neutrinos} \\ 10^{25} m_{\rm GeV}^{-2} & \text{heavy neutrinos} \end{cases} \qquad (2.7.36)$$

applicable for neutrino lifetimes $\tau \gtrsim 3 \times 10^{17}$ sec. The forbidden region of the mass–lifetime plane is shown in Fig. 2.7.4.

$t_{rec} \simeq 6 \times 10^{12} (\Omega_0 h^2)^{-\frac{1}{2}} \textbf{sec} \leq \tau \leq t_U$ If neutrinos decay after recombination, but before the present epoch, then the decay-produced photons will not interact and should appear today in the diffuse photon background. Again, for simplicity, assume that each neutrino decay produces one photon of energy $m/2$. Then the present flux of such photons is

$$\frac{d\mathscr{F}_\gamma}{d\Omega} = \frac{n_\nu c}{4\pi}$$

$$\simeq 3 \times 10^{11} \text{ cm}^{-2}\text{sr}^{-1}\text{s}^{-1} \qquad \text{light neutrinos}$$

$$\simeq 4 \times 10^4 m_{\rm GeV}^{-3} \text{ cm}^{-2}\text{sr}^{-1}\text{s}^{-1} \quad \text{heavy neutrinos} \qquad (2.7.37)$$

where we have assumed that when the neutrino species decays, it is non-relativistic, so that each decay-produced photon today has energy $E \simeq m/2(1 + z_D)$, where $(1 + z_D) \simeq 3.5 \times 10^{11} (\Omega_0 h^2)^{-\frac{1}{3}} \tau_{\rm sec}^{-\frac{2}{3}}$. Comparing these flux estimates to our rough estimate of the diffuse background flux we obtain the constraints,

$$m \lesssim 2 \times 10^6 (\Omega_0 h^2)^{-\frac{1}{3}} \tau_{\rm sec}^{-\frac{2}{3}} \text{ eV} \qquad \text{light neutrinos}$$

$$m \gtrsim 8 \times 10^{-3} (\Omega_0 h^2)^{\frac{1}{6}} \tau_{\rm sec}^{\frac{1}{3}} \text{ GeV} \quad \text{heavy neutrinos} \qquad (2.7.38)$$

applicable for neutrino lifetimes in the range $3.5 \times 10^{11} (\Omega_0 h^2)^{-\frac{1}{3}}$ sec $\lesssim \tau \lesssim 3 \times 10^{17}$ sec. For very light neutrino species, the assumption that the species decays when it is nonrelativistic breaks down. If the species decays after $t = t_{\rm therm}$ and before the present epoch, and is relativistic when it decays, the decay-produced photons will be comparable in energy and in number to the CMBR photons and will cause significant distortions to the CMBR. Thus a neutrino species that decays while relativistic in the time interval $10^6 \lesssim t \lesssim 3 \times 10^{17}$ sec is forbidden. The excluded region is $200 \lesssim t_{\rm sec}/m_{\rm eV} \lesssim 4 \times 10^{20} (\Omega_0 h^2)^{\frac{1}{3}}$, for

$$m_{\rm eV} \lesssim \begin{cases} 3.5 \times 10^8 (\Omega_0 h^2)^{-\frac{1}{3}} t_{\rm sec}^{-\frac{2}{3}} & t_{\rm sec} \gtrsim 4.4 \times 10^{10} (\Omega_0 h^2)^{-2} \\ 4.6 \times 10^6 t_{\rm sec}^{-\frac{1}{2}} & t_{\rm sec} \lesssim 4.4 \times 10^{10} (\Omega_0 h^2)^{-2}. \end{cases} \qquad (2.7.39)$$

The forbidden region of the mass–lifetime plane is shown in Fig. 2.7.4.

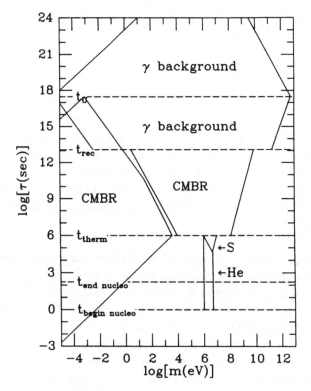

Fig. 2.7.4   Cosmological limits to the mass and lifetime of an unstable neutrino species that decays radiatively (from [KOL 90]).

$t_{therm} \simeq 10^6\,\text{sec} \leq \tau \lesssim t_{rec}$ For neutrino decays that occur during this epoch, the decay-produced photons can scatter with electrons, which can in turn scatter with cosmic microwave background radiation (CMBR) photons, thereby changing the spectral shape of the CMBR [SAT 77]. However, during this epoch, processes that can alter the number of photons in the CMBR – for example, the double Compton process, $\gamma + e \rightarrow \gamma + \gamma + e-$ are not effective (i.e., $\Gamma < H$). Therefore, the result of dumping significant amounts of electromagnetic energy density from neutrino decays is a Bose–Einstein spectrum (with $\mu_\gamma \neq 0$) for the CMBR. The CMBR is to a very good precision a black body. Thus, any electromagnetic energy density resulting from neutrino decays during this epoch must be much less than that in the CMBR itself. Recalling that

$$\frac{\rho_\nu}{\rho_\gamma} = \frac{mY_\infty s}{\rho_\gamma}$$

$$\simeq 0.1 m/T \qquad \text{light neutrinos}$$

$$\simeq 2 \times 10^{-8} m_{\text{Gev}}^{-3} m/T \quad \text{heavy neutrinos} \qquad (2.7.40)$$

and requiring that $\rho_\nu/\rho_\gamma \lesssim 1$, we obtain the following limits for a neutrino species that decays during this epoch:

$$m \lesssim 10^7 \tau_{\text{sec}}^{-\frac{1}{2}} \text{eV} \qquad \text{light neutrinos}$$

$$m \gtrsim 4 \times 10^{-3} \tau_{\text{sec}}^{\frac{1}{4}} \text{GeV} \quad \text{heavy neutrinos} \qquad (2.7.41)$$

where we have taken $t_{\text{sec}} \simeq T_{\text{MeV}}^{-\frac{1}{2}}$. These limits are applicable for neutrino lifetimes in the range $10^6 \text{sec} \lesssim \tau \lesssim 10^{13} \text{sec}$. The forbidden region of the mass–lifetime plane is shown in Fig. 2.7.4. (A neutrino species that decays after nucleosynthesis and produces photons of energy greater than 30 MeV can lead to photofission of the light elements produced during nucleosynthesis; additional, more stringent bounds result [LIN 79].)

$t_{\text{end nucleo}} \simeq 3\,\text{min} \leq \tau \leq t_{\text{therm}}$ For neutrino decays that occur during this epoch, the decay-produced photons can be thermalized into the CMBR because both Compton and double Compton scattering are effective ($\Gamma > H$). However, in so doing the entropy per comoving volume is increased. This has the effect of decreasing the present value of η relative to the standard scenario. It is known that luminous matter (necessarily baryons) provides $\Omega_{LUM} \sim 0.01$, and thus provides direct evidence that today $\eta \gtrsim 4 \times 10^{-11}$. On the other hand, primordial nucleosynthesis indicates that at the time of nucleosynthesis η corresponded to a present value of $(3-10) \times 10^{-10}$ [YAN 84]. Thus any entropy production after the epoch of nucleosynthesis must be less than a factor of $\sim 10^{-9}/4 \times 10^{-11} \sim 30$. This leads to the limits

$$10^9 \gtrsim m_{\text{eV}} \tau_{\text{sec}}^{\frac{1}{2}} \qquad \text{light neutrinos}$$

$$10^7 \gtrsim m_{\text{GeV}}^{-2} \tau_{\text{sec}}^{\frac{1}{2}} \qquad \text{heavy neutrinos} \qquad (2.7.42)$$

applicable for neutrino lifetimes in the range $200\,\text{sec} \lesssim \tau \lesssim 10^6 \text{sec}$. This bound too is shown in Fig. 2.7.4 (also see [SAT 77]).

$t_{\text{begin nucleo}} \simeq 1\,\text{sec} \leq \tau \leq t_{\text{end nucleo}}$ If the neutrino lifetime is longer than about a second, then massive neutrinos can contribute significantly to the mass density of the universe during nucleosynthesis, potentially leading to an increase in ^{4}He production. Only the equivalent of 1 additional neutrino species can be tolerated without overproducing ^{4}He. One additional neutrino species is about equivalent to the energy density contributed by photons. Since the crucial epoch is when the neutron-to-proton ratio freezes out ($t \sim 1$ sec, $T \sim 1$ MeV), the constraint that follows is $(\rho_\nu/\rho_\gamma)_{T \simeq \text{MeV}} \lesssim 1$. This results in the mass limit

$$m \gtrsim 5 \times 10^{-3} \text{GeV} \quad \text{heavy neutrinos.} \qquad (2.7.43)$$

Note there is no corresponding limit for a light species because a light species is just one additional relativistic neutrino species. This limit, which is applicable to a heavy neutrino species with lifetime greater than about 1 sec is shown in Fig. 2.7.4.

$\tau \ll 1$ sec A neutrino species that decays earlier than about 1 sec after the bang disappears without leaving much of a cosmological trace. Its decay products thermalize before primordial nucleosynthesis, and its only effect is to increase the entropy per comoving volume. If we understood the origin of the baryon-to-entropy ratio in great detail, and could predict its "prenucleosynthesis" value, then we could use entropy production by the decaying neutrino species to obtain constraints for very short lifetimes.

Astrophysical implications Neutrino decay into visible modes can have "astrophysical" effects too [COW 77]. As the detection of neutrinos from SN 1987A dramatically demonstrated, type II supernovae are a copious source of neutrinos. The integrated flux of neutrino-decay-produced photons from type II supernovae that have occurred throughout the history of the universe can be used to obtain a very stringent bound to acceptable neutrino masses and lifetimes.

Each type II supernova releases about 3×10^{53} erg of energy in thermal neutrinos with average energy about 12 MeV – or about $N_{\nu\bar{\nu}} \simeq 5 \times 10^{57}$ neutrinos and antineutrinos of each species. The historical (last 1,000 years) type II rate in our own galaxy is about 1 per 30 years (give or take a factor of 3), and the observed extragalactic rate is roughly $1.1h^2$ per 100 years per $10^{10}L_{B\odot}$. Using the measured mean blue luminosity density of the universe, $L_{B\odot} \sim 2.4\,h \times 10^8\,\mathrm{Mpc}^{-3}$, this translates into a present type II rate (per volume) of $\Gamma_{SN} \simeq 2.5\,h^3 \times 10^{-85}\,\mathrm{cm}^{-3}\,\mathrm{s}^{-1}$. Assuming that the type II rate has been constant over the history of the universe (a bold assumption), the differential photon number flux is

$$\frac{d\mathscr{F}_\gamma}{d\Omega\,dE} = \frac{9}{5\sqrt{2}}\frac{\gamma_{SN}t_U^2 N_{\nu\bar{\nu}}}{4\pi\langle E_\nu\rangle\tau/m}\frac{1}{\langle E_\nu\rangle^{\frac{1}{2}}E^{\frac{1}{2}}} \qquad (2.7.44)$$

where for simplicity we have assumed that the supernovae neutrinos are mono-energetic, with $E_\nu = \langle E_\nu\rangle \simeq 12$ MeV, that each decay-produced photon carries half the energy of the parent neutrino, and a flat universe. For this energy spectrum $\langle E_\gamma\rangle = \langle E_\nu\rangle/6 \simeq 2$ MeV. Comparing the expected photon number flux at $\langle E_\gamma\rangle$,

$$\langle E_\gamma\rangle\frac{d\mathscr{F}_\gamma}{d\Omega\,dE} \simeq \frac{1}{2}\frac{\Gamma_{SN}t_U^2 N_{\nu\bar{\nu}}m}{4\pi\langle E_\nu\rangle\tau} \qquad (2.7.45)$$

with the measured diffuse γ-ray flux at a few MeV, $3 \times 10^{-3}\,\mathrm{cm}^{-2}\mathrm{sr}^{-1}\mathrm{s}^{-1}$, we obtain the following constraint:

$$\tau_{\mathrm{sec}} \gtrsim 5 \times 10^{12}(\Gamma_{SN}/3 \times 10^{-85}\,\mathrm{cm}^{-3}\,\mathrm{s}^{-1})m_{\mathrm{eV}}. \qquad (2.7.46)$$

Of course, this bound only applies to neutrino species light enough to have been produced in supernovae ($m \lesssim 10$ MeV) and that decay outside the envelope of the

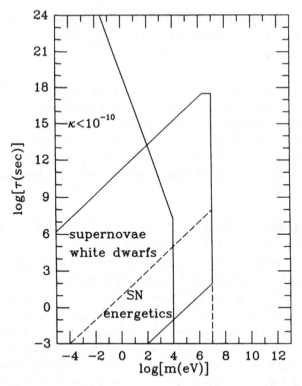

Fig. 2.7.5 Astrophysical limits to the mass and lifetime of an unstable neutrino that decays radiatively (from [KOL 90]).

exploding star ($\tau_{sec} \gtrsim 10^{-5} m_{eV}$) by the present epoch ($t \lesssim 10^{11} m_{eV}$ sec). (Based upon γ-ray observations of SN 1987A made by the SMM spacecraft a similar, slightly more restrictive bound obtains. Furthermore, the lack of observed ionization around SN 1987A by $\nu \to \nu_e^+ e^\pm$ limits this mode for $m > 1$ MeV (see [COW 77]).) This constraint is shown in Fig. 2.7.5.

For a neutrino species that decays within the envelope of the exploding star, and thereby deposits energy in the envelope, a different bound can be derived. Any energy deposited by neutrino decays in the envelope will be thermalized and radiated in the visible part of the spectrum. The energy radiated by SN 1987A in the visible was only about 10^{47} erg, while each neutrino species carries off about 10^{53} ergs! The energy that is deposited in the envelope by a hypothetical, unstable neutrino species is

$$E_{DEP} \simeq N_{\nu\nu} \langle E_\nu \rangle \min[1, R_{BSG}/\tau_{LAB}] \qquad (2.7.47)$$

$$\simeq \min[10^{53} \text{ erg}, 10^{48} m_{eV}/\tau_{sec} \text{ erg}] \qquad (2.7.48)$$

where $R_{BSG} \sim 3 \times 10^{12}$ cm is the radius of the envelope of the progenitor blue super giant (Sanduleak -69 202, by name), and $\tau_{LAB} = \langle E_\nu \rangle \tau/m$ is the neutrino lifetime in the rest frame of the supernova. Comparing this to the observed energy of 10^{47} ergs, we obtain the bound

$$m_{eV}/\tau_{sec} \lesssim 0.1 \quad (t_{sec} \gtrsim 10^{-5} m_{eV}) \tag{2.7.49}$$

$$m_{eV} \gtrsim 10^7 \quad (t_{sec} \lesssim 10^{-5} m_{eV}). \tag{2.7.50}$$

This constraint too is shown in Fig. 2.7.5.

A neutrino species that can decay radiatively, $\nu_j \rightarrow \nu_i + \gamma$, necessarily has an electromagnetic coupling that may be quantified as a transition magnetic moment, $\mu_{ij} = \kappa_{ij}(e/2m_e)$. The transition magnetic moment and neutrino mass and lifetime are related by

$$\tau^{-1} = \alpha_{EM}\kappa^2 m^3/8m_e^2$$

$$\kappa = 0.44 \tau_{sec}^{-\frac{1}{2}} m_{eV}^{-\frac{3}{2}} \tag{2.7.51}$$

where we have assumed $m_j \gg m_i$. The transition moment leads to an electromagnetic correction to $\nu - e$ scattering. Laboratory limits to $\nu - e$ scattering through the transition moment leads to the bound $\kappa_{e\mu} \lesssim 10^{-8}$, or

$$\tau_{sec} \gtrsim 2 \times 10^{15} m_{eV}^{-3} \quad (\nu_\mu \rightarrow \nu_e + \gamma). \tag{2.7.52}$$

Further, such a transition moment leads to neutrino pair emission from white dwarfs and red giants through the process *plasmon* $\rightarrow \nu_i \nu_j$. For $\kappa \sim 10^{-10} - 10^{-11}$ plasmon $\nu\bar{\nu}$ emission can be a very significant cooling mechanism for these objects, and can effect their evolution. Based upon this, a limit of $\kappa_{ij} \lesssim 10^{-10}$ or so has been derived for neutrinos of less mass than 10 keV (see, e.g., the paper of Beg *et al.* [COW 77]). This translates to the limit

$$\tau_{sec} \gtrsim 2 \times 10^{19} m_{eV}^{-3} \quad (m \lesssim 10 \, \text{keV}). \tag{2.7.53}$$

All of the astrophysical and cosmological constraints just discussed are summarized in Figs. 2.7.4. and 2.7.5. These constraints serve to illustrate how a large variety of cosmological and astrophysical observations can be used to probe particle properties in regimes beyond the reach of the terrestrial laboratory.

2.7.3. Limits to the number of families

Perhaps the most significant area to date where cosmological and astrophysics constraints have affected particle physics is in limiting the number of neutrino families, N_ν. This is the first time that a prediction made about elementary particle physics from cosmological arguments [STE 77] has actually been confirmed by measurements done with accelerators [LEP 90]. This important cosmological bound came from big bang nucleosynthesis. A second very different, but less stringent, bound comes from SN 1987A [SCH 87b]. Let us first look at the big bang nucleosynthesis argument.

It might be noted that when the big bang nucleosynthesis arguments on N_ν were first put forth [STE 77] in the mid-1970s, particle accelerators had just found charm, bottom, and tau. Thus is appeared almost as if each new detector would find a new fundamental particle. The cosmological argument that N_ν had to be small thus seemed to fly in the face of the conventional wisdom of the time. The confirmation of small N_ν has clearly become a cornerstone for the whole astroparticle connection.

The power of big bang nucleosynthesis comes from the fact that essentially all of the input physics is well determined. The relevant temperatures, 0.1 to 0.05 MeV, are well explored in nuclear physics labs. Thus, what nuclei do under such conditions is not a matter of guesswork but is precisely known. In fact, the nuclear physics is known far better for these temperatures than it is known in the centers of star like our sun. The temperature at the center of the sun is only a little over 1 keV. This energy is below the energy at which nuclear reaction rates yield significant results in laboratory experiments, and only the long times and higher densities available in stars enable anything to take place at all! Unfortunately, for stellar astrophysics this means that nuclear reaction rates must be extrapolated to many orders of magnitude below their laboratory-measured values. The big bang laboratory does not have this problem. The reactions occur at temperatures and densities at which cross sections and the like are known and well studied in the laboratory.

To calculate what happens, all one has to do is follow a gas of baryons with density ρ_b as the universe expands and cools. As far as nuclear reactions are concerned, the important epoch begins a little above 1 MeV and ends a little below 100 keV. At higher temperatures, no complex nuclei other than single neutrons and protons can exist, and the ratio of neutrons to protons, n/p, is just determined by thermodynamic equilibrium, $n/p = e^{-Q/T}$, where $Q = 1.3$ MeV is neutron–proton mass difference. Equilibrium applies because the weak interaction rates are much faster than the expansion of the universe at temperatures much above 1 MeV. At temperatures much below 0.1 MeV, the electrostatic repulsion of nuclei prevents nuclear reactions from proceeding as fast as the cosmological expansion separates the particles.

After the weak interaction drops out of equilibrium, around 1 MeV, the ratio of neutrons to protons changes more slowly, by free neutrons decaying to protons and similar transformations of neutrons to protons via interactions with the ambient leptons. By the time the universe reaches 0.1 MeV, the ratio is slightly below $\frac{1}{7}$. For temperatures above 0.1 MeV, the high entropy of the universe suppresses the abundance of nuclei. Once the temperature drops to about 0.1 MeV, nuclei begin to be present in significant amounts, starting with ^{2}D adding neutrons and protons, making ^{3}H and ^{3}He. These, in turn, capture neutrons and protons to produce ^{4}He or ^{3}H and ^{3}He can collide to also yield ^{4}He. Since ^{4}He is the most tightly bound nucleus (in this region of the periodic table), the flow of reactions converts almost all the neutrons that exist at 0.1 MeV into ^{4}He (for neutron/proton ratios less than unity).

The two-body chain essentially ceases there, because there are no stable nuclei at either mass-5 or mass-8. Since the baryon density at big bang nucleosynthesis is relatively low (much less than $1 \, g/cm^3$) only reactions involving two-particle collisions occur. It can be seen that combining the most abundant nuclei neutrons, protons, and ^4He via two-body interactions always leads to unstable mass-5. Even when one combines ^4He with rarer nuclei like ^3H or ^3He, we still only get to mass-7, which when hit by a proton, the most abundant nucleus around, yields mass-8. Eventually, ^3H radioactively decays to ^3He, and any mass-7 made, radioactively decays to ^7Li. Thus, big bang nucleosynthesis makes ^4He with traces of ^2D, ^3He, and ^7Li. (Also, all the protons left over that did not capture neutrons remain as hydrogen.) All other chemical elements are made later in stars and in related processes. (Stars jump the mass-5 and -8 instability by having gravity compress the matter to sufficient densities that three-body collisions can occur and jump the mass-5 and -8 gaps.) A neutron/proton ratio of $\sim \frac{1}{7}$ yields a resultant ^4He primordial mass fraction, $Y = 2(n/p)/(n/p + 1) = 0.25$.

The only cosmological parameter in such calculations is the density of the baryon gas at a given temperature. From the thermodynamics of the expanding universe we know that $\rho_b \propto T^3$, thus we can relate the baryon density at 10^{11} K to the baryon density today, when the temperature is about 2.75 K. The problem is, we don't know ρ_b today, so the calculation must be carried out for a range in ρ_b. The cosmological expansion rate depends on the total mass–energy density. For cosmological temperatures much above 1 eV the energy density of radiation exceeds the mass–energy density of the baryon gas. Thus, during big bang nucleosynthesis, we need the radiation density as well as the baryon density. The baryon density determines the density of the nuclei and thus their interaction rates, and the radiation density controls the expansion rate of the universe at those times. The density of radiation is just proportional to the number of "types" of radiation. Thus, the density of radiation is not a free parameter provided we know how many types of relativistic particles exist at temperatures $\sim 0.1 - 1.0 \, MeV$.

Assuming that the relativistic particles at 1 MeV are photons, e, μ, and τ neutrinos (and their antiparticles) and electrons (and positrons), the big bang nucleosynthetic yields have been calculated for a range in present ρ_b (more precisely the baryon to photon ratio), going from less than that observed in galaxies to greater than that allowed by the observed large-scale dynamics of the universe. The ^4He yield is almost independent of the baryon density, with a very slight rise in the density due to the decreasing entropy per baryon, which enables nucleosynthesis to start slightly earlier, when the neutron/proton ratio was higher. No matter what assumptions one makes about the baryon density, it is clear that ^4He is predicted by big bang nucleosynthesis to be around 25 percent of the mass of the universe. This was first noted by Hoyle and Taylor [HOY 64] and later found by Peebles [PEE 66] and by Wagoner, Fowler, and Hoyle [WAG 67]. The current results do not differ in any qualitative way from Wagoner, Fowler, and Hoyle's original detailed calculations.

The fact that the observed helium abundance in all objects is about 20–30 percent is certainly a nice confirmation of these ideas. Since stars produce a yield of only 2 percent in all the heavy elements combined, stars cannot easily duplicate such a large ^{4}He yield. While the predicted big bang yields of the other light elements were also calculated in the 1960s, they were not considered important at that time, since it was assumed in the 1960s that these nuclei were made in more significant amounts in stars [FOW 62]. However, work by our group at Chicago [YAN 84], and others, thoroughly established big bang nucleosynthesis and turned it into a tool for probing the universe, by showing that other light element abundances had major contributions from the big bang and that the effects of any stellar contributions could be removed by appropriate techniques. Today the big bang predictions for all four light isotopes are used to test the model and use it as a probe of conditions at early times.

In particular, it was demonstrated in the early 1970s that contrary to the ideas of the 1960s, deuterium could not be made in any significant amount by *any* realistic contemporary astrophysical process [EPS 76]. The big bang deuterium yield decreases rapidly with increasing ρ_b. At high densities deuterium gets more completely converted to ^{4}He; quantitatively this means that the present density of baryons must be below $\sim 5 \times 10^{-31}$ g/cm^3 in order for the big bang to have produced enough deuterium to explain the observed abundance. Similar, though more complex, arguments [KAW 87] were also developed for ^{3}He, and most recently for ^{7}Li, so that it can be said that only if the baryon density is between 2×10^{-31} g/cm^3 and 5×10^{-31} g/cm^3 are all the observed light element abundances consistent with the big bang yields. If the baryon density were outside of this narrow range, a significant disagreement between the big bang predictions and the observed abundances would result. To put this in perspective, it should be noted that for this range in densities, the predicted abundances for the four separate species cover a range from 25 percent to one part in $\sim 10^{10}$. The big bang yields all agree with only one freely adjustable parameter, ρ_b.

Recently, several nonstandard scenarios of primordial nucleosynthesis have been proposed [SCH 87a]; however, these scenarios with their additional adjustable parameters seem to be unable to account for the abundances of the four light isotopes, especially ^{7}Li. This speaks to the remarkable success of the standard scenario of big bang nucleosynthesis.

This narrow range in baryon density for which concordance occurs is very interesting. Let us convert it into units of the critical cosmological density for the allowed range of Hubble expansion rates. From big bang nucleosynthesis [EPS 76; KAW 87], it follows that the baryon density Ω_B is less than 0.12 and greater than 0.03 (once one includes [FRE 84] age constraints on a flat universe); that is the universe *cannot be closed with baryonic matter*. If the universe is truly at critical density, then nonbaryonic matter is required. This argument has led to one of the major areas of research at the particle–cosmology interface, namely, the search for nonbaryonic dark matter.

Another important conclusion regarding the allowed range in baryon density is that it is in very good agreement with the density implied from the dynamics of galaxies, *including their dark halos.* An early version of this argument, using only deuterium, was described over 15 years ago [GOT 74]. As time has gone on, the argument has strengthened and the fact remains that galactic dynamics and nucleosynthesis both suggest densities of about 10 percent of the critical density. Thus, if the universe is indeed at critical density, as many believe, it requires that the bulk of the matter not be associated with galaxies and their halos, as well as being nonbaryonic.

With the growing success of big bang nucleosynthesis, the predictions came under more scrutiny. In particular, the ^{4}He yield was examined in detail since it is the most abundant of the nuclei, and thus in principle it is the one that observers should be able to measure to highest accuracy. In addition, it is very sensitive to the n/p ratio.

In the standard calculation it is assumed that photons, electrons, and the three known neutrino species (and their antiparticles) are present in the universe at the time of nucleosynthesis. However, by doing the calculation with additional species of neutrinos we can see when ^{4}He yields exceed observational limits. The bound on ^{4}He comes from observations of helium in many different objects in the universe. However, since ^{4}He is not only produced in the big bang but in stars as well, it is important to estimate what part of the helium in some astronomical object is primordial, from the big bang, and what part is due to stellar production after the big bang. To do this we [GAL 89] have found that the carbon content of the object is well suited for tracking the additional helium produced. Carbon is made in the same mass stars that also produce ^{4}He; thus as the carbon abundance increases, so must the helium. (Other heavy elements such as oxygen have been used previously, but these elements are not produced in the same mass stars as those that produce the bulk of the helium.) The extrapolation of helium to zero carbon content in an object should be a good estimate of the primordial helium. We obtain ~ 0.235 as our best estimate for the mass fraction of helium produced in the big bang. The upper bound is what is important here. We formally estimate a three standard deviation bound as 0.247. In particular, it seems clear that the primordial ^{4}He was at least a little less than 25 percent. Since objects have heavy elements and possibly some associated extrastellar-produced helium and still have helium abundances of 25 percent, this certainly seems like a *very* safe upper bound. In fact, if anything our estimates are on the high side due to possible systematic errors; for example, Pagel [PAG 87] finds collisional excitation reduces the 0.235 to 0.233.

We find (see Fig. 2.7.6) that three (or two) types of neutrinos fit the data well, and a fourth is only marginally allowed if helium slightly exceeds the 3σ upper bound; any more neutrinos are strictly prohibited. Since each family contains a neutrino, we are saying that the total number of families is probably three. $N_\nu = 3$ has now been confirmed at LEP SLC [LEP 90]. Of course, the cosmological argument assumes that the neutrinos are "light," that is, less massive than ~ 10 MeV, and LEP includes all ν's up to ~ 45 GeV.

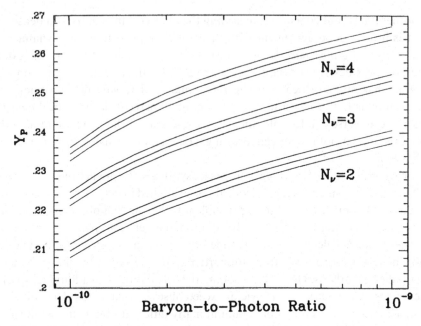

Fig. 2.7.6 Helium mass fraction versus the baryon-to-photon ratio η. The lower bound of 2×10^{-10} derives from the ^{3}He + D and ^{7}Li constraints, and the upper bound of 7×10^{-10} from the D and ^{7}Li constraints. The three lines for each neutrino family correspond to neutron half-lives of 10.4, 10.5, and 10.6 minutes (from [YAN 84]).

2.7.4 Supernova 1987A and neutrino counting

Let us now compare this bound with the supernova constraint. As is now well appreciated, neutrinos were detected from SN 1987A by both Kamiokande [HIR 87] and IMB [BIO 87]. Both of these H_2O Cherenkov detectors are most sensitive to $\bar{\nu}_e + p \rightarrow n + e^+$ because of its larger cross section.

If the $\bar{\nu}_e$ flux is assumed to come from a Fermi–Dirac (F-D) distribution at temperature T and total $\bar{\nu}$ energy, $\epsilon_{\bar{\nu}_e}$, both IMB and Kamiokande are simultaneously fit with $T \sim 4$ to $4.5 \, \text{MeV}$ and $\epsilon_{\bar{\nu}_e} \sim 3$ to $4.5 \times 10^{52} \, \text{erg}$. These figures are in remarkable agreement with the Standard Model [MAY 87] for gravitational core collapse of a massive star, if $N_\nu = 3$. Thus, we have confidence that we have witnessed such a core collapse, and that we have a good understanding of its physics. Let us now turn the argument around and see how sensitive our expected fluxes are to N_ν.

In a collapse to a neutron star, the binding energy, ϵ_B, must be radiated as neutrinos. The initial neutronization burst of ν_e's carries away a fraction $f_n \lesssim 10$ percent of ϵ_B on a time scale of $\gtrsim 10 \, \text{ms}$. The remaining energy comes out in thermal $\nu\bar{\nu}$ pairs from reactions like

$$e^+ e^- \rightarrow \nu \bar{\nu} \tag{2.7.54}$$

where through neutral currents all species of neutrinos with $m_\nu \lesssim 10\,\text{MeV}$ are emitted.

Since electron scattering rates are small compared to $\bar{\nu}_e$ capture, even with five times more free electrons than protons, at most we expect one or two scattering events in the detectors for a SN at 50 kpc (distance to LMC). Thus, the detectable fraction of ϵ_B is $\epsilon_{\bar{\nu}_e}$ where

$$\epsilon_{\bar{\nu}_e} \approx \frac{(1 - f_n)}{2N_\nu} \epsilon_B \tag{2.7.55}$$

assuming an equipartition of energy entitled in the various neutrino species, as is found in the detailed models. (While average energy per neutrino is higher for ν_μ and ν_τ, their flux is correspondingly lower.) The number of counts, n, one expects in a detector of mass M_D, is

$$n \approx \frac{\epsilon_{\bar{\nu}_e}}{\langle E_{\bar{\nu}_e} \rangle} \frac{\langle \sigma \rangle}{4\pi R^2} \frac{2M_D}{18m_p} \tag{2.7.56}$$

where m_p is the proton mass, $R \approx 50\,\text{kpc}$ is the distance to LMC, $\langle E_{\bar{\nu}_e} \rangle$ is the average $\bar{\nu}_e$ energy, and $\langle \sigma \rangle$ is the cross section appropriately averaged over an F-D distribution with appropriate threshold factors and efficiencies taken into account. The temperature of ν_e's is found to be $\sim 3.2\,\text{MeV}$ ($\langle E_\nu \rangle \approx 10\,\text{MeV}$) to good accuracy. Temperatures are very insensitive to model parameters being determined by microphysics at the neutrinosphere [SCH 87]. The temperature for $\bar{\nu}_e$'s is somewhat higher due to the smaller opacities at late times as protons disappear in the core, thereby minimizing charged-current interactions. This enables the $\bar{\nu}_e$'s to come from deeper in the star. Mayle et al. [MAY 87] find $T_{\bar{\nu}_e} \sim 4\,\text{MeV}$ in good agreement with the temperature inferred from the observations. (They do find a higher than thermal high energy tail to the distribution that can effect the high threshold IMB but not Kamiokande.) For detectors like Kamiokande where the threshold is well below the peak of the cross-sectional weighted distribution, it is reasonable to use

$$\langle \sigma \rangle \approx \sigma_0 (g_\nu^2 + 3g_A^2) 12 T_{\bar{\nu}_e}^2. \tag{2.7.57}$$

(For IMB a more careful procedure must be applied due to its high threshold.) Substituting into Eq. (2.7.56) yields

$$n = \frac{5.2}{(N_\nu / 3)} \left(\frac{1 - f_n}{2 \times 10^{53}\,\text{erg}} \right) \left(\frac{T_{\bar{\nu}_e}}{4\,\text{MeV}} \right) \left(\frac{M_D}{\text{ktons}} \right) \left(\frac{50\,\text{kpc}}{R} \right)^2 \tag{2.7.58}$$

which for $M_D = 2.14\,\text{ktons}$ (Kamioka) gives a prediction of 11 counts for $N_\nu = 3$. While they actually observe 11, one should weigh their counts by efficiency effects to obtain 16.5 ± 5. Solving for N_ν yields

$$N_\nu = (2 \pm 0.6) \left[\left(\frac{T_{\bar{\nu}_e}}{4\,\text{MeV}} \right) \left(\frac{\epsilon_B}{2 \times 10^{53}\,\text{erg}} \right) \left(\frac{1 - f_n}{0.9} \right) \left(\frac{50\,\text{kpc}}{R} \right) \right]. \tag{2.7.59}$$

Let us now see how high we can push this. While models can be found with $f_n > 0.1$, it is obvious that $1 - f_n$ can never exceed unity. The effective $T_{\bar{\nu}_e}$, as used above, varies by $\lesssim 25$ percent. The binding energy for 1.4 M neutron stars (the mass of the collapsing core) is found to vary from 1.5 to 3×10^{53} ergs for a wide range of equation-of-state [ARN 77]. Thus, we choose 3×10^{53} erg (4×10^{53} erg) as an (extreme) upper bound. The distance to the LMC varies in the astronomical literature by < 7 percent. We'll adopt an extreme limit of 10 percent consistent with current SN 1987A determination of the distance [WAG 88]. Combining all these extreme values yields

$$N_\nu < 6.6(8.9). \tag{2.7.60}$$

A more careful calculation taking into account different thresholds for both IMB and Kamiokande to obtain measured $\epsilon_{\bar{\nu}_e}$ for predicted yields at the $T_{\bar{\nu}_e}$ inferred from the data yields essentially the same result ($N_\nu \lesssim 6.7(9.0)$) as given above. Thus, SN 1987A gives a limit to N_ν comparable to accelerator experiments but not as strong as the big bang nucleosynthesis limits.

2.7.5 Other constraints from SN 1987A

SN 1987 A has proven to be an amazing neutrino laboratory. In addition to the previously mentioned limits, it has placed limits to the charge and magnetic moment of the neutrino that exceed current laboratory limits and its constraint to the mass of ν_e is comparable to the best laboratory limits. Let us briefly review these bounds.

2.7.6 Magnetic moment

Barbieri and Mohapatra [BAR 88a,b], and Lattimer and Cooperstein have shown that the observation of $\bar{\nu}_e$'s from SN 1987A constrains the value of the magnetic moment of the neutrino to $\lesssim 10^{-11} \mu_B$. The argument is twofold, involving in a crucial way the fact the interaction cross section of right-handed neutrinos must be significantly weaker than those of left-handed neutrinos. (Right-handed Dirac neutrinos must interact more weakly so that they do not get counted in the big bang nucleosynthesis arguments [OLI 81].) First, there is the limit from cooling the proto-neutron star too rapidly if ν_L's can change to ν_R's as a result of magnetic moment interactions in the protoneutron star core. Second, there is the effect that a flipped ν_R can escape from the higher temperature inner core and then get flipped back to a ν_L by the intergalactic magnetic field. This latter situation could yield 70 MeV $\bar{\nu}_e$'s which were definitely not detected. It is argued that these processes limit the magnetic moment to $\lesssim 10^{-13}$ with $10^{-11} \mu_B$ as an extreme upper limit. However, Okun [OKU 88] has argued that these arguments can be circumvented if the magnetic moment is not static but is a Majoron transition moment or if an appropriate MSW mixing [MIK 86] of neutrino species also occurs in the supernova.

2.7.7 Neutrino mass

Since the observed neutrino burst was relatively narrow ($\lesssim 10$ sec), despite energies that spanned a range of about a factor of two, it is obvious that any neutrino rest mass must be very small. While the relationship between mass, timespread, and energy is a simple one, the key here is to decide on the significance of the time and energy spread, and to estimate what the intrinsic spread was in the neutrino burst in the absence of finite masses.

The crucial, but simple, relationship at the heart of any analysis to constrain the ν_e mass from the IMB and Kamiokande data is that for the time delay suffered by a neutrino during its flight to earth:

$$\Delta t \cong \frac{1}{2}\frac{R}{c}\frac{m^6}{E^2} \simeq 2.6\,\mathrm{sec}\,\frac{(m/10\,\mathrm{eV})^2}{(E/10\,\mathrm{MeV})^2}. \tag{2.7.61}$$

From this simple equation for Δt, it is clear that any mass constraint which follows will be in the general range of about 20 eV, or so, which is comparable to existing laboratory limits. Given the sparseness of the data set (19 events in total), the subtleties of the detectors (response, thresholds, etc.) and the absence of a very specific, well-accepted Standard Model of the initial cooling, it is not surprising that many authors "derived" limits (and even values!) for the ν_e mass ranging from a few eV to 30 eV. The most extensive and careful analyses to date [LAM 89] provide limits of around 20–25 eV. While SN 1987A has not really improved existing bounds, it is interesting that the constraint that is found is comparable to the present laboratory limits.

2.7.8 Neutrino mixing

Neutrino mixing has been proposed as a solution to the solar neutrino problem [MIK 86], and the Homestake and Kamiokande observations of solar neutrinos place constraints on allowed mixing parameters [KOL 87b]. A supernova could potentially also test neutrino mixing [WAL 87]. If neutrino mixing occurs between supernova emission and detection, it can obviously alter the detected neutrino signal.

If MSW mixing is indeed the solution to the solar neutrino problem, then only $\nu_e \nu_\mu(\nu_\tau)$ mixing is possible and not $\bar{\nu}_e \to \bar{\nu}_\mu(\bar{\nu}_\tau)$. Thus, the solar neutrino solution would not affect the $\bar{\nu}_e$ flux. However, it could deplete the initial neutronization burst. Unfortunately, there is no conclusive evidence that even a single $\nu_e + e^- \to \nu_e + e^-$ scattering event associated with the neutronization burst was seen.

If we drop the solar neutrino solution and go to general MSW mixing, then we can mix $\bar{\nu}_\mu(\bar{\nu}_\tau)$ into $\bar{\nu}_e$, which might enhance the energy slightly, but would otherwise do little. No effect would occur for the electron scattering ν_e's. Thus no definitive statement can be made from SN 1987A about neutrino mixing and oscillations.

2.7.9 Secret interactions

Precious little is known about any interactions that neutrinos may have beyond the standard weak interactions, for example, additional neutrino–neutrino interactions as in the Majoron model. Since neutrinos from the supernova traversed 170 000 light-years through the cosmic seas of relic neutrinos (and other particles such as Majorons) without apparent attenuation, any unknown (i.e., secret) interactions they might have with neutrinos (or other particles in the sea of relics) can be constrained:

$$\sigma_{\text{secret}} \lesssim 10^{-25} \, \text{cm}^2. \tag{2.7.62}$$

2.7.10 Radiative decays

The fluence of neutrinos from SN 1987A was enormous, $\sim 10^{10} \, \text{cm}^{-2}$ per species (integrated over the observed burst). On the other hand there was no observation (above instrument background) of any high energy γ-rays: based upon the data of the gamma ray spectrometer aboard the Solar Maximum Mission and γ-ray detectors on the Pioneer Venus Orbiter a γ-ray fluence limit for the same time period of $\lesssim 1 \, \text{cm}^{-2}$ follows. This means that less than about 1 in 10^{10} of the supernova neutrinos could have decayed producing a γ-ray. From these non-observations of γ-rays a limit of

$$\tau_{\text{sec}}/m_{\text{eV}} \gtrsim 2 \times 10^{15} B_\gamma \tag{2.7.63}$$

can be set to the radiative decay of any neutrino species. Here B_γ is the branching ratio for the radiative decay mode.

Note added in proof: Recent LEP (90) results have severely constrained heavy (cold) weakly interacting particles as dark matter. In particular, heavy neutrinos with masses $\lesssim 406 \, \text{GeV}$ are now excluded.

For a recent update on big-bang nucleosynthesis see D. Schramm and M. Turner [SCH 98].

References

[ABE 87]	K. Abe *et al.*, *Phys. Rev. Lett.* 58 (1987) 636.
[ACH 95]	B. Achkar *et al.*, *Nucl. Phys.* B434 (1995) 503.
[ADL 72]	S. Adler, *Phys. Rev.* 177 (1969) 2426; D. J. Gross and R. Jackiw, *Phys. Rev.* D6 (1972) 477; C. Bouchiat, J. Iliopoulos and P. Meyer, *Phys. Lett.* B38 (1972) 519.
[AGI 89]	M. Aglietta *et al.*, *Europhys. Lett.* 8 (1989) 611.
[AKH 88]	E.Kh. Akhmedov, preprint Moscow (1988) unpublished.
[ALB 92]	H. Albrecht *et al.* (ARGUS), *Phys. Lett.* B292 (1992) 221.
[ALE 94]	A. Alessandrello *et al.*, *Phys. Lett.* B335 (1994) 519.
[ALE 96]	G. Alexander *et al.* (OPAL), *Z. Phys.* C72 (1996) 231.
[ALEPH 89]	D. Decamp *et al.* (ALEPH Coll.), *Phys. Lett.* B231 (1989) p. 519.
[ALEPH 92]	see for example ALEPH Coll., *Phys. Rep.* 216 (1992) 253.
[ALEPH 93]	ALEPH Coll., *Phys. Lett.* B313 (1993) 520.
[ALEPH 95]	ALEPH Coll., *Phys. Lett.* B346 (1995) 379.

[ALEPH 97]   ALEPH Coll., *Phys. Lett.* B414 (1997) 362.
[ALEPH 98]   ALEPH Coll., *Eur. Phys. J.* C2 (1998) 395.
[ALL 97]   W. W. M. Allison *et al.*, *Phys. Lett.* B391 (1997) 491.
[ALS 93]   M. Alston-Garnjost *et al.*, *Phys. Rev. Lett.* 71 (1993) 831.
[ALT 97a]   G. Altarelli, J. Ellis, G. F. Guidice, S. Lola and M. L. Mangano, preprint hep–ph/ 9703276.
[ALT 97b]   G. Altarelli, R. Barbieri, F. Caravaglios CERN-th/97-290, to appear in *J. Mod. Phys.* A (1998).
[ALV 40]   L. W. Alvarez and R. Cornog, *Phys. Rev.* 56 (1939) 613 and *Phys. Rev.* 57 (1940) 248.
[APO 98]   CHOOZ Collab., M. Appollonio *et al.*, *Phys. Lett.* B420 (1998) 397.
[ARA 86a]   J. Arafune, N. Koga, K. Morokuma and T. Watanabe, *J. Phys. Soc. Japan* 55 (1986) 3806.
[ARA 86b]   J. Arafune and T. Watanabe, *Phys. Rev.* C34 (1986) 336.
[ARN 77]   W. Arnett and R. Bowers, *Ap. J.* 33 (1977) 33.
[ASP 87]   ASP Coll., C. Hearty *et al.*, *Phys. Rev. Lett.* 58 (1987) 1711.
[ASS 94]   K. Assamagan *et al.*, *Phys. Lett.* B335 (1994) 231.
[ASS 96]   K. Assamagan *et al.*, *Phys. Rev.* D53 (1996) 6065.
[ATH 96]   C. Athanassopoulos *et al.*, LSND Collab., *Phys. Rev.* C54 (1996) 2685.
[AUL 82]   C. S. Aulakh and R. N. Mohapatra, *Phys. Lett.* B119 (1982) 136.
[AYR 95]   MINOS Collab., D. Ayres *et al.*, *Fermilab. reports: NUMI–L–63, NUMI–L–375* (1995).
[BAB 87]   K. S. Babu and V. S. Mathur, *Phys. Lett.* B196 (1987) 218.
[BAB 95]   K. S. Babu and R. N. Mohapatra, *Phys. Rev. Lett.* 75 (1995) 2276.
[BAB 97]   K. S. Babu *et al.*, preprint hep–ph/9703299 (March 1997).
[BAH 88]   J. Bahcall and R. Ulrich, *Rev. Mod. Phys.* 60 (1988) 297.
[BAK 96]   H. Backe *et al.* in *Proc. 17th Int. Conference on Neutrino Physics and Astrophysics (Helsinki 1996)*, ed. K. Enqvist *et al.* (World Scientific, Singapore, 1996).
[BAM 95a]   P. Bamert, C. P. Burgess and R. N. Mohapatra, *Nucl. Phys.* B438 (1995) 3.
[BAM 95b]   P. Bamert, C. P. Burgess and R. N. Mohapatra, *Nucl. Phys.* B449 (1995) 25.
[BAR 72]   W. A. Bardeen, R. Gastmans and B. Lautrip, *Nucl. Phys.* B46 (1972) 319.
[BAR 88a]   R. Barbieri and R. N. Mohapatra, *Phys. Rev. Lett.* 61 (1988) 27.
[BAR 88b]   R. Barbieri, R. N. Mohapatra, J. Lattimer and J. Cooperstein, *Phys. Rev. Lett.* 61 (1988) 23.
[BAR 88c]   A. S. Barabash, ITEP preprint, Moscow (1988).
[BAR 88d]   S. Barr and A. Halprin, *Phys. Lett.* B202 (1988) 279.
[BAR 89]   V. Barger, G. F. Guidice and T. Han, *Phys. Rev.* D40 (1989) 2987.
[BAU 97]   L. Baudis, J. Hellmig, H. V. Klapdor–Kleingrothaus, A. Müller, F. Petry, Y. Ramachers and H. Strecker, *Nucl. Instrum. Methods* A385 (1997) 265.
[BAV 95]   E. Baver and M. Leurer, *Phys. Rev.* D51 (1995) 260.
[BEC 92]   IBM Collab., R. Becker-Szendy *et al.*, *Phys. Rev.* D46 (1992) 3270.
[BEC 95]   R. Becker-Szendy *et al.*, *Nucl. Phys. B (Proc. Suppl.)* 38 (1995) 331.
[BEG 78]   M. A. B. Bèg, W. J. Marciano and M. Ruderman, *Phys. Rev.* D17 (1978) 1395.
[BEG 82]   M. A. B. Bèg and A. Sirlin, *Phys. Rep.* 88 (1982) 1.
[BEH 82]   H. Behrens and W. Bühring, *Electron Radial Wave Functions and Nuclear Beta-Decay* (Clarendon Press, Oxford, 1982).
[BEL 95a]   A. I. Belesev *et al.*, *Phys. Lett.* B350 (1995) 263.
[BEL 95b]   G. Belanger, F. Boudjema, D. London and H. Nadeau, *Phys. Rev.* D53 (1996) 6292.
[BEN 81]   C. L. Bennett *et al.*, *Phys. Lett.* B107 (1981) 19.
[BER 63]   J. Bernstein, M. Ruderman and G. Feinberg, *Phys. Rev.* 132 (1963) 1227.
[BER 71]   K. E. Bergkvist, *Physica Scripta* 4 (1971) 23.
[BER 72]   K. E. Bergkvist, *Nucl. Phys.* B39 (1972) 317, *ibid.*, 371.
[BER 92]   Z. G. Berezhiani, A.Yu. Smirnov and J. W. F. Valle, *Phys. Lett.* B291 (1992) 99.
[BER 95]   Z. Berezhiani and R. N. Mohapatra, *Phys. Rev.* D52 (1995) 6607.
[BET 35]   H. A. Bethe, *Proc. Cambridge Phil. Soc.* 32 (1935) 108.

[BIL 78]   S. M. Bilenky and B. Pontecorvo, *Phys. Rep.* 41 (1978) 225.
[BIL 80]   S. Bilenky, J. Hosek and S. Petcov, *Phys. Lett.* B94 (1980) 495.
[BIL 84]   S. Bilenky, N. Nedelcheva and S. Petcov, *Nucl. Phys.* B247 (1984) 61.
[BIL 87]   S. M. Bilenky and S. T. Petcov, *Rev. Mod. Phys.* 59 (1987) 671.
[BIL 96]   S. M. Bilenky *et al.*, *Phys. Lett.* B356 (1995) 273; *Phys. Rev.* D54 (1996) 1881.
[BIO 87]   R. M. Bionta *et al.*, *Phys. Rev. Lett.* 58 (1987) 1494.
[BJO 64]   J. D. Bjorken and S. D. Drell, *Relativistic Quantum Mechanics* (McGraw-Hill, 1964).
[BOE 96]   F. Boehm *et al.*, *The Palo Verde experiment* (1996) http://www.cco.caltech.edu/songhoon/Palo-Verde.html.
[BOR 87]   S. Boris *et al.*, *Phys. Rev. Lett.* 58 (1987) 2019.
[BOR 92]   L. Borodovsky *et al.*, *Phys. Rev. Lett.* 68 (1992) 274.
[BUC 87]   W. Buchmüller, R. Rückl and D. Wyler, *Phys. Lett.* B191 (1987) 442.
[BUR 93]   C. P. Burgess and J. M. Cline, *Phys. Lett.* B298 (1993) 141; *Phys. Rev.* D49 (1994) 5925.
[BUS 95]   D. Buskulic *et al.* (ALEPH), *Phys. Lett.* B349 (1995) 585.
[BUT 93]   J. Butterworth and H. Dreiner, *Nucl. Phys.* B397 (1993) 3; H. Dreiner and P. Morawitz, *Nucl. Phys.* B428 (1994) 31.
[CAL 93]   D. O. Caldwell and R. N. Mohapatra, *Phys. Rev.* D48 (1993) 3259.
[CAM 98]   L. Camilleri in *Proc. XVIII Int. Conf. on Neutrino Physics and Astrophysics*, Takayama, Japan, 4–9 June 1998, ed. by Y. Suzuki and Y. Totsuka. *Nucl. Phys. B (Proc. Suppl.)* 77 (1999) 3.
[CAR 93]   C. D. Carone, *Phys. Lett.* B308 (1993) 85.
[CAU 96]   E. Caurier, F. Nowacki, A. Poves and J. Retamosa, *Phys. Rev. Lett.* 67 (1996) 1954.
[CDF 91]   CDF Coll., F. Abe *et al.*, *Phys. Rev.* D44 (1991) 29.
[CDF 95]   CDF Coll., *Phys. Rev. Lett.* 74 (1995) 2626.
[CELLO 88]   CELLO Coll., H. J. Behrend *et al.*, *Phys. Lett.* B215 (1988) 186.
[CEN 94]   ICARUS Collab., P. Cennini, LNGS-94/99-I (1994).
[CHA 86]   CHARM Collab., P. Vilain *et al.*, *Phys. Lett.* B179 (1986) 301.
[CHI 81]   Y. Chikashige, R. N. Mohapatra and R. D. Peccei, *Phys. Lett.* B98 (1981) 265; *Phys. Rev. Lett.* 45 (1980) 265.
[CHO 97]   D. Choudhury and S. Raychaudhuri, preprint hep–ph/9702392.
[CIN 93]   M. Cinabro *et al.* (CLEO), *Phys. Rev. Lett.* 70 (1993) 3700.
[CIS 71]   A. Cisneros, *Astro Space Sci.* 10 (1971) 87.
[COP 96]   C. J. Copi, D. N. Schramm and M. S. Turner, *Phys. Rev.* D55 (1997) 3389.
[COW 72]   R. Cowsik and J. McClelland, *Phys. Rev. Lett.* 29 (1972) 669; G. Gerstein, Ya. B. Zeldovich, *Zh. Eksp. Teor. Fiz, Pisma Red4* (1966) 174; G. Marx, A. Szalay, *Neutrino 72* (Budapest) 123.
[COW 77]   R. Cowsik, *Phys. Rev. Lett.* 39 (1977) 784; S. W. Falk and D. N. Schramm, *Phys. Lett.* 79B (1978) 511; M. A. Beg, W. J. Marciano and M. Ruderman, *Phys. Rev.* D17 (1978) 1395; M. Fukugita and S. Yazaki, *Phys. Rev.* D36 (1987) 3817. G. A. Tammann, *Ann. N. Y. Acad. Sci.* 302 (1977) 61; S. van den Bergh, R. D. McClure and R. Evans, *Ap. J.* 323 (1987) 44; E. Kolb and M. S. Turner, *Phys. Rev. Lett.* 62 (1989) 509; R. Cowsik, P. Hoflich and D. N. Schramm, *Phys. Lett.* B (1990) in press.
[CUR 49]   S. C. Curran, J. Angus and A. L. Cockroft, *Phil. Mag.* 40 (1949) 53.
[D0 95]   D0 Coll., *Phys. Rev. Lett.* 74 (1995) 2632.
[DAN 62]   G. Danby *et al.*, *Phys. Rev. Lett.* 9 (1962) 36.
[DAN 95]   F. A. Danevich *et al.*, *Phys. Lett.* B344 (1995) 72.
[DAU 95]   K. Daum *et al.*, *Z. Phys.* C66 (1995) 417.
[DAV 86]   R. Davis, *Seventh GUT Workshop, Tokyo* (1988) 237.
[DAV 94]   S. Davidson, D. Bailey and A. Campbell, *Z. Phys.* C61 (1994) 613; M. Leurer, *Phys. Rev. Lett.* 71 (1993) 1324; *Phys. Rev.* D50 (1994) 536.
[DEG 88]   G. Degrassi, W. Marciano and A. Sirlin, *Brookhaven Report* (1988), unpublished.
[DELPHI 89]   DELPHI Coll., P. Aarnio *et al.*, *Phys. Lett.* B231 (1989) 539.

[DELPHI 97]   DELPHI Coll., *Z. Phys.* C74 (1997) 577.
[DEN 89]   D. Denegri, B. Sadoulet and M. Spiro, *Rev. Mod. Phys.* A (1989).
[DEU 90]   J. Deutsch, M. Lebrun and R. Prieels, *Nucl. Phys.* A518 (1990) 149.
[DOI 81]   M. Doi *et al.*, *Phys. Lett.* B102 (1981) 323.
[DOI 85]   M. Doi, T. Kotani and E. Takasugi, *Progr. Theor. Phys. Suppl.* 83 (1985) 1.
[DOI 93]   M. Doi and T. Kotani, *Progr. Theor. Phys.* 89 (1993) 139.
[DOL 72]   A. D. Dolgov, L. B. Okun and V. I. Z. Zacharov, *Nucl. Phys.* B41 (1972) 197.
[DOL 81]   A. D. Dolgovich and Ya. B. Zeldovich, *Rev. Mod. Phys.* 53 (1981) 1.
[DONUT 98]   *SCIENCE* Magazine, 17 July, 1998.
[DRU 87]   E. G. Drukarev and M. I. Strikman, *Phys. Lett.* B186 (1987) 1.
[ELL 87]   S. R. Elliot, A. A. Hahn and M. K. Moe, *Phys. Rev. Lett.* 59 (1987) 1649.
[ELL 92]   S. R. Elliott *et al.*, *Phys. Rev.* C46 (1992) 1535.
[ELL 76]   J. R. Ellis, 'Zedology', in *Proc. LEP Summer Study*, Les Houches 1978, *CERN Yellow Report* 79-01 (1979) 618.
[EPS 76]   R. Epstein, J. Lattimer and D. Schramm, *Nature* 263 (1976) 198.
[ERL 98]   J. Erler and D. M. Pierce, *SLAC Pub. 7634* (1998).
[FAC 85]   Fackler, B. Jeziorski, W. Kolos, H. J. Monkhorst and K. Szalewicz, *Phys. Rev. Lett.* 55 (1985) 1388.
[FAL 94]   T. Falk, A. Olive and M. Srednicki, *Phys. Lett.* B339 (1994) 248.
[FOG 95]   G. L. Fogli *et al.*, *Phys. Rev.* D52 (1995) 5334.
[FOW 62]   W. Fowler, J. Greenstein and F. Hoyle, *Geophys. J. R. A.S.* 6 (1962) 148.
[FRA 96]   A. Franklin, *Rev. Mod. Phys.* 67 (1995) 457.
[FRE 83]   K. Freese, E. W. Kolb and M. S. Turner, *Phys. Rev.* D27 (1983) 1689; G. Steigman and M. S. Turner, *Nucl. Phys.* B253 (1985) 375.
[FRE 84]   K. Freese and D. Schramm, *Nucl. Phys.* B233 (1984) 167.
[FRI 86]   M. Fritschi, E. Holzschuh, W. Kündig, J. W. Petersen, R. E. Pixley and H. Stüssi, *Phys. Lett.* B173 (1986) 485.
[FRI 91]   M. Fritschi, E. Holzschuh, W. Kündig and H. Stüssi, *Nucl. Phys. B (Proc. Suppl.)* 19 (1991) 205.
[FRI 95]   H. Fritzsch and Zhi-Zhong Xing, *Phys. Lett.* B372 (1996) 265.
[FRO 96]   P. Froelich and A. Saenz, *Phys. Rev. Lett.* 77 (1996) 4724.
[FUJ 80]   K. Fujikawa and R. Shrock, *Phys. Rev. Lett.* 55 (1980) 963.
[FUK 87a]   M. Fukugita and S. Yazaki, *Phys. Rev.* D36 (1987) 3817.
[FUK 87b]   M. Fukugita, T. Yanagida, *Phys. Rev. Lett.* (1987) 1807.
[FUK 94]   Y. Fukuda *et al.*, *Phys. Lett.* B335 (1994) 237.
[FUK 98]   Y. Fukuda *et al. Evidence for oscillation of atmospheric neutrinos*, hepex/9807003, 3 July 1998.
[FUR 39]   W. H. Furry, *Phys. Rev.* 56 (1939) 1184.
[GAI 96]   T. K. Gaiser *et al.*, *Phys. Rev.* D54 (1996) 5578.
[GAL 89]   L. Gallagher, G. Steigman and D. Schramm, *Com. Astron. Astrophys.* (1989).
[GEL 79]   M. Gell-Mann, P. Ramond and R. Slansky in *Supergravity*, ed. D. Freedman and P. van Nieuwenhuizen (North Holland, Amsterdam 1979) 315.
[GEL 81]   G. B. Gelmini and M. Roncadelli, *Phys. Lett.* B99 (1981) 411.
[GEL 95]   G. Gelmini and E. Roulet, *Rep. Prog. Phys.* 58 (1995) 1207.
[GEO 81]   H. M. Georgi, S. L. Glashow and S. Nussinov, *Nucl. Phys.* B193 (1981) 297.
[GER 96]   G. Gervasio, in [KLA 96b].
[GOE 35]   M. Goeppert–Mayer, *Phys. Rev.* 48 (1935) 512.
[GOM 98]   NOMAD Collab., J. J. Gomes-Cadenas in *Proc. XVIII Int. Conf. on Neutrino Physics and Astrophysics*, Takayama, Japan, 4–9 June 1998, ed. by Y. Suzuki and Y. Totsuka. *Nucl. Phys. B (Proc. Suppl.)* 77 (1999) 3.
[GOT 74]   J. Gott, J. Gunn, D. Schramm and B. Tinsley, *Ap. J.* 194 (1974) 543.
[GRO 85]   K. Grotz and H. V. Klapdor, *Phys. Lett.* B157 (1985) 242.
[GRO 86]   K. Grotz and H. V. Klapdor, *Nucl. Phys.* A460 (1986) 395.
[GRO 90]   K. Grotz and H. V. Klapdor, *The Weak Interaction in Nuclear, Particle and Astrophysics* (Adam Hilger, Bristol, Philadelphia, 1990).

[H1 95]	H1 Collab., *Phys. Lett.* B353 (1995) 578; *Z. Phys.* C64 (1994) 545.
[H1 96]	H1 Collab., S. Aid *et al.*, *Phys. Lett.* B369 (1996) 173.
[HAL 84]	L. Hall and M. Suzuki, *Nucl. Phys.* B231 (1984) 419.
[HAM 53]	D. R. Hamilton, W. P. Alford and L. Gross, *Phys. Rev.* 92 (1953) 1521.
[HAN 49]	G. C. Hanna and B. Pontecorvo, *Phys. Rev.* 75 (1949) 983.
[HAX 84]	W. Haxton and G. Stephenson, Jr., *Prog. Part. Nucl. Phys.* 12 (1984) 409.
[HEL 96]	J. Hellmig, *PhD Thesis*, Univ. of Heidelberg, 1996.
[HIR 87]	K. Hirata *et al.*, *Phys. Rev. Lett.* 58 (1987) 1490.
[HIR 92]	K. S. Hirata *et al.*, *Phys. Lett.* B280 (1992) 146.
[HIR 95a]	M. Hirsch, H. V. Klapdor-Kleingrothaus and S. G. Kovalenko, *Phys. Rev. Lett.* 75 (1995) 17.
[HIR 95c]	J. Hirsch *et al.*, *Nucl. Phys.* A582 (1995) 124.
[HIR 95d]	M. Hirsch, H. V. Klapdor-Kleingrothaus and S. Kovalenko, *Phys. Lett.* B352 (1995) 1.
[HIR 96a]	M. Hirsch, H. V. Klapdor-Kleingrothaus and S. G. Kovalenko, *Phys. Lett.* B378 (1996) 17; *Phys. Rev.* D54 (1996) R4207.
[HIR 96b]	M. Hirsch, H. V. Klapdor-Kleingrothaus, S. G. Kovalenko and H. Päs, *Phys. Lett.* B372 (1996) 8.
[HIR 96c]	M. Hirsch, H. V. Klapdor-Kleingrothaus and S. Kovalenko, *Phys. Rev.* D53 (1996) 1329.
[HIR 96e]	M. Hirsch, H. V. Klapdor-Kleingrothaus and S. G. Kovalenko, *Phys. Lett.* B372 (1996) 181.
[HIR 97]	M. Hirsch, H. V. Klapdor-Kleingrothaus and S. G. Kovalenko, *Phys. Lett.* B398 (1997) 311; *Phys. Lett.* B403 (1997) 291.
[HMC 94]	Heidelberg–Moscow Collab., *Phys. Lett.* B336 (1994) 141.
[HMC 95]	Heidelberg–Moscow Collab., *Phys. Lett.* B356 (1995) 450.
[HMC 96]	Heidelberg–Moscow Collab., *Phys. Rev.* D54 (1996) 3641.
[HMC 97]	Heidelberg–Moscow Collab., *Phys. Rev.* D55 (1997) 54; *Phys. Lett.* B407 (1997) 219.
[HOL 92a]	E. Holzschuh, *Rep. Prog. Phys.* 55 (1992) 1035.
[HOL 92b]	E. Holzschuh, M. Fritschi and W. Kündig, *Phys. Lett.* B287 (1992) 381.
[HOY 64]	F. Hoyle and R. Taylor, *Nature* 203 (1964) 1108.
[ICA 94]	ICARUS Coll., *LNGS-94/99-I*, May 1994.
[IMB 87]	IMB Coll., R. M. Bionta *et al.*, *Phys. Rev. Lett.* 58 (1987) 1494.
[IOA 94]	A. Ioanissyan and J. W. F. Valle, *Phys. Lett.* B322 (1994) 93.
[JAC 75]	J. D. Jackson, *Classical Electrodynamics*, 2nd Edition (Wiley, New York, 1975)
[JEC 94]	B. Jeckelmann, P. F. A. Goudsmit and H. J. Leisi, *Phys. Lett.* B335 (1994) 326.
[JUN 96]	G. Jungmann, M. Kamionkowski and K. Griest, *Phys. Rep.* 267 (1996) 195.
[JÖR 94]	V. Jörgens *et al.*, *Nucl. Phys. (Proc. Suppl.)* B35 (1994) 378.
[KAJ 98]	SUPER-KAMIOKANDE Collab., T. Kajita *et al.*, in *Proc. XVIII Int. Conf. on Neutrino Physics and Astrophysics*, Takayama, Japan, 4–9 June 1998, ed. by Y. Suzuki and Y. Totsuka. *Nucl. Phys. B (Proc. Suppl.)* 77 (1999) 3.
[KAM 87]	K. Hirata *et al.*, *Phys. Rev. Lett.* 58 (1987) 1490.
[KAP 82]	I. G. Kaplan, V. N. Smutny and G. V. Smelov, *Phys. Lett.* B112 (1982) 417.
[KAP 83]	I. G. Kaplan, V. N. Smutny and G. V. Smelov, *Sov. Phys. JETP* 57 (1983) 483.
[KAP 85]	I. G. Kaplan, G. V. Smelov and V. N. Smutny, *Phys. Lett.* B161 (1985) 389.
[KAP 86]	I. G. Kaplan and G. V. Smelov in *Nuclear Beta Decays and Neutrino*, eds. T. Kotani, H. Ejiri and E. Takasugi (World Scientific, Singapore, 1986).
[KAP 88]	I. G. Kaplan and V. N. Smutny, *Adv. Quantum Chem.* 19 (1988) 289.
[KAW 87]	L. Kawano, D. Schramm and G. Steigmann, *Ap. J.* 327 (1987) 750.
[KAY 82]	B. Kayser, *Phys. Rev.* D26 (1982) 1662.
[KAY 83]	B. Kayser and A. S. Goldhaber, *Phys. Rev.* D28 (1983) 2341.
[KAY 84]	B. Kayser, *Phys. Rev.* D30 (1984) 1023.
[KAY 87]	B. Kayser, in: *New and Exotic Phenomena*, eds. O. Fackler and J. Tran Thanh Van (Editions Frontières, Gif-sur-Yvette, France, 1987) p. 349.

[KAY 88a] B. Kayser, F. Gibrat-Debu and F. Perrier, *The Physics of Massive Neutrinos* (World Scientific, Singapore, 1988).

[KAY 88b] B. Kayser, in: *Fifth Force-Neutrino Physics*, eds. O. Fackler and J. Tran Thanh Van (Editions Frontières, Gif-sur-Yvette, France, 1988) p. 109. This article discusses an illustration of this difficulty.

[KAY 88c] B. Kayser, in: *CP Violation*, ed. C. Jarlskog (World Scientific, Singapore, 1988) to be published.

[KAY 89] B. Kayser, S. Petcov and S. P. Rosen, in preparation.

[KLA 84] H. V. Klapdor and K. Grotz, *Phys. Lett.* B142 (1984) 323.

[KLA 94] H. V. Klapdor-Kleingrothaus, *Progr. Part. Nucl. Phys.* 32 (1994) 261.

[KLA 95] H. V. Klapdor-Kleingrothaus and A. Staudt, *Non-Accelerator Particle Physics* (IOP Publ., Bristol, Philadelphia, 1995).

[KLA 96a] H. V. Klapdor-Kleingrothaus, in *Proc. Int. Workshop on Double Beta Decay and Related Topics*, Trento, 1995, eds. H. V. Klapdor-Kleingrothaus and S. Stoica (World Scientific, Singapore, 1996).

[KLA 96b] H. V. Klapdor-Kleingrothaus and S. Stoica, eds., *Proc. Int. Workshop on Double Beta Decay and Related Topics*, Trento, 1995 (World Scientific, Singapore, 1996).

[KLA 97a] H. V. Klapdor-Kleingrothaus, Invited talk at *NEUTRINO 96*, Helsinki, 1996 (World Scientific, Singapore, 1997).

[KLA 97b] H. V. Klapdor-Kleingrothaus and M. Hirsch, *Z. Phys.* A359 (1997) 361.

[KLA 98] H. V. Klapdor-Kleingrothaus and H. Päs, eds., *Proc. Int. Workshop Beyond the Desert – Accelerator and Non-Accelerator Approaches*, Castle Ringberg, 1997 (IOP Publ., Bristol, Philadelphia, 1998).

[KLA 98a] H. Klapdor-Kleingrothaus, talk presented at the conference *Neutrino 98*, held in Takayama, Japan, June 1998.

[KLE 96] J. Kleinfeller, *Nucl. Phys. B (Proc. Suppl)* 48 (1996) 207.

[KOB 73] M. Kobayashi and T. Maskawa, *Prog. Theor. Phys.* 49 (1973) 652.

[KOB 80] I. Kobzarev *et al.*, *Sov. J. Nucl. Phys.* 32 (1980) 823.

[KOL 85] W. Kolos, B. Jeziorski, K. Szalewicz and H. J. Monkhorst, *Phys. Rev.* A31 (1985) 551.

[KOL 87] E. Kolb and M. Turner, *Phys. Rev. Lett.* D36 (1987) 2895.

[KOL 90] E. Kolls and M. Turner, *The Early Universe* (Addison Wesley, Redwood City, 1989).

[KUZ 90] V. Kuzmin, V. Rubakov and M. Shaposhnikov, *Phys. Lett.* B185 (1985) 36; M. Fukugita and T. Yanagida, *Phys. Rev.* D42 (1990) 1285; G. Gelmini and T. Yanagida, *Phys. Lett.* B294 (1992) 53; B. Campbell *et al.*, *Phys. Lett.* B256 (91) 457.

[KYU 84] A. V. Kyuldjiev, *Nucl. Phys.* B243 (1984) 387.

[L3 89] L3 Coll., B. Adeva *et al.*, *Phys. Lett.* B231 (1989) p. 509.

[L3 98] L3 Collab., *Phys. Lett.* B431 (1998) 199.

[LAM 89] D. Lamb, T. Loredo and F. Melia, *Phys. Rev.* D (1989).

[LAN 52] L. M. Langer and R. J. D. Moffat, *Phys. Rev.* 88 (1952) 689.

[LAN 88] P. Langacker, in *Neutrinos* ed. H. V. Klapdor (Springer, Heidelberg, New York, 1988) p. 71.

[LAN 92] P. Langacker, M. Luo and A. K. Mann, *Rev. Mod. Phys.* 64 (1992) 87.

[LAT 88] J. M. Lattimer *et al.*, *Phys. Rev. Lett.* 61 (1988) 23.

[LAV 96] M. Laveder, *Nucl. Phys. B (Proc. Suppl.)* 48 (1996) 188.

[LEE 77a] B. Lee and R. Shrock, *Phys. Rev.* D16 (1977) 1444.

[LEE 77b] B. W. Lee and S. Weinberg, *Phys. Rev. Lett.* 39 (1977) 165, although it was discovered independently by many people: P. Hut, *Phys. Lett.* 69B (1977) 85; K. Sato and H. Kobayashi, *Prog. Theor. Phys.* 58 (1977) 1775; M. I. Vysotskii, A. D. Dolgov and Ya. B. Zel'dovich, *JETP Lett.* 26 (1977) 188; and the basic argument was first noted by Ya. B. Zel'dovich, *Adv. Astron. Astrophys.* 3 (1965) 241; and H. Y. Chiu, *Phys. Rev. Lett.* 17 (1966) 712.

[LEE 94] D. G. Lee and R. N. Mohapatra, *Phys. Lett.* B329 (1994) 463.

[LEP 84]   *LEP Design Report*, CERN-LEP/84-01 (1984).

[LEP 86]   *Physics at LEP* (CERN 86-02, Geneva, 1986).

[LEP 90]   B. Adeva *et al.*, L3 Collab., *Phys. Lett.* B231 (1989) 509 and D. Decamp *et al.*,
            ALEPH Collab., *Phys. Lett.* B231 (1989) 519; M. Z. Akrawy *et al.*, OPAL Collab.,
            Phys. Lett. B240 (1990) 497; P. Aarnio *et al.*, DELPHI Collab., *Phys. Lett.* B241
            (1990) 435; D. Decamp *et al.*, ALEPH Collab., *Phys. Lett.* B235 (1990) 399. For
            most recent results see Section 2. 5.

[LEP 98a]   The LEP collab., summary prepared for 1998 winter conferences, LEPEWWG/
            98-01 (1998), http://www.cern.ch/LEPEWWG/stanmod/

[LEP 98b]   DELPHI Coll., *E. Phys. J.* C1 (1998) 1; L3 Collab., *Phys. Lett.* B415 (1997) 299;
            OPAL Coll., *E. Phys. J.* C2 (1998) 607.

[LEU 94]   M. Leurer, *Phys. Rev.* D49 (1994) 333.

[LIM 88]   C. S. Lim and W. J. Marciano, *Phys. Rev.* D37 (1988) 1368.

[LIN 79]   D. Lindley, *MNRAS* 188 (1979) 15.

[LIU 87]   J. Liu, *Phys. Rev.* D35 (1987) 3447.

[LOB 96]   V. M. Lobashev *et al.* in *Proc. 17th Int. Conf. on Neutrino Physics and Astrophysics
            (Helsinki 1996)*, ed. K. Enqvist *et al.* (World Scientific, Singapore, 1996).

[LOP 88]   J. L. Lopez and L. Durand, *Phys. Rev.* C37 (1988) 535.

[LUB 80]   V. A. Lubimov, E. G. Novikov, V. Z. Nozik, E. F. Tretyakov and V. S. Kosik, *Phys.
            Lett.* B94 (1980) 266.

[LUB 81]   V. A. Lubimov, E. G. Novikov, V. Z. Nozik, E. F. Tretyakov, V. S. Kozik and
            N. F. Myasoedov, *Sov. Phys. JETP* 54 (1981) 616.

[LUC 85]   J. C. Lucio, A. Rosado and A. Zepeda, *Phys. Rev.* D31 (1985) 1091.

[MA 78]   E. Ma and J. Okada, *Phys. Rev. Lett.* 41 (1978) 287.

[MAC 85]   MAC Coll., E. Fernandez *et al.*, *Phys. Rev. Lett.* 54 (1985) 1118; W. T. Ford *et al.*,
            Phys. Rev. D33 (1986) 3472.

[MAC 96]   D. Macina, *Nucl. Phys. B (Proc. Suppl.)* 48 (1996) 183.

[MAK 62]   Z. Maki, M. Nakagawa and S. Sakata, *Prog. Theor. Phys.* 28 (1962) 247.

[MAR 69]   R. E. Marshak *et al.*, *Theory of Weak Interaction in Particle Physics* (Wiley-
            Interscience, 1969) 71.

[MAR 77]   W. J. Marciano *et al.*, *Phys. Lett.* B67 (1977) 303.

[MAR 80]   W. J. Marciano and A. Sirlin, *Phys. Rev.* D22 (1980) 2695.

[MAR 86]   W. J. Marciano and Z. Parsa, *Ann. Rev. Nucl. Part. Sci.* 36 (1986) 171.

[MAR 97]   K. Martens, *Proc. Int. Europhys. Conf. on High Energy Physics*, 19–26 August 1997,
            Jerusalem. Israel.

[MARKII 89a] MarkII Coll., G. S. Abrams *et al.*, *Phys. Rev. Lett.* 63 (1989) p. 724.

[MARKII 89b] MarkII Coll., G. S. Abrams *et al.*, *Phys. Rev. Lett.* 63 (1989) p. 2173.

[MAY 87]   R. Mayle, J. Wilson and D. Schramm, *Ap. J.* 318 (1987) 288; D. Schramm,
            J. Wilson and R. Mayle, in *Proc. 1st Int. Symp. Underground Physics* (1985).

[MCK 82]   B. McKellar, *Los Alamos Report LA-UR-82-1197* (1982), unpublished.

[MIK 86]   S. P. Mikheyev and A.Yu. Smirnov, *Nuovo Cimento* C9 (1986) 17 and
            L. Wolfenstein, *Phys. Rev.* D17 (1978) 2364.

[MIN 95]   MINOS Coll., NUMI-L-63, February 1995.

[MOE 91]   M. K. Moe, *Phys. Rev.* C44 (1991) R931.

[MOE 94]   M. K. Moe, *Prog. Part. Nucl. Phys.* 32 (1994) 247; *Nucl. Phys. (Proc. Suppl.)* B38
            (1995) 36.

[MOH 80]   R. Mohapatra and G. Senjanovic, *Phys. Rev. Lett.* 44 (1980) 912.

[MOH 81]   R. Mohapatra and G. Senjanovic, *Phys. Rev.* D23 (1981) 165.

[MOH 86a]   R. N. Mohapatra, *Phys. Rev.* D34 (1986) 3457.

[MOH 86b]   R. N. Mohapatra, *Phys. Rev.* D34 (1986) 909.

[MOH 88]   R. N. Mohapatra and E. Takasugi, *Phys. Lett.* B211 (1988) 192.

[MOH 91]   R. N. Mohapatra, P. B. Pal, *Massive Neutrinos in Physics and Astrophysics* (World
            Scientific, Singapore, 1991).

[MOH 92]   R. N. Mohapatra, *Unification and Supersymmetry* (Springer, Heidelberg, New York,
            1986 and 1992).

[MOH 94]	R. N. Mohapatra, *Progr. Part. Nucl. Phys.* 32 (1994) 187.
[MOH 95]	R. N. Mohapatra and S. Nussinov, *Phys. Lett.* B346 (1995) 75.
[MOH 96]	R. N. Mohapatra, *Proc. Neutrino 96*, Helsinki, 1996 (World Scientific, Singapore, 1997) p. 290.
[MOH 97]	R. N. Mohapatra, *Proc. Int. School on Neutrinos*, Erice, Italy, 1997; *Progr. Part. Nucl. Phys.* 40 (1998), and private communication.
[MUT 88]	K. Muto and H. V. Klapdor, in *Neutrinos*, ed. H. V. Klapdor (Springer, Heidelberg, New York, 1988) p. 183.
[MUT 89]	K. Muto, E. Bender and H. V. Klapdor, *Z. Phys.* A334 (1989) 177, 187;
[MUT 91]	K. Muto, E. Bender and H. V. Klapdor-Kleingrothaus, *Z. Phys.* A339 (1991) 435.
[NEM 94]	NEMO Collaboration, *Nucl. Phys. (Proc. Suppl.)* B35 (1994) 369.
[NIE 82]	J. Nieves, *Phys. Rev.* D26 (1982) 3152.
[NIS 97]	K2K Collab., K. Nishikawa *et al.*, *Nucl. Phys. B (Proc. Suppl.)* 59 (1997) 289.
[NOT 88]	D. Nötzold, Max Planck Institute, München 1988 (unpublished).
[OKU 86]	L. B. Okun *et al.*, *Sov. Phys. JETP* 64 (1986) 446.
[OKU 88]	L. Okun, *Proc. Neutrino 88*, ed. J. Schereps, (World Scientific, 1989) 828.
[OLI 81b]	K. A. Olive, D. N. Schramm, G. Steigman, M. S. Turner and J. Yang, *Ap. J.* 246 (1981) 557.
[OLI 95]	K. A. Olive and G. Steigman, *Phys. Lett.* B354 (1995) 357.
[OPAL 89]	OPAL Coll., M. Z. Akrawy *et al.*, *Phys. Lett.* B231 (1989) p. 530.
[OPAL 95]	OPAL Coll., *Z. Phys.* C65 (1995) 47.
[OTT 95]	E. W. Otten, *Nucl. Phys. B (Proc. Suppl.)* 38 (1995) 26.
[PAG 87]	B. Pagels, *Proc. 1987 Rencontre de Moriond, Astrophys.* (1987) ed. J. Audouze.
[PDG 96]	Particle Data Group, *Phys. Rev.* D 54 (1996) 1.
[PDG 98]	The 1998 Review of Particle Physics, C. Caso *et al.*, *Euro. Phys. J.* C3 (1998) 1.
[PEE 66]	P. J. E. Peebles, *Phys. Rev. Lett.* 16 (1966) 410.
[PEL 93]	J. Peltoniemi and J. Valle, *Nucl. Phys.* B406 (1993) 409.
[PET 77]	S. T. Petcov, *Sov. J. Nucl. Phys.* 25 (1977) 340 and 641.
[PET 86]	S. Petcov, *Phys. Lett.* B178 (1986) 57.
[PET 94]	S. T. Petcov and A.Yu. Smirnov, *Phys. Lett.* B322 (1994) 109.
[PET 98]	SOUDAN-2 Collab., E. Peterson in *Proc. XVIII Int. Conf. on Neutrino Physics and Astrophysics*, Takayama, Japan, 4–9 June 1998, ed. Y. Suzuki and Y. Totsuka. *Nucl. Phys. B (Proc. Suppl.)* 77 (1999) 3.
[PIE 98]	F. Pietropaolo in *Proc. XVIII Int. Conf. on Neutrino Physics and Astrophysics*, Takayama, Japan, 4–9 June 1998, ed. Y. Suzuki and Y. Totsuka. *Nucl. Phys. B (Proc. Suppl.)* 77 (1999) 3.
[PLA 70]	M. A. Player and P. Sanders, *J. Phys.* 133 (1970) 1620.
[PON 57]	B. Pontecorvo, *Zh. Eksp. Theor. Phys.* 33 (1957) 549. (*Sov. Phys. JETP* 6 (1958) 429).
[PON 58]	B. Pontecorvo, *Zh. Eksp. Theor. Phys.* 34 (1958) 247.
[PON 67]	B. Pontecorvo, *Zh. Eksp. Theor. Phys.* 53 (1967) 1717 (*Sov. Phys. JETP* 26 (1967) 989).
[PON 71]	B. Pontecorvo, *JETP Lett.* 13 (1971) 199.
[PRI 95]	J. R. Primack, J. Holtzman, A. Klypin and D. O. Caldwell, *Phys. Rev. Lett.* 74 (1995) 2160.
[RAD 85]	E. Radescu, *Phys. Rev.* D32 (1985) 1266.
[RAF 88]	G. Raffelt and D. Dearborn, *Phys. Rev.* D37 (1988) 549.
[RAF 96]	G. Raffelt and J. Silk, *Phys. Lett.* B366 (1996) 429.
[RAG 94]	R. S. Raghavan, *Phys. Rev. Lett.* 72 (1994) 1411.
[REI 53]	F. Reines and C. L. Cowan, *Phys. Rev.* 92 (1953) 830; C. L. Cowan *et al.*, *Science* 124 (1956) 103.
[REI 76]	F. Reines, H. Gurr and H. Sobel, *Phys. Rev.* D37 (1976) 315.
[ROB 91]	R. G. H. Robertson, T. J. Bowles, G. J. Stephenson, D. J. Wark, J. F. Wilkerson and D. A. Knapp, *Phys. Rev. Lett.* 67 (1991) 957.
[ROM 97]	A. Romosan *et al.*, *Phys. Rev. Lett.* 78 (1997) 2912.

[RON 98]   MACRO Collab., F. Ronga in *Proc. XVIII Int. Conf. on Neutrino Physics and Astrophysics*, Takayama, Japan, 4–9 June 1998, ed. Y. Suzuki and Y. Totsuka. *Nucl. Phys. B (Proc. Suppl.)* 77 (1999) 3.

[ROY 92]   D. P. Roy, *Phys. Lett.* B283 (1992) 270.

[RUB 96]   C. Rubbia, *Proc. TAUP 95*, Toledo, 1995, *Nucl. Phys. B (Proc. Suppl.)* 48 (1996) 172.

[RUJ 81]   A. De Rújula, *Nucl. Phys. B* 188 (1981) 414.

[SAE 97]   A. Saenz and P. Froelich, *Phys. Rev.* B (1997) 2045.

[SAR 83]   S. Sarantakos, A. Sirlin and W. J. Marciano, *Nucl. Phys.* B217 (1983) 84.

[SAT 77]   K. Sato and H. Kobayashi, *Prog. Theor. Phys.* 58 (1977) 1775; D. A. Dicus, E. W. Kolb, V. L. Teplitz and R. V. Wagoner, *Phys. Rev.* D17 (1978) 1529; S. Miyama and K. Sato, *Prog. Theor. Phys.* 60 (1977) 1703.

[SAT 98]   CHORUS Collab., O. Sato in *Proc. XVIII Int. Conf. on Neutrino Physics and Astrophysics*, Takayama, Japan, 4–9 June 1998, ed. Y. Suzuki and Y. Totsuka. *Nucl. Phys. B (Proc. Suppl.)* 77 (1999) 3.

[SCH 80]   J. Schechter and J. Valle, *Phys. Rev.* D22 (1980) 2227.

[SCH 81]   J. Schechter and J. Valle, *Phys. Rev.* D24 (1981) 1883.

[SCH 82]   J. Schechter and J. Valle, *Phys. Rev.* D25 (1982) 2951.

[SCH 87a]   R. Scherrer, J. Applegate and C. Hogan, *Phys. Rev.* D35 (1987) 1151; C. Alcock, G. Fuller and G. Mathews, *Ap. J.* 320 (1987) 439; S. Dimopoulos, R. Esmailzadeh, L. Hall and G. Starkman, *Phys. Rev. Lett.* 60 (1988) 7.

[SCH 87b]   D. Schramm, *Comm. Nucl. Part. Physics* A17 (1987) 239.

[SCH 81]   J. Schechter and J. W. F. Valle, *Phys. Rev.* D25 (1982) 2951.

[SCH 91]   S. Schafroth, *Thesis, Univ. of Zurich*; T. A. Claxton, S. Schafroth and P. F. Meier, *Phys. Rev. A* 45 (1991) 6209.

[SCH 98]   D. N. Schramm and M. S. Turner, *Rev. Mod. Phys.* 70 (1998) 319.

[SHR 82]   R. Shrock, *Nucl. Phys.* B206 (1982) 359.

[SIL 85]   J. Silk, K. Olive and M. Srednicki, *Phys. Rev. Lett.* 55 (1985) 257; T. Gaisser, G. Steigman and S. Tilav, *Phys. Rev.* D34 (1986) 2206; K. Ng, K. Olive and M. Srednicki, *Phys. Lett.* B188 (1987) 138.

[SIM 96]   F. Simkovic *et al.*, *Proc. TAUP 95*, Toledo, 1995; *Nucl. Phys. B (Proc. Suppl.)* 48 (1996) 257.

[SIM 97]   F. Simkovic *et al.*, *Phys. Lett.* B393 (1997) 267.

[SIR 80]   A. Sirlin, *Phys. Rev.* D22 (1980) 971.

[SOU 92]   I. A. D'Souza and C. S. Kalman, *Preons, Models of Leptons, Quarks and Gauge Bosons as Composite Objects* (World Scientific, Singapore, 1992).

[STA 90]   A. Staudt, K. Muto and H. V. Klapdor-Kleingrothaus, *Europhys. Lett.* 13 (1990) 31.

[STE 77]   G. Steigman, D. Schramm and J. Gunn, *Phys. Lett.* B66 (1977) 202. G. Steigman, K. Olive, M. Turner and D. Schramm, *Phys. Lett.* B175 (1986) 33.

[STE 91]   J. Steinberger, *Phys. Rep.* 203 (1991) 345.

[STE 93]   R. I. Steinberg, *Proc. 5th Int. Workshop on Neutrino Telescopes*, Venezia, March 1993.

[STO 95]   W. Stoeffl and D. J. Decman, *Phys. Rev. Lett.* 75 (1995) 3237.

[SUT 76]   P. Sutherland *et al.*, *Phys. Rev.* D31 (1976) 2700.

[SUZ 96]   Y. Suzuki, *Proc. Neutrino 96*, Helsinki, June 1996.

[SUZ 97]   A. Suzuki, private communication 1997, and KAMLAND proposal (in Japanese).

[SUZ 98]   KAM-LAND Collab., A. Suzuki in *Proc. XVIII Int. Conf. on Neutrino Physics and Astrophysics*, Takayama, Japan, 4–9 June 1998, ed. Y. Suzuki and Y. Totsuka. *Nucl. Phys. B (Proc. Suppl.)* 77 (1999) 3.

[SWI 96]   A. M. Swift *et al.* in *Proc. 17th Int. Conf. on Neutrino Physics and Astrophysics (Helsinki 1996)*, ed. K. Enqvist *et al.* (World Scientific, Singapore, 1996).

[TAK 84]   E. Takasugi, *Phys. Lett.* B149 (1984) 372.

[TOM 87]   T. Tomoda and A. Faessler, *Phys. Lett.* B199 (1987) 475.

[TRE 75]   E. F. Tret'yakov, *Izv. Akad. Nauk SSSR, Ser. Fiz.* 39 (1975) 583, (English trans. *Bull. USSR Acad. Sci. Phys. Ser.* 39 (1975) 102).

[TRE 95]   V. I. Tretyak and Yu. Zdesenko, *At. Data Nucl. Data Tables* 61 (1995) 43.

[UA1 87a]   UA1 Coll., C. Albajar *et al.*, *Phys. Lett.* B185 (1987) 241.

[UA1 87b]   UA1 Coll., C. Albajar *et al.*, *Phys. Lett.* B198 (1987) 271.

[UA2 87]   UA2 Coll., R. Ansari *et al.*, *Phys. Lett.* B194 (1987) 158.

[VID 94]   G. S. Vidyakin *et al.*, *JETP Lett.* 59 (1994) 390.

[VID 96]   For a review of universality in the charged current, see H. Videau, *Proc. ICHEP96*, Warsaw, ed. S. Adjuk (World Scientific, 1998) 1013.

[VIL 93]   For a discussion of universality of neutrino coupling, see in particular: CHARMII coll., P. Vilain *et al.*, *Phys. Lett.* B320 (1993) 203.

[VOG 86]   P. Vogel and M. R. Zirnbauer, *Phys. Rev. Lett.* 57 (1986) 3148.

[VUI 93]   J.-C. Vuilleumier *et al.*, *Phys. Rev.* D48 (1993) 1009.

[VYS 77]   M. I. Vysotsky, Ya. B. Zel'dovich, M.Yu. Khlopov and V. M. Chechetkin, *Zh. Eksp. Teor. Fiz. Pis'ma* 26 (1977) 200; 27 (1978) 533; K. Sato and H. Kobayashi, *Prog. Theor. Phys.* 58 (1977) 1775; D. A. Dicus, E. W. Kolb and V. L. Teplitz, *Phys. Rev. Lett.* 39 (1977) 168; *Ap. J.* 221 (1978) 327; T. Goldman and G. J. Stephenson, *Phys. Rev.* D16 (1977) 2256.

[WAG 67]   R. Wagoner, W. Fowler and F. Hoyle, *Ap. J.* 148 (1967) 3.

[WAL 87]   T. Walker and D. Schramm, *Phys. Lett.* B195 (1987) 331.

[WEI 72]   S. Weinberg, in *Gravitation and Cosmology* (John Wiley and Sons Inc., New York, 1972).

[WEI 93]   Ch. Weinheimer *et al.*, *Phys. Lett.* B300 (1993) 210.

[WIL 83]   R. D. Williams and S. E. Koonin, *Phys. Rev.* C27 (1983) 1815.

[WIN 88]   K. Winter, *Proc. Neutrino 88* (World Scientific, 1988) 403, *Phys. Lett.* 345 (1995) 115.

[WOJ 98]   S. Wojcicki in *Proc. XVIII Int. Conf. on Neutrino Physics and Astrophysics*, Takayama, Japan, 4–9 June 1998, ed. Y. Suzuki and Y. Totsuka. *Nucl. Phys. B (Proc. Suppl.)* 77 (1999) 3.

[WOL 78]   L. Wolfenstein, *Phys. Rev.* D17 (1978) 2369.

[WOL 81]   L. Wolfenstein, *Phys. Lett.* B107 (1981) 77.

[YAN 79]   T. Yanagida, in: *Proc. Workshop on Unified Theory and Baryon Number in the Universe*, eds. O. Sawada and A. Sugamoto (KEK, Tsukuba, Japan, 1979).

[YAN 84]   J. Yang, M. Turner, G. Steigman, D. Schramm and K. Olive, *Ap. J.* 281 (1984) 493.

[YAS 94]   S. Yasumi *et al.*, *Phys. Lett.* B334 (1994) 229.

[YOU 95]   Ke You *et al.*, *Phys. Lett.* B265 (1995) 53.

[ZEI 98]   KARMEN Collab., B. Zeitnitz in *Proc. XVIII Int. Conf. on Neutrino Physics and Astrophysics*, Takayama, Japan, 4–9 June 1998, ed. Y. Suzuki and Y. Totsuka. *Nucl. Phys. B. (Proc. Suppl.)* 77 (1999) 3.

[ZOR 63]   J. C. Zorn, G. E. Chamberlin and V. W. Hughes, *Phys. Rev.* 129 (1963) 2566.

3

Theory of the interaction of neutrinos with matter*

3.1 Introduction

In his letter to the "Dear Radioactive Friends," W. Pauli identified neutrinos as extremely penetrating particles, perhaps as penetrating as gamma rays.

A few years later, with the Fermi theory of β-decays, the interaction of neutrinos with matter took a more precise form. Neutrinos turned out to be more penetrating than gamma rays but, otherwise, the characteristics of neutrinos envisaged by Pauli were quite confirmed. In more recent years, the elusive neutrino has turned out to be one of the most powerful tools to explore particle interactions.

The theory of neutrino interaction with matter, from Fermi to our times, has evolved in several respects.

The four-fermion interaction was eventually recognized as the low-energy manifestation of a more basic, gauge-theoretical interaction, mediated by massive intermediate vector bosons. A renormalizable theory, unifying the weak and the electromagnetic interactions, has been formulated and found to agree, as far as we can tell, with the experimental data. The asymmetry between the massless photon and the massive vector bosons has been interpreted as the manifestation of spontaneous symmetry breaking.

Only left-handed neutrinos seem to be coupled to matter. We have direct experimental proof of the existence of two different species of neutrinos (ν_e and ν_μ) and very strong arguments for a third species, ν_τ, associated with the τ lepton.

The main elements of the present theory of neutrino interactions with matter are reviewed in this chapter.

3.2 The basic elements of the standard electroweak theory

3.2.1 The minimal gauge group: SU(2) × U(1)

It is quite straightforward to identify the minimal gauge group of the observed weak and electromagnetic interactions.

We begin with the electron–neutrino system. The β-transitions

$$\mu^+ \rightarrow \bar{\nu}_\mu + e^+ + \nu_e$$
$$P \rightarrow N + e^+ + \nu_e$$

* L. Maiani, Physics Department, Universita' di Roma "La Sapienza," Roma, Italy, and INFN-Sezione di Roma, Italy.

are well described by the assumption that the $e^+-\nu_e$ pair is created by the operator:[1]

$$J_\mu = \tfrac{1}{2}\bar\nu_e\gamma_\mu(1+\gamma_5)e$$

$$= \bar\nu_{eL}\gamma_\mu e_L. \qquad (3.2.1)$$

The charges corresponding to J_μ and to its Hermitian conjugate, J_μ^+, act as raising and lowering operators on the doublet:

$$1_L = \begin{bmatrix} \nu_{eL} \\ e_L \end{bmatrix} \qquad (3.2.2)$$

In fact, if we define

$$T = \int d^3x J_0(x,t)$$

the canonical equal-time anticommutation rules for the electron and neutrino field lead to the commutation relations

$$[T, e_L^+] = \nu_{eL}^+; [T, \nu_{eL}^+] = 0$$

(we assume that currents are conserved, so that T is time-independent).

The electromagnetic current and charge for the $e-\nu_e$ system have the form:

$$J_\mu^{\text{e.m.}} = -\bar e\gamma_\mu e$$

$$= -[\bar e_L\gamma_\mu e_L + \bar e_R\gamma_\mu e_R]$$

$$Q = -\int d^3x[e_L^+e_L + e_R^+e_R] \qquad (3.2.3)$$

and bring in the right-handed electron field.

With only electrons and neutrinos it is not possible to obtain a closed algebra with T, T^+, and Q only.[2] In fact, the commutator

$$[T, T^+] = 2T^3 = \int d^3x[\nu_{eL}^+\nu_{eL} - e_L^+e_L] \qquad (3.2.4)$$

which completes an $SU(2)$ algebra with T and T^+, does not coincide with Q.

It is immediately seen, however, that the difference $Q - T^3$ commutes with T, T^+, and Q. If we call this difference Y, then T, T^+, T^3, and Y form an $SU(2) \times U(1)$ algebra, thus identified as the minimal algebra that includes the weak and electromagnetic charges [GLA 61]. The electric charge and the weak hypercharge

---

[1] We use the Bjorken–Drell γ-matrices and the metric $g_{\mu\nu} = \text{diag}(-1, -1, -1, +1)$. Left- and right-handed components of a Dirac field are projected by $(1+\gamma_5)/2$ and $(1-\gamma_5)/2$, respectively.
[2] This could be obtained by adding new, positively charged and neutral lepton fields as, e.g., in the Georgi–Glashow O(3) model [GEO 72].

are given explicitly by

$$Q = T^3 + Y \tag{3.2.5}$$

$$Y = \int d^3x \{ -\tfrac{1}{2} [\nu_{eL}^+ \nu_{eL} - e_L^+ e_L] - e_R^+ e_R \}. \tag{3.2.6}$$

$SU(2) \times U(1)$ charges obey the commutation rules:

$$[T^i, T^j] = i\epsilon^{ijk} T^k \tag{3.2.7}$$

$$[T^i, Y] = 0 \tag{3.2.8}$$

with

$$T = T^1 + iT^2$$
$$T^+ = T^1 - iT^2.$$

Finally, the currents corresponding to T^i and Y are

$$J_\mu^i = 1_L \gamma_\mu \frac{\tau^i}{2} 1_L \tag{3.2.9}$$

$$Y_\mu = -\tfrac{1}{2} 1_L \gamma_\mu 1_L - \bar{e}_R \gamma_\mu e_R \tag{3.2.10}$$

with 1_L defined in Eq. (3.2.2) and τ^i the three Pauli matrices.

Besides the left-handed lepton doublet, we had to introduce the right-handed electron field, a singlet under $SU(2)$, with $Y = 1$. A right-handed neutrino field could also be considered. At the present stage, this would be quite immaterial as such a field has vanishing $SU(2)$ charges and also vanishing $U(1)$ charges, because of Eq. (3.2.5), hence no gauge interaction at all.

3.2.2 The symmetric gauge interaction

We can now write the symmetric Lagrangian describing the interaction of the $e-\nu_e$ system with the gauge bosons of $SU(2) \times U(1)$. The interaction is determined by the so-called minimal substitution, well known from quantum electrodynamics, which is, in our case,

$$\partial_\mu 1_L \rightarrow \left[\partial_\mu - ig \frac{\tau^i}{2} W_\mu^i - ig' \left(-\frac{1}{2} \right) B_\mu \right] 1_L$$

for the left-handed doublet, 1_L, and

$$\partial_\mu e_R \rightarrow [\partial_\mu - ig'(-1) B_\mu] e_R$$

for the right-handed electron field. In correspondence to the four generators of the algebra (3.2.7-8) we have four gauge fields, classified according to their electric charge. W_μ^3 and B_μ are neutral, while W_μ^1 and W_μ^2 are the real components of the complex field describing the charged intermediate bosons. Defining

$$W_\mu = \frac{1}{\sqrt{2}}(W_\mu^1 + iW_\mu^2) \tag{3.2.11}$$

W_μ annihilates a W^- and creates a W^+. Also, the two simple factors of the gauge group require two independent couplings, g and g'.

From the kinetic energy term in the $e - \nu_e$ free Lagrangian density, we obtain in conclusion the following gauge-symmetric interaction Lagrangian:

$$L_{\text{fermions}} = L_0 + L_{\text{int}}$$

$$L_0 = +i1_L\partial^\mu\gamma_\mu 1_L + i\bar{e}_R\partial^\mu\gamma_\mu e_R$$

$$L_{\text{int}} = gW_\mu^i 1_L \frac{\tau^i}{2}\gamma_\mu 1_L + g'B_\mu\left[-\bar{e}_R\gamma_\mu e_R - \frac{1}{2}1_L\gamma_\mu 1_L\right]. \tag{3.2.12}$$

The interaction terms involving the charged vector boson correspond to $i = 1, 2$ in Eq. (3.2.12). Rewritten in terms of the complex field W_μ, they read:

$$L_W = g\frac{1}{\sqrt{2}}W_\mu\bar{e}_L\gamma_\mu\nu_{eL} + \text{Hermitian conjugate} \tag{3.2.13}$$

which shows that indeed the charged gauge field is coupled to the current J_μ we started with in Section 3.2.1, as required by the observation.

The Lagrangian (3.2.12) must be completed with the trilinear and quadrilinear term in the W fields, arising from the gauge-invariant Yang–Mills Lagrangian [YAN 54], and representing the self-interaction of the $SU(2)$ gauge fields:

$$L_{Y-M} = -\frac{1}{4}(W_{\mu\nu}^i W^{\mu\nu i} + B_{\mu\nu}B^{\mu\nu})$$

$$W_{\mu\nu}^i = \partial_\mu W_\nu^i - \partial_\nu W_\mu^i + g\epsilon^{ijk}W_\mu^j W_\nu^k$$

$$B_{\mu\nu} = \partial_\mu B_\nu - \partial_\nu B_\mu.$$

The complete Lagrangian density is, in conclusion,

$$L_{\text{tot}} = L_{\text{fermions}} + L_{Y-M}$$

$$= L_0 + L_{\text{int}} + L_{Y-M} \tag{3.2.14}$$

and L_{tot} is fully symmetric under the $SU(2) \times U(1)$ gauge transformations:

$$\delta 1_L(x) \rightarrow \frac{i}{2}[\tau^i \epsilon^i(x) - \epsilon(x)] 1_L(x)$$

$$\delta e_R(x) \rightarrow -i\epsilon(x) e_R(x)$$

$$\delta W_\mu^i(x) \rightarrow \frac{1}{g}\partial_\mu \epsilon^i(x) - \epsilon^{ijk} \epsilon^j(x) W_\mu^k(x)$$

$$\delta B_\mu(x) \rightarrow \frac{1}{g'}\partial_\mu \epsilon(x).$$

3.2.3 Breaking SU(2) × U(1) to U(1)$_Q$

In the real world, the $SU(2) \times U(1)$ symmetry is broken by the gauge boson and by the electron masses (at least) to the $U(1)_Q$ gauge symmetry associated with the electric charge.

In the original formulation of Glashow [GLA 61], appropriate symmetry-breaking terms were added to L_{tot}, to represent just these effects. In the next paragraph we discuss the more modern formulation of Weinberg [WEI 67] and Salam [SAL 68] where the reduction to $U(1)_Q$ is achieved by spontaneous symmetry breaking. In any case, it is useful to analyze the constraints imposed on the gauge boson mass terms and interactions by the exact $U(1)_Q$ gauge invariance, in a way that is independent from the origin of symmetry breaking itself.

To be compatible with the electric charge conservation, the gauge boson mass matrix must have the form

$$L_{Mass} = \tfrac{1}{2}M_W^2(W_\mu^1 W_\mu^1 + W_\mu^2 W_\mu^2) + \tfrac{1}{2}[W_\mu^3, B_\mu]M^2 \begin{bmatrix} W_\mu^3 \\ B_\mu \end{bmatrix}.$$

M_W is the mass of the charged particle associated with the complex field combination, Eq. (3.2.11), and M^2 is the 2×2 real symmetric matrix that mixes the neutral fields.

Local $U(1)_Q$ gauge invariance requires M^2 to have a vanishing eigenvalue, that is, vanishing determinant, corresponding to the massless photon. Hence, M^2 has the form

$$M^2 = \begin{bmatrix} (M_3)^2 & -M_3 M_0 \\ -M_3 M_0 & (M_0)^2 \end{bmatrix} \tag{3.2.15}$$

with arbitrary M_3 and M_0.

We define the eigenstates of the mass matrix according to

$$A_\mu = \cos\theta_W B_\mu + \sin\theta_W W_\mu^3$$
$$Z_\mu = -\sin\theta_W B_\mu + \cos\theta_W W_\mu^3 \tag{3.2.16}$$

with A_μ the massless photon field and Z_μ the massive neutral boson. The weak mixing angle θ_W and the Z mass can be obtained directly from Eq. (3.2.15)

$$\tan \theta_W = \frac{M_0}{M_3}$$

$$(M_Z)^2 = (M_0)^2 + (M_3)^2$$

$$= \frac{(M_3)^2}{\cos^2 \theta_W}. \tag{3.2.17}$$

In addition, it is convenient to introduce the so-called ρ parameter:

$$\rho = \frac{(M_W)^2}{(M_3)^2}$$

so that

$$M_W = (\rho)^{1/2} M_Z \cos \theta_W. \tag{3.2.18}$$

In all there are three independent parameters. We take them to be the W and Z masses and the weak mixing angle θ_W.

The angle θ_W specifies also the Z-matter coupling. Substituting the definition (3.2.16) in the interaction Lagrangian (3.2.12), the terms corresponding to the interaction of the neutral bosons read:

$$(g \sin \theta_W J_\mu^3 + g' \cos \theta_W Y_\mu) A_\mu + (g \cos \theta_W J_\mu^3 - g' \sin \theta_W Y_\mu) Z_\mu.$$

We have to impose that the massless particle associated with A_μ couples precisely to the e.m. current Eq. (3.2.3). This implies the further relations (e is the absolute value of the electron electric charge):

$$g \sin \theta_W = e$$

$$\frac{g'}{g} = \tan \theta_W \tag{3.2.19}$$

and the Z-matter coupling

$$L_Z = \frac{g}{\cos \theta_W} (J_\mu^3 - \sin^2 \theta_W J_\mu^{\text{e.m.}}) Z_\mu. \tag{3.2.20}$$

The interaction brings in only one more parameter, since the ratio between g and g' is fixed by Eq. (3.2.19) in terms of θ_W. In conclusion, four parameters (M_W, M_Z, θ_W, e) specify completely the massive boson masses and interactions in a $U(1)_Q$ invariant theory.

3.2.4 Spontaneous symmetry breaking

In field theory, it may happen that no quantum state exists that would correspond to a symmetric vacuum state. The real, stable, vacuum state is not symmetric under the symmetry. In this case, where a symmetric Lagrangian does not admit a symmetric vacuum, we speak of a spontaneously broken symmetry. If scalar (nonsinglet) fields are present, spontaneous breaking of the gauge symmetry is possible even for a weak, perturbative, interaction (for a more extended discussion see, e.g., [COL 85]).

The minimal possibility that gives rise to a realistic vector boson and fermion spectrum is to introduce a scalar $SU(2)$ doublet, with $Y = +\frac{1}{2}$ (see the comments at the end of this section):

$$\phi = \begin{bmatrix} \phi^+ \\ \phi^0 \end{bmatrix}. \tag{3.2.21}$$

The most general, renormalizable, and global $SU(2) \times U(1)$ symmetric Lagrangian density for ϕ is:

$$L_\phi = \partial^\mu \phi^+ \partial_\mu \phi - V(\phi^+ \phi)$$
$$V(\phi^+ \phi) = \mu^2 \phi^+ \phi + \lambda (\phi^+ \phi)^2 \tag{3.2.22}$$

with the Hamiltonian density:

$$H_\phi = \partial^0 \phi^+ \partial^0 \phi + \nabla \phi^+ \nabla \phi V(\phi^+ \phi).$$

Positivity of the Hamiltonian requires $\lambda > 0$.

For $\mu^2 > 0$, the Hamiltonian has an absolute minimum at $\phi = 0$. In the quantum theory, this field configuration corresponds to the ground, that is, vacuum state, and it is obviously symmetric under the full set of $SU(2) \times U(1)$ transformations. Quantizing the field ϕ around the field configuration $\phi = 0$ gives rise to a theory where the exact symmetry of the Lagrangian is faithfully reproduced in the physical states.

For $\mu^2 < 0$, the extremum at $\phi = 0$ corresponds to a local maximum of the potential V, rather than a minimum. Absolute minima are located on the surface:

$$\phi^+ \phi = -\frac{\mu^2}{2\lambda}. \tag{3.2.23}$$

A stable theory is obtained by developing the field ϕ around one (otherwise arbitrary) point of this surface, which we may take to be the point:

$$\langle \phi \rangle_0 = \eta \begin{bmatrix} 0 \\ 1 \end{bmatrix}$$
$$\eta = \sqrt{-\mu^2/2\lambda} \tag{3.2.24}$$

with no loss of generality. In the quantum theory, $\langle\phi\rangle_0$ is the vacuum expectation value (v.e.v.) of the field ϕ. Its nonvanishing value indicates that the ground state is not symmetric under $SU(2) \times U(1)$, the signal of spontaneous symmetry breaking.

The choice made in (3.2.24) is such that only ϕ^0 has a nonvanishing v.e.v. and the vacuum is $U(1)_Q$ invariant. The choice (3.2.24) is always possible and we see here a nice feature of the one-doublet model: The electric charge is automatically conserved.

Field fluctuations around $\langle\phi\rangle_0$ can be analyzed in terms of particles. A direct inspection of (3.2.22) shows that a massive neutral scalar particle is present, the Higgs boson σ, together with three massless scalar particles with electric charges $+1$, -1, and 0, the Goldstone bosons corresponding to the broken generators [GOL 62]. The Goldstone bosons are described by the fields $\xi^i(x)$ in the following parameterization of ϕ:

$$\phi = U(x) \begin{bmatrix} 0 \\ \eta + \frac{\sigma(x)}{\sqrt{2}} \end{bmatrix}$$

$$U(x) = \exp\left[i\xi^i(x)\frac{\tau^i}{\sqrt{2\eta}}\right].$$

(3.2.25)

Equation (3.2.25) makes it explicit that the fields ξ^i are associated with degrees of freedom that correspond to a local (gauge) $SU(2)$ transformation on the real, down, $SU(2)$ spinor:

$$\hat\phi = \begin{bmatrix} 0 \\ \eta + \frac{\sigma(x)}{\sqrt{2}} \end{bmatrix}.$$

We now extend the Lagrangian (3.2.22) to make it invariant under local $SU(2) \times U(1)$ transformations. The minimal substitution for ϕ is

$$\partial_\mu \phi \rightarrow \left[\partial_\mu + ig\frac{\tau^i}{2}W_\mu^i + ig'\left(+\frac{1}{2}\right)B_\mu\right]\phi.$$

(3.2.26)

After this substitution, the Lagrangian (3.2.22) can be added to L_{tot}, Eq. (3.2.14).

The further, and last, term to be included in the total Lagrangian describes the interaction of ϕ with the fermion fields. It is here that the question of the right-handed neutrino comes up again.

With no ν_{eR} field, the only possible Yukawa-type coupling of ϕ to the $e-\nu_e$ system is

$$L_{\phi-e} = g_e \bar{l}_L \phi e_R + \text{h.c.}$$

(3.2.27)

If a ν_{eR} field also exists, a further coupling can be constructed, with the $Y = -\frac{1}{2}$ field ϕ_c, the charge conjugate of ϕ:

$$\phi_c = i\tau^2\phi^* = \begin{bmatrix} \phi^{0*} \\ -\phi^- \end{bmatrix}.$$

The additional coupling is

$$L_{\phi_c-\nu} = g_\nu \bar{l}_L \phi_c \nu_{eR} + \text{h.c.} \tag{3.2.28}$$

and the complete Yukawa Lagrangian is

$$L_{\text{Yukawa}} = L_{\phi-e} + L_{\phi_c-\nu}. \tag{3.2.29}$$

In conclusion, the full, local $SU(2) \times U(1)$ symmetric Lagrangian is

$$L = L_{\text{fermions}} + L_\phi + L_{\text{Yukawa}} + L_{Y-M}. \tag{3.2.30}$$

3.2.5 The particle spectrum

We are now in a position to illustrate the mass spectrum resulting from the Lagrangian (3.2.30) in the spontaneously broken case $\mu^2 < 0$ and therefore $\eta \neq 0$.

1 Scalar particles   According to Eq. (3.2.25), the fields $\xi^i(x)$ can be removed by the local $S(2)$ gauge transformation associated with the $SU(2)$ matrix $U^{-1}(x)$. Thus, the Goldstone degrees of freedom disappear in the gauge invariant theory.[3] This is the well-known Higgs phenomenon, [HIG 64; ENG 64; KIB 67]. There remains only one degree of freedom in the scalar field, associated with the field $\sigma(x)$ and corresponding to a neutral particle, the Higgs boson, with a mass:

$$(M_\sigma)^2 = 2\lambda\eta^2. \tag{3.2.31}$$

The gauge in which

$$\xi^i(x) = 0 \tag{3.2.32}$$

is called the "unitary gauge," since in this gauge only the physical degrees of freedom appear. We shall work in the unitary gauge.

2 Vector bosons   After the minimal substitution Eq. (3.2.26), a mass term for the gauge bosons is generated by the term in L_ϕ which is quadratic in the gauge fields. Explicitly,

$$(L_{\text{mass}})_{\text{gauge bosons}} = (0, \eta) \left[-ig\frac{\tau^1}{2} W_\mu^i - ig'\left(+\frac{1}{2}\right)B_\mu \right]$$

$$\times \left[ig\frac{\tau^k}{2} W_\mu^k + ig'\left(+\frac{1}{2}\right)B_\mu \right]\begin{pmatrix} 0 \\ \eta \end{pmatrix}.$$

---

[3] Rather, they are transferred to the gauge sector and provide the longitudinal components of the massive vector fields.

Working out this expression, one finds that the mass matrix has the form given in Section 3.2.3, with

$$(M_W)^2 = \tfrac{1}{2}g^2\eta^2 \tag{3.2.33}$$

$$(M_3)^2 = \tfrac{1}{2}g^2\eta^2 = (M_W)^2 \tag{3.2.34}$$

$$(M_0)^2 = \tfrac{1}{2}g'^2\eta^2. \tag{3.2.35}$$

Equation (3.2.19) is reproduced automatically, but a new relation comes out, which translates into the relation between W and Z mass:

$$M_W = M_Z \cos\theta_W. \tag{3.2.36}$$

Equation (3.2.36) is quite well obeyed by the data (see Chapter 4). It is valid whenever only Higgs doublets and singlets are present. In more general cases, M_W and M_Z are independent and Eq. (3.2.18) defines the ρ parameter.

The W and Z couplings to matter coincide with those found previously, Eqs. (3.2.13) and (3.2.20).

In conclusion, the gauge boson masses and interactions are described by three independent parameters, in the Higgs doublet case: η, g, and g', or, equivalently, M_W, θ_W, and e.

3 Fermions The fermion mass arises from the Yukawa coupling. Replacing ϕ with its v.e.v. in Eq. (3.2.27), we read the electron mass directly:

$$M_e = g_e\eta. \tag{3.2.37}$$

A nonvanishing neutrino mass is obtained if there is a right-handed neutrino field, in which case Eq. (3.2.28) gives:

$$M_\nu = g_\nu\eta. \tag{3.2.38}$$

Neutrinos are, in this case, massive Dirac particles with $V-A$ weak coupling. The mass term and the interaction respect lepton number conservation.

With a doublet Higgs field, this is in fact the only way one can get a massive neutrino. Conversely, the absence of ν_{eR} and the restriction of Higgs scalars to $SU(2)$ doublets provides a natural way to obtain an exactly massless neutrino. A Majorana mass for the neutrino (see Chapter 2) and the corresponding lepton-number violation is possible without right-handed neutrino fields, if scalar $SU(2)$ triplets exist (see, e.g., [GEL 81]).

As we have seen in this section, the scalar field ϕ fulfils a double role: It gives a mass to the gauge bosons and to the fermions. Although these effects are both related to the breaking of the $SU(2) \times U(1)$ symmetry, they could arise from different sources, in principle. In this respect, the single-doublet model is really minimal. The electron mass term

$$\bar{e}_L e_R$$

behaves under $SU(2) \times U(1)$ as a doublet with $Y = -\frac{1}{2}$. The field (3.2.21) is thus uniquely selected to transform this term into the invariant Yukawa coupling, Eq. (3.2.27). A Higgs doublet with $Y = -\frac{1}{2}$ is also required to generate the coupling Eq. (3.2.28), and is needed in any case to provide a mass for the down-type quarks (Section 3.3.2). Again the one-doublet model is minimal, in that the $Y = -\frac{1}{2}$ field can be just the charge-conjugate of ϕ. This is not always possible. For example, in the supersymmetric extensions of the standard theory (see, e.g., [NIL 84]) two independent Higgs doublets with $Y = \pm\frac{1}{2}$ are required.

It is quite remarkable that the doublet choice needed for fermion masses is confirmed in the gauge sector by the (independent) fact that the relation Eq. (3.2.26) is experimentally well obeyed.

Although introduced for very good reasons, it must be stressed that we lack any direct evidence for the existence of the Higgs boson at present. The experimental search for the scalar Higgs boson is one of the primary goals of modern particle physics.

3.3 Lepton and quark families

3.3.1 Lepton families

The considerations in Section 3.2 can be extended to include further lepton multiplets, such as $\mu - \nu_\mu$ and $\tau - \nu_\tau$.

Assuming left-handed doublets,

$$\begin{pmatrix} \nu_{\mu L} \\ \mu_L \end{pmatrix} \quad \begin{pmatrix} \nu_{\tau L} \\ \tau_L \end{pmatrix}$$

and right-handed singlets:

$$\mu_R, \ \tau_R$$

the total weak and e.m. currents are simply obtained by adding muon and tau terms analogous to the electron ones. For example,

$$J_\lambda = \bar{\nu}_{eL}\gamma_\lambda e_L + \bar{\nu}_{\mu L}\gamma_\lambda \mu_L + \bar{\nu}_{\tau L}\gamma_\lambda \tau_L. \tag{3.3.1}$$

Equation (3.3.1) embodies the time-honored concept of electron-muon universality [PON 47; PUP 48; KLE 48; LEE 49; TIO 49], extended to the τ lepton.

A more interesting situation arises for the Yukawa couplings. The possibility of nondiagonal and intrinsically complex couplings may give rise to lepton flavor nonconservation, neutrino oscillations [PON 58; GRI 69; BIL 77] and CP violation,[1] if right-handed neutrino fields do exist.

---

[1] CP violating neutrino oscillations have been studied in [CAB 78].

The generalization of Eqs. (3.2.27–28) to the case of three lepton families is

$$L_{\phi-E} = (\bar{L}_L)_\alpha(g_E)_{\alpha\beta}\phi(E_R)_\beta + \text{h.c.} \tag{3.3.2}$$

$$L_{\phi_c-E} = (\bar{L}_L)_\alpha(g_N)_{\alpha\beta}\phi_c(N_R)_\beta + \text{h.c.} \tag{3.3.3}$$

$\alpha, \beta = e, \mu, \tau$, run over the lepton families $(L_L)_\alpha$, $(E_R)_\alpha$, and $(N_R)_\alpha$ denote the corresponding left-handed doublets and the charged or neutral right-handed singlets, respectively.

When the v.e.v. of ϕ is substituted into Eqs. (3.3.2–3), nondiagonal, complex mass matrices are generated for all leptons:

$$(L_{\text{mass}})_{\text{lepton}} = (\bar{E}_L)_\alpha(M_E)_{\alpha\beta}(E_R)_\beta + (\bar{N}_L)_\alpha(M_N)_{\alpha\beta}(N_R)_\beta + \text{h.c.}$$

$$M_E = g_E\eta \tag{3.3.4}$$

$$M_N = g_N\eta.$$

Several nondiagonal terms and phases in the mass matrices can be eliminated by field redefinitions. In fact, we may perform independent, unitary transformations on the doublets and on the electronlike singlets without affecting the gauge interactions:

$$L_L \rightarrow UL_L \tag{3.3.5}$$

$$E_R \rightarrow VE_R \tag{3.3.6}$$

with U and V 3×3 matrices in flavor space (the above transformations are usually referred to as belonging to the "horizontal" group). Under these transformations,

$$M_E \rightarrow U^+M_EV$$

and we are free to choose U and V such that the resulting M_E is diagonal, with real and positive entries.

If $M_N = 0$ (equivalently, if there are no right-handed neutrino fields), this is the whole story. The charged lepton mass matrix is real-diagonal and so are the couplings to the Higgs boson field, $\sigma(x)$. All leptonic interactions are exactly CP and lepton flavor conserving.

If M_N is nonvanishing, we have to diagonalize it with a further transformation on the neutrino fields. In the basis where M_E is real diagonal, we send

$$N_L \rightarrow U'N_L \tag{3.3.7}$$

$$N_R \rightarrow V'N_R. \tag{3.3.8}$$

The transformation (3.3.7) is done on neutrino fields alone, so that it breaks $SU(2) \times U(1)$. Furthermore, (3.3.7) breaks the conservation of the lepton flavors (if U' is nondiagonal) and CP (if it is complex). Expressed in terms of the physical charged lepton and neutrino fields, the charged current (3.3.1) now reads:

$$J_\lambda = \bar{N}_L\gamma_\lambda(U') + E_L. \tag{3.3.9}$$

Equation (3.3.9) is the leptonic counterpart of the Cabibbo–Kobayashi–Maskawa weak current [CAB 63; KOB 73] to be discussed in Section 3.4. On the other hand, the neutral current J_λ^3, Eq. (3.2.20), is always flavor diagonal because U' is unitary:

$$J_\mu^3 = \tfrac{1}{2}(\bar{N}_L(U')^+\gamma_\lambda U'N_L - \bar{E}_L\gamma_\lambda E_L)$$

$$= \tfrac{1}{2}(\bar{N}_L\gamma_\lambda N_L - \bar{E}_L\gamma_\lambda E_L). \qquad (3.3.10)$$

The same applies to the e.m. current, so that the photon and Z coupling to the leptons are flavor conserving, the leptonic counterpart of the Glashow–Iliopoulos–Maiani (GIM) mechanism [GLA 70].

3.3.2 Quarks

Quarks and gluons are the basic degrees of freedom of hadronic matter [GEL 64; and see, e.g., LEE 81; OKU 82]. In this section we describe the electroweak interactions of quarks. The observed semileptonic decays indicate a quite remarkable lepton–hadron universality. This is accounted for by classifying left- and right-handed quark fields in $SU(2)$ doublets and singlets, respectively:

$$\begin{pmatrix} u_L \\ d_L \end{pmatrix} \begin{pmatrix} c_L \\ s_L \end{pmatrix} \begin{pmatrix} t_L \\ b_L \end{pmatrix} \cdots$$

$$u_R, c_R, t_R, \ldots, d_R, s_R, b_R, \ldots .$$

Quarks of each flavor come in three colors. The corresponding $SU(3)$ gauge symmetry generates the basic strong interactions, and it is assumed to commute with the full electroweak symmetry, $U(1)$ included. Thus, quarks of the same flavor and different color have the same weak hypercharge and electric charge.[2] Uplike quarks $(u, c, t, \ldots)$ have electric charge $Q = \tfrac{2}{3}$ and downlike quarks $(d, s, b, \ldots)$ have $Q = -\tfrac{1}{3}$. To be consistent with Eq. (3.2.5), we have to assign

$$Y(Q_L) = \tfrac{1}{6}$$

$$Y(U_R) = \tfrac{2}{3}$$

$$Y(D_R) = -\tfrac{1}{3}$$

---

[2] Schemes wherein color generators do not commute with the electric charge and quarks have integral charges have been considered by many authors, in particular by Pati and Salam [PAT 73], starting from the pioneering work of Han and Nambu [HAN 65]. Deep inelastic scattering data do favor fractionally charged quarks and neutral gluons. The factored structure $SU(3)_{colour} \times SU(2) \times U(1)$ is also required by exact color symmetry and permanent quark confinement.

where Q_L, U_R, and D_R denote the doublets and uplike and downlike singlets, respectively.

The hadronic parts of the $SU(2) \times U(1)$ currents are, in conclusion,

$$(J_\mu^i)_{\text{hadr}} = \bar{Q}_L \gamma_\mu \frac{\tau^i}{2} Q_L$$
$$(Y_\mu)_{\text{hadr}} = \tfrac{1}{6} \bar{Q}_L \gamma_\mu Q_L + \tfrac{2}{3} \bar{U}_R \gamma_\mu U_R - \tfrac{1}{3} \bar{D}_R \gamma_\mu D_R$$

(here and in the following, summation of color indices is understood).

Quark masses break the $SU(2) \times U(1)$ symmetry and they arise from the quark–Higgs field Yukawa couplings, in the same way as lepton masses. The analysis done in Section 3.3.1 can be repeated word by word. The invariant quark–Higgs couplings are

$$L_{\phi-U} = (\bar{Q}_L)_\alpha (g_U)_{\alpha\beta} \phi_c (U_R)_\beta + \text{h.c.} \qquad (3.3.11)$$

$$L_{\phi_c-D} = (\bar{Q}_L)_\alpha (g_D)_{\alpha\beta} \phi (D_R)_\beta + \text{h.c.} \qquad (3.3.12)$$

and they give rise to the mass terms:

$$(L_{\text{mass}})_{\text{quark}} = (\bar{Q}_L)_\alpha (M_U)_{\alpha\beta} (U_R)_\beta + (\bar{Q}_L)_\alpha (M_D)_{\alpha\beta} (D_R)_\beta + \text{h.c.}$$
$$M_{U,D} = g_{U,D} \eta. \qquad (3.3.13)$$

We set ourselves in the basis where the M_U is already diagonal, with real and positive entries, and diagonalize M_D with the further transformations:

$$D_L \to U_{\text{CKM}} D_L \qquad (3.3.14)$$

$$D_R \to V D_R. \qquad (3.3.15)$$

In terms of the physical fields, the charged weak hadronic current now contains nondiagonal, flavor-changing terms:

$$(J_\mu)_{\text{hadr}} = \bar{U}_L \gamma_\mu U_{\text{CKM}} D_L \qquad (3.3.16)$$

with U_{CKM} the Cabibbo–Kobayashi–Maskawa matrix [CAB 63; KOB 73]. For three- quark weak doublets, the mixing matrix depends on three real angles and one CP-violating phase (see Section 3.4 and Chapter 4 for more details).

Flavor violation via quark mixing is a well-studied phenomenon, known since the observation of strange particle weak decays and the formulation of the Cabibbo theory. The fact that the observed CP violation in K decays is induced by a complex phase in the mixing matrix has recently received an important confirmation, with the observation of a nonvanishing value for the ratio ϵ'/ϵ [BUR 88], which measures the amount of direct CP violation in the weak transition $K_L \to \pi\pi$. However, the matter has not been settled yet, since the result of a more recent experiment is compatible with vanishing ϵ'/ϵ [PAT 90].

Unlike the charged current, the neutral hadronic weak and e.m. currents are flavor diagonal, because of the unitarity of U_{CKM} (GIM mechanism [GLA 70]).

3.3.3 Anomalies

The proof that nonabelian spontaneously broken gauge theory is renormalizable and unitary [HOO 71; LEE 72; ABE 73] makes essential use of gauge invariance, in particular of the conservation of the currents to which the gauge bosons are coupled. Current conservation in a quantum field theory is, however, quite a subtle problem because of the possible occurrence of anomalies, that is, quantum corrections that spoil the conservation of a classically conserved current.

The origin of quantum anomalies has to do with the regularization procedure, needed to carry over the renormalization of the theory in a meaningful way. If the regularization does not respect a given symmetry, it may happen that the corresponding currents, although conserved at the classical level, are not conserved in the full quantum theory.

The typical example is that of the axial current

$$J_\mu^5 = \bar{e}\gamma_\mu\gamma_5^e$$

in quantum electrodynamics. In the Pauli–Villars regularization, one has to introduce one (or more) fictitious, heavy, negative metric, fermion(s), E, to cut off the divergences in the electron quantum loops. The total axial current associated to the chiral transformations over all fields is now

$$(J_\mu^5)_{\text{tot}} = \bar{e}\gamma_\mu\gamma_5 e + \bar{E}\gamma_\mu\gamma_5 E.$$

Even in the limit where the physical electron is massless, the axial current is not conserved, owing to the presence of the heavy particle.[3] When the mass of the latter is sent to infinity and the cutoff dependence removed by renormalization, the hard breaking of chiral symmetry introduced by the regularization remains, in the form of the Adler–Bell–Jackiw anomaly [ADL 69; BEL 69] arising from the triangular diagrams of Fig. 3.3.1.

The result is that the axial current is not conserved even for a massless electron:

$$\partial^\mu J_\mu^5 = Ce^2 F_{\mu\nu}\bar{F}^{\mu\nu} + O(m_e) \qquad (3.3.17)$$

---

[3] A similar, hard breaking of the chiral symmetry occurs in lattice gauge theories with fermions [NIE 81; KAR 81; BOC 85].

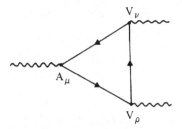

Fig. 3.3.1   The basic triangle diagram responsible for the Adler–Bell–Jackiw anomaly.

with:

$$\bar{F}^{\mu\nu} = \frac{1}{2}\epsilon^{\mu\nu\rho\sigma}F_{\rho\sigma}$$

$$C = \frac{1}{8\pi^2}.$$

The role of anomalies in the electroweak gauge theory has been studied in [BOU 72] and [GRO 72], with the conclusion that, indeed, renormalizability is spoiled if the Adler–Bell–Jackiw fermion anomalies of the currents coupled to the gauge fields do not vanish.

To work out the conditions posed on the theory by anomaly cancellation, it is convenient to shift to an alternative, although equivalent, notation to the one used until now. We replace all right-handed fields, leptons, and quarks with the corresponding charge-conjugate left-handed fields; for example, we replace the field e_R, which annihilates a right-handed electron and creates a left-handed positron, with the positron left-handed field, which annihilates a left-handed positron and creates a right-handed electron:

$$e_R \rightarrow (e_C)_L$$

In this way, the two fields e_L and $(e_C)_L$ are associated with the same set of physical states as e_L and e_R, with the advantage that all fermion fields now have the same chirality.[4] With this notation, there are only left-handed chiral currents, which we denote by

$$J_\mu^A = \bar{f}_L\gamma_\mu X^A f_L \tag{3.3.18}$$

with A running over the group generators and f_L a row vector including all fermion fields.

The currents (3.3.18) are conserved in the symmetric theory, at the classical level. At the quantum level, current conservation is endangered by the ABJ anomaly arising from triangle diagrams like those in Fig. 3.3.1, with photons replaced by the

---

[4] This form of field bookkeeping is used in the construction of grand-unified theories where $SU(2)$ singlets and doublets are included in the same multiplet and therefore must have the same chirality.

gauge bosons. Taking into account the different factors at the vertices, one derives the result [BAR 69]

$$\partial^\mu J_\mu^A = -\tfrac{1}{12} C D^{ABC} g^B g^C F_{\mu\nu}^B \bar{F}^{C\mu\nu} \tag{3.3.19}$$

with

$$D^{ABC} = \mathrm{Tr}(X^A \{X^B, X^C\})$$

where $g^A = g$ or g' for A belonging to $SU(2)$ or $U(1)$, respectively, and the trace goes over all fermions.

For definiteness, we shall restrict to fermion fields that are either doublets or singlets under $SU(2)$, so that

$$
\begin{aligned}
(2T^3)^2 &= 1, \quad \text{for } SU(2) \text{ doublets} \\
&= 0, \quad \text{for singlets.}
\end{aligned} \tag{3.3.20}
$$

The identity

$$\mathrm{Tr}(\tau^i \{\tau^j, \tau^k\}) = 0$$

implies that the anomaly vanishes when all indices in (3.3.19) belong to $SU(2)$.[5]

To analyze the other cases, it is convenient to replace Y with the electric charge, using Eq. (3.2.5). We need to consider three cases.

1 $\mathrm{Tr}(Q^3)$. This quantity vanishes separately for leptons and quarks, since electrically charged particles are always present in conjugated pairs, for example e_L and e_{CL}.

2 $\mathrm{Tr}[(Q^2)T^3]$. Again using Eq. (3.2.5), we obtain

$$
\begin{aligned}
\mathrm{Tr}[(Q^2)T^3] &= \mathrm{Tr}\{[Y^2 + 2Y\,T^3 + (T^3)^2]T^3\} \\
&= 2\mathrm{Tr}[Y(T^3)^2] = \tfrac{1}{2}\mathrm{Tr}_{\mathrm{doub}}(Y) \\
&= \tfrac{1}{2}\mathrm{Tr}_{\mathrm{doub}}(Q)
\end{aligned}
$$

where we have used (3.3.20), and $\mathrm{Tr}_{\mathrm{doub}}$ means that the trace goes only over $SU(2)$ doublets.

3 $\mathrm{Tr}[Q(T^3)^2]$. This gives again the trace of Q over the doublets, because of Eq. (3.3.20).

In conclusion, the presence of anomalies is determined by

$$D = \mathrm{Tr}_{\mathrm{doub}}(Q). \tag{3.3.21}$$

---

[5] In fact, this is true for any set of $SU(2)$ multiplets, because any $SU(2)$ representation, R, is equivalent to its complex conjugate, $R*$. From $R = R*$, it follows that $D^{ABC}(R) = D^{ABC}(R*)$. On the other hand, $D^{ABC}(R) = -D^{ABC}(R*)$ in general, since D contains an odd number of generators and therefore is odd under conjugation. Thus, $D^{ABC}(R) = 0$ for $SU(2)$ and, in general, for any group with only real representations. In the usual terminology, $SU(2)$ is anomaly-free, albeit in a trivial way.

The contributions of a single lepton or quark multiplet, for example, the ν_e-e or the $u-d$ doublets are

$$D_{\text{lept}} = -1$$
$$D_{\text{quark}} = 3\left(\tfrac{2}{3} - \tfrac{1}{3}\right) = +1$$

(the factor 3 comes from color); thus a single quark and lepton family has a vanishing anomaly by itself.

The interplay between quarks and leptons implied by the vanishing of the anomaly is a most convincing indication of a more unified scheme, where the strong and electroweak interactions originate from a simple gauge group and, correspondingly, quarks and leptons make part of a single, anomaly-free, structure. An example is given by the grand-unified $SO(10)$ scheme [FRI 75; CHA 77; GEO 79] whereby quarks and leptons of each generation fill up an irreducible, 16-dimensional, multiplet. The cancellation of quark versus lepton anomalies follows from the fact that $SO(10)$ itself is anomaly-free; that is, ABJ anomalies vanish for any representation.

The contribution to the anomaly of each multiplet is mass independent. One may wonder what happens when the mass of a given doublet, the $b-t$ doublet, say, becomes large, for example, with respect to the W and Z masses.[6] In this case, we may integrate out the b and t quark fields in the Feynman path integral that defines the correlation functions. After this is done, we are left with an expression of the form

$$Z = \int [\delta\Omega] \det[iD^\mu \gamma_\mu + g_{\phi-b}\phi + g_{\phi_c-t}\phi_c] e^{i(S_{\text{light}})}$$

where D^μ is the gauge-covariant derivative for the b and t fields and the functional integral is done over the light degrees of freedom only (the other quark and the lepton, Higgs and gauge fields). After setting

$$\phi = \langle\phi\rangle_0 + \begin{pmatrix} 0 \\ \dfrac{\sigma(x)}{\sqrt{2}} \end{pmatrix}$$

one can develop the determinant in inverse powers of $M_{b,t} = (g_{\phi-b}, g_{\phi_c-t})\eta$. In this way, we obtain a nonpolynomial, nonrenormalizable, effective interaction of the Wess–Zumino type, involving vector and scalar fields. This interaction gives precisely the same contribution to the anomaly as the b and t pair did before. We have traded an anomaly-free renormalizable theory, including the $b-t$ doublet, with an equally anomaly-free, but nonrenormalizable, effective theory involving

---

[6] This can be done by tuning the gauge and the b and t Yukawa couplings without necessarily leaving the domain of validity of perturbation theory.

light particles only. The nonrenormalizable effective theory contains in an essential way the mass scales $M_{b,t}$. Once we get to momentum scales larger than $M_{b,t}$, we "see" the elementary $b-t$ doublet and the original renormalizable theory comes back again.

3.4 Low-energy interactions

3.4.1 Low-energy limit and the parameters of the standard theory

In the very low energy limit, the massive vector bosons disappear from the theory described in the previous sections. Only the light fermions, the photon, and, possibly, the Higgs boson remain. To the lowest order, fermions are subject to the familiar QED interaction, to the interaction with the Higgs boson, σ, and to the residual weak interaction due to W and Z exchange. The latter takes the familiar form of a current × current, four-fermion interaction.

The muon-decay effective Lagrangian due to W-exchange is obtained from Eq. (3.2.13). It has the canonical $V-A$ form [FEY 58; MAR 58; SAK 58]:

$$L_{\text{eff}} = \frac{G}{\sqrt{2}} [\bar{e}\gamma_\lambda(1+\gamma_5)\nu_e][\bar{\nu}_\mu\gamma_\lambda(1+\gamma_5)\mu] \tag{3.4.1}$$

with the Fermi constant G related to the gauge coupling by

$$\frac{G}{\sqrt{2}} = \frac{g^2}{8M_W^2}. \tag{3.4.2}$$

Neutral-current neutrino processes are induced by Z-exchange, with an effective Lagrangian obtained from Eqs. (3.2.18–20). In the case of ν_μ,

$$L_{\text{eff}} = \rho\frac{G}{\sqrt{2}} [\bar{\nu}_\mu\gamma_\lambda(1+\gamma_5)\nu_\mu]$$
$$[\bar{f}\gamma_\lambda(1+\gamma_5)(T^3 - Q\sin^2\theta_W)f - \sin^2\theta_W\bar{f}\gamma_\lambda(1-\gamma_5)Qf] \tag{3.4.3}$$

where f is any fermion ($f \neq \mu$, ν_μ) and $\rho = 1$ with Higgs doublets only.

The electromagnetic and weak interactions in the low-energy limit are already sufficient to determine all couplings of the theory, but for the Higgs boson mass. We review here one convenient way to fix the independent couplings from the corresponding number of physical quantities.[1]

---

[1] The choice of the physical quantities that determine the independent couplings of the theory is obviously nonunique. The invariance of physical quantities from the choice of the normalization conditions is expressed by the renormalization group equations (see, e.g., [LEE 81] or [COL 85]). In practice, since it is wiser to use those quantities that can be measured more precisely, the choice is considerably restricted.

Table 3.4.1. *Basic parameters of the standard electroweak theory*

Parameter	Value	Source
α^{-1}	137.035963(15)	Josephson effect
$\sin^2\theta_W$	0.234 ± 0.013	Neutral-current neutrino vs. antineutrino scattering off $I=0$ nuclei
ρ	1.002 ± 0.015	Neutral-current vs. charged-current neutrino scattering off $I=0$ nuclei
G	$1.16634(2) \cdot 10^{-5}\,\text{GeV}^{-2}$	Muon decay

1 The gauge boson sector The vector boson mass and interactions are determined by g, g', M_W, and ρ. Using Eqs. (3.2.19), we can replace g and g' with the electron charge (one uses, rather, the fine-structure constant α) and $\sin\theta_W$. The dimensionful parameter M_W can be replaced by the Fermi constant.

The fine-structure constant is determined from the Josephson effect; $\sin^2\theta_W$ can be obtained from the ratio of antineutrino versus neutrino neutral-current cross section off a given target (e.g., $I=0$ nuclei or electrons). The ratio of neutral-current versus charged-current neutrino cross section is directly related to ρ, once $\sin^2\theta_W$ is given. Finally, the value of the Fermi constant can be obtained from muon decay.

The present values of $\sin^2\theta_W$ and ρ, as obtained from neutral-current neutrino scattering off $I=0$ nuclei are reported in Table 3.4.1 (the determination of $\sin^2\theta_W$ is discussed in more detail in Chapter 4). The measured value of ρ supports the doublet Higgs model quite well.

When expressed in terms of the basic parameters, the W mass, the Higgs field v.e.v., and the electroweak gauge couplings have the form

$$(M_W)^2 = \frac{\pi\alpha}{\sqrt{2G}}\frac{1}{\sin^2\theta_W} = (37.28\,\text{GeV})^2\frac{1}{\sin^2\theta_W} \tag{3.4.4}$$

$$\eta = (2\sqrt{2G})^{-1/2} = 174\,\text{GeV} \tag{3.4.5}$$

$$\alpha_W = \frac{g^2}{4\pi} \cong \frac{1}{32} \tag{3.4.6}$$

$$\alpha'_W = \frac{g'^2}{4\pi} \cong \frac{1}{105}.$$

The W-mass is now a prediction of the theory, whose value agrees quite well with the experimental determination (see Chapter 4).

2 The lepton sector The Yukawa couplings of charged leptons are directly related to their physical masses. From Eqs. (3.3.2–4) we read:

$$L_{\sigma-1} = \sum_{1}\frac{g_1}{\sqrt{2}}\sigma(x)\bar{1}(x)1(x)$$

$$= \frac{1}{\eta\sqrt{2}}\sum_{1}m_1\sigma(x)\bar{1}(x)1(x) \tag{3.4.7}$$

and from (3.4.5) we obtain

$$\frac{g_e^2}{4\pi} = \frac{G}{\pi\sqrt{2}} m_e^2 \cong 6.9 \times 10^{-13}$$

$$\frac{g_\mu^2}{4\pi} \cong 2.9 \times 10^{-8}$$

$$\frac{g_\tau^2}{4\pi} \cong 8.6 \times 10^{-6}. \tag{3.4.8}$$

Limits to neutrino masses and mixing angles are discussed in Chapter 2.

The smallness of the Yukawa versus the gauge couplings, and the wide disparities of the Yukawa couplings among themselves, constitute perhaps the most unsatisfactory feature of the standard theory.

3 The quark sector Following Eqs. (3.3.13–14), we can express the downtype quark mass, or Yukawa coupling, matrix according to

$$M_D = g_D \eta = U_{\text{CKM}} m_D V^+$$

with m_D real, diagonal, and positive. The matrix V is eliminated from the very beginning by the transformation (3.3.15), which is just an $SU(2) \times U(1)$ invariant field redefinition. Therefore the mass matrix and the corresponding Yukawa couplings are determined by the unitary matrix U_{CKM} and by the physical quark masses m_d, m_s, m_b.

After the transformation (3.3.15), the coupling of the neutral Higgs boson to the physical quarks is diagonal in flavor and it is determined by the diagonal quark masses themselves, analogously to the lepton case:

$$L_{\sigma-q} = \frac{1}{\eta\sqrt{2}} \sum_q m_q \sigma(x) \bar{q} q(x).$$

The following set of quark masses is commonly quoted,[2] derived from hadron spectroscopy (see [GAS 82]) and from the nonobservation of the t quark in P–P collisions [ABE 90]:

$$m_u = m_d \cong 6 \,\text{MeV}$$

$$m_c \cong 1.8 \,\text{GeV}; \quad m_s \cong 170 \,\text{MeV}$$

$$m_b \cong 5.0 \,\text{GeV}; \quad m_t > 77 \,\text{GeV}. \tag{3.4.9}$$

Nondiagonal couplings associated with the U_{CKM} matrix appear in the interaction of the would-be Goldstone boson fields $\xi^i(x)$ – see Eq. (3.2.25) – which,

[2] Quark confinement makes the definition of quark mass ambiguous, especially for light quarks. The values quoted here correspond to the "current" quark masses, related to the pion and kaon masses by the partial conservation of the axial current. This mass still depends (although only logarithmically) on the momentum scale where the scalar quark density $\bar{q}q(x)$ is normalized to the free value.

however, disappear completely in the unitary gauge. In the unitary gauge, nondiagonal couplings remain only in the weak charged current through the CKM matrix – see Eq. (3.3.16) – and are determined by the flavor-changing hadronic β-decays.

A convenient parameterization of U_{CKM}, in terms of three real angles and one *CP*-violating phase, is the following [MAI 77]:

$$U_{CKM} = \begin{pmatrix} c_\beta c_\theta & c_\beta s_\theta & s_\beta e^{i\phi} \\ -c_\gamma s_\theta - c_\theta s_\gamma s_\beta e^{-i\phi} & c_\gamma c_\theta - s_\gamma s_\beta s_\theta e^{-i\phi} & c_\beta s_\gamma \\ s_\theta s_\gamma - c_\gamma c_\theta s_\beta e^{-i\phi} & -c_\theta s_\gamma - c_\gamma s_\theta s_\beta e^{-i\phi} & c_\gamma c_\beta \end{pmatrix} \qquad (3.4.10)$$

where we have used the notation $c_\beta = \cos\beta$, and so on. The smallness of the mixing angles (see the next section) and the empirical ordering:

$$\theta \gg \gamma \gg \beta$$

has suggested the simplified parameterization [WOL 83]

$$U_{CKM} = \begin{pmatrix} 1 - \lambda^2/2 & \lambda & A\rho\lambda^3 e^{i\phi} \\ -\lambda & 1 - \lambda^2/2 & A\lambda^2 \\ A\lambda^3(1 - \rho e^{-i\phi}) & -A\lambda^2 & 1 \end{pmatrix}$$

which corresponds to

$$s_\theta = \lambda; \quad s_\gamma = A\lambda^2; \quad s_\beta = A\rho\lambda^3$$

up to terms of higher order than λ^3.

The angles θ, γ, β are determined by the beta-decay amplitudes

$$u \rightarrow s + e^+ + \nu_e$$
$$b \rightarrow c + e^- + \bar{\nu}_e$$
$$b \rightarrow u + e^- + \bar{\nu}_e$$

respectively. θ coincides with the Cabibbo angle. The experimental information about the mixing angles is discussed in Chapter 4. Commonly accepted values are

$$\theta = 0.221 \pm 0.002$$
$$\gamma = 0.051 \pm 0.009 \quad (\text{i.e., } A = 1.05 \pm 0.17). \qquad (3.4.11)$$

At present we have only an upper bound to the third angle:

$$\beta < 1.0 \times 10^{-2} \quad (\text{i.e., } \rho < 1.0). \qquad (3.4.12)$$

As a result of the strong interaction uncertainties, very little can be said at present about the *CP*-violating phase, ϕ.

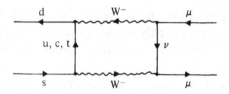

Fig. 3.4.1   Box diagram for the $K_L \to \mu^+\mu^-$ amplitude in lowest order.

3.4.2 Flavor-changing neutral currents

There is a well-obeyed selection rule in weak decays, namely, that decays involving flavor-changing neutral currents (FCNC) are suppressed to a very high degree. The first and typical example is that of the transition (this and the following experimental values are taken from [PAR 86]):

$$K_L \to \mu^+\mu^- \tag{3.4.13}$$

known to occur with a very low rate:

$$\frac{\Gamma(K_L \to \mu^+\mu^-)}{\Gamma(K^+ \to \mu^+\nu)} = (3.4 \pm 0.7) \times 10^{-9}. \tag{3.4.14}$$

Other known cases are:

$$\frac{\Gamma(K^+ \to \nu\bar{\nu}\pi^+)}{\Gamma(K^+ \to e^+\nu\pi^0)} < 1.2 \times 10^{-5} \tag{3.4.15}$$

$$\frac{\Gamma(B^0 \to e^+e^- + \text{anything})}{\Gamma(B^0 \to \text{all})} < 0.8 \times 10^{-2} \tag{3.4.16}$$

$$\frac{\Gamma(B^0 \to \mu^+\mu^- + \text{anything})}{\Gamma(B^0 \to \text{all})} < 0.7 \times 10^{-2}$$

$$\frac{\Gamma(\mu \to ee^+e^-)}{\Gamma(\mu \to \nu e\bar{\nu})} < 1.9 \times 10^{-9} \tag{3.4.17}$$

$$\frac{\Gamma(\tau \to 3 \text{ charged leptons})}{\Gamma(\tau \to \text{all})} :$$

$$< 4.0 \times 10^{-4} \ (ee^+e^-)$$

$$< 3.3 \times 10^{-4} \ (e\mu^+\mu^-)$$

$$< 4.9 \times 10^{-4} \ (\mu\mu^+\mu^-)$$

$$< 4.4 \times 10^{-4} \ (\mu e^+e^-). \tag{3.4.18}$$

We have already seen that the Z coupling is flavor diagonal to the lowest order. At the one-loop level, the GIM cancellation of the up versus charmed quark contributions to the box diagrams in Fig. 3.4.1 makes $K_L \rightarrow \mu^+\mu^-$ amplitude of order

$$A(K_L \rightarrow \mu^+\mu^-)_{\text{GIM}} \cong \frac{\alpha(m_c^2 - m_u^2)}{M_W^2} A(K^+ \rightarrow \mu^+\nu). \qquad (3.4.19)$$

That is, it gives the further suppression factor $(m_c^2 - m_u^2)/M_W^2$, in addition to the factor α, which reduces in general one-loop versus tree-level amplitudes. It is precisely this further suppression, of about 6×10^{-4}, that makes the theoretical prediction agree with the experimental ratio, Eq. (3.4.14). A simple factor of α with respect to $A(K^+ \rightarrow \mu^+\nu)$ would not be enough.

What happens in the general, spontaneously broken, gauge theory has been analyzed by Glashow and Weinberg [GLA 77], and can be illustrated as follows. We expand the amplitude (3.4.19) according to

$$A(K_L \rightarrow \mu^+\mu^-) = \left[a_Z^{(0)} + a_H^{(0)} + \alpha a^{(1)} + \frac{\alpha(m_c^2 - m_u^2)}{M_W^2} b^{(1)} + \cdots \right] A(K^+ \rightarrow \mu^+\nu)$$

$$(3.4.20)$$

where the first two terms arise from tree-level Z and Higgs exchange. Dots represent higher-order terms in powers of α and of $1/M_W^2$. The expansion Eq. (3.4.20), with the appropriate combination of quark or lepton masses, is valid for any FCNC transition.

Natural GIM cancellation means that $a_Z^{(0)}$ and $a_{(1)}$ vanish and that the Higgs exchange term gives a contribution not larger than the term proportional to $b^{(1)}$ in (3.4.20), for all FCNC processes and for generic values of the independent coupling constants in the Lagrangian, that is, without requiring special cancellations of otherwise independent contributions.

1 Z coupling at the tree level   Consider left- and right-handed fermions, f_{QL} and f_{QR}, of electric charge, Q, and denote by T_L^3 and T_R^3 the diagonal matrices that represent the $SU(2)$ generator T^3 in the two subspaces spanned by f_{QL} and f_{QR}, respectively (in the standard case, $T_L^3 = -\frac{1}{2}$ and $T_R^3 = 0$ for $Q = -1$ and $Q = -\frac{1}{3}$ fermions). After spontaneous breaking, the fermions of electric charge Q will mix arbitrarily. To diagonalize the mass matrix, we make the transformations

$$\begin{aligned} f_{QL} &\rightarrow U_L f_{QL} \\ f_{QR} &\rightarrow U_R f_{QR}. \end{aligned} \qquad (3.4.21)$$

As a result, the contribution of the fermions to the neutral current is

$$(J_\mu^3)_f = \bar{f}_{QL}\gamma_\mu U_L^+ T_L^3 U_L f_{QL} + \bar{f}_{QR}\gamma_\mu U_R^+ T_R^3 U_R f_{QR}.$$

This expression can be flavor diagonal for generic unitary matrices $U_{L,R}$ only if $T_{L,R}^3$ are multiples of the unit matrix. This is the first conclusion: In order to have a flavor-diagonal Z coupling at the tree level, *fermions of the same electric charge and helicity must all have the same value of T^3.*

2 One-loop corrections FCNC transitions are induced by the familiar box diagrams, Fig. 3.4.1, which involve the exchange of a pair of W^+ and W^- bosons. The leading term of order α involves the products of the charged generators, $T_{L,R}$ and $T_{L,R}^+$, and the condition for these terms to be flavor diagonal is that the matrices

$$U_L^+(T_L T_L^+)U_L, \quad U_L^+(T_L^+ T_L)U_L$$

be multiples of the unit matrix (and similarly for the right-handed matrices). This can be obtained, for generic unitary matrices $U_{L,R}$, if and only if *fermions of the same electric charge and helicity all have the same value of the total weak isospin T*[21]. (Note that this condition is experimentally required only in the K_L and in the $\mu \to 3e$ case.)

3 Higgs couplings The general order of magnitude of the Higgs coupling to the fermions – see Eqs. (3.3.4) and (3.3.12) – is

$$g_{\phi - f} \cong e \frac{m_f}{M_W}.$$

Thus, the coefficient $a_H^{(0)}$ in Eq. (3.4.20), which represents the tree-level Higgs exchange contribution, is of the order

$$a_H^{(0)} \cong e^2 \frac{m_s m_\mu}{M_W^2} \frac{M_W^2}{M_\sigma^2}.$$

As a crude estimate, we take all fermion masses to be of the same order ($m_s \cong m_\mu \cong m_c \cong m_f$). Then, $a_H^{(0)}$ is of the same order or larger than the GIM suppression factor (3.4.19), provided

$$\frac{M_W^2}{M_\sigma^2} = \left(\frac{e}{4\lambda}\right)^2 > e^2 > \alpha$$

which is possible, in general, since $\lambda < 1$ (λ is the Higgs self-interaction coupling, Section 3.2). The conclusion is that tree-level Higgs exchange can be dangerous if nondiagonal Higgs couplings are of the same order of magnitude as the diagonal ones.

In the general case, the coupling of the neutral components of the Higgs scalar multiplets to the fermion of electric charge Q takes the form

$$L_{\phi^0 - f} = \bar{f}_L g^\alpha \phi_\alpha^0 f_R + \text{h.c.} \tag{3.4.22}$$

where g^α are matrices in the space of the flavors associated with f_{QL} and f_{QR} and α labels the different Higgs fields present in the theory. Denoting by η_α the vacuum expectation value of ϕ_α^0, the complex mass matrix of the fermions f is

$$M_f = g^\alpha \eta_\alpha.$$

The transformation Eq. (3.4.21) diagonalizes M_f and gives rise to the fermion–neutral Higgs couplings

$$L_{\chi^0_f} = \bar{f}_{QL}(U_L^+ g^\alpha U_R)\frac{\chi^0_\alpha}{\eta\sqrt{2}} f_{QR} + \text{h.c.}$$

where we have set

$$\phi^0_\alpha(x) = \eta_\alpha + \frac{1}{\sqrt{2}}\chi^0_\alpha(x).$$

It is quite clear that, in general, the transformation (3.4.21) leaves some of the coupling matrices g^α nondiagonal. The only case in which M_f and the coupling matrices are simultaneously diagonal for generic unitary matrices $U_{L,R}$ is when there is only one Higgs field, which gives M_f proportional to g. This is the third conclusion: Higgs couplings are naturally flavor diagonal if *fermions of the same electric charge are coupled to only one scalar Higgs multiplet.*

In conclusion, the natural GIM suppression of *all* flavor-changing neutral current processes provides quite a strong constraint to the structure of any $SU(2)\times U(1)$ gauge theory. Since beta decays indicate that the left-handed (right-handed) electron is an $SU(2)$ doublet (singlet), one is led uniquely to the standard theory, sequential, quantum numbers for the other leptons. The only possible allowed variation is to have two distinct Higgs doublets with $Y=\pm\frac{1}{2}$, rather than only one, as in Section 3.2.3.

Quarks are similarly frozen, once the u–d family is given the left-handed doublet, right-handed singlet structure required by nuclear β-decays.

The absence of ABJ anomaly completes the picture by fixing the ratio of the u-quark to the electron electric charge, once the neutrino has been given a vanishing charge.

Although historically constructed piece by piece, the standard electroweak theory is quite unique.

3.4.3 Is there a τ-neutrino?

There is no direct experimental evidence, as yet, that the neutrino emitted in τ decay

$$\tau^- \rightarrow \nu_\chi + e^- + \bar{\nu}_e \tag{3.4.23}$$

is different from both ν_e and ν_μ. As we shall see, however, the observed τ decays already prove that the τ lepton cannot be an $SU(2)$ singlet. At least one new particle must come with the τ to complete a nontrivial multiplet. The simplest choice is, of course, the left-handed τ-neutrino, to complete the third, sequential, lepton family. This choice is compatible with all that is known about the τ properties. More recently, the assignment of τ_L to a weak doublet has been confirmed by the observation of the backward–forward asymmetry in τ pair production from e^+e^- annihilation.

We suppose that both τ_L and τ_R are $SU(2)$ singlets, so that the lepton scheme is

$$\begin{pmatrix} \nu_{eL} \\ e_L \end{pmatrix} \quad \begin{pmatrix} \nu_{\mu L} \\ \mu_L \end{pmatrix} = \text{doublets}$$

$$\tau_L, e_R, \mu_R, \tau_R = \text{singlets}. \qquad (3.4.24)$$

We assume only left-handed, massless neutrinos. This scheme contradicts the conditions we have derived in Section 3.4.2, so that unnatural parameter tuning will be necessary to avoid contradictions with flavor-changing neutral currents. This is not the point, however, which is rather to see whether it is at all possible to fix the parameters so as to make the τ-singlet hypothesis compatible with data (what follows is an updated version of the arguments given in [ALT 77] and [HOR 77]).

After symmetry breaking, the two neutrinos are coupled, in the charged weak current, to two orthogonal linear combinations of e_L, μ_L and τ_L. Since neutrinos are massless, therefore degenerate, we have the further liberty of a unitary transformation over the neutrino fields. We define ν_{eL} so that it does not couple to μ_L. In conclusion, the charged leptonic current takes the form

$$J_\mu = (\bar{\nu}_{eL}, \bar{\nu}_{\mu L}) \begin{pmatrix} \cos\theta_1 & 0 & \sin\theta_1 \\ -\sin\theta_2\sin\theta_1 & \cos\theta_2 & \sin\theta_2\cos\theta_1 \end{pmatrix} \begin{pmatrix} e_L \\ \mu_L \\ \tau_L \end{pmatrix} \qquad (3.4.25)$$

and the neutral current J_μ^3 is

$$J_\mu^3 = \tfrac{1}{2}[\bar{\nu}_{eL}\gamma_\mu\nu_{eL} + \bar{\nu}_{\mu L}\gamma_\mu\nu_{\mu L}] - \tfrac{1}{2}\bar{l}_L\gamma_\mu U + U1_L \qquad (3.4.26)$$

where U is the 2×3 matrix given in Eq. (3.4.25). To the lowest order in the angles $\theta_{1,2}$ (we will see shortly that these angles must be small) the charged lepton contribution to J_μ^3 is

$$(J_\mu^3)_{\text{ch.lept}} = -\tfrac{1}{2}[\bar{e}_L\gamma^\lambda e_L + \bar{\mu}_L\gamma^\lambda\mu_L + (\theta_1^2 + \theta_2^2)\bar{\tau}_L\gamma_\lambda\tau_L$$
$$+ (-\theta_1\theta_2\bar{e}_L\gamma^\lambda\mu_L + \theta_1\bar{e}_L\gamma_\lambda\tau_L + \theta_2\bar{\mu}_L\gamma^\lambda\tau_L + \text{h.c.})]. \qquad (3.4.27)$$

The angles $\theta_{1,2}$ are constrained by (1) the absence of $\mu \to 3e$ transitions; (2) the observed e–μ universality in τ decay; and (3) the decay rate of τ into one charged lepton plus neutrinos.

1. *The $\mu \to e$ coupling in Eq. (3.4.27) gives rise to the $\mu \to 3e$ transition via Z exchange.* The effective Lagrangian is

$$L_{\mu \to 3e} = 2\frac{G}{\sqrt{2}}\theta_1\theta_2(\bar{e}_L\gamma_\lambda\mu_L)[(1 - 2\sin^2\theta_W)\bar{e}_L\gamma^\lambda e_L + 2\sin^2\theta_W\bar{e}_R\gamma^\lambda e_R]$$

$$(3.4.28)$$

It gives

$$\frac{\Gamma(\mu \to ee^+e^-)}{\Gamma(\mu \to \nu e \bar{\nu})} = \frac{1}{2}(\theta_1\theta_2)^2[(1 - 2\sin^2\theta_W)^2 + 2\sin^4\theta_W]. \tag{3.4.29}$$

2. *The leptonic τ decay modes can be classified according to the number of charged leptons in the final states.* The corresponding effective Lagrangians, computed from Eqs. (3.4.25–26) are

$$L_{\tau \to 1\text{ch.lept.}} = 4\frac{G}{\sqrt{2}}(\theta_1\bar{\nu}_{eL} + \theta_2\bar{\nu}_{\mu L})\gamma_\lambda\tau_L(\bar{e}_L\gamma^\lambda\nu_{eL} + \bar{\mu}_L\gamma^\lambda\nu_{\mu L})$$
$$- 2\frac{G}{\sqrt{2}}(\theta_1\bar{e}_L + \theta_2\bar{\mu}_L)\gamma_\lambda\tau_L(\bar{\nu}_{eL}\gamma^\lambda\nu_{eL} + \bar{\nu}_{\mu L}\gamma^\lambda\nu_{\mu L}) \tag{3.4.30}$$

where the first (second) term arises from W- (Z-)exchange, while

$$L_{\tau \to 3\text{ch.lept.}} = 2\frac{G}{\sqrt{2}}(\theta_1\bar{e}_L + \theta_2\bar{\mu}_L)\gamma_\lambda\tau_L[(1 - 2\sin^2\theta_W)\bar{e}_L\gamma^\lambda e_L$$
$$+ 2\sin^2\theta_W\bar{e}_R\gamma^\lambda e_R + (e \to \mu)]. \tag{3.4.31}$$

Defining

$$\Gamma^{(0)} = \Gamma(\mu \to \nu e\bar{\nu})\left(\frac{m_\tau}{m_\mu}\right)^5 \cong 6.2 \times 10^{11} \text{ sec}^{-1}$$

we find, from (3.4.30)

$$\Gamma(\tau \to e\nu\bar{\nu}) = \left[\tfrac{1}{2}(\theta_1)^2 + (\theta_2)^2\right]\Gamma^{(0)}$$
$$\Gamma(\tau \to \mu\nu\bar{\nu}) = \left[(\theta_1)^2 + \tfrac{1}{2}(\theta_2)^2\right]\Gamma^{(0)} \tag{3.4.32}$$

and from (3.4.31):

$$\Gamma(\tau \to ee^+e^-) = \theta_1^2\left[\sin^4\theta_W + \tfrac{1}{2}(1 - 2\sin^2\theta_W)^2\right]\Gamma^{(0)}$$
$$\Gamma(\tau \to \mu e^+e^-) = \theta_2^2\left[\sin^4\theta_W + \tfrac{1}{4}(1 - 2\sin^2\theta_W)^2\right]\Gamma^{(0)} \tag{3.4.33}$$

the decay rates of the channels obtained by $e \leftrightarrow \mu$ exchange can be obtained from (3.4.33) by the exchange $\theta_1 \leftrightarrow \theta_2$.

With the experimental limit (3.4.17) and $\sin^2\theta_W \cong 0.23$, one obtains from Eq. (3.4.29):

$$(\theta_1\,\theta_2)^2 < 0.96 \times 10^{-8}. \tag{3.4.34}$$

On the other hand, Eq. (3.4.32) and the close equality of the experimental τ branching ratio into one electron or one muon plus neutrinos:

$$B(\tau \to e\nu\bar{\nu})_{\text{expt}} = (16.5 \pm 0.9) \times 10^{-2}$$
$$B(\tau \to \mu\nu\bar{\nu})_{\text{expt}} = (18.5 \pm 1.1) \times 10^{-2} \qquad (3.4.35)$$

imply

$$(\theta_1)^2 \cong (\theta_2)^2$$

to within, say, 10 percent. Combining with (3.4.34), we derive the strong bound

$$\theta_1^2 \cong \theta_2^2 < 0.98 \times 10^{-4}. \qquad (3.4.36)$$

The leptonic rate is accordingly reduced with respect to the universality value, $\Gamma^{(0)}$. From Eqs. (3.4.32) and (3.4.36) we predict

$$\Gamma(\tau \to e\nu\bar{\nu}) < 1.5 \times 10^{-4}\, \Gamma^{(0)} \cong 0.91 \times 10^8\, \text{sec}^{-1}$$

while the measured τ lifetime combined with the branching ratio (3.4.35) gives

$$\Gamma(\tau \to e\nu\bar{\nu})_{\text{expt}} = (4.9 \pm 0.8) \times 10^{11}\, \text{sec}^{-1}. \qquad (3.4.37)$$

Eq. (3.4.37) agrees with $\Gamma^{(0)}$, and rules out the singlet hypothesis.

An independent, equally strong, inconsistency is provided by the ratio of the decay rates into 3 versus 1 charged lepton. If we add up the rates in (3.4.33) and the related muonic ones, and compare with the total rate into 1 charged lepton, we predict, independently from the values of $\theta_{1,2}$:

$$R_{\frac{3}{1}} = \frac{\Gamma(\tau \to 3 \text{ charged leptons})}{\Gamma(\tau \to 1 \text{ charged lepton})}$$
$$= \tfrac{1}{6}[8\sin^4\theta_W + 3(1 - 2\sin^2\theta_W)^2] \cong 0.21. \qquad (3.4.38)$$

On the other hand, adding up all the experimental upper limits (3.4.18) and comparing with the branching ratios (3.4.35), one finds:

$$\left(R_{\frac{3}{1}}\right)_{\text{expt}} < 0.4 \times 10^{-2}$$

which is completely inconsistent with (3.4.38).

We may add, for completeness, that the latter argument runs for the b quark [BRA 78] in exactly the same way. If b_L and b_R were both $SU(2)$ singlets, the d, s, and b coupling to the u and c quark would be analogous to (3.4.25), with the rows of U replaced by the first two rows of the CKM matrix, Section 3.4.2. Irrespective of the values of the mixing parameters, one predicts a certain amount of FCNC transitions in b-quark decay. In particular, one easily obtains

$$R(b)_{\text{FCNC}} = \frac{\Gamma(b \to e^+ e^- + \cdots)}{\Gamma(b \to e\nu + \cdots)}$$

$$= \left[\sin^4\theta_W + \tfrac{1}{4}(1 - 2\sin^2\theta_W)^2 \right] \cong 0.13. \qquad (3.4.39)$$

The upper bound (3.4.16), combined with the observed semileptonic branching ratio

$$B(b \to e\nu + \cdots) \cong 0.13 \qquad (3.4.40)$$

gives instead

$$R(b)_{\text{FCNC, expt}} < 0.06 \qquad (3.4.41)$$

which is inconsistent with (3.4.39) and with the b-singlet hypothesis.

3.5 Neutrino–lepton cross sections

In this section we discuss the neutrino cross sections of leptonic systems, as given by the standard theory. Although difficult to observe, neutrino–lepton scattering provides the opportunity to test the standard theory in a very clean situation, free from complicated strong interaction effects. In particular, a very precise measurement of $\sin^2\theta_W$ can be obtained from ν_μ–e scattering.

We assume massless, purely left-handed neutrinos, thus the exact conservation of lepton flavors, and restrict to the lowest order in the electroweak interaction.

3.5.1 A prototype case

As a prototype of neutrino–lepton scattering we take the process

$$\nu_\mu + e \to \nu_\mu + e \qquad (3.5.1)$$

induced by Z-exchange. Initially we work in the center of mass frame, denoting by θ and E the neutrino-scattering angle and energy and by E_e and m the electron energy and mass, respectively. In terms of the familiar Mandelstam variable s,

$$E_e = (E^2 + m^2)^{1/2} = (s + m^2)/2\sqrt{s}$$
$$E = (s - m^2)/2\sqrt{s}. \qquad (3.5.2)$$

We also introduce the electron velocity in the c.m. frame, β:

$$\beta = (s - m^2)/(s + m^2). \qquad (3.5.3)$$

Finally, the invariant four-momentum transfer between the two neutrinos is given by

$$t = -2E^2(1 - \cos\theta). \tag{3.5.4}$$

The process (3.5.1) is conveniently described in terms of helicity amplitudes [JAC 64]. The neutrino helicity is fixed to be $-\frac{1}{2}$. We have a total of four helicity amplitudes, which we label with the sign of the outgoing and ingoing electron helicity. Helicity amplitudes are normalized so that the differential cross section for unpolarized electrons is

$$\frac{d\sigma}{d\cos\theta} = \frac{1}{2}2\pi\{|f_{-,-}(\cos\theta)|^2 + |f_{+,+}(\cos\theta)|^2 + |f_{-,+}(\cos\theta)|^2 + |f_{+,-}(\cos\theta)|^2\}. \tag{3.5.5}$$

We write the S-matrix element according to

$$S_{fi} = {}_iT_{fi} = (2\pi)^4\delta^4(P_f - P_i)(2\pi)^{-6}(E_eE)^{-1}iM_{fi}. \tag{3.5.6}$$

The Feynman invariant matrix element M_{fi} is obtained directly from the Z-exchange Feynman diagram and the helicity amplitudes are then given by[1]

$$f_{fi} = \frac{1}{2\pi(s)^{1/2}}M_{fi}. \tag{3.5.7}$$

In the unitary gauge, the Z-propagator takes the form

$$D(q)_{\mu\nu} = -i\left(g_{\mu\nu} - \frac{q_\mu q_\nu}{M_Z^2}\right)\frac{1}{q^2 - M_Z^2} \tag{3.5.8}$$

and we find, from Eqs. (3.2.20) and (3.4.3):

$$M_{fi} = -2\sqrt{2}\rho G_Z(t)[(J_\nu)^\lambda(J_e)^\lambda]_{fi}$$
$$G_Z(t) = \frac{G}{1 - t/M_Z^2} \tag{3.5.9}$$

G is the Fermi constant and $\rho = 1$ in the Higgs doublet model. The weak current matrix elements in (3.5.9) are given by

$$[(J_\nu)^\lambda]_{fi} = \frac{1}{2}\bar{u}_\nu(k')\gamma^\lambda(1 + \gamma_5)u_\nu(k)$$
$$[(J_e)^\lambda]_{fi} = \frac{1}{2}\bar{u}_e(p')[g_L\gamma^\lambda(1 + \gamma_5) + g_R\gamma^\lambda(1 - \gamma_5)]u_e(p) \tag{3.5.10}$$

---

[1] In the case where final and initial state masses are not equal, there is an additional factor $(p'/p)^{\frac{1}{2}}$ in Eq. (3.5.7), where $p(p')$ is the c.m. momentum of the initial (final) state. This applies, e.g., to the inverse μ-decay considered in Section 3.5.3.

where, for the electron,

$$g_L = -\tfrac{1}{2} + \sin^2\theta_W$$
$$g_R = \sin^2\theta_W. \tag{3.5.11}$$

The explicit calculation with the appropriate helicity spinors leads to the result

$$f_{-,-}(\nu_\mu e; \theta) = -\rho\sqrt{2}\frac{G_Z(t)}{2\pi(s)^{1/2}}EE_e[2(1+\beta)g_L - (1-\beta)(1-\cos\theta)g_R]$$

$$f_{+,+}(\nu_\mu e; \theta) = -\rho\sqrt{2}\frac{G_Z(t)}{2\pi(s)^{1/2}}EE_e(1+\beta)(1+\cos\theta)g_R \tag{3.5.12}$$

$$f_{+,-}(\nu_\mu e; \theta) = -f_{-,+}(\nu_\mu e; \theta) = -\rho\sqrt{2}\frac{G_Z(t)}{2\pi(s)^{1/2}}Em\sin\theta\, g_R$$

and to the differential and total cross sections

$$d\sigma(\nu_\mu e)/d\cos\theta = (2\rho^2 G_Z(t)^2/(\pi s))\{(EE_e)^2[(1+\beta)^2 g_L^2$$
$$+ (1+\beta\cos\theta)^2 g_R^2] - (Em)^2(1-\cos\theta)g_L g_R\} \tag{3.5.13}$$

$$\sigma(\nu_\mu e) = (4\rho^2 G_Z(t)^2/(\pi s))\{(EE_e)^2[(1+\beta)^2 g_L^2$$
$$+ (1+\beta^2/3)g_R^2] - (Em)^2 g_L g_R\}. \tag{3.5.14}$$

The amplitudes for antiparticle scattering are easy to work out. First, *CP* symmetry implies that switching particles into antiparticles and reversing the helicities leaves the amplitude invariant

$$CP: f_{a,b}(\bar{\nu}_\mu e^+; \theta) = f_{-a,-b}(\nu_\mu e; \theta)$$
$$f_{a,b}(\bar{\nu}_\mu e; \theta) = f_{-a,-b}(\nu_\mu e^+; \theta). \tag{3.5.15}$$

Next, we rewrite the electron current in terms of the positron field

$$(J_e)^\lambda = \bar{e}[g_L\gamma^\lambda(1+\gamma_5) + g_R\gamma^\lambda(1-\gamma_5)]e$$
$$= -\bar{e}_C[g_L\gamma^\lambda(1-\gamma_5) + g_R\gamma^\lambda(1+\gamma_5)]e_C \tag{3.5.16}$$

which shows that we go from the electron to the positron with the simple exchange

$$e \to e^+: \quad g_L \leftrightarrow -g_R \tag{3.5.17}$$

so that, in conclusion, we find:

$$f_{-,-}(\bar{\nu}_{\mu}e; \theta) = \rho\sqrt{2}\frac{G_Z(t)}{2\pi(s)^{1/2}}EE_e(1+\beta)(1+\cos\theta)g_L$$

$$f_{+,+}(\bar{\nu}_{\mu}e; \theta) = \rho\sqrt{2}\frac{G_Z(t)}{2\pi(s)^{1/2}}EE_e[2(1+\beta)g_R - (1-\beta)(1-\cos\theta)g_L]$$

$$f_{+,-}(\bar{\nu}_{\mu}e; \theta) = -f_{-,+}(\bar{\nu}_{\mu}e; \theta)$$

$$= -\rho\sqrt{2}\frac{G_Z(t)}{2\pi(s)^{1/2}}Em\sin\theta\, g_L.$$

(3.5.18)

Antineutrino–positron cross sections are obtained directly from Eqs. (3.5.13–14) with the substitution (3.5.17).

Equations (3.5.12) and (3.5.18) show that helicity is conserved in the limit of vanishing electron mass. The factors $(1+\cos\theta)$ and $(1-\cos\theta)$, which appear in $f_{+,+}(\nu_{\mu}e; \theta)$ and in $f_{-,-}(\bar{\nu}_{\mu}e; \theta)$, is related to helicity and angular momentum conservation. In these cases, the initial state has a nonvanishing angular momentum along the direction of the electron and helicity conservation requires the final state to have the opposite value, for the scattering of 180°. Therefore, the amplitude must vanish in the backward direction.

We conclude the exercise by giving the differential cross section in the laboratory frame. The commonly used variables are the neutrino energy, ω, and the final electron kinetic energy

$$T = \frac{(pp') - m^2}{m}$$

(3.5.19)

or the scaled variable

$$y = \frac{T}{\omega}.$$

(3.5.20)

After a simple calculation, one finds:

$$d\sigma(\nu_{\mu}e)/dy = 2m\omega\rho^2\frac{G_Z(t)^2}{\pi}\left[g_L^2 + (1-y)^2g_R^2 - \frac{my}{\omega}g_Lg_R\right].$$

(3.5.21)

3.5.2 ν_e and $\bar{\nu}_e$–e elastic scattering

With respect to the previous case, for example, (3.5.1), the reaction

$$\nu_e(\bar{\nu}_e) + e \rightarrow \nu_e(\bar{\nu}_e) + e$$

(3.5.22)

is special because of the additional contribution from W-exchange. The effective Lagrangian for W-exchange is given by Eq. (3.4.1), with μ, ν_{μ} replaced by e and ν_e, respectively, and the appropriate W-propagator restored. Next, we use the Fierz-rearrangement identity

$$[\bar{e}\gamma_\lambda(1+\gamma_5)\nu_e][\bar{\nu}_e\gamma_\lambda(1+\gamma_5)e] = +[\bar{\nu}_e\gamma_\lambda(1+\gamma_5)\nu_e][\bar{e}\gamma_\lambda(1+\gamma_5)e] \qquad (3.5.23)$$

to bring the effective Lagrangian for W-exchange in the form of the neutral-current Lagrangian, Eq. (3.4.3). The amplitudes for ν_e–e scattering take the form given in (3.5.12) and

$$g(\nu_e-e)_L = g_L + \frac{G_W(u)}{\rho G_Z(t)}$$
$$g(\nu_e-e)_R = g_R. \qquad (3.5.24)$$

For $\bar{\nu}_e$–e scattering, we have to use Eq. (3.5.18), with

$$g(\bar{\nu}_e-e)_L = g_L + \frac{G_W(s)}{\rho G_Z(t)}$$
$$g(\bar{\nu}_e-e)_R = g_R. \qquad (3.5.25)$$

For the presently available neutrino energies, we may simplify the formulae given above. Since

$$-t < s = 2m\omega \cong (0.3\,\text{GeV})^2(\omega/100\,\text{GeV}) \ll M_Z^2 \qquad (3.5.26)$$

we can safely neglect the momentum dependence of the Z- and W-propagators, up to neutrino energies of the order of 100 GeV. Moreover, for accelerator neutrinos the final electron is quite relativistic and we may set $m=0$. In this case, we find:

$$d\sigma(\nu e)/dT = 2m\rho^2\frac{G^2}{\pi}[g_L^2 + (1-y)^2 g_R^2]$$
$$d\sigma(\bar{\nu} e)/dT = 2m\rho^2\frac{G^2}{\pi}[g_R^2 + (1-y)^2 g_L^2] \qquad (3.5.27)$$

with $0 < y < 1$, and

$$\sigma(\nu e) = 2m\omega\rho^2\frac{G^2}{\pi}[g_L^2 + g_R^2/3] \qquad (3.5.28)$$

$$\sigma(\bar{\nu} e) = 2m\omega\rho^2\frac{G^2}{\pi}[g_R^2 + g_L^2] \qquad (3.5.29)$$

where $g_{L,R}$ are given by Eq. (3.5.11) or (3.5.24).

A further consequence of the small electron mass is that the final electron is emitted with a very small angle in the laboratory system. The cross section being quite isotropic in the c.m. system, most of the electrons are projected by the Lorentz transformation within an angle such that

$$\tan \theta_{\text{lab}} \cong \gamma^{-1} \cong 2m/\sqrt{s} \cong 10^{-2}(10 \text{ GeV}/\omega)^{1/2}. \tag{3.5.30}$$

A separate measurement of the total cross sections (3.5.28–29) gives g_L and g_R, up to a fourfold sign ambiguity, provided that the ratio

$$R(\nu e) = \sigma(\nu e)/\sigma(\bar{\nu} e) \tag{3.5.31}$$

is between $\frac{1}{3}$ and 3. Only one of the four points thus determined in the $g_L - g_R$ plane can fall on the segment representing Eq. (3.5.11) or (3.5.24), so we can determine $\sin^2 \theta_W$ unambiguously in this way.

3.5.3 Charged-current scattering

The inverse μ-decay processes

$$\nu_\mu + e \rightarrow \mu^- + \nu_e \tag{3.5.32}$$

$$\bar{\nu}_e + e \rightarrow \mu^- + \bar{\nu}_\mu \tag{3.5.33}$$

are mediated by W-exchange only. We give directly the differential cross sections.[2] In the c.m. frame, we find:

$$d\sigma(\nu_\mu e \rightarrow \mu^- \nu_e)/d\cos\theta = \frac{G_W(t)^2}{2\pi} s \left(1 - \frac{m_\mu^2}{s}\right)^2 \tag{3.5.34}$$

$$d\sigma(\bar{\nu}_e e \rightarrow \mu^- \bar{\nu}_\mu)/d\cos\theta = \frac{G_W(t)^2}{2\pi} E_e E_\mu \cdot (1 - m_\mu^2/s)^2(1 + \beta_e \cos\theta)(1 + \beta_\mu \cos\theta). \tag{3.5.35}$$

In the laboratory frame, it is convenient to use the incoming neutrino energy, ω, and the variable analogous to the one in (3.5.19) or (3.5.20), namely,

$$T = \frac{[(pp') - m_e^2]}{m_e} = \omega y.$$

---

[2] We neglect terms smaller by a factor $(m_e m_\mu / M_W^2)$, contributed by the longitudinal part of the W-propagator.

One finds:

$$d\sigma(\nu_\mu e \to \mu^- \nu_e)/dy = 2m_e\omega \frac{G_W(t)^2}{\pi}\left[1 - \frac{(m_\mu^2 - m_e^2)}{2\omega m_e}\right] \qquad (3.5.36)$$

$$d\sigma(\bar\nu_e e \to \mu^- \bar\nu_\mu)/dy = 2m_e\omega \frac{G_W(t)^2}{\pi}(1-y)\left[1 + \frac{(m_\mu^2 - m_e^2)}{2\omega m_e} - y\right]. \qquad (3.5.37)$$

Of course, in the high-energy limit where we neglect $m_{e,\mu}$ with respect to all energies, the above formulae coincide with Eqs. (3.5.27–28), with $g_L = 1$ and $g_R = 0$. In particular for the total cross sections we find, in this limit and neglecting W-propagator effects:

$$\sigma(\nu_\mu e \to \mu^- \nu_e) = 2m_e\omega G^2/\pi \qquad (3.5.38)$$

$$\sigma(\bar\nu_e e \to \mu^- \bar\nu_\mu) = 2m_e\omega G^2/3\pi \qquad (3.5.39)$$

$$2m_e G^2/\pi \cong 15.33 \times 10^{-42}\ \mathrm{cm^2/GeV}. \qquad (3.5.40)$$

The factor of $\frac{1}{3}$ in (3.5.39) with respect to (3.5.38) is typical of the antifermion–fermion scattering in a $V-A$ theory. As noticed already, because of helicity and angular momentum conservation the differential c.m. cross section is isotropic for fermion–fermion scattering, while it is proportional to $(1 + \cos\theta)^2$ for antifermion–fermion scattering. The integration over $\cos\theta$ leads then to the $3:1$ ratio in the total cross sections.

3.6 Neutrino–hadron scattering

Neutrino–hadron interactions have been observed over a wide energy range, from a few MeV corresponding to the energies of neutrinos provided by nuclear reactors (or by the sun), up to energies of the order of 10–100 GeV, obtained with accelerator neutrinos.

The study of neutrino–hadron interactions has been of great value in understanding the structure of the fundamental interactions. The observation of semileptonic neutral-current processes has given crucial support to the standard electroweak theory. Deep inelastic neutrino–hadron processes have confirmed that the parton picture emerged from deep inelastic electron–proton scattering and have been crucial in determining the quantum numbers of the hadron constituents, which led, eventually, to the determination of QCD as the basic theory of the strong interactions. Finally, neutrino production of like-sign dimuons has given the first experimental evidence for the existence of charmed hadrons.

In this section, we focus on processes that are at the two ends of the energy region explored thus far, namely neutrino–nucleon elastic and deep inelastic scattering. Intermediate energy reactions, such as baryon resonance formation or pion production, have also provided a great deal of information on the basic hadron properties, but we shall have to put this subject aside for the lack of space.

To lowest order, the basic semileptonic processes

$$\nu_l + \alpha \rightarrow l^- + \beta, \tag{3.6.1a}$$

$$\nu_l + \alpha \rightarrow \nu_l + \beta, \tag{3.6.1b}$$

(α and β are hadronic states) are described by W- and Z-exchange, respectively. Defining

$$S_{fi} = iT_{fi} = (2\pi)^4 \delta^4 (P_f - P_i)(2\pi)^{-3}(EE')^{-1/2} iM_{fi} \tag{3.6.2}$$

where E and E' are the initial and final lepton energies, the amplitudes corresponding to (3.6.1) and (3.6.2) are

$$M_{fi}^{(W)} = -2\sqrt{2}G_W(t)(J_1)^\lambda \langle \beta | J_\lambda | \alpha \rangle, \tag{3.6.3a}$$

$$M_{fi}^{(Z)} = -2\sqrt{2}\rho G_Z(t)(J_\nu)^\lambda \langle \beta | J_\lambda^{(3)} - \sin^2\theta_W J_\lambda^{(\text{e.m.})} | \alpha \rangle. \tag{3.6.3b}$$

$$(J_1)^\lambda = \tfrac{1}{2}\bar{u}_1(k')\gamma^\lambda(1 + \gamma_5)u_\nu(k), \tag{3.6.4a}$$

$$(J_\nu)^\lambda = \tfrac{1}{2}\bar{u}_\nu(k')\gamma^\lambda(1 + \gamma_5)u_\nu(k). \tag{3.6.4b}$$

J_λ, and $J_\lambda^{(3)}$ are the hadronic weak (charged and neutral) currents, $J_\lambda^{(\text{e.m.})}$ the electromagnetic current (see Section 3.3.2). $G_{W,Z}(t)$ are defined as in (3.5.9) and

$$t = (p_\alpha - p_\beta)^2 = (q)^2. \tag{3.6.5}$$

3.6.1 Elastic scattering

We consider first the charged-current processes

$$\nu_e + N \rightarrow e^- + P, \tag{3.6.6a}$$

$$\bar{\nu}_e + P \rightarrow e^+ + N. \tag{3.6.6b}$$

We shall neglect the proton–neutron mass difference and the electron mass.

The relevant component of the hadronic current in (3.6.6a) is the $\Delta S = 0$ part:

$$J_\lambda = \cos\theta_C J_\lambda^{(\Delta S=0)} + \cdots + J_\lambda^{(\Delta S=0)} = \bar{u}_L \gamma_\lambda d_L \tag{3.6.7}$$

θ_C is the Cabibbo angle, and we have approximated $\cos\beta$ with unity in the CKM matrix (Section 3.4.1).

The current is divided into a vector and an axial vector part:

$$J_\lambda^{(\Delta S=0)} = \tfrac{1}{2}(V_\lambda + A_\lambda) \tag{3.6.8}$$

and the nucleon matrix elements are expanded in form-factors.[1] For the vector current we write:

$$(2\pi)^3(E_P E_N)^{1/2}\langle P|V_\lambda|N\rangle = \bar{u}_P(p')\left[g_V(t)\gamma_\lambda + i\frac{1}{2M}\kappa(t)\sigma_{\lambda\mu}q^\mu + g_S(t)q_\lambda\right]u_N(p)$$

(3.6.9)

where

$$q = p - p'$$

and M is the nucleon mass. In the following, it will be convenient to use the so-called charge and magnetic form-factors:

$$g_E(t) = g_V(t) - \frac{t}{4M^2}\kappa(t)$$
$$g_M(t) = g_V(t) + \kappa(t).$$

(3.6.10)

For the axial current, we have two form-factors:

$$(2\pi)^3(E_P E_N)^{1/2}\langle P|A_\lambda|N\rangle = \bar{u}_P(p')[g_A(t)\gamma_\lambda\gamma_5 + g_P(t)\gamma_5 q_\lambda]u_N(p).$$ (3.6.11)

The currents (3.6.8) are among the Noether currents associated with the hadron chiral symmetry $SU(2)_L \times SU(2)_R$, which includes the familiar isotopic spin symmetry (see, e.g., [LEE 81; COL 85]).

In the limit of equal u and d quark masses, the isospin symmetry is exact, and the vector current is conserved. This statement coincides with the CVC hypothesis of Feynman and Gell-Mann [FEY 58] and relates the weak current to the isovector part of the electromagnetic current.

The axial current is conserved for vanishing u and d quark mass. The masses given in (3.4.9) are so small as to make the nonconservation of the axial current of the same order as that of the vector current. However, unlike isospin, the axial generators are broken spontaneously by the QCD interaction. The corresponding Goldstone particle is the pion, which should have a vanishing mass in the limit of massless quarks.

The CVC hypothesis, translated into the matrix elements (3.6.9), relates the vector form-factors to the e.m. isovector form-factors:

$$g_E(t) = [G_E(t)]_P - [G_E(t)]_N$$
$$g_M(t) = [G_M(t)]_P - [G_M(t)]_N$$
$$g_S(t) = 0$$

(3.6.12)

---

[1] With the present convention, the left-handed quark in J_λ corresponds to a V + A interaction. We define $\sigma_{\lambda\mu} = 1/(2i)[\gamma_\mu, \gamma_\mu]$. A term proportional to $\sigma_{\lambda\mu}\gamma_5 q^\mu$ in the axial current is excluded by time-reversal invariance.

and:

$$g_E(0) = g_V(0) = Q_P - Q_N = 1$$
$$g_M(0) - 1 = \kappa(0) = \kappa_P - \kappa_N \cong +3.7. \tag{3.6.13}$$

The conservation equation for the axial current reads:

$$q^\lambda \bar{u}_P(p')[g_A(t)\gamma_\lambda\gamma_5 + g_P(t)\gamma_5 q_\lambda]u_N(p)$$
$$= \bar{u}_P(p')[-g_A(p)2M + tg_P(t)]\gamma_5 u_N(p) = 0. \tag{3.6.14}$$

$g_P(t)$ receives a contribution from the diagram where a pion is created from vacuum by the current. This contribution is singular at $t = 0$, in the massless quark limit where (3.6.14) applies:

$$[g_P(t)]_\pi = f_\pi \frac{g_{P\pi N}}{t - m_\pi^2} \rightarrow f_\pi \frac{g_{P\pi N}}{t}.$$

In the same limit, Eq. (3.6.14) gives rise to the Goldberger–Treiman relation

$$g_A(0) = \frac{f_\pi g_{P\pi N}}{2M} \tag{3.6.15}$$

where f_π and $g_{P\pi N}$ are the pion decay constant ($f_\pi = 132\,\text{MeV}$) and the strong pion–nucleon constant ($g_{P\pi N}^2/8\pi \approx 15$), respectively. The Goldberger–Treiman relation agrees well with the experimental value of $g_A(0)$ obtained from the neutron β-decay:

$$g_A(0) \cong +1.25. \tag{3.6.16}$$

As Eq. (3.6.15) shows, the value of g_A is not determined by the symmetry, like g_V. Rather, the symmetry gives a relation between vertices with zero and one pion emitted (i.e., a soft-pion theorem).

In conclusion, the vector part of the current is known completely, at all values of t, in terms of the e.m. form-factors for all momentum transfers. At $t = 0$, all the matrix elements are known, Eqs. (3.6.13) and (3.6.16).

The cross section for (3.6.6a) is obtained from the spin-averaged matrix element squared:

$$\frac{1}{2}\sum |M_{fi}|^2 = \frac{G^2 \cos^2\theta_C}{2}\frac{1}{2}L^{\mu\nu}T_{\mu\nu}$$
$$L^{\mu\nu} = 2[k^\mu k'^\nu + k'^\mu k^\nu - g^{\mu\nu}(kk') + i\epsilon^{\mu\alpha\nu\beta}k_\alpha k'_\beta] \tag{3.6.17}$$

where k and k' are the initial and final lepton momenta.

In $T_{\mu\nu}$ we can drop terms proportional to q_μ and/or q_ν, which give vanishing contributions when contracted with $L^{\mu\nu}$, in the limit of massless lepton. In particular, we may drop the contribution of the induced pseudoscalar form-factor.

We find:[2]

$$T_{\mu\nu} = 2[p_\mu p_\nu f_2 - M^2 g_{\mu\nu} f_1 + \tfrac{1}{2} \epsilon_{\mu\alpha\nu\beta} p^\alpha p'^\beta f_3] \tag{3.6.18}$$

with

$$f_1(t) = g_A(t)^2 - \frac{t}{4M^2}[g_A(t)^2 + g_M(t)^2]$$

$$f_2(t) = \frac{g_E(t)^2 - \dfrac{t}{4M^2} g_M(t)^2}{1 - \dfrac{t}{4M^2}} + g_A(t)^2 \tag{3.6.19}$$

$$f_3(t) = 2g_M(t) g_A(t).$$

We denote by ω, ω', and θ_{lab} the initial and final lepton energies and the scattering angle in the laboratory frame:

$$\sin^2(\theta_{\text{lab}}/2) = \frac{-tM^2}{(s - M^2)(s + t - M^2)}. \tag{3.6.20}$$

The cross section computed from (3.6.17–19) takes the form[3]

$$\frac{d\sigma}{d\omega'} = M \frac{G^2 \cos^2\theta_C}{\pi} \frac{\omega'}{\omega} \left\{ \cos^2(\theta_{\text{lab}}/2) f_2(t) + 2\sin^2(\theta_{\text{lab}}/2) f_1(t) \right.$$

$$\left. + \frac{\omega + \omega'}{M} \sin^2(\theta_{\text{lab}}/2) f_3(t) \right\} \tag{3.6.21}$$

$$\frac{d\sigma}{dt} = \frac{1}{2M} \frac{d\sigma}{d\omega'}. \tag{3.6.22}$$

A few comments are in order.

3.6.1.1 The antineutrino cross section (3.6.6b) is obtained by changing the sign of the V, A interference (see Section 3.3).

$$\nu \to \bar\nu : f_3(t) \to -f_3(t). \tag{3.6.23}$$

3.6.1.2 The cross sections for the neutral current processes

$$\nu + P(N) \to \nu + P(N), \tag{3.6.24}$$

$$\bar\nu + P(N) \to \bar\nu + P(N) \tag{3.6.25}$$

---

[2] The calculation is greatly simplified if, in the vector current, one eliminates the term in $\sigma_{\lambda\mu} q^\mu$ using the identity: $i1/2M\sigma_{\sigma\mu}q\mu = -1/2M(p + p')_\mu + \gamma_\mu$ valid when taken between positive energy spinors.

[3] It is easy to verify that for a pointlike, minimally coupled particle, $g_E = g_M = 1$, $g_A = $ constant, we get back the result (3.5.21), with $2 g_{L,R} = (1 \pm g_A)^2$ and $y = \omega - \omega'/\omega$.

are obtained by a suitable redefinition of the form-factors. From Eq. (3.6.4b) we find (the upper or lower sign refers to proton or neutron):[4]

$$g_{E,M}^{(0)}(t) = \pm\tfrac{1}{2}g_{E,M}(t) - \sin^2\theta_W[G_{E,M}(t)]_{P/N}$$
$$g_A^{(0)}(t) = \pm\tfrac{1}{2}g_A(t). \tag{3.6.26}$$

3.6.1.3 The same results apply to the muonic processes

$$\nu_\mu + N \rightarrow \mu^- + P \tag{3.6.27}$$
$$\bar{\nu}_\mu + P \rightarrow \mu^+ + N \tag{3.6.28}$$

provided we are well above the muon-production threshold (i.e., for $\omega \gg 100\,\text{MeV}$).

We discuss the result Eq. (3.6.21) in the two extreme cases.

1 Nonrelativistic limit   In this case,

$$\omega = \omega' \ll M; \quad t = 0.$$

In terms of $\cos(\theta_{\text{lab}})$ we find the simple result

$$
\begin{aligned}
\frac{d\sigma(\nu_e N \rightarrow e^- P)}{d\cos\theta_{\text{lab}}} &= \frac{d\sigma(\bar{\nu}_e P \rightarrow e^+ N)}{d\cos\theta_{\text{lab}}} \\
&= \frac{G^2\cos^2\theta_C}{2\pi}\omega^2\{g_V(0)^2 + 3g_A(0)^2 \\
&\quad + \cos\theta_{\text{lab}}[g_V(0)^2 - g_A(0)^2]\}
\end{aligned} \tag{3.6.29}
$$

$$\sigma = \frac{G^2\cos^2\theta_C}{\pi}\omega^2[g_V(0)^2 + 3g_A(0)^2]. \tag{3.6.30}$$

Neutrino and antineutrino cross sections coincide and are determined by the same couplings that appear in the neutron β-decay.

For neutrinos of a few MeV (including, e.g., the high-energy tail of solar neutrino and Supernova neutrinos) the cross section in Eq. (3.6.30) is not much larger than the corresponding neutrino–electron cross sections given in (3.5.30), the nucleon to electron cross section ratio being of the order of $\omega/2m_e$.

2 High-energy limit   The differential cross section is dominated, in this limit,[5] by the form-factors. Since the e.m. form-factors are known, we can use the elastic neutrino–nucleon scattering to determine the axial form-factor.

---

[4] Neglecting the contribution to the axial coupling of strange and charmed quark pairs in the nucleon.
[5] We are assuming that the c.m. energy is much larger than the nucleon mass, but still smaller than the W mass. The behavior for $s > M_W^2$ is discussed in Section 3.7.

The e.m. form-factors have been found to show, approximately, the dipole behavior:

$$\frac{g_E(t)}{g_E(0)} \simeq \frac{g_M(t)}{g_M(0)} \simeq \frac{1}{\left(1 - \dfrac{t}{m_V^2}\right)^2} \tag{3.6.31}$$

with

$$m_V \simeq 0.84 \, \text{GeV}. \tag{3.6.32}$$

The axial current form-factor can be obtained by making the difference between neutrino and antineutrino cross sections

$$\frac{d\sigma^\nu}{d\omega'} - \frac{d\sigma^{\bar\nu}}{d\omega'} = 4M \frac{G^2 \cos^2\theta_c}{\pi} \frac{\omega'}{\omega} \frac{\omega + \omega'}{M} \sin^2(\theta_{\text{lab}}/2) g_M(t) g_A(t).$$

A result similar to (3.6.31) is obtained for the axial current, with

$$m_A \simeq 0.9 \, \text{GeV}. \tag{3.6.33}$$

Owing to the rapid fall-off of the form-factors at large t, the elastic cross section approaches a constant limit. As seen from Eqs. (3.6.21–22), the differential cross section at zero momentum transfer is energy-independent:

$$\left(\frac{d\sigma}{dt}\right)_{t=0} = \frac{G^2 \cos^2\theta_c}{2\pi} [g_V(0)^2 + g_A(0)^2]. \tag{3.6.34}$$

At high energy, the form-factors cut the differential cross section for values of $|t| > m_{V,A}^2$, so that

$$\sigma \simeq m_V^2 \left(\frac{d\sigma}{dt}\right)_{t=0} \simeq \frac{G^2 m_V^2}{\pi} \simeq 10^{-38} \, \text{cm}^2. \tag{3.6.35}$$

3.6.2 Inelastic processes

For energies above 1 GeV, the interaction of neutrinos with matter is dominated by the inelastic processes:

$$\nu_1 + \text{nucleon} \rightarrow 1^- + \text{hadrons}, \tag{3.6.36}$$

$$\nu_1 + \text{nucleon} \rightarrow \nu_1 + \text{hadrons}. \tag{3.6.37}$$

In the completely inclusive situation, where only the final lepton is observed,[6] the cross section depends upon three variables, which we may take as the initial and final lepton (laboratory) energies and scattering angle, ω, ω', and θ_{lab}. The latter two

---

[6] In the charged-current process (3.5.36) one can reconstruct completely the kinematics, also for nonmonochromatic incident neutrinos, by measuring the final lepton energy and angle and the final hadron total energy.

variables are often replaced by the lepton momentum transfer and by the energy transmitted to the hadrons:

$$Q^2 = -t = (k' - k)^2$$
$$\nu = \omega - \omega' = -\frac{1}{M}(qp) \tag{3.6.38}$$

or by the corresponding Bjorken scaling variables:

$$x = \frac{Q^2}{2M\nu}$$
$$y = \frac{\omega - \omega'}{\omega} = -\frac{(qp)}{(kp)}. \tag{3.6.39}$$

The inclusive differential cross section is accordingly written as

$$\frac{d^2\sigma}{d\cos\theta_{\mathrm{lab}}d\omega'} = 2\omega\omega'\frac{d^2\sigma}{dQ^2d\nu} = \frac{1}{M}\frac{1-y}{y}\frac{d^2\sigma}{dxdy} \tag{3.6.40}$$

with s the c.m. energy squared in the lepton–nucleon system:

$$s = 2\omega M + M^2.$$

For definiteness, we specialize to neutrino–proton scattering. The inelastic cross section for unpolarized protons is determined by the spin-averaged matrix element squared:

$$\sum_\beta (2\pi)^4 \delta^4(p_\beta - p_\alpha + q)\langle P|J(0)_\mu^+|\beta\rangle\langle\beta|J(0)_\nu|P\rangle|_{\mathrm{spin\ av}}.$$
$$= (2\pi)^{-3}(E_P)^{-1}(2\pi W_{\mu\nu}). \tag{3.6.41}$$

In turn, the hadronic tensor $W_{\mu\nu}$ can be expressed in terms of inelastic structure functions, according to

$$W_{\mu\nu} = \frac{p_\mu p_\nu}{M}W_2(\nu, Q^2) - Mg_{\mu\nu}W_1(\nu, Q^2) - \frac{1}{2M}i\epsilon_{\mu\alpha\nu\beta}p^\alpha q^\beta W_3(\nu, Q^2) + \cdots. \tag{3.6.42}$$

Dots represent terms proportional to q_μ and/or q_ν, which vanish when contracted with the lepton tensor, Eq. (3.6.17).

The inelastic cross section takes a form very similar to Eq. (3.6.21), namely,

$$\frac{d\sigma}{dQ^2 d\nu} = \frac{G^2\cos^2\theta_C}{2\pi}\frac{\omega'}{\omega}\left\{\cos^2(\theta_{\mathrm{lab}}/2)W_2(\nu, Q^2) + 2\sin^2(\theta_{\mathrm{lab}}/2)W_1(\nu, Q^2)\right.$$
$$\left. + \frac{\omega + \omega'}{M}\sin^2(\theta_{\mathrm{lab}}/2)W_3(\nu, Q^2)\right\}. \tag{3.6.43}$$

Among the final states there is, of course, the single nucleon state. We can immediately identify the elastic contribution to the cross section or to the structure functions:

$$\frac{d^2\sigma^{(el)}}{dQ^2 d\nu} = \delta\left(\nu - \frac{Q^2}{2M}\right)\frac{d\sigma}{dt}$$

$$W^{(el)}_{1,2,3}(\nu, Q^2) = \delta\left(\nu - \frac{Q^2}{2M}\right)f_{1,2,3}(-Q^2). \tag{3.6.44}$$

The deep inelastic region is defined by the conditions

$$Q^2 \gg M^2; \ \nu \gg M; \ \ x = \text{fixed}. \tag{3.6.45}$$

The properties of neutrino deep inelastic structure functions in QCD are discussed in detail in Chapter 5. In the following, we give a brief illustration of the parton picture of deep inelastic processes [FEY 72] and relate deep inelastic cross sections to the elementary neutrino–quark cross sections.

3.6.3 The parton description

Let us describe the deep inelastic process in the Breit frame of reference, where the momentum transfer has the space component only:

$$q_0 = q_x = q_y = 0; \ \ q_z = -\sqrt{Q^2}.$$

In this frame, the proton has a very large momentum along the z-axis:

$$p_z = E_p = \frac{\nu M}{\sqrt{Q^2}} \ \rightarrow \ \text{infinity}$$

Following Feynman, we assume that the proton can be described as a cloud of pointlike objects, the partons [FEY 72]. The parton distribution inside the proton is specified by a set of probability functions, $f_i(z)$, called parton densities, such that $f_i(z)$ is the probability density to find a parton of type i with a fraction z of the proton longitudinal momentum.

Experiment shows a definite cut in the transverse-momentum distribution of the final particles in hadronic collisions. Accordingly, we assume that partons have a limited transverse momentum distribution ($\Delta_{pT} \simeq 0.1$ GeV) and treat them as if they were collinear with the proton. Thus, neglecting the proton and the parton mass, the parton four-momentum is

$$(p_i)_\mu = z p_\mu \tag{3.4.46}$$

and we may define parton variables analogous to (3.6.38–39); that is,

$$\nu_i = \frac{1}{M}(qp_i) = z\nu$$

$$y_i = -\frac{(qp_i)}{(kp_i)} = y.$$

The deep inelastic process arises from the incoherent elastic scattering of the neutrino over each parton. After the collision, the hadronic system ends up in a configuration quite different from the initial one, with one parton scattered away from the others. This highly excited state will then evolve in the multihadronic state observed in the final state of process (3.6.36–37).

We may express the inclusive cross section in terms of the elementary neutrino–parton cross sections, according to

$$\frac{d\sigma}{dQ^2 d\nu} = \sum_i \int dz f_i(z) \frac{\partial \nu_i}{\partial \nu} \frac{d\sigma_i}{dQ^2 d\nu_i}$$

$$= \sum_i \int dz f_i(z) z \delta \left(z\nu - \frac{Q^2}{2M} \right) \frac{d\sigma_i}{dQ^2}$$

$$= \sum_i f_i(x) \frac{x}{\nu} \frac{d\sigma_i}{dQ^2}. \tag{3.6.47}$$

Thus the Bjorken scaling variable x in (3.6.39) is identified with the fractional parton momentum.

For dimensional reasons, the neutrino cross section for a massless, point-like parton has the form

$$\frac{d\sigma_i}{dq^2} = \frac{G^2}{\pi} g_i(y_i) = \frac{G^2}{\pi} g_i(y)$$

so that, in terms of adimensional variables, we obtain[7]

$$\frac{d\sigma}{dxdy} = \sum_i f_i(x) \frac{d\sigma_i}{dy}(xs, y) = \frac{G^2}{\pi} s \sum_i x f_i(x) g_i(y). \tag{3.6.48}$$

With a view to the experimental results, we identify the weakly interacting partons with quarks. The parton cross section to be used in Eq. (3.6.48) is the one appropriate to massless, pointlike, spin $\frac{1}{2}$ particles and it can be taken from Section 3.5, Eq. (3.5.27), so that

$$\frac{d\sigma}{dxdy} = \frac{G^2}{\pi} 2M\omega \sum_i x f_i(x) [(g_L^{(i)})^2 + (1-y)^2 (g_R^{(i)})^2]. \tag{3.6.49}$$

---

[7] Replacing $d\sigma_i(xs, y)/dy$ with the corresponding electron–parton cross section, one obtains the cross section for deep inelastic electron–proton scattering. We are as usual neglecting W or Z propagator effects, which can be easily restored in G^2.

As for the structure functions, we find, in the scaling limit (3.6.45):

$$MW_1(\nu, Q^2) = F_1(x) = \frac{1}{2}\sum_i f_i(x)[(g_L^{(i)})^2 + (g_R^{(i)})^2]$$

$$\nu W_2(\nu, Q^2) = F_2(x) = \sum_i x f_i(x)[(g_L^{(i)})^2 + (g_R^{(i)})^2] \qquad (3.6.50)$$

$$\nu W_3(\nu, Q^2) = F_3(x) = \sum_i f_i(x)[(g_L^{(i)})^2 - (g_R^{(i)})^2]$$

which shows the relation $F_2(x) = 2x F_1(x)$, typical of spin $\frac{1}{2}$ partons [CAL 69; BJO 69b; GRO 69]. Equations (3.6.48–50) embody what is called the scaling property of deep inelastic processes.

Although innocent looking, the validity of scaling is far from trivial. One assumption we had to make is that the hadrons can be described in terms of a certain number of elementary fields. Given the success of the quark model in describing the hadron spectroscopy, we can take this for granted. However, there is another, crucial, assumption that we can illustrate as follows.

In the Breit system we are considering, the typical time scale of the neutrino–parton collision is given by $1/\sqrt{Q^2}$, which is much smaller than the typical strong interaction time, of order $1/M$. If we can neglect the action of the strong interaction during the collision time, we can argue that the neutrino indeed probes the bare quanta of the elementary fields and we can derive the pointlike cross section and scaling.

It is a fact, however, that the hypothesis is not verified in a general, renormalizable, field theory. Any field theory contains a hidden, high-energy scale, namely, the scale of the ultraviolet cutoff Λ, required to make quantum corrections finite before renormalization. In this situation, there are strong interaction processes that occur with any time scale longer than $1/\Lambda$. Hence also at time scales of the order of $1/\sqrt{Q^2}$ there may occur significant strong-interaction modifications to the parton cross section. The only way in which the scaling behavior can be resurrected is that processes at very short time scales (or short distances) are suppressed by an attenuation of the interaction strength, namely, if the theory is asymptotically free.[8]

The same conclusion is obtained following the Bjorken approach [BJO 69a]. The matrix element squared in Eq. (3.6.41) is recognized to be the Fourier transform of

---

[8] One can be suspicious about the above argument, since the cutoff disappears in the renormalized theory, but the argument is correct. Indeed, in the renormalized theory of cutoff is replace by a finite scale μ, representing the momentum scale where we define the renormalized couplings. Letting all masses go to zero, scale invariance is in general violated by logarithmic corrections of the form $\log(Q^2/\mu^2)$. It is recovered only if these corrections conspire to give a decreasing strength at large Q^2, namely, if the theory is asymptotically free.

the current–current product at different space-time points:

$$\sum (2\pi)^4 \delta^4 (p_\beta - p_\alpha + q) \langle P|J(0)^+_\mu |\beta\rangle \langle \beta|J(0)_\nu|P\rangle|_{\text{spin av}}$$

$$= \int dx^4 e^{iqx} \langle P|J(x)^+_\mu J(0)_\nu|P\rangle|_{\text{spin av}}.$$

In the limit (3.6.45), the integral is dominated by the light-cone singularities of the current product. Scaling requires these singularities to coincide with those of the free quark model, namely, that strong interactions become free at short distances, $x^2 \cong 1/Q^2 < 1/M^2$.

The fact that the scaling laws Eqs. (3.6.48–50) are approximately obeyed in the real world has therefore to be seen as highly nontrivial information on the strong interaction dynamics. In fact, it selects non-abelian gauge theories, the only ones that are asymptotically free [GRO 73; POL 73; COL 73; this result has also been reported in an unpublished work of G. t'Hooft (1972)] and leads directly to QCD.

Scaling cannot be exact, since freedom sets in only asymptotically. The logarithmic scaling violations expected in QCD are considered in Chapter 5.

3.6.4 Deep inelastic cross sections

To conclude the section we give the formulae for (anti-)neutrino scattering off an equal mixture of protons and neutrons (i.e., $I = 0$ nuclei). We denote by $f_{d,u,s}(x)$ and $f_{\bar{u},\bar{d},\bar{s}}(x)$ the parton densities of the proton. For the neutron,

$$f_{d,u}(x)|_N = f_{u,d}(x); \quad f_s(x)|_N = f_s(x)$$

$$f_{\bar{u},\bar{d}}(x)|_N = f^x_{\bar{d},\bar{u}}; \quad f_{\bar{s}}(x)|_N = f_{\bar{s}}(x). \qquad (3.6.51)$$

1 Charged currents We set $\cos\theta_C = 1$ at first and ignore heavy quark production. The parton reactions are

$$\nu_1 + d \to 1^- + u; \; \nu_1 + \bar{u} \to 1^- + \bar{d}$$

$$\bar{\nu}_1 + u \to 1^+ + d; \; \bar{\nu}_1 + \bar{d} \to 1^+ + \bar{u}. \qquad (3.6.52)$$

We find:

$$\frac{d\sigma(\nu_1 \to 1^-)}{dxdy} = \frac{1}{2}\left[\frac{d\sigma(\nu_1 \to 1^-)}{dxdy}\bigg|_P + \frac{d\sigma(\nu_1 \to 1^-)}{dxdy}\bigg|_N\right]$$

$$= \frac{G^2}{\pi} M\omega \sum_i x\{[f_d(x) + f_u(x)] + (1-y)^2[f_{\bar{u}}(x) + f_{\bar{d}}(x)]\} \qquad (3.6.53)$$

$$\frac{d\sigma(\bar{\nu} \to 1^+)}{dx\,dy} = \frac{1}{2}\left[\frac{d\sigma(\bar{\nu} \to 1^+)}{dx\,dy}\bigg|_P + \frac{d\sigma(\bar{\nu} \to 1^+)}{dx\,dy}\bigg|_N\right]$$

$$= \frac{G^2}{\pi}M\omega\sum_i x\{[f_{\bar{d}}(x) + f_{\bar{u}}(x)] + (1-y)^2[f_u(x) + f_d(x)]\} \qquad (3.6.54)$$

and

$$\sigma(\nu_1 \to 1^-) = \frac{G^2}{\pi}M\omega[\langle x\rangle_q + \tfrac{1}{3}\langle x\rangle_{\bar{q}}] \qquad (3.6.55)$$

$$\sigma(\nu_1 \to 1^+) = \frac{G^2}{\pi}M\omega[\tfrac{1}{3}\langle x\rangle_q + \langle x\rangle_{\bar{q}}] \qquad (3.6.56)$$

$$\frac{G^2}{\pi}M = 1.41 \times 10^{-38}\,\text{cm}^2/\text{GeV}$$

where $\langle x\rangle_{q,\bar{q}}$ are the fractional momentum carried by quarks and antiquarks.

The cross sections (3.6.55–6) are much larger than the corresponding neutrino–electron cross section, the ratio being of the order of $M/m_e \cong 1.8 \times 10^3$. Experiments support very well the linear rise of the cross section with energy and give a ratio of about 3 between neutrino and antineutrino, as obtained from (3.6.55–6), if we neglect the antiquark contribution to the proton momentum, the so-called valence approximation. Experimentally, one finds:

$$\sigma(\nu_1 \to 1^-)|_{\text{expt}} = \frac{G^2}{\pi}M\omega(0.44 \pm 0.01)$$

$$\sigma(\bar{\nu} \to 1^+)|_{\text{expt}} = \frac{G^2}{\pi}M\omega(0.21 \pm 0.01) \qquad (3.6.57)$$

which gives: $\langle x\rangle_q + \langle x\rangle_{\bar{q}} \cong 0.49$, $\langle x\rangle_{\bar{q}} \cong 0.07$. Quarks do not carry the total proton momentum, the missing one has to be carried by the gluons (this result was first found in electron–nucleon deep inelastic scattering).

Neutrinos can produce heavy quark flavors on nucleons, due to the non-diagonal terms in the charged weak current. In the charm case, for example, we have the following processes:

$$\nu_1 + d \to 1^- + c\,(U_{cd} \cong -\sin\theta_C);$$

$$\nu_1 + s \to 1^- + c\,(U_{cs} \cong \cos\theta_C);$$

$$\bar{\nu}_1 + \bar{d} \to 1^+ + \bar{c}\,(U_{cd} \cong -\sin\theta_C);$$

$$\bar{\nu}_1 + \bar{s} \to 1^+ + \bar{c}\,(U_{cs} \cong \cos\theta_C).$$

We can determine the matrix elements of the CKM-matrix, U_{cd} and U_{cs}, from the observation of charm production on nucleons. The contribution to the cross section can be obtained by using the cross sections for the inverse μ-decay, Eqs. (3.5.36–37), with the appropriate changes:

$$\frac{d\sigma(\nu_1 \to c)}{dx\,dy} = \frac{G}{\pi} 2M\omega x[|U_{cd}|^2 f_d(x) + |U_{cs}|^2 f_s(x)] \left(1 - \frac{m_c^2}{xs}\right)$$

$$\frac{d\sigma(\bar{\nu}_1 \to \bar{c})}{dx\,dy} = \frac{G}{\pi} 2M\omega x[|U_{cd}|^2 f_{\bar{d}}(x) + |U_{cs}|^2 f_{\bar{s}}(x)](1 - y) \left(1 + \frac{m_c^2}{sx} - y\right).$$

Charm is produced by neutrinos both on valence d-quarks (but with the small coupling, U_{cd}) and nonvalence, sea, s-quarks, with a large coupling, U_{cs}. Only sea quarks are involved in antineutrino reactions.

At not too large values of $s \cong 2M\omega \cong m_c^2$, heavy flavor production is a source of scaling violations, which adds to the logarithmic scaling violations mentioned above.

2 Neutral currents We obtain:

$$\frac{d\sigma(\nu_1 \to \nu_1)}{dx\,dy}$$

$$= \frac{1}{2}\left[\frac{d\sigma(\nu_1 \to \nu_1)}{dx\,dy}\bigg|_P + \frac{d\sigma(\nu_1 \to \nu_1)}{dx\,dy}\bigg|_N\right]$$

$$= \frac{G^2}{\pi} M\omega \sum_i x\{[f_d(x) + f_u(x)][A + B(1-y)^2] + 2f_s(x)[C + D(1-y)^2]$$

$$+ [f_{\bar{u}}(x) + f_{\bar{d}}(x)][B + A(1-y)^2] + 2f_s(x)[D + C(1-y)^2]\} \qquad (3.6.58)$$

$$A = (g_L^{(u)})^2 + (g_L^{(d)})^2$$

$$B = (g_R^{(u)})^2 + (g_R^{(d)})^2$$

$$C = (g_L^{(s)})^2; \ D = (g_R^{(s)})^2$$

with

$$g_L^{(u,d)} = \pm \tfrac{1}{2} - Q^{(u,d)} \sin^2\theta_W; \ g_R^{(u,d)} = -Q^{(u,d)} \sin^2\theta_W$$

s and d couplings are equal.

The neutral current is flavor diagonal in the standard theory. Hence we expect no single production of heavy quarks, as observed experimentally to a good precision.

3.7 High-energy behavior and unitarity limits

In a renormalizable field theory, unitarity of the scattering matrix (S-Matrix) is satisfied order by order in perturbation theory and for any center-of-mass energy, at least until the energy is so large that $g^2 \log(s/M^2)$ becomes of order unity and perturbation theory loses its meaning.[1] This behavior is to be contrasted with that of nonrenormalizable theories. In the latter case, amplitudes grow like polynomials in the energy, and unitarity is violated at a finite energy (i.e., not exponentially large), no matter how small the coupling constant is.

It is interesting to study in detail the way perturbative unitarity is satisfied [WEI 71] in the standard theory. In the unitary gauge (Section 3.2), the theory is superficially nonrenormalizable and, in fact, individual Feynman diagrams give rise to dangerously increasing terms. Unitarity is enforced by delicate cancellations between different amplitudes.

One could even try to go the other way round [LLE 73; COR 74]; that is one could consider a general theory involving fermions, scalars, and massive vector bosons and work out the general conditions for which unitarity in all scattering channels is satisfied by the lowest-order amplitudes, up to exponentially large energies. It turns out that the conditions are so tight as to lead *almost* to a proof that the theory must be a spontaneously broken gauge theory. In the case of a semisimple group, in fact, the proof can be carried to the end. This result gives strong support to the idea that there are no renormalizable alternatives to the Higgs mechanism, to describe non-abelian massive vector bosons in the weak coupling regime.

The constraints posed by the unitarity of the S-Matrix on two-particle into two-particle scattering are formulated easily in terms of partial-wave amplitudes.

The helicity amplitudes are developed in generalized partial waves, according to (see, e.g., [JAC 64; BER 72])

$$f_{\lambda3,\lambda4;\lambda1,\lambda2}(\theta, s) = \sum_j (2j + 1) d^{(j)}_{\mu,\lambda}(\theta) \langle \lambda_3, \lambda_4 | f^{(j)}(s) | \lambda_1, \lambda_2 \rangle \qquad (3.7.1)$$

λ_1 to λ_4 are the initial and final helicities, $\lambda = \lambda_1 - \lambda_2$, $\mu = \lambda_3 - \lambda_4$ and $d^{(j)}_{\mu,\lambda}(\theta)$ is the polar-angle-dependent part of the finite rotation matrices, for angular

---

[1] To investigate what happens at such large energies, one has to improve on perturbation theory, using renormalization-group techniques. If the theory is asymptotically free, the large logarithms actually add up in such a way as to reduce the scattering amplitudes, and perturbative unitarity is obeyed also at very high energy.

momentum j. We note the explicit expression of the first few d-matrices:

$$d_{0,0}^{(0)}(\theta) = 1$$

$$d^{(1)}1, 1(\theta) = d^{(1)} - 1, - 1^{(\theta)} = \frac{1}{2}(1 + \cos\theta)$$

$$d^{(1)}0, 0(\theta) = \cos\theta$$

$$d^{(1)}1, 0(\theta) = -d^{(1)}0, 1^{(\theta)} = d^{(1)}0, - 1^{(\theta)}$$

$$= -d^{(1)} - 1, 0^{(\theta)} = \frac{1}{\sqrt{2}}\sin\theta$$

$$d_{1,-1}^{(1)}(\theta) = d^{(1)} - 1, 1^{(\theta)} = \frac{1}{2}(1 - \cos\theta). \tag{3.7.2}$$

In turn, the partial-wave amplitudes can be obtained from the expansion (3.7.1), using the orthogonality property:

$$\frac{1}{2}\int d\cos\theta d_{\mu,\lambda}^{(j)}(\theta)d_{\mu,\lambda}^{(j')}(\theta) = \frac{1}{2j+1}\delta_{j,j'}.$$

The partial-wave matrix $f^{(j)}$ is related to the S-Matrix by the relation

$$\langle\lambda_3, \lambda_4|f^{(j)}(s)|\lambda_1, \lambda_2\rangle = \frac{1}{2q}\langle\lambda_3, \lambda_4|T^{(j)}(s)|\lambda_1, \lambda_2\rangle$$

$$= \frac{1}{2iq}\langle\lambda_3, \lambda_4|(S^{(j)}(s) - 1)|\lambda_1, \lambda_2\rangle = \frac{\eta e^{i2\delta} - 1}{2iq} \tag{3.7.3}$$

q is the center-of-mass momentum, δ the phase shift, and ρ the inelasticity parameter. Unitarity of the S-Matrix implies that δ is real and

$$0 < \eta < 1$$

so that the partial-wave amplitudes are bounded by $1/q$:

$$|\langle\lambda_3, \lambda_4|f^{(j)}(s)|\lambda_1, \lambda_2\rangle| < \frac{1}{q}. \tag{3.7.4}$$

Let us apply the above considerations, first to $\nu_\mu{-}e$ elastic scattering.

1 $\nu_\mu - e$ elastic scattering The helicity amplitudes have been given in Section 3.5, Eq. (3.5.12). To leading order in the energy and reinserting the neutrino helicity in the notation, the helicity amplitudes are

$$f_{-\frac{1}{2},-\frac{1}{2};-\frac{1}{2},-\frac{1}{2}}(s,\theta) = -\rho\sqrt{2}\frac{G_Z(t)}{2\pi}2E_{g_L}$$

$$f_{-\frac{1}{2},+\frac{1}{2};-\frac{1}{2},+\frac{1}{2}}(s,\theta) = -\rho\sqrt{2}\frac{G_Z(t)}{2\pi}2E_{g_R}\frac{1+\cos\theta}{2}$$

$$f_{-\frac{1}{2},-\frac{1}{2};-\frac{1}{2},+\frac{1}{2}}(s,\theta) = f_{-\frac{1}{2},+\frac{1}{2};-\frac{1}{2},-\frac{1}{2}}(s,\theta)$$

$$= -\rho\sqrt{2}\frac{G_Z(t)}{4\pi}m\sin\theta g_R.$$

(3.7.5)

a. Fermi interaction. If we neglect the t-dependence in $G_Z(t)$, that is, in the case of a local Fermi interaction, the amplitudes in (3.7.5) correspond to $j=0$, 1 partial-wave amplitudes.

The unitarity bound is violated by the leading, helicity-nonflip, amplitudes and by the next-to-leading, helicity-flip, amplitudes.

Numerically, the $j=0$ amplitude exceeds the unitarity bound for

$$\sqrt{2}\rho g_L\frac{GE^2}{\pi} > 1; \quad E > (\rho g_L)^{-\frac{1}{2}}(436\,\text{GeV})$$

(3.7.6)

b. Z-exchange. Taking into account the t-dependence of G_Z, the $j=0$ amplitude is given by

$$f^{(0)}(s) = -\sqrt{2}\rho g_L\frac{G}{\pi}E\int d\cos\theta\frac{1}{2}\frac{1}{1-t/M_Z^2}$$

$$= -\sqrt{2}\rho g_L E\frac{G}{\pi}\frac{M_Z^2}{s}\log\left(\frac{1+s}{M_Z^2}\right)$$

$$= \frac{g^2}{16\pi(\cos\theta_W)^2}\frac{\rho}{E}\log\left(1+\frac{s}{M_Z^2}\right)$$

where g is the $SU(2)$ gauge coupling (see Section 3.2). Unitarity is now obeyed up to exponentially large energies:

$$s \cong M_Z^2\exp[16\pi(\cos\theta_W)^2/g^2]$$

The same holds for the $j=1$ partial-wave amplitudes.

The softening due to the vector boson exchange, as compared to the local Fermi interaction, is sufficient to satisfy tree-level unitarity for fermion–fermion scattering. It is not so for amplitudes involving the real production of vector bosons [GEL 69]. The polarization vector of a longitudinal (i.e., vanishing helicity) massive vector boson, with space momentum $\mathbf{k}$ and energy ω, is

$$\epsilon^\mu = \frac{\omega}{M}\left(\frac{\mathbf{k}}{|k|}, \frac{|k|}{\omega}\right)$$

(3.7.7)

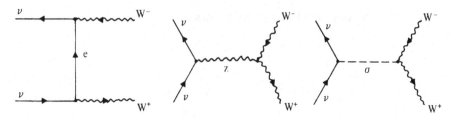

Fig. 3.7.1   Lowest order Feynman diagrams for the process: $\nu + \bar{\nu} \to W^+ - W^-$.

The components of ϵ^μ increase linearly with energy.[2] The situation is similar to the one encountered in the case of the local four-fermion interaction. For dimensional reasons, the helicity amplitudes increase (at most) like ω/M^2. The leading and next-to-leading amplitudes violate unitarity, and have to cancel out in a renormalizable theory.

2 W-pair production   We consider the amplitude for the process

$$\nu_e(p) + \bar{\nu}_e(p') \quad \to \quad W^+(k) + W^-(k') \tag{3.7.8}$$

with two longitudinally polarized bosons, determined by the diagrams of Fig. 3.7.1. For illustrative reasons, we shall not neglect the neutrino mass. We work in the c.m. frame, take $\mathbf{p}$ along the positive z-axis and denote by E, m, the neutrino energy and mass, respectively, and by v the W velocity ($v = k\omega = k/E$).

The electron-exchange diagram gives rise to the invariant Feynman amplitude:

$$M^{(e)} = -\frac{g^2}{4} \frac{1}{(p-k)^2 - m_e^2} \epsilon^\mu(k')\epsilon^\nu(k)$$
$$\{\bar{v}(p')[\gamma_\mu(p'-k')\gamma_\nu](1+\gamma_5)u(p)\}. \tag{3.7.9}$$

Reducing the product of three gamma matrices,[3] $M^{(e)}$ can be evaluated using Eq. (3.7.7) and the current matrix elements:

$$J^\alpha = \tfrac{1}{2}\bar{v}(p')\gamma^\alpha(1+\gamma_5)u(p)$$

for the different neutrino helicity states.

For the case where the neutrino (antineutrino) has negative (positive) helicity, one finds:

$$(Jx, y, z, t)_{-\frac{1}{2},+\frac{1}{2}} = \frac{1}{2}E(1+\beta)(1, -i, 0, 0) \tag{3.7.10}$$

---

[2] The linear rise in energy of the polarization vector is, of course, the origin of the irregular behavior of the massive boson propagator, Eq. (3.5.8), whose longitudinal part approaches a constant limit as the components of q_μ go to infinity.

[3] With the identity: $\gamma_\mu\gamma_\alpha\gamma_\nu = g_{\mu\alpha}\gamma_\nu + \gamma_\mu g_{\alpha\nu} - g_{\mu\nu}\gamma_\alpha + i\epsilon_{\mu\alpha\nu\beta}\gamma^\beta\gamma_5$.

while in the equal-helicity case,

$$(Jx, y, z, t)_{+\frac{1}{2},+\frac{1}{2}} = \tfrac{1}{2}m(0, 0, -1, -1)$$
$$(Jx, y, z, t)_{-\frac{1}{2},+\frac{1}{2}} = \tfrac{1}{2}m(0, 0, +1, -1).$$

(3.7.11)

The electron mass has disappeared from the numerator of Eq. (3.7.9), because of the $(1 + \gamma_5)$ projectors. Furthermore, in the opposite-helicity case, Eq. (3.7.10), the neutrino mass appears only through β, namely only in terms of order m^2/E^2 with respect to the leading terms. Therefore, we may neglect the electron mass altogether and, to compute the opposite-helicity amplitude, we may also set $m = 0$. On the other hand, the equal-helicity amplitude is itself proportional to m, Eq. (3.7.11), and it represents the next-to-leading contribution in E.

We find

$$f^{(e)}(s, \theta)_{0,0;-\frac{1}{2},+\frac{1}{2}} = \frac{1}{4\pi(s)^{\frac{1}{2}}} (v)^{\frac{1}{2}} M^{(e)}_{-\frac{1}{2},+\frac{1}{2};0,0}$$

$$= \frac{g^2}{16\pi M^2} v^{\frac{1}{2}} \omega \sin\theta \frac{2\omega^2 \left(1 - \dfrac{1}{v}\cos\theta\right) + M^2}{2\omega^2(1 - v\cos\theta) - M^2 - m^2 + m_e^2}$$

$$= \frac{G}{2\pi\sqrt{2}} \omega \sin\theta[1 + O(\omega^{-2})] \quad \rightarrow \quad \frac{G}{2\pi} \omega[d^{(1)}_{0,-1}(\theta)]$$

(3.7.12)

and

$$f^{(e)}(s, \theta)_{0,0;+\frac{1}{2},+\frac{1}{2}} = -f^{(e)}(s, \theta)_{-\frac{1}{2},-\frac{1}{2};0,0}$$

$$= -\frac{Gm}{4\pi\sqrt{2}}(1 + \cos\theta)[1 + O(\omega^{-2})]$$

$$\rightarrow -\frac{Gm}{4\pi\sqrt{2}}[d^{(0)}_{0,0}(\theta) + d^{(1)}_{0,0}(\theta)].$$

(3.7.13)

Evidently, the amplitude (3.7.12) is dominated by the $j = 1$ partial wave, which violates unitarity at an energy of the same order as in (3.7.6). The next-to-leading amplitude (3.7.13) also violates unitarity in the $j = 0$ and 1 partial waves.

The singular behavior of the $j = 1$ partial wave is neatly cured by the Z-exchange diagram. A tedious but straightforward calculation leads to the results:

$$f^{(Z)}(s,\theta)_{0,0;-\frac{1}{2},+\frac{1}{2}} = -\frac{g^2}{16\pi M^2} v^{1/2} \omega \sin\theta \frac{4\omega^2 + 2M^2}{4\omega^2 - M_Z^2}$$

$$= -\frac{G}{2\pi\sqrt{2}} \omega \sin\theta[1 + O(\omega^{-2})]$$

$$f^{(Z)}(s,\theta)_{0,0;+\frac{1}{2},+\frac{1}{2}} = \frac{Gm}{4\pi\sqrt{2}} \cos\theta[1 + O(\omega^{-2})]. \qquad (3.7.14)$$

It is in the $j=0$ partial waves that the Higgs mechanism shows its power. The neutrino and W couplings of the Higgs boson are completely determined by particle masses, as discussed in Sections 3.2 to 3.4. In particular,

$$L_{\sigma-W^-W^+} = gM(\sigma W_\mu^+ W^\mu)$$

$$L_{\sigma-\nu\bar\nu} = \frac{g_\nu}{\sqrt{2}} \bar\nu\nu$$

$$g_\nu = g \frac{m}{\sqrt{2M}}$$

so that the Higgs-boson-exchange diagram is given in terms of the same constants which appear in the previous contributions. In fact, one easily finds:

$$f^{(\sigma)}(s,\theta)_{0,0;+\frac{1}{2},+\frac{1}{2}} = \frac{Gm}{4\pi\sqrt{2}} v^{1/2} \frac{1+v^2}{2} \frac{4\omega^2}{4\omega^2 - M_\sigma^2}$$

$$\rightarrow \frac{Gm}{4\pi\sqrt{2}}. \qquad (3.7.15)$$

The overall amplitude is well behaved at large energy.

3.8 Coherent effects in neutrino propagation through matter

In this section we discuss the coherent effects in neutrino propagation through matter. No such effect has been established, as yet. Effects of this kind have been discussed in connection with the (thus far unsuccessful) effort to detect the cosmic 2K background neutrinos. More recently, resonant neutrino oscillations in matter have been advocated as the origin of the (still controversial) reduction of the solar neutrino flux. Coherent effects during propagation of the neutrino in matter may be important for neutrinos in stars and during a supernova explosion.

We limit ourselves to a discussion of the main ingredients and to a presentation of the basic formulae, as the subject is raised in other parts of this book. Except for Section 3.8.3, we consider the propagation through homogeneous matter, with no net polarization.

3.8.1 The refraction index

Neutrinos undergo refraction while moving in a medium; namely, a plane wave propagating in the (homogeneous) medium has the space dependence

$$\Psi(\mathbf{x}) = \Psi(0)e^{+i n k x}$$

$\mathbf{k}$ is the space momentum, related in the usual way to the energy, ω. For example,

$$|\mathbf{k}| = \omega$$

for massless neutrinos.

A simple way to compute the refraction index is to observe that the interaction with a homogeneous medium adds a constant, V, to the neutrino energy, with respect to the energy in vacuum. The wave has a momentum k in vacuum and a momentum nk in the medium, where n is such that the total energy is the same in the two cases:

$$\omega(nk) + V = \omega(k). \tag{3.8.1}$$

For n close to unity:

$$\frac{d\omega}{dk} k(n-1) = -V \qquad n = 1 - \frac{V}{vk}. \tag{3.8.2}$$

The current–current interaction Hamiltonian produced by W- and Z-exchange, leads to the very simple form:

$$V(\nu) = \frac{G}{\sqrt{2}} \sum_1 N^{(i)} g^{(i)}(\nu) \tag{3.8.3}$$

where $N^{(i)}$ is the number density of the ith particle species present in the medium and $g^{(i)}(\nu)$ are effective couplings for the given neutrino type, which we shall determine presently.

Equivalently, for a disordered and not-too-dense medium, the refraction index n is related to the forward-scattering amplitude on each particle species, according to (see, e.g., [NEW 66]):

$$n = 1 + \sum_1 \frac{2\pi N^{(i)}}{k^2} f^{(i)}(\omega, \cos\theta = 0) \tag{3.8.4}$$

where $f^{(i)}$ is normalized so that

$$\frac{d\sigma}{d\Omega} = |f|^2.$$

The energy in Eq. (3.8.3) is very small. A convenient unit is

$$V^{(0)} = \frac{G}{\sqrt{2}} \rho^{(0)} = 3.81 \times 10^{-14} \, \text{eV} \qquad (3.8.5)$$

with $\rho^{(0)} = 1$ Avogadro number/cm^3.

1 Normal matter For neutrinos propagating in nonrelativistic, unpolarized matter, $f^{(i)}$ is the average of the helicity nonflip amplitudes computed in Section 3.5. From (3.5.12) we find (massless neutrinos, we set $\rho = 1$ throughout):

$$f(\omega, \cos\theta = 0)^{(i)} = \tfrac{1}{2}(f_{+,+} + f^{(i)}_{-,-}) = -\frac{G}{2\pi\sqrt{2}}\omega[2(g_L + g_R)^{(i)}]$$

so that

$$n = 1 - \left(\frac{G}{\sqrt{2}}\frac{1}{\omega}\right) \sum_1 N^{(i)} 2(g_L + g_R)^{(i)}. \qquad (3.8.6)$$

Comparing with Eqs. (3.8.2) and (3.8.4), we see that

$$g^{(i)}(\nu) = 2(g_L + g_R)^{(i)}. \qquad (3.8.7)$$

For massive neutrinos, the factor ω^{-1} in Eq. (3.8.6) is replaced by $(1 + v)/(2vk)$, where v is the neutrino velocity.

We have to distinguish three cases.

– ν_μ, ν_τ, *etc.* Adding the quark vector charges (no approximation is involved here since the vector current is conserved) one finds:

$$g^{(\text{Proton})}(\nu_\mu) = -g^{(\text{Electron})}(\nu_\mu)$$

$$= 1 - 4\sin^2\theta_W$$

$$g^{(\text{Neutron})}(\nu_\mu) = -1. \qquad (3.8.8)$$

For electrically neutral matter, the electron and proton contributions cancel exactly and we are left with the simple result

$$V(\nu_\mu) = -\left(\frac{G}{\sqrt{2}}\right) N_{\text{Atoms}}\, g(\nu_\mu)$$

$$g(\nu_\mu) = (A - Z). \qquad (3.8.9)$$

Notice that $n > 1$, which corresponds to an attractive interaction between the neutrino and the atoms of the medium. See Eq. (3.8.2). Indeed, the residual neutrino–neutron interaction is attractive, since they have opposite weak charges.

− ν_e As discussed in Section 3.5, this is a special case, because of the additional contribution from W-exchange.
W-exchange gives a repulsive contribution (nondiagonal weak charges are equal) of a strength twice as great as that of the neutron. See Eqs. (3.2.13), (3.2.20), and (3.3.15). Therefore

$$V(\nu_e) = -\left(\frac{G}{\sqrt{2}}\right) N_{\text{Atoms}}\, g(\nu_e)$$

$$g(\nu_e) = (A - 3Z).$$

(3.8.10)

− *Antineutrinos*   It follows from Eqs. (3.3.15–17) that the antineutrino refraction index is obtained by the sign switch:

$$n(\bar{\nu}) - 1 = -[n(\nu) - 1].$$

(3.8.11)

A further consequence is that $n = 1$ for Majorana neutrinos.

2 Supernova core   The above equations describe neutrino propagation in stellar environments, except for the case of the supernova core. In this case, one has to take into account the interaction with the other neutrinos, which are not a negligible fraction of the particles present in the core.

Neutrinos interact among themselves via Z-exchange. This gives rise to a repulsive interaction. The contribution of unequal-flavor neutrinos is twice that of the neutrons found previously, because neutrinos are purely left-handed. For equal-flavor neutrinos, there is an additional factor of two, due to equal-particle interference. In conclusion, we find for ν_e, for example,

$$V(\nu_e) = -\frac{G}{\sqrt{2}}\left[N^{(N)} - 4N^{(\nu_e)} - 2\sum_{l \neq e} N^{(\nu_l)}\right].$$

(3.8.12)

3.8.2 Mechanical effects of unpolarized bodies

Neutrinos traversing a body with refractive index n, exchange a momentum of the order of:

$$\Delta k \cong k|n - 1|.$$

Thus, under a stationary neutrino flux,

$$\Phi_\nu = \rho_\nu \upsilon$$

we may expect the body to receive a force of first order in the Fermi constant, G:

$$\text{Force} \cong S\Phi_\nu k|n-1| = SV\rho_\nu$$

$$= S\left(\frac{G}{\sqrt{2}} N_{\text{Atoms}}\right) g(v)\rho_\nu \tag{3.8.13}$$

where S is the surface of the body. In contrast, forces transmitted by nonforward collisions are quadratic in G.

The force in (3.8.13) is peculiar. One can easily prove that [CAB 82; SMI 83]:

The net force of order G vanishes if the spatial density of neutrinos is constant, irrespective of the neutrino momentum distribution.

The net force acting on a body is given by the negative of the time derivative of the neutrino total momentum. By definition,

$$\mathbf{P}_\nu = -i\int d^3x \nu^+(x)\nabla\nu(x)$$

and

$$\mathbf{F} = -\frac{d\mathbf{P}_\nu}{dt} = -i[H,\mathbf{P}_\nu] = -i[H_{\text{int}},\mathbf{P}_\nu]$$

$$= -\frac{G}{\sqrt{2}}g(\nu)\int d^3x N_{\text{Atoms}}(x)\nabla\rho_\nu(x).$$

Evidently, there is no net force of order G, for a spatially homogeneous distribution.

Forces of order $(G)^{3/2}$ can be generated by total reflection. Neutrinos arriving on the surface under a grazing angle, α, smaller than the limiting angle α_0 [OPH 74]:

$$\alpha_0 = (2|n-1|)^{\frac{1}{2}} \tag{3.8.14}$$

exert a force per unit surface:

$$\frac{F}{S} = 4V\rho\nu\left(\frac{\alpha}{\alpha_0}\right)^2.$$

For an uncollimated beam, with angular spread $\Delta\theta$, one could expect a net force due to total reflection equal to

$$\frac{F}{S} = \frac{4}{3}V\rho_\nu\left(\frac{\alpha_0}{\Delta\theta}\right) \propto (G)^{3/2}.$$

However, one has to consider that neutrinos arriving on the other side of the surface are also reflected and give an opposite force. Taking all into account, one obtains also in this case a negative result [CAB 82], namely,

> *Unless the neutrino flux is collimated with α_0, the net force associated with total reflection vanishes.*

The above results eliminate refraction and/or total reflection as a means of detecting the neutrinos associated with the cosmic 2K background (as considered, e.g., in [OPH 74; LEW 80]). Even assuming Dirac particles with a substantial neutrino–antineutrino asymmetry, we expect the flux to be homogeneous in space and certainly not collimated within the very small angle α_0, so that the previous results apply.

The macroscopic force due to refraction is nonvanishing only to the order G^2. Numerically, it is of the order of 10^{-23} dyne, in the most favorable case of massive, galaxy-bound neutrinos [CAB 82], much too small to be detectable with foreseeable techniques.

3.8.3 The effect of polarization

We consider the interaction energy of a single fermion at rest (electron, proton, or neutron) in a stationary flux of neutrinos. To be definite, we consider massless, muon neutrinos. The spin dependent Hamiltonian due to Z-exchange is:

$$
\begin{aligned}
H^{(Z,i)} &= +\frac{G}{2\sqrt{2}}\bar{\nu}(0)\gamma_\mu(1+\gamma_5)\nu(0)[V^\mu - A^\mu]^{(i)} \\
&= \frac{G}{\sqrt{2}}\rho_\nu[2(g_L - g_R)^{(i)}]\int\frac{d^3k}{\omega}f_\nu(\mathbf{k})(\mathbf{v}\cdot\boldsymbol{\sigma})
\end{aligned}
\tag{3.8.15}
$$

where $\mathbf{k}$ and $\mathbf{v}$ are the neutrino momentum and velocity ($|\mathbf{v}|=1$) and $\boldsymbol{\sigma}/2$ is the target-particle spin. We have also introduced the neutrino momentum-distribution function, normalized so that

$$
\int\frac{d^3k}{\omega}f_\nu(\mathbf{k}) = 1.
$$

For an isotropic neutrino momentum distribution, as in the rest frame of the cosmic background neutrinos, the interaction energy vanishes. Owing to the earth's motion through the galaxy, however, the neutrino distribution is asymmetric, and a nonvanishing value of the spin-associated interaction results [STO 74]. Notice the difference with respect to the previous case, where the momentum anisotropy did not matter.

To first order in the earth velocity in the galaxy, $\mathbf{V}$, it is not difficult to see that

$$\int \frac{d^3k}{\omega} f_\nu(\mathbf{k})\mathbf{v} = -\frac{2}{3}\mathbf{V}. \tag{3.8.16}$$

To this end, one uses the fact that the distribution is isotropic in the cosmic-background rest frame, that is,

$$\frac{d^3k}{\omega} f_\nu(\mathbf{k}) = \frac{d^3k_{rf}}{\omega_{rf}} g(|\mathbf{k}_{rf}|)$$

$$\mathbf{k}_{rf} = \mathbf{k} + \omega\mathbf{V}$$

$$\omega_{rf} = \omega + \mathbf{k}\cdot\mathbf{V}$$

so that

$$\int \frac{d^3k}{\omega} f_\nu(\mathbf{k})\mathbf{v} = \int \frac{d^3k_{rf}}{\omega_{rf}} g(|\mathbf{k}_{rf}|)[\mathbf{v}_{rf}(1 + \mathbf{v}_{rf}\cdot\mathbf{V}) - \mathbf{V}] = -\tfrac{2}{3}\mathbf{V}.$$

The interaction energy of each particle is thus

$$H^{(Z,i)} = -\frac{G}{\sqrt{2}}\rho_\nu \frac{2}{3}(\boldsymbol{\sigma}\cdot\mathbf{V})[2(g_L + g_R)^{(i)}]. \tag{3.8.17}$$

In principle, the presence of the energy splitting (3.8.17) can be detected from the observation of the spin precession. In the case of a macroscopically polarized body – for example, a ferromagnet – the energy splitting gives rise to a torque acting on the body, which, in fact, is the only nonvanishing effect to order G.

The energy in Eq. (3.8.17) is, at any rate, very small indeed. With respect to the value $\mathbf{V}^{(0)}$ in Eq. (3.8.5), we have lost a factor of $|\mathbf{V}|\rho_\nu/\rho^{(0)}$. Using

$$|\mathbf{V}| = 10^{-3}$$

$$\rho_\nu = 10^2\,\mathrm{cm}^{-3}$$

as appropriate for the earth motion through the galaxy and for the cosmic background neutrino density, the splitting between the two spin levels is of the order of[1]

$$\Delta H \cong 6 \times 10^{-39}\,\mathrm{eV}. \tag{3.8.18}$$

---

[1] The smallest value given in [STO 74] is still larger than ours, since it corresponds to a neutrino density of about $10^6\,\mathrm{cm}^{-3}$.

It is clear that the measurement of such tiny energy differences is out of the question, for the time being.

3.8.4 Phonon excitation

For completeness, we consider also the excitation of phonons in a gravitational antennalike system[2] of mass M, volume V, and length L. For definiteness, we take the parameters of the Roma-CERN antenna [AMA 79]:

$$M = \rho_0 V = 2.3 \times 10^6 \, g$$
$$L = 3 \times 10^2 \, cm \tag{3.8.19}$$

where ρ_0 is the average mass density (in $g \, cm^{-3}$), and focus on the mode with frequency

$$\omega_{\alpha 0} = 0.54 \times 10^4 \, sec^{-1}. \tag{3.8.20}$$

To describe the antenna vibrational modes, one introduces the displacement vector $\mathbf{u}(\mathbf{x}, t)$ according to

$$\rho = \rho_0 (1 + \nabla \cdot \mathbf{u}) \tag{3.8.21}$$

and develops $\mathbf{u}$ in normal coordinates:

$$\mathbf{u}(\mathbf{x}, t) = (V)^{1/2} \sum_\alpha Q_\alpha(t) \mathbf{u}_\alpha(\mathbf{x}). \tag{3.8.22}$$

The wave functions are normalized so that

$$\int d^3 x \, \mathbf{u}_\alpha \cdot \mathbf{u}_\beta = \delta_{\alpha\beta}.$$

The unperturbed Hamiltonian of the antenna, in terms of the canonical normal coordinates, is

$$H = \frac{1}{2M} \sum_\alpha P_\alpha(t)^2 + \frac{M}{2} \sum_\alpha \omega_\alpha^2 Q_\alpha(t)^2. \tag{3.8.23}$$

In turn, normal coordinates are expressed in terms of the annihilation and creation operators of a phonon of frequency ω_α according to

$$Q_\alpha = \frac{1}{(2M\omega_\alpha)^{1/2}} (a_\alpha + a_\alpha^+).$$

---

[2] The material of this paragraph is derived from unpublished work by N. Cabibbo and myself.

To obtain the cross section for the process

$$\nu(p) + \text{ground state} \rightarrow \nu(p') + 1 \text{ phonon} \tag{3.8.24}$$

one considers the effective interaction Hamiltonian:

$$H_{\text{int}} = -\left(\frac{G}{\sqrt{2}}\right)\frac{Y}{m}\int d^3x \rho_\nu(\mathbf{x}, t)\rho(\mathbf{x}, t) \tag{3.8.25}$$

where $Y = g(\nu)/A$, m is the nucleon mass. Then

$$\sigma_\alpha = \frac{2\pi}{\nu(p)}\int\frac{d^3p'}{(2\pi)^3}[1 - n(p')]\delta(E' - E - \omega_\alpha)|\langle f|H_{\text{int}}|i\rangle|^2 \tag{3.8.26}$$

$\nu(p)$ is the velocity of the incoming neutrino. We consider a nondegenerate situation, where the occupation numbers are small, $n(p') \ll 1$. The matrix element in (3.8.26) is determined by the adimensional form-factor

$$\int d^3x e^{i\mathbf{kx}}(\nabla \cdot \mathbf{u}_\alpha) = \frac{(V)^{1/2}}{L}g_\alpha(\mathbf{k})$$

which we can assume to be of order unity for $\mathbf{k}$ small.

We restrict to nonrelativistic neutrinos for simplicity and obtain the cross section:

$$\sigma_\alpha = \frac{E'^2}{\pi}|\langle f|H_{\text{int}}|i\rangle|^2$$

$$= \frac{\langle|g_\alpha|^2\rangle}{16\pi}(Gm^2)^2(2Y)^2\frac{E'^2}{m^2}\frac{M}{m}\frac{1}{(L\omega_\alpha)^2}\frac{\omega_\alpha}{m^3} \tag{3.8.27}$$

and the phonon production rate:

$$\frac{dP_\alpha}{dt} = \langle v\rangle\rho_\nu\sigma_\alpha = \omega_\alpha f_\alpha. \tag{3.8.28}$$

The quantity $\langle|g_\alpha|^2\rangle$ denotes the angular average of the form-factor squared. The pure number f_α characterizes the transition rate of (3.8.24). We find, finally:

$$f_{\alpha0} = \frac{\langle|g_{\alpha0}|^2\rangle}{16\pi}(Gm^2)^2(2Y)^2\frac{E'^2}{m^2}\frac{M}{m}\frac{1}{(L\omega_{\alpha0})^2}\frac{\rho_\nu}{m^3}\langle v\rangle. \tag{3.8.29}$$

The most favorable case is that of massive, galaxy-bound neutrinos, with a mass, density, and velocity of the order of $10^6\,\text{eV}$, $10^6\,\text{cm}^{-3}$, and 10^{-3}, respectively.

We obtain:

$$f_{\alpha 0} = 1.0 \times 10^{-27} \langle |g_{\alpha 0}|^2 \rangle (2Y)^2 \left(\frac{m_\nu}{10\,\text{eV}}\right)^2 \frac{\rho_\nu}{10^6\,\text{cm}^{-3}}. \tag{3.8.30}$$

Needless to say, this is a very small effect, corresponding to the average creation of one phonon of this particular mode every 10^{24} sec.

3.8.5 Neutrino oscillations in matter

Neutrino oscillations, if they exist, are affected by the presence of matter, as pointed out by Wolfenstein [WOL 78, 79]. Here we consider the amplification of neutrino oscillations, originally discussed by Mikheyev and Smirnov [MIK 86].

Following Eq. (3.8.1), we write the energy of an ultrarelativistic neutrino propagating inside matter according to

$$H = p + \frac{m^2}{2p} + V(\nu) \tag{3.8.31}$$

$p = nk$ is the neutrino momentum inside matter and m^2 the neutrino mass-squared matrix. We consider the case of two-neutrino mixing – for example, $\nu_e-\nu_\mu$ – where m^2 and V are two-by-two matrices. In the $\nu_e-\nu_\mu$ basis we write:

$$m^2 = \begin{bmatrix} m_{e-e}^2 & m_{e-\mu}^2 \\ m_{e-\mu}^2 & m_{\mu-\mu}^2 \end{bmatrix}$$

$$V = \begin{bmatrix} V(\nu_e) & 0 \\ 0 & V(\nu_\mu) \end{bmatrix}$$

$V(\nu_e)$ and $V(\nu_\mu)$ are the interaction energies for electron and muon neutrinos, respectively, considered previously. They differ because of W-exchange. See Section 3.8.1, which gives

$$V(\nu_e) - V(\nu_\mu) = \Delta V = \sqrt{2}GZN_{\text{Atoms}}$$
$$= +\sqrt{2}GY\frac{\rho}{M} \tag{3.8.32}$$

$Y = Z/A$ is the number of active particles (electrons, in this case) per nucleon, ρ the mass density, and M the nucleon mass.

Discarding trivial contributions proportional to the unit matrix, we have to diagonalize the matrix

$$H' = -\epsilon\sigma_3 + \beta\sigma_1 \tag{3.8.33}$$

with

$$2p\epsilon = \tfrac{1}{2}(m_{\mu-\mu}^2 - m_{e-e}^2) - p\Delta V$$

$$2p\beta = m_{e-\mu}^2.$$

We define the mixing angle according to

$$\sin^2 2\theta = \frac{\beta^2}{\beta^2 + \epsilon^2}. \qquad (3.8.34)$$

The probability to observe a ν_μ at time t, if we started with ν_e at time $t = 0$, is then

$$P(\nu_e \rightarrow \nu_\mu) = \sin^2\theta\tfrac{1}{2}(1 - \cos 2\pi t/L)$$
$$L = \frac{2\pi}{\Delta h}. \qquad (3.8.35)$$

where Δh is the difference of the eigenvalues of the matrix (3.8.33):

$$\Delta h = 2(\beta^2 + \epsilon^2)^{1/2} = 2\epsilon(\cos 2\theta)^{-1}. \qquad (3.8.36)$$

We append a subscript vac or mat for the quantities in vacuum and in matter, respectively. In vacuum, Δh is related to the difference of the mass-squared of the physically massive neutrinos:

$$\Delta h_{\text{vac}} = \frac{\Delta m^2}{2p} \qquad (3.8.37)$$

and L_{vac} is the familiar oscillation length:

$$L_{\text{vac}} = \frac{4\pi p}{\Delta m^2}. \qquad (3.8.38)$$

Defining further,

$$L_0 = \frac{2\pi}{\Delta V}$$

we find, from Eq. (3.8.36):

$$\sin^2 2\theta_{\text{mat}} = \frac{\sin^2 2\theta_{\text{vac}}}{1 - 2\cos 2\theta_{\text{vac}} \dfrac{L_{\text{vac}}}{L_0} + \left(\dfrac{L_{\text{vac}}}{L_0}\right)^2} \tag{3.8.39}$$

$$L_{\text{mat}} = \frac{L_{\text{vac}}}{\left[1 - 2\cos 2\theta_{\text{vac}} \dfrac{L_{\text{vac}}}{L_0} + \left(\dfrac{L_{\text{vac}}}{L_0}\right)^2\right]^{\frac{1}{2}}}. \tag{3.8.40}$$

The interesting fact is that the mixing angle in matter exhibits a resonant behavior: $\sin^2 2\theta_{\text{mat}}$ reaches unity, no matter how small θ_{vac} is, if

$$\frac{L_{\text{vac}}}{L_0} = \cos 2\theta_{\text{vac}}. \tag{3.8.41}$$

The reason for the resonance is clear from Eq. (3.8.33). If parameters are such that ϵ vanishes, the mixing angle will be maximal, no matter how small the off-diagonal matrix elements are. In fact, the resonance condition (3.8.41) is equivalent to

$$\Delta V = \frac{1}{2p}(m_{\mu-\mu}^2 - m_{e-\epsilon}^2). \tag{3.8.42}$$

It is quite easy to derive the conditions under which the oscillations of neutrinos emitted in the sun will be amplified by the occurrence of the resonance.

1 Resonance condition   We see from Eq. (3.8.41) that the critical density at which the resonance occurs is

$$\rho_{\text{crit}} = \frac{M}{\sqrt{2}GY} \frac{\Delta M^2 \cos 2\theta_{\text{vac}}}{2p}. \tag{3.8.43}$$

Given that the density inside the sun decreases from the center, if the density at the place where neutrinos are emitted is larger than ρ_{crit}, neutrinos escaping from the sun will find a critical layer where the resonance condition is met. Thus, the existence of the critical layer requires ρ_{crit} to be smaller than the maximum density at the center of the sun, namely,

$$\Delta m^2 \cos 2\theta_{\text{vac}} < p\frac{2\sqrt{2}GY\rho_{\text{max}}}{M}$$

$$= 2V^{(0)}p\frac{2Z}{A}\frac{\rho_{\text{max}}}{1\,\text{g cm}^{-3}}$$

$$= 2.2 \times 10^{-4}(\text{eV})^2 \frac{2Z}{A}\frac{\rho_{\text{max}}}{300\,\text{g cm}^{-3}}\frac{p}{10\,\text{MeV}} \tag{3.8.44}$$

The numerical value is approximately correct for the sun and for the high-energy neutrinos that give rise to the chlorine–argon transition in the Davies experiment.

2 Adiabatic condition   The mixing angle in Eq. (3.8.39) remains large in a shell where the density is close to ρ_{crit}. Considering, for instance, the half-width at half-maximum of the expression (3.8.39), we see that

$$\tfrac{1}{2} < \sin^2 2\theta_{mat} < 1$$

for

$$\rho_{crit} - \Delta\rho < \rho < \rho_{crit} + \Delta\rho$$
$$\Delta\rho = \rho_{crit} \tan 2\theta_{vac}. \tag{3.8.45}$$

It is not difficult to see that the resonance is effective only if the radial extension of the shell, $\Delta\rho(d\rho/dr)^{-1}$, is much larger than the neutrino oscillation length. If this is the case, a ν_e created in the large density region, $\rho > \rho_{crit}$, will switch adiabatically to ν_μ as it goes through the critical layer. Since at resonance

$$(L_{mat})_{res} = \frac{L_{vac}}{|\sin 2\theta_{vac}|}$$

the adiabatic condition is

$$\Delta m^2 \sin 2\theta_{vac} \tan 2\theta_{vac} > 4\pi p \frac{(d\rho/dr)_{crit}}{\rho_{crit}}$$
$$= 4.1 \ 10^{-8} \frac{p}{10 \,\text{MeV}} (R_{Sun})(d \log \rho/dr)_{crit} \tag{3.8.46}$$

We see from Eqs. (3.8.44) and (3.8.46) that values of $\Delta m^2 \cong 10^{-4} \,\text{eV}^2$ and $\sin 2\theta_{vac} \cong 10^{-4}$ lead to large oscillations for the neutrinos detected in the Davies experiment. The reason why this explanation of the (possible) neutrino deficit is attractive is that these values are much more natural, in grand unified theories, than the maximal mixing required to get the same effect from vacuum oscillations. In particular, since we know that $\Delta V > 0$, we read from Eq. (3.8.42) that a necessary condition to have resonant oscillations for neutrinos, not antineutrinos, coming from the sun is that

$$m^2_{\mu-\mu} > m^2_{e-e}$$

a condition that is fulfilled in most grand unified schemes.

References

[ABE 73]   E. S. Abers and B. W. Lee, *Phys. Reports* C9 (1973) 1.
[ABE 90]   F. Abe *et al.*, *Phys. Rev. Lett.*, 64 (1990) 143.

[ADL 69]   S. Adler, *Phys. Rev.* 177 (1969) 2426.
[ALT 77]   G. Altarelli, N. Cabibbo, L. Maiani, and R. Petronzio, *Phys. Lett.* 67B (1977) 463.
[AMA 79]   E. Amaldi and G. Pizzella, "Search for Gravitational Waves," in *Relativity, Quanta and Cosmology*, ed. F. De Finis, Johnson Reprint, New York, vol. 1 (1979) 9.
[BAR 69]   W. Bardeen, *Phys. Rev.* 184 (1969) 1848.
[BEL 69]   J. S. Bell and R. Jackiw, *Nuovo Cimento* 60 (1969) 47.
[BER 72]   V. Berestetski, E. Lifshitz, and L. Pitayevzki, *Theorie Quantique Relativiste*, MIR, Moscow (1972).
[BIL 77]   S. M. Bilenky and B. M. Pontecorvo, *Usp. Fiz. Nauk* 23 (1977) 181.
[BJO 69a]   J. D. Bjorken, *Phys. Rev.* 179 (1969) 1547.
[BJO 69b]   J. D. Bjorken and E. A. Paschos, *Phys. Rev.* 185 (1969) 1975.
[BOC 85]   M. Bochicchio, L. Maiani, G. Martinelli, G. C. Rossi, and M. Testa, *Nucl. Phys.* B262 (1985) 331.
[BOU 72]   C. Bouchiat, J. Iliopoulos, and Ph. Meyer, *Phys. Lett.* 38B (1972) 519.
[BRA 78]   G. C. Branco and R. N. Mohapatra, *Phys. Rev.* D18 (1978) 4246.
[BUR 88]   H. Burkhardt *et al.*, *Phys. Lett.* 206B (1988) 163.
[CAB 63]   N. Cabibbo, *Phys. Rev. Lett.* 10 (1963) 531.
[CAB 78]   N. Cabibbo, *Phys. Lett.* 72B (1978) 333.
[CAB 82]   N. Cabibbo and L. Maiani, *Phys. Lett.* 114B (1982) 115.
[CAL 69]   C. G. Callan and D. J. Gross, *Phys. Rev. Lett.* 22 (1969) 156.
[CHA 77]   M. Chanowitz, J. Ellis, and M. K. Gaillard, *Nucl. Phys.* B218 (1977) 506.
[COL 73]   S. Coleman and D. J. Gross, *Phys. Rev. Lett.* 31 (1973) 851.
[COL 85]   S. Coleman, *Aspects of Symmetry*, Cambridge University Press (1985).
[COR 74]   J. M. Cornwall, D. N. Levin, G. Tiktopoulos, *Phys. Rev.* D10 (1974) 1145.
[ENG 64]   F. Englert and R. Brout, *Phys. Rev. Lett.* 13 (1964) 321.
[FEY 58]   R. P. Feynman and M. Gell-Mann, *Phys. Rev.* 109 (1958) 193.
[FEY 72]   R. P. Feynman, *Photon–Hadron Interactions*, Benjamin (1972).
[FRI 75]   H. Fritzsch and P. Minkowski, *Ann. of Phys.* 93 (1975) 193.
[GAS 82]   J. Gasser and H. Leutwyler, *Phys. Rep.* 87 (1982) 77.
[GEL 64]   M. Gell-Mann, *Phys. Lett.* 8 (1964) 214.
[GEL 69]   M. Gell-Mann, M. L. Goldberger, N. M. Kroll, and F. E. Low, *Phys. Rev.* 179 (1969) 1518.
[GEL 81]   G. Gelmini and M. Roncadelli, *Phys. Lett.* 99B (1981) 411.
[GEO 72]   H. Georgi and S. L. Glashow, *Phys. Rev. Lett.* 28 (1972) 1494.
[GEO 79]   H. Georgi and D. V. Nanoupolos, *Nucl. Phys.* B155 (1979) 152.
[GLA 61]   S. L. Glashow, *Nucl. Phys.* 22 (1961) 579.
[GLA 70]   S. L. Glashow, J. Iliopoulos, and L. Maiani, *Phys. Rev.* D2 (1970) 1285.
[GLA 77]   S. L. Glashow and S. Weinberg, *Phys. Rev.* D15 (1977) 1958.
[GOL 62]   J. Goldstone, A. Salam, and S. Weinberg, *Phys. Rev.* 127 (1962) 965.
[GRI 69]   V. Gribov and B. M. Pontecorvo, *Phys. Lett.* 28B (1969) 493.
[GRO 69]   D. J. Gross and C. H. Llewellyn Smith, *Nucl. Phys.* B14 (1969) 337.
[GRO 72]   D. Gross, *Phys. Rev.* D6 (1972) 477.
[GRO 73]   D. J. Gross and F. Wilczeck, *Phys. Rev. Lett.* 30 (1973) 1343.
[HAN 65]   H. Y. Han and Y. Nambu, *Phys. Rev.* 139 (1965) B1006.
[HIG 64]   P. Higgs, *Phys. Lett.* 12 (1964) 132.
[HOO 71]   G. W. 't Hooft, *Nucl. Phys.* B33 (1971) 167.
[HOR 77]   D. Horn and G. G. Ross, *Phys. Lett* 67B (1977) 460.
[JAC 64]   M. Jacob and G. C. Chew, *Strong Interaction Physics*, W. A. Benjamin, New York (1964).
[KAR 81]   L. H. Karsten and J. Smit, *Nucl. Phys.* B183 (1981) 103.
[KIB 67]   T. W. Kibble, *Phys. Rev.* 155 (1967) 1554.
[KLE 48]   O. Klein, *Nature* 161 (1948) 897.
[KOB 73]   M. Kobayashi and K. Maskawa, *Progr. Theor. Phys.* 49 (1973) 652.
[LEE 49]   T. D. Lee, R. Rosenbluth, and C. N. Yang, *Phys. Rev.* 75 (1949) 9905.
[LEE 72]   B. W. Lee and J. Zinn-Justin, *Phys. Rev.* D5 (1972) 3121, 3137.

[LEE 81]   T. D. Lee, *Particle Physics and Introduction to Field Theory*, Harwood, London (1981).

[LEW 80]   R. R. Lewis, *Phys. Rev.* D21 (1980) 663.

[LLE 73]   C. H. Llewellyn Smith, *Phys. Lett.* 46B (1973) 283.

[MAI 77]   L. Maiani, *Proc. Int. Symp. on Lepton and Photon Interaction at High Energy*, Hamburg, 1977.

[MAR 58]   R. E. Marshak and E. C. G. Sudarshan, *Phys. Rev.* 109 (1958) 1860.

[MIK 86]   S. P. Mikheyev and A. Yu. Smirnov, *Nuovo Cimento* 9 (1986) 17.

[NEW 66]   R. G. Newton, *Scattering Theory of Waves and Particles*, McGraw-Hill (1966).

[NIE 81]   H. B. Nielsen and M. Ninomiya, *Nucl. Phys.* B185 (1981) 20; B195 (1982) 541; B193 (1981) 173.

[NIL 84]   H. Nilles, *Phys. Rep.* 110 (1984) 1.

[OKU 82]   L. B. Okun, *Leptons and Quarks*, North-Holland, Amsterdam (1982).

[OPH 74]   R. Opher, *Astron. Astrophys.* 37 (1974) 135.

[PAR 86]   Particle Data Group, M. Aguile-Betuitez *et al.*, *Phys. Lett.* 170B (1986) 1.

[PAT 90]   J. R. Patterson *et al.*, *Phys. Rev. Lett.* 64 (1990) 1491.

[PAT 73]   J. C. Pati and A. Salam, *Phys. Rev.*, D8 (1973) 1240.

[POL 73]   D. Politzer, *Phys. Rev. Lett.* 30 (1973) 1346.

[PON 47]   B. M. Pontecorvo, *Phys. Rev.* 72 (1947) 246.

[PON 58]   B. M. Pontecorvo, *JETP* 6 (1958) 429.

[PUP 48]   G. Puppi, *Nuovo Cimento* 5 (1948) 505.

[SAK 58]   J. J. Sakurai, *Nuovo Cimento* 7 (1958) 649.

[SAL 68]   A. Salam, *Proc. 8th Nobel Symp.*, ed. N. Svartholm, Almqvist and Wicksell, Stockholm (1968).

[SMI 83]   P. F. Smith and J. D. Lewin, *Phys. Lett.* 127B (1983) 185.

[STO 74]   L. Stodolski, *Phys. Rev. Lett.* 34 (1974) 110.

[TIO 49]   J. Tiomno and J. A. Wheeler, *Revs. Mod. Phys.* 21 (1949) 153.

[WEI 67]   S. Weinberg, *Phys. Rev. Lett.* 19 (1967) 1264.

[WEI 71]   S. Weinberg, *Phys. Rev. Lett.* 27 (1971) 1688.

[WOL 78]   L. Wolfenstein, *Phys. Rev.* D17 (1978) 2369.

[WOL 79]   L. Wolfenstein, *Phys. Rev.* D20 (1979) 2634.

[WOL 83]   L. Wolfenstein, *Phys. Rev. Lett.* 51 (1983) 1945.

[YAN 54]   C. N. Yang and R. Mills, *Phys. Rev.* 96 (1954) 191.

4

Experimental studies of the weak interaction

4.1 Structure of the charged weak current*

4.1.1 Space-time structure of the weak Lagrangian

The scattering of neutrinos from nucleons or electrons is a unique tool for probing the internal quark structure of these nucleons and for testing the properties of the weak interaction in a large range of momentum transfers. Since neutrinos are only subject to the weak interaction, these reactions offer the possibility of studying the weak interaction alone. This interaction proceeds through the exchange of intermediate vector bosons, the $W^{\pm}$ bosons for charged currents, and the Z^0 boson for neutral currents. Neutral currents were discovered in muon–neutrino interactions in 1973, and the vector bosons were discovered 10 years later at the antiproton–proton collider at CERN. This chapter is devoted to the charged-current interactions; the neutral ones are treated in Chapter 5.

The first weak reaction observed was β decay, that is, the emission of an electron and an antineutrino from a nucleus. Fifty years of experimentation were needed to establish that this process could be described by a Lagrangian of the current–current type with the W propagator. In the low-energy approximation this can be written as

$$L = \frac{G}{\sqrt{2}} \sum_i (\bar{e}\Gamma_i \nu_e)(\bar{n}\Gamma_i(C_i + \gamma_5 C_i')p) + \text{h.c.} \tag{4.1.1}$$

In the most general case, the operators Γ_i between the electron (e) and neutrino (ν) spinors and between the neutron (n) and proton (p) spinors can be of five different types: 1 for scalar (S), γ^μ for vector (V), $\sigma^{\mu\nu}$ for tensor (T), $\gamma^5\gamma^\mu$ for axial vector (A), and γ^5 for pseudoscalar (P) interaction.

Experiments in nuclear beta decay were found to be consistent with the $V–A$ structure for the operators in the lepton current, and with a V and A structure for the nucleon current [FEY 58; SUD 58]. If this Lagrangian is formulated in the quark language, it is simplified to

$$L = \frac{G}{\sqrt{2}} J^{+\alpha} \cdot j_\alpha^- + \text{h.c.} \tag{4.1.2}$$

* K. Kleinknecht, Institut für Physik, Johannes Gutenberg-Universität, Mainz, Germany.

where j_α^- is the leptonic charged current:

$$j_\alpha^- = \bar{e}\gamma_\alpha(1+\gamma_5)\nu_e$$

and $J^{+\alpha}$ is the hadronic charged current between two quarks, u and d:

$$J^{+\alpha} = u\gamma^\alpha(1+\gamma_5)d \cdot V_{ud}$$

V_{ud} is the coupling strength of the transition of d to u compared with that of a purely leptonic process, as for example,

$$\mu^+ \to e^+ + \nu_e + \bar{\nu}_\mu.$$

The problems of this phenomenological Fermi Lagrangian for a pointlike four-fermion interaction at high-momentum transfers are cured by the intermediate vector boson theory, where two currents of two fermions are interacting through the exchange of a charged heavy $W^\pm$ boson. The Lagrangian then reads:

$$L = -\frac{g}{\sqrt{2}}W^\alpha(x)J_\alpha(x) + \text{herm. conj.} \tag{4.1.3}$$

where g is a universal $SU(2)_L$ gauge coupling of the W field to the fermion charged current. Since the intermediate boson propagator

$$D^{\mu\nu}(q) = \frac{q^\mu q^\nu/M_{W^2} - g^{\mu\nu}}{q^2 - M_{W^2}}$$

becomes $g^{\mu\nu}/M_{W^2}$, in the limit $|q^2| \ll M_{W^2}$, the phenomenological Fermi Lagrangian and the intermediate boson theory coincide at low q^2 with the coupling constants being related

$$\frac{G}{\sqrt{2}} = \frac{g^2}{8M_{W^2}}. \tag{4.1.4}$$

If the fermion charged current is generalized to include all leptons and quarks, it reads:

$$J_\alpha(x) = \sum_{i=e,\mu,\tau} \bar{\nu}_{iL}\gamma_\alpha l_L + \sum \bar{q}_L\gamma_\alpha V_{qq'}q'_L \tag{4.1.5}$$

where the quarks with $\frac{2}{3}$ charge are called q, and the ones with $-\frac{1}{3}$ charge q'. Here the V–A structure is built in by taking only the left-handed spinor components of fermions, $\psi_L = (1 - \gamma_5)\,\psi/2$ [WEY 29; SAL 57; LAN 57; LEE 57], and the flavor-dependent weak coupling of quarks is parameterized in a unitary 3×3 matrix, as proposed by Kobayashi and Maskawa [KOB 73]. This flavor structure of the weak interaction is treated in Section 4.1.2.

4.1.1.1 Inverse β-decay

Historically, this was the first reaction observed with neutrinos. Antineutrinos in the energy range up to 8 MeV from β-decays in a power reactor are detected in an experimental apparatus containing protons, using the reaction

$$\bar{\nu}_e p \to e^+ n. \tag{4.1.6}$$

In the original Savannah River experiment [REI 59], 1400 liters of organic liquid scintillator with an admixture of cadmium octoate was exposed to a $\bar{\nu}_e$ flux of $1.3 \times 10^{13}\,\mathrm{cm}^{-2}\,\mathrm{sec}^{-1}$. The positron was detected by a first pulse at the time of the reaction, while the neutron was moderated by the protons in the scintillator and then captured in cadmium. A delayed signal from γ rays created in the capture process following the e^+ pulse in a time interval from 0.75 to 25.75 µsec was used as signature for the neutrino-induced process. Detection efficiencies were estimated at 85 ± 5 percent for the positron and 10 ± 2 percent for the neutron. In order to separate the extremely small neutrino signal from the background, at a five times larger rate, induced by cosmic radiation, measurements were taken with the reactor on and the reactor off. The difference of rates is then 34 ± 4 events/h, leading to a cross section of $(11 \pm 2.6)10^{-44}\,\mathrm{cm}^2$. This result is consistent with the value expected for a two-component neutrino, $(10.7 \pm 0.7)\,10^{-44}\,\mathrm{cm}^2$ [NEZ 66]. The inverse β-decay reaction has been used recently in searches for neutrino oscillations. Again, nuclear power reactors are used as a source of electron antineutrinos of energies up to 8 MeV. Energy spectra of positrons in delayed coincidence with neutrons are measured with the detector at different distances from the reactor core.

As an example, we discuss the experiment of the Caltech-SIN-TUM collaboration [GAB 84]. Here five planes of liquid scintillator of 377 l total volume serve as proton target, positron detector, and neutron moderator, while the neutron is detected in four ^{3}He wire chambers interleaved with the scintillator. Position information from both the positron signal and the neutron detectors, and the requirement of spatial correlation between both reduces the accidental background considerably. Positron energy spectra for reactor on/off are shown in Fig. 4.1.1, as well as the difference. The smooth curve shown is the calculated spectrum, based on measured β spectra and composition of fission fragments, assuming no oscillation. The ratio of integrated experimental yield to that predicted on the basis of a massless two-component neutrino is 1.05 ± 0.02 (stat) ± 0.06 (syst) for a distance of $L = 37.9\,\mathrm{m}$ and $1.06 \pm 0.02 \pm 006$ for $L = 45.9\,\mathrm{m}$. This shows that the measured cross section for the inverse β-decay is consistent with the two-component neutrino theory.

In addition, the ratio of integrated rates at the two distances, $R = 1.01 \pm 0.03$ (stat) ± 0.02 (syst), is consistent with the absence of $\bar{\nu}_e$ oscillations, and can be used to set limits on oscillation parameters (see Section 2.4).

4.1.1.2 Inverse muon decay

The second neutrino type, the muon neutrino, has been discovered in reactions with nucleons [DAN 62]. See Chapter 1 for a reprint. At a much smaller cross section, the reaction

$$\nu_\mu e^- \to \mu^- \nu_e \tag{4.1.7}$$

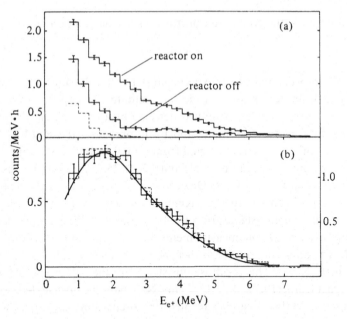

Fig. 4.1.1   (a) Positron energy spectra from reactor-induced $\bar{\nu}_e$ interactions and from background events for a detector at 45.9-m distance from reactor core (dashed line: accidental coincidences). (b) Experimental positron spectra (reactor on minus reactor off) for distances 45.9 m (solid, l.h. scale) and 37.9 m (dashed, r.h. scale). Smooth curve is predicted spectrum for $V-A$ interaction without neutrino oscillations [GAB 84].

is a purely leptonic process mediated by charged currents. It can be used [JAR 70] to complement the precise measurements of muon decay, $\mu^- \rightarrow e^- \nu_\mu \bar{\nu}_e$ [BUR 85; FET 86], in order to test the $V-A$ structure of leptonic weak interactions and the two-component theory of neutrinos.

In the helicity projection form of the charged weak current, there are couplings of left-handed (L) and right-handed (R) fermions of the types LL, RR, LR and RL. Experiments on muon decay show that at least one of the couplings, the scalar coupling g_{LL}^S or the vector coupling g_{LL}^V does not vanish. They also show that the two couplings of the RR type and six couplings of the LR and RL type are consistent with zero within tight limits [FET 86].

For the LL couplings, there is a theoretical bound

$$|g_{LL}^V|^2 + |g_{LL}^S|^2/4 \leq 1. \tag{4.1.8}$$

While muon decay experiments cannot distinguish between g_{LL}^V and g_{LL}^S couplings, this distinction is measurable in inverse muon decay experiments because the incoming neutrino is left-handed [BAC 61]. Therefore the only coupling contributing is g_{LL}^V, and the asymptotic cross section is $\sigma = \sigma_{as} E_\nu |g_{LL}^V|^2$. The $V-A$ interaction predicts $\sigma_{as} = 2m_e G_F^2/\pi$, $g_{LL}^V = 1$, and $g_{LL}^S = 0$.

The cross section for this inverse muon decay has been measured at CERN [ARM 79c, BER 83, VIL 95]. We discuss the most recent experiment of the CHARM II

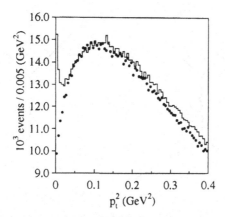

Fig. 4.1.2 $p_\perp^2$ distribution for events with $E_\mu > 10.9\,\text{GeV}$ and hadronic energy $< 1.5\,\text{GeV}$ for μ^- from the neutrino exposures (solid line) and for μ^+ from antineutrino exposures (dots). The μ^+ distribution is normalized to the μ^- distribution in the range $0.05\,\text{GeV}^2 < p_\perp^2 < 0.1\,\text{GeV}^2$.

collaboration [VIL 95]. The detector consists of a glass-scintillator/streamertube sampling calorimeter measuring deposited energy, followed by a toroidal magnetic iron core spectrometer measuring muon momenta [GEI 93]. The detector was exposed to a horn-focused wide-band neutrino beam produced by 400 GeV protons from the CERN SPS, with a mean energy of 24 GeV for neutrinos, 19 GeV for antineutrinos.

The reaction (4.1.7) is characterized by a threshold at $E_\nu^* = 10.8\,\text{GeV}$, the absence of detectable recoil energy at the reaction point, and the forward emission of a muon at small angles, $(\theta_\mu < 10\,\text{mrad})$, that is, at small $q^2 = 4E_\nu E_\mu \sin^2\theta_\mu/2$. The selection criteria in the experiment require small hadronic recoil energy deposition $(< 1.5\,\text{GeV})$ around the reaction point and a $\mu^-(\mu^+)$ of more than 10.9 GeV energy for neutrino (antineutrino) exposure. The antineutrino data are used to determine the background to reaction (4.1.7) consisting of quasi-elastic scattering on a nucleon,

$$\nu_\mu n \rightarrow \mu^- p \qquad (4.1.9)$$

by studying the corresponding one induced by antineutrinos,

$$\bar{\nu}_\mu n \rightarrow \mu^+ n. \qquad (4.1.10)$$

The $p_\perp^2$ distribution of 10^6 neutrino-induced and antineutrino-induced events is shown in Fig. 4.1.2.

While for μ^- events a peak near $p_\perp = 0$ indicates the presence of inverse muon decay, the μ^+ data show a smooth distribution. The $p_\perp^2$-dependence of backgrounds to reaction (4.1.7), quasi-elastic scattering and one-pion production, $\nu_\mu N \rightarrow \mu^- \pi N$, is expected to be the same for ν and $\bar{\nu}$ on an isoscalar target. Subtracting the background in the region $p_\perp^2 < 0.1\,\text{GeV}^2$ according to a simulated quasi-elastic distribution based on the μ^+ data, the signal for reaction (4.1.7) amounts to $N = (18.13 \pm 0.37\,(\text{stat}) \pm 0.57\,(\text{syst})) \times 10^3$ events. From this and the measured

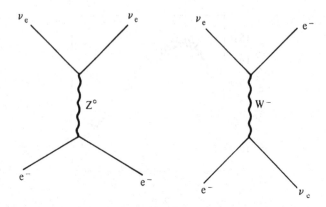

Fig. 4.1.3   Feynman diagrams for ν_e scattering from electrons.

neutrino flux one obtains the asymptotic cross section slope $\sigma_{as} = (16.51 \pm 0.93)10^{-42}\,\mathrm{cm}^2\,\mathrm{GeV}^{-1}$.

Another experiment of similar sensitivity was done by the CCFR collaboration [MIS 90]. Here, an iron/scintillator calorimeter with toroidal magnets was exposed to the Tevatron quadrupole triplet neutrino beam. The inverse muon decay signal again shows up as a peak of 2015.6 ± 98.5 events at $p_\perp^2 < 0.2\,\mathrm{GeV}^2$ in neutrino-induced events. The cross-section slope derived from these is $\sigma_{as} = (16.93 \pm 8.85$ (stat) ± 0.52 (syst)$) \times 10^{-42}\,\mathrm{cm}^2\,\mathrm{GeV}^{-1}$.

If we take these two most precise experiments together, we obtain the cross section slope $(16.71 \pm 0.68) \times 10^{-42}\,\mathrm{cm}^2\,\mathrm{GeV}^{-1}$. The Standard Model prediction for σ_{as} is $17.23 \times 10^{-42}\,\mathrm{cm}^2\,\mathrm{GeV}^{-1}$, and the ratio of experimental slope to the predicted one is $S = 0.970 \pm 0.040$. This result shows perfect agreement with the Standard Model. From the constraint on the sum of the vector and scalar LL couplings above, one obtains a limit on the scalar coupling:

$$|g_{LL}^S|^2 \le 4(1 - S) < 0.39 \text{ at } 90\% \text{ CL.} \tag{4.1.11}$$

4.1.1.3 The reaction $\nu_e e \rightarrow \nu_e e$

While the scattering of muon neutrinos on electrons can only proceed through neutral currents (see Section 4.2.4.1), electron neutrinos can scatter from electrons both by neutral and charged-current reactions (Fig. 4.1.3). The total rate in the Weinberg–Salam model is determined by destructive interference of these contributions. Indeed, the cross section for $\nu_e e^-$ scattering in the Standard Model is given by [HOO 71]

$$\sigma = \frac{G^2 s}{4\pi}[(v_e + a_e + 2)^2 + (v_e - a_e)^2/3] \tag{4.1.12}$$

where $s = 2ME_\nu$ is the center-of-mass energy squared in the reaction, and the neutral current couplings of electrons are $a_e = -\frac{1}{2}$ and $v_e = (-\frac{1}{2} + 2\sin^2\theta_W)$. The integrated

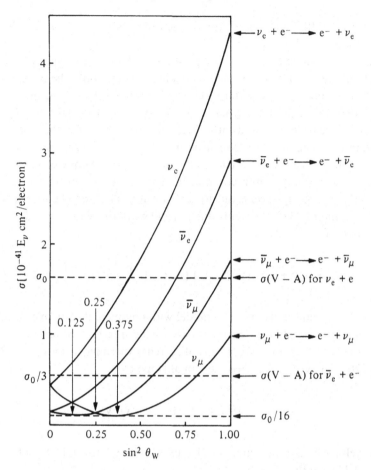

Fig. 4.1.4 Cross sections for neutrino–electron scattering processes, as calculated in the standard electroweak theory, as a function of the Weinberg mixing parameter, $\sin^2\theta_W$ ['tHO 71].

cross section is then

$$\sigma = \frac{G^2 s}{\pi}\left[\frac{4}{3}\sin^4\theta_W + \sin^2\theta_W + \frac{1}{4}\right].$$

The dependence of σ from $\sin^2\theta_W$ is shown in Fig. 4.1.4, together with other neutrino–electron cross sections. If $\sin^2\theta_W = 0.23$ is inferred from other experiments, the presence of the destructive interference predicted in the standard electroweak model can be tested.

The experiment has been done at Los Alamos by a UCI-LANL-Maryland collaboration [CHE 86; ALL 90]. The reaction $\nu_e e \to \nu_e e$ is identified by detecting the recoil electron with energy less than the maximum neutrino energy (52.8 MeV) in the forward $10°$ cone as required by kinematics. Neutrinos are produced at the beam stop of the LAMPF 780 MeV proton accelerator from $\pi^+ \to \mu^+ \nu_\mu$ decay at rest, followed by $\mu^+ \to e^+ \nu_e \bar{\nu}_\mu$ decay at rest.

The detector is a 15 t sandwich of plastic scintillators (for energy and timing information) and flash tube planes (for tracking information). The beam has a flux of 4×10^7 neutrinos/(cm^2 sec) at a distance of 9 m from the target. In a four-year exposure 1983–6, the total electron-neutrino flux was $(9.16 \pm 0.76) \, 10^{14} \, \nu_e \, \text{cm}^{-2}$ at the detector. The beam is pulsed with a cycle time of 8.3 ms, such that beam-on and beam-off rates can be measured concurrently. The cosmic background is subtracted using the data with beam off. The total sample of neutrino–electron scattering events after background subtraction is 295 ± 35 events. From this, $27.4 \pm 4.7 \nu_\mu e$ events and $33.5 \pm 5.9 \, \bar{\nu}_\mu e$ events due to neutral currents have to be subtracted. The remaining signal of 234 ± 35 events due to $(\nu_e e)$ scatterings corresponds to a cross section slope of

$$\sigma/E_\nu = (9.9 \pm 1.5 \text{ (stat)} \pm 1.0 \text{ (syst)}) 10^{-42} \, \text{cm}^2 \text{GeV}^{-1}. \tag{4.1.13}$$

Assuming the standard electroweak model with complete destructive interference between neutral and charged currents (size of the interference term $I = -1$) and inferring the Weinberg angle from neutral current experiments ($\sin^2\theta_W = 0.23$), one expects 226 ± 23 events or a cross-sectional slope of

$$\sigma/E_\nu = (9.7 \pm 1.0) 10^{-42} \, \text{cm}^2 \text{GeV}^{-1}. \tag{4.1.14}$$

Alternatively, leaving the value of the interference term I free and inferring $\sin^2\theta_W = 0.23$, one finds

$$I = -1.07 \pm 0.17 \text{ (stat)} \pm 0.12 \text{ (syst)} \tag{4.1.15}$$

to be compared with the value $I = -1.08$ in the Standard Model. This experiment therefore demonstrates a nonvanishing interference term at a 5 standard deviation level.

4.1.1.4 *V–A structure of the quark current in inelastic neutrino reactions*

Inelastic neutrino scattering can be used to search for the existence of right-handed weak quark currents coupling to the left-handed weak lepton current. In the quark parton model the reactions

$$\nu_\mu N \to \mu^- X$$
$$\bar{\nu}_\mu N \to \mu^+ X \tag{4.1.16}$$

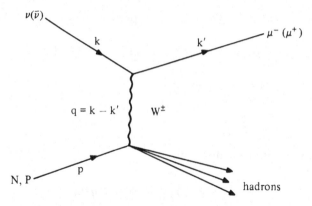

Fig. 4.1.5   Feynman diagram for deep inelastic neutrino–nucleon scattering.

in the deep inelastic region can be described as the scattering of neutrinos from the quark (and antiquark) constituents of the nucleon. The usual kinematic variables are the following (see Fig. 4.1.5):

$$q = k - k' \quad Q^2 = -q^2 = 4E_\nu E_\mu \sin^2 \theta/2 \quad x = Q^2/\nu$$
$$\nu = p \cdot q/M = E_\nu - E_\mu = E_h \quad y = \nu/E_\nu.$$

In the scattering of a left-handed neutrino from a left-handed quark via a left-handed $V-A$ interaction, the total angular momentum vanishes in the c.m. system, such that the differential cross section is flat in the angular variable $\cos \theta^*$, with θ^* being the c.m. scattering angle between neutrino and muon directions. Since $\cos \theta^*$ is related to the Bjorken scaling variable $y = \nu/E_\nu$ by

$$y = (1 - \cos \theta^*)/2 \tag{4.1.17}$$

this leads to a flat distribution in y between 0 and 1. In particular, scattering by $180°$ ($y = 1$) is possible. On the other hand, neutrino scattering from a right-handed quark involves an angular momentum of 1, and the angular distribution is of the form $(1 - y)^2$. If the Lagrangian is purely left-handed, namely,

$$L = \frac{G}{\sqrt{2}} (\bar\mu\gamma_\mu(1 + \gamma_5)\nu)(\bar u\gamma^\mu(1 + \gamma_5)d)$$

then the cross section in terms of quark and antiquark structure functions, $q(x)$ and $\bar q(x)$, becomes

$$\frac{d^2\sigma^\nu}{dx\,dy} = \sigma^0(q(x) + (1 - y)^2\bar q(x))$$

$$\frac{d^2\sigma^\nu}{dx\,dy} = \sigma^0(\bar q(x) + (1 - y)^2 q(x)).$$

If, however, right-handed currents are also present, that is,

$$L = \frac{G}{\sqrt{2}}(\bar{\mu}\gamma_\mu(1+\gamma_5)\nu)(\bar{u}\gamma^\mu[C_L(1+\gamma_5) + C_R(1-\gamma_5)]d)$$

then the y distributions are of the form

$$\frac{d^2\sigma}{dx\,dy} = \sigma^0(q_L(x) + (1-y)^2 q_R(x)) \tag{4.1.18}$$

and conversely for antineutrinos:

$$\frac{d^2\sigma}{dx\,dy} = \sigma^0(q_R(x) + (1-y)^2 q_L(x)). \tag{4.1.19}$$

Here,

$$q_L = q(x) + \rho^2\bar{q}(x)$$
$$q_R = \bar{q}(x) + \rho^2 q(x) \tag{4.1.20}$$

are linear combinations of the quark and antiquark structure functions, and $\rho = |C_R/C_L|$ is the ratio of right-handed to left-handed couplings.

The analysis of the y distribution has been used by the CDHS collaboration to search for right-handed currents. This analysis is based on a sample of 175 000 $\bar{\nu}$ and 90 000 ν events collected in wide-band and narrow-band beams at the CERN SPS [ABR 82b]. The average momentum transfer squared is $\langle Q^2 \rangle = 33\,(\text{GeV}/c)^2$. Evidence for right-handed currents could be seen at y near 1 and large x in antineutrino reactions, where the contribution of sea antiquarks is negligible and where the scattering of antineutrinos from valence quarks vanishes like $(1-y)^2$. Figure 4.1.6 shows the ratio $R = (d\sigma^{\bar{\nu}}/dy)/(d\sigma^{\nu}/dy)$ as a function of y for two regions of x. At large x, $x > 0.4$, the ratio vanishes as y approaches 1, showing that the contribution of right-handed quark currents is very small. An upper limit on ρ^2 can be found by using the ratio (for $y \to 1$ and $x > 0.4$)

$$\frac{q_R}{q_L} = \frac{(d^2\sigma^{\bar{\nu}}/dx\,dy) - (1-y)^2 d^2\sigma^{\nu}/dx\,dy}{(d^2\sigma^{\nu}/dx\,dy) - (1-y)^2 d^2\sigma^{\bar{\nu}}/dx\,dy}.$$

A limit of $|\rho|^2 < 0.009$ was obtained, for 90 percent CL. This measurement puts limits on left–right symmetric models [BEG 77; GEL 78] based on the gauge group $SU(2)_R \times SU(2)_L \times U(1)$, where two sets of intermediate bosons, M_L and M_R are mixed to yield two mass eigenstates, W_1 (mass m_1) and W_2 (mass m_2). The limit from this measurement on m_1^2/m_2^2 and on the mixing angle θ is shown in Fig. 4.1.7. Also shown are results from muon decay. In contrast to the muon decay result, the limit obtained here is also valid if the right-handed neutrino is heavy. If the right-handed neutrino is lighter than the muon, the most sensitive result comes from very precise measurements [CAR 83; STR 84] of the muon decay rate near the endpoint of the positron momentum spectrum. Polarized muons from pion decay were stopped in

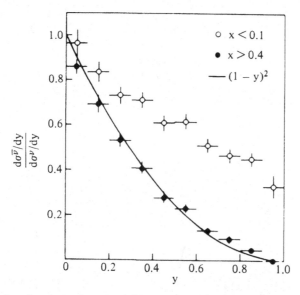

Fig. 4.1.6   Ratio of antineutrino to neutrino cross sections, as function of y, for small $x(<0.1)$ and large $x(>0.4)$ [ABR 82b].

(nondepolarizing) pure metal foils. The stopping process occurred either in a spin-holding magnetic field B of 1.1 T parallel to the muon polarization, or in a transverse field B of 70 gauss inducing muon spin rotation. In the first case, a $V-A$ interaction forces the positron rate to vanish at the endpoint; $V+A$ would maximize this rate. In the second case, $V-A$ produces a maximum muon spin rotation. The results of both experiments give the very precise 9 percent C.L. contour as the allowed range around the origin corresponding to pure $V-A$ interaction (or left-handed W_L boson) in Fig. 4.1.7.

4.1.1.5 Conservation of helicity

In inelastic neutrino–nucleon interactions it is possible to test experimentally whether the helicity carried by the incident neutrino is transferred to the outgoing muon. This can help to clear up the helicity structure of the weak charged-current interaction.

The most general Lagrangian for the interaction [KIN 74] including V, A, S, T, P terms, neglecting antiquarks in the nucleon, leads to a distribution in the inelasticity y for the process $\nu + N \to \mu^+ + X$ of the form

$$\frac{d\sigma}{dy} \sim 2(g_V - g_A)^2 + 2(g_V + g_A)^2(1-y)^2$$

$$+ (|g_S|^2 + |g_P|^2)y^2 + 32|g_T|^2(1 - y/2)^2$$

$$+ 8Re(g_T(g_S^* + g_P^*))y(1 - y/2) \tag{4.1.21}$$

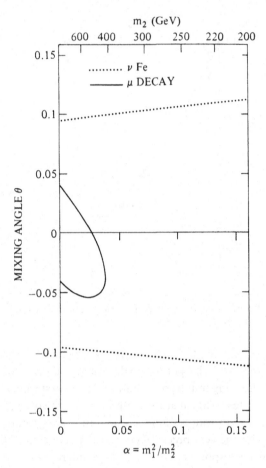

Fig. 4.1.7   Limits on the mixing angle θ between W_R and W_L, as a function of the mass ratio m_1^2/m_2^2 [STR 84; ABR 82a].

where the g_i are the coupling constants of the corresponding Lorentz invariant operators. The experimental data on these y distributions [BOS 78; GRO 79; JON 82] are perfectly consistent with constant and $(1-y)^2$ terms only, that is, with a V–A interaction (see 4.1.1.4). However, the same distribution could be obtained with a suitable combination of S, T, and P terms. This confusion [KAY 74; KIN 74] can be resolved by measurements of the outgoing muon [CHE 71] because, for massless leptons, V and A currents preserve the helicity at the lepton vertex, while S, T, and P currents change sign. Since muon antineutrinos in the beam from π or K decays are experimentally known to have positive helicity [BAC 61], the outgoing μ^+ leptons are expected to have the same helicity H = +1 if only A or V currents are present, and H = −1 for any combination of S, T, or P currents. The experiment was carried out [JON 83] in the wide-band beam at the CERN SPS using the combined detectors of the two big neutrino experiments CDHS [HOL 78a] and CHARM [DID 80]. Here the massive (1500 t) CDHS

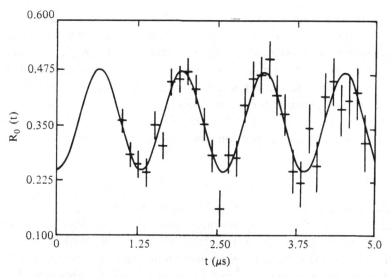

Fig. 4.1.8   Observed time dependence of the oscillating forward-backward positron asymmetry from polarized muons, $R_0(t)$ [JON 83].

iron-scintillator calorimeter served as an instrumented target focusing with its toroidal magnetic field the positive muons toward the detector (and beam) axis. The CHARM calorimeter made of marble ($CaCO_3$) and scintillator was embedded in a weak transverse magnetic field of 58 G. It was used as a polarimeter for the outgoing positive muons by using their decay $\mu^+ \rightarrow e^+ \nu_e \bar{\nu}_\mu$. The measured quantity was the asymmetry

$$R(t) = (N^b(t) - N^f(t))/(N^b(t) + N^f(t))$$

of muon stops in the CHARM detector with positrons emitted forward (f) or backward (b) at a time t after the muon stop. Since the muon spin is precessing in the weak transverse magnetic field with a frequency of 4.9 MHz, the asymmetry oscillates with time t at a period of 1.3 μs:

$$R(t) = R_0 \cos(\omega t + \phi) + R_1.$$

   Experimental results on R from 17 000 detected μ^+ decays are shown in Fig. 4.1.8. The best fit to the data yields

$$R_0 = 0.116 \pm 0.010 \text{ and}$$

$$\phi = -3.02 \pm 0.08 \text{ rad.}$$

This measured phase at $t = 0$ is consistent with the value $\phi = -\pi$ predicted if only V or A currents are present, and in contradiction to the value $\phi = 0$ expected for S, T, or P type currents. The measured oscillation amplitude R_0 of the asymmetry is related to the magnitude of the muon polarization P, by

$$R_0 = \alpha P$$

where α is the analyzing power of the polarimeter. The value of α was obtained from a Monte Carlo simulation of the μ^+ path and decay, and of the positron passage through matter. The depolarization of the μ^+ during the stopping process in marble was determined experimentally with muons from π decay at $140\,\text{MeV}/c$ by comparing a marble analyzer to one made of carbon, known to have no depolarizing effect. The resulting value for the magnitude of the original μ^+ polarization is $P = 0.82 \pm 0.07\,(\text{stat}) \pm 0.12\,(\text{syst})$. Since the phase ϕ is known, the main interaction types must be V and A, and limits on possible admixtures from S, T, or P currents can be derived from the value of P:

$$\sigma_{S.T.P}/\sigma_{\text{tot}} < 20\% \text{ at } 95\% \text{ C.L.}$$

These limits become more stringent if, as additional information, the inelasticity y measured in the CDHS target calorimeter is used. Average values are $\langle y \rangle = 0.404 \pm 0.01$, $\langle E^\nu \rangle = 30.3 \pm 0.12\,\text{GeV}$, and $\langle Q^2 \rangle = 4.0 \pm 0.04\,\text{GeV}^2/c^2$. If contributions from tensor currents are ignored, in the high y ($y > 0.5$) region S and P contributions, proportional to y^2, would dominate over the ones from V and A. Conversely, at low y ($y < 0.2$) S and P contributions are negligible. See Eq. (4.1.21). By comparing the asymmetry R_0 in these two regions of inelasticity y, a polarization for $y > 0.5$ of $P = 1.10 \pm 0.24$ is obtained. This gives the upper limit

$$\sigma_{S.P.}/\sigma_{\text{tot}} < 7\% \text{ at } 95\% \text{ C.L.} \tag{4.1.22}$$

It can be concluded that the weak leptonic charged current remains dominantly composed of vector and axial-vector currents at momentum transfers of several $(\text{GeV}/c)^2$.

4.1.1.6 V–A structure in dimuon production reactions

The $V-A$ structure of the charm-changing quark current ($\Delta C = 1$) can be studied by measuring the inelasticity (y) distribution of neutrino-induced opposite-sign dimuon events. Such events of the type

$$\nu_\mu + N \rightarrow \mu^- \mu^+ X \tag{4.1.23}$$

are due, apart from a small background caused by μ^+ from π^+ decay, to charm production via the quark reactions

$$\nu_\mu d \rightarrow \mu^- c \quad \text{rate} \sim |V_{cd}|^2$$
$$\nu_\mu s \rightarrow \mu^- c \quad \text{rate} \sim |V_{cs}|^2$$

with subsequent semileptonic weak decay of the charmed hadron according to

$$c \rightarrow s\mu\nu$$
$$\rightarrow d\mu\nu.$$

The observed flavor structure of the charged currents in such that processes with $\Delta C = \Delta S = \Delta Q = 1$ are favored over those with $\Delta C = 1$, $\Delta S = 0$. The suppression of these latter currents is similar to the one of strangeness – changing currents, as described by the Cabibbo angle, $\sin\theta_c \sim \frac{1}{5}$. Indeed, $|V_{cd}|/|V_{cs}| \sim \frac{1}{5}$ (see Sections 4.1.2.1–4).

The charm origin of the opposite-sign dimuon events is established by the small transverse momentum of the wrong-sign muon (μ^+ in neutrino reactions) relative to the axis of the hadron shower and by the observed back-to-back emission of the two muons in the plane perpendicular to the neutrino direction. In addition, the hadron shower in opposite-sign dimuon events contains more strange particles than the one in single muon neutrino events, in line with the expectation of the favored charm decay to strange hadrons.

The distribution of dimuon events in the inelasticity variable y is expected to be flat for a purely left-handed $V-A$ coupling (see Section 4.1.1.4) since it is due to neutrino–quark scattering. Likewise, also for the corresponding antineutrino reaction

$$\bar{\nu}_\mu + N \to \mu^+ \mu^- X$$

the basic quark reactions are

$$\bar{\nu}_\mu + \bar{s} \to \bar{c} + \mu^+$$
$$\bar{\nu}_\mu + \bar{d} \to \bar{c} + \mu^+$$

and here, too, a flat y distribution is expected. However, a right-handed $V+A$ coupling would lead to a $(1-y)^2$ dependence in both cases.

This y distribution has been studied in several neutrino experiments [ABR 82c; JON 81]. The largest event sample contains 10 381 neutrino-induced dimuon events [ABR 82c], recorded in the CDHS detector [HOL 78] in wide-band and narrow-band beams at CERN. The observed $d\sigma/dy$ distributions are distorted by the requirement that both muons have momenta above $5\,\text{GeV}/c$. This distortion is less severe at high neutrino energy. The simulation of the expected y distribution takes into account the measured charm fragmentation function in the variable $z = E_D/E_c$ with E_D the energy of the charmed hadron and E_c the energy of the charmed quark. The measured average z is $\langle z \rangle = 0.68 \pm 0.08$.

The observed y-distributions, together with the expected shape for $V-A$ coupling and $V+A$ coupling are shown in Fig. 4.1.9. There is good agreement with the $V-A$ prediction. A parameterization of the form $d\sigma/dy \propto \beta(1-y)^2 + (1-\beta)$ gives the following upper limit for the strength of the right-handed coupling:

$$\beta < 0.07 \text{ at 95 percent C.L.}$$

In a similar experiment, the CHARM collaboration [JON 81] finds $\beta = 0.15 \pm 0.10$ from a sample of 285 $\bar{\nu}_\mu$-induced dimuon events.

If the interaction Lagrangian has the form

$$L^{\Delta C=1} = \frac{G}{\sqrt{2}} (\bar{\mu}\gamma_\alpha(1+\gamma_5)\nu)(g_L J_\alpha^{\Delta C=1}(V-A) + g_R J_\alpha^{\Delta C=1}(V+A))$$

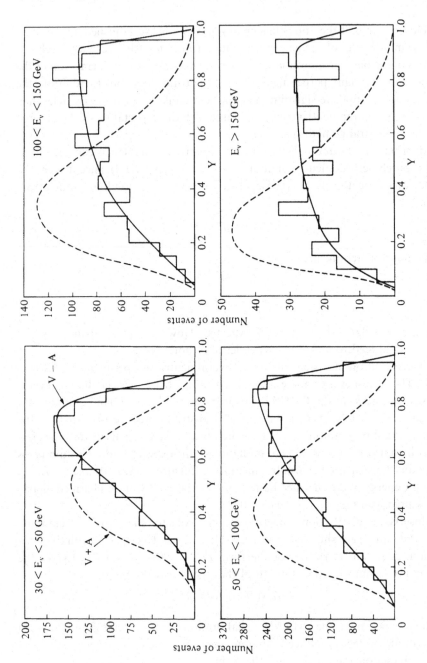

Fig. 4.1.9 Distribution in the inelasticity y of neutrino-induced opposite-sign dimuon events, in different neutrino energy intervals. Solid line is prediction for left-handed $(V-A)$ coupling of charm producing charged current, dashed line for $V+A$ coupling [ABR 82c].

then this limit transforms to a limit on the square of the coupling constants

$$g_R^2/(g_L^2 + g_R^2) < 0.07 \text{ (95 percent C.L.)}. \qquad (4.1.24)$$

4.1.2 Flavor structure of the weak quark current

4.1.2.1 The Cabibbo model

Most of the hadronic weak processes considered above in order to determine the space-time structure of weak currents were dealing with the light quarks u and d. The most prominent weak process involving these quark flavors is nuclear beta decay. In the quark picture, we consider this decay $n \rightarrow pe^- \bar{\nu}$ as a transformation of a valence d quark in the neutron (udd) into a u quark, thereby emitting a W^- boson. The process can be depicted by the diagram of Fig. 4.1.10. From the previous part of this chapter we infer that the Lagrangian for this reaction is

$$L = \frac{G}{\sqrt{2}} J^{+\alpha} \cdot j_\alpha^- + \text{h.c.}$$

where j_α^- is the leptonic charged current

$$j_\alpha^- = \bar{e}\gamma_\alpha(1 + \gamma_5)\nu_e$$

and $J^{+\alpha}$ is the hadronic charged current

$$J^{+\alpha} = \bar{u}\gamma^\alpha(1 + \gamma_5)d \cdot V_{ud}$$

V_{ud} is the coupling strength of the transition of d to u compared to that of a purely leptonic process.

Analogously we consider β-decay of Λ hyperon (quark state uds) to be the transformation of the s quark into an u quark, with the two other constituent quarks, u and d, again being spectators (Fig. 4.1.11). Experimentally it was observed that the coupling strength of the corresponding piece of the hadronic current

$$J^{+\alpha} = \bar{u}\gamma^\alpha(1 + \gamma_5) \cdot V_{us} \qquad (4.1.25)$$

is about 5 times smaller than in neutron decay. This led Cabibbo to postulate that the eigenstates of the then-known quarks with $-\frac{1}{3}$ charge are not the flavor eigenstates d and s but a linear combination, rotated by an angle θ, the Cabibbo angle: $d_c = d\cos\theta + s\sin\theta$. When it became experimentally clear that strangeness-changing neutral currents in the decay $K_L \rightarrow \mu^+\mu^-$ are strongly suppressed (the branching ratio is $(8.1^{+2.8}_{-1.8})10^{-9}$ [SHO 79]), Glashow, Iliopoulos, and Maiani [GLA 70] realized that this contradicted the expectation derived from the weak-electro-magnetic diagram (Fig. 4.1.12): In order to cancel this diagram by another one of the same magnitude but opposite sign, they postulated the existence of another

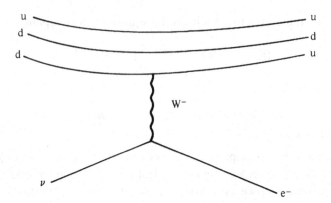

Fig. 4.1.10   Quark spectator diagram for neutron β-decay.

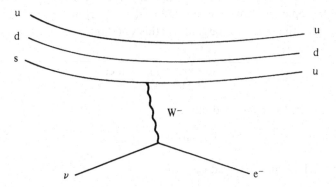

Fig. 4.1.11   Quark spectator diagram for Λ hyperon β-decay.

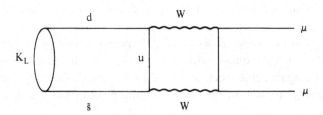

Fig. 4.1.12   Feynman diagram for the decay $K_L \to \mu\mu$.

quark with $+\frac{2}{3}$ charge, the charm quark. As the corresponding $-\frac{1}{3}$ charge weak eigenstate they took the orthogonal state to d_c, that is,

$$s_c = -d \sin\theta + s \cos\theta$$

If the mass of this quark is not too different from the one of the up-quark ($m_c \simeq 2\,\mathrm{GeV}$), the diagram (Fig. 4.1.12) and the one with the u quark replaced by a c quark cancel nearly exactly, being proportional to $G \sin\theta \cos\theta$ and $G \cos\theta(-\sin\theta)$ respectively.

For four quarks then, the hadronic weak-charged current is

$$J^{+\alpha} = (\bar{u}\bar{c}) \begin{bmatrix} \cos\theta & \sin\theta \\ -\sin\theta & \cos\theta \end{bmatrix} \gamma^\alpha(1+\gamma_5) \begin{bmatrix} d \\ s \end{bmatrix} \qquad (4.1.26)$$

where the 2×2 unitary (and in effect real) matrix is simply a rotation matrix with one rotation angle θ. There is no room here for a violation of time-reversal invariance, or *CP* violation.

4.1.2.2 The Kobayashi–Maskawa quark-mixing scheme

With the observation of the fifth quark, bottom b, and the presumed existence of another $+\frac{2}{3}$ charge quark, top, the current has to be extended. Kobayashi and Maskawa [KOB 73] realized that if one enlarges the mixing matrix correspondingly, there are three rotation angles and one complex phase as free parameters of the theory. The current is then

$$J^{+\alpha} = (\bar{u}\bar{c}\bar{t}) \begin{bmatrix} V_{ud} & V_{us} & V_{ub} \\ V_{cd} & V_{cs} & V_{cb} \\ V_{td} & V_{ts} & V_{tb} \end{bmatrix} \gamma^\alpha(1+\gamma_5) \begin{bmatrix} d \\ s \\ b \end{bmatrix}. \qquad (4.1.27)$$

The very important point noted by Kobayashi and Maskawa (KM) was the presence of a nontrivial complex phase in this matrix. They suggested that this could be the origin of *CP* violation.

Their original parameterization of the matrix in terms of the three angles $\theta_1, \theta_2, \theta_3$ and the phase δ is given in Table 4.1.1. In this parameterization, the angle θ_1 corresponds to couplings between the first and second generation of quarks, while couplings between second and third generations and between first and third are connected to both θ_2 and θ_3. A more useful parameterization would be such that each angle corresponds to one pairing of generations. Such formulations have been given by Maiani [MAI 77] and Chau and Keung [CHA 83] for three generations, and by Harari and Leurer [HAR 86] and Fritzsch and Plankl [FRI 87] in a form generalizable for an arbitrary number of generations. We choose here the "standard parameterization," as advocated in [GIL 86b], and shown in Table 4.1.2 for three generations. The experimental determination of the elements of the KM matrix has received a great deal of attention in the last few years [KLE82a,b, 83; PAK 82; PAS 82; CHA 83; BUR 84) because these elements and the three angles plus phase that can be deduced from them are constants of nature that are not explained in the Standard Model. Any theory that goes beyond the Standard Model has to explain these angles. There is the conjecture [FRI 79; STE 83] that these angles are related to the ratios of quark masses. Of course, if there are four generations, the analysis has to be extended, and parameterizations have been given for this case [BOS 80; GRO 85; TUR 85] including the extension of the standard parameterization above [HAR 86; FRI 87]. Experimental information on the weak quark couplings comes from measurements of weak decays of light and heavy quarks

Table 4.1.1. *Kobayashi–Maskawa parameteriza-tion of mixing matrix*

$$V = \begin{pmatrix} c_1 & s_1 c_3 & s_1 s_3 \\ -s_1 c_2 & c_1 c_2 c_3 - e^{i\delta} s_2 s_3 & c_1 c_2 c_3 + e^{i\delta} s_2 c_3 \\ s_1 s_2 & -c_1 s_2 c_3 - e^{i\delta} c_2 s_3 & -c_1 s_2 s_3 + e^{i\delta} c_2 c_3 \end{pmatrix}$$

Table 4.1.2. *Quark-mixing matrix in standard parameterization*

$$V = \begin{pmatrix} c_{12} c_{13} & s_{12} c_{13} & s_{13} e^{-i\delta_{13}} \\ -s_{12} c_{23} - c_{12} s_{23} s_{13} e^{i\delta_{13}} & c_{12} c_{23} - s_{12} s_{23} s_{13} e^{i\delta_{13}} & s_{23} c_{13} \\ s_{12} s_{23} - c_{12} c_{23} s_{13} e^{i\delta_{13}} & -c_{12} s_{23} - s_{12} c_{23} s_{13} e^{i\delta_{13}} & c_{23} c_{13} \end{pmatrix}$$

and from neutrino production of charm quarks. The next paragraph contains these experimental constraints, and the following one proceeds to derive bounds on the mixing angles in the case of three or four generations.

4.1.2.3 *Measurements of weak quark-coupling constants*

Light quark couplings

Coupling V_{ud}   This parameter is obtained by comparing decay rates of nuclear beta decays and muon decays. A new analysis of radiative corrections in the nuclear beta decays has been presented by Marciano and Sirlin [MAR 86a], and more recently the inconsistencies between ft-values of low Z and high $Z 0 \rightarrow 0$ Fermi transitions have been revised and removed by Sirlin and Zucchini [SIR 86].

New data have been reported on superallowed $0 \rightarrow 0$ decays [SAV 95; ORM 95]. Averaging these two ft-values, and keeping the same error as for one of the results, we obtain $|V_{ud}| = 0.9735 \pm 0.0005$. Considering the argument [SAI 95] that the change in charge-symmetry violation for quarks inside nucleons that are embedded in nuclei increases that ft-value by 0.08 to 0.2%, we increase ft by 0.1±0.1% and obtain:

$$|V_{ud}| = 0.9740 \pm 0.0010. \tag{4.1.28}$$

Coupling V_{us}   Two kinds of experimental information on the coupling of strange and up-quarks exist: One is the self-consistent analysis of weak semileptonic hyperon decays, as done by the WA2 collaboration [BOU 83], with the result $V_{us} = 0.231 \pm 0.003$. The other comes from an analysis of K_{e3} decays, $K_L \rightarrow \pi e \nu$ and

$K^+ \to \pi e \nu$. This result, $V_{us} = 0.2196 \pm 0.0023$, differs from the one obtained in hyperon decays; Leutwyler and Roos [LEU 84] argue that $SU(3)$ breaking effects are larger in hyperon decays than in K meson decays, and that therefore the results from K_{e3} are more reliable from a theoretical point of view. We therefore use the value

$$V_{us} = 0.2196 \pm 0.0023. \qquad (4.1.29)$$

Charm–quark couplings

Coupling V_{cd} This coupling has been determined from measurements of single charm production in neutrino and antineutrino reactions. The coupling parameter is obtained from the measured ratio of dimuon to single muon production cross sections in neutrino reactions $(R^\nu = \sigma^\nu_{\mu^-\mu^+}/\sigma^\nu_{\mu^-})$ and antineutrino reactions $(R^{\bar\nu} = \sigma^{\bar\nu}_{\mu^+\mu^-}/\sigma^{\bar\nu}_{\mu^+})$.

The differential cross sections for neutrino charm production on isoscalar targets are

$$\frac{d\sigma^\nu}{dx\,dy} = \frac{G^2 M E_\nu x}{\pi} [V^2_{cd}(u(x) + d(x)) + |V_{cs}|^2 2s(x)] \qquad (4.1.30)$$

$$\frac{d\sigma^{\bar\nu}}{dx\,dy} = \frac{G^2 M E_{\bar\nu} x}{\pi} [V^2_{cd}(\bar u(x) + \bar d(x)) + |V_{cs}|^2 2\bar s(x)] \qquad (4.1.31)$$

where $u(x)$, $d(x)$, and $s(x)$ are the quark density distributions in the proton, G is the Fermi coupling constant, M the nucleon mass, E_ν the neutrino laboratory energy, and x and y the Bjorken scaling variables.

Experimentally, the observation of charm production has been done mainly by three methods: (1) direct observation of the short-lived decay of charmed hadrons in emulsions; (2) observation of semileptonic charm decay $c \to s + \mu^+ + \nu_\mu$ in neutrino-induced dimuon events, $\nu N \to \mu^- \mu^+ X$; and (3) observation of semileptonic charm decay $c \to s + e^+ + \nu_e$ in dilepton events, $\nu N \to \mu^- e^+ X$. By far the largest event samples have been collected using the second method.

In order to obtain the coupling parameter V_{cd}, the contribution of charm production from the strange sea s and $\bar s$ quarks has to be eliminated. According to the cross sections given in (4.1.30) and (4.1.31), this can be done by using the weighted difference of neutrino and antineutrino cross sections:

$$\beta V^2_{cd} = \left(\frac{R^\nu - R R^{\bar\nu}}{1 - R}\right) \frac{2}{3} \qquad (4.1.32)$$

where β is the semileptonic branching ratio of the mixture of charmed particles produced in the neutrino reactions, and $R = \sigma^{\bar\nu}/\sigma^\nu$ is the ratio of total neutrino cross sections. Figure 4.1.13 shows R^ν and $R^{\bar\nu}$ as obtained by the CDHS neutrino experiment [ABR 82c].

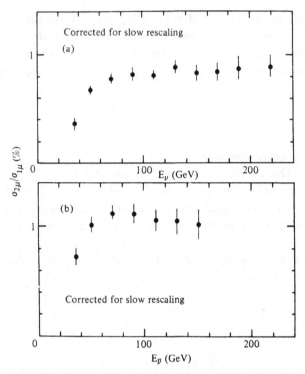

Fig. 4.1.13   Ratio of dimuon production and single muon production cross sections (a) for neutrinos (b) for antineutrinos. Rates are corrected for detector acceptance and for charm production threshold effects ("slow rescaling") [ABR 82c].

The result of the CDHS collaboration,

$$\beta V_{cd}^2 = (0.41 \pm 0.07)10^{-2} \tag{4.1.33}$$

has been used to extract V_{cd}.

The corresponding value from the Tevatron neutrino experiment CCFR [BAZ 95] was obtained with a next-to-leading order QCD analysis. The result is $0.534 \pm 0.021^{+0.025}_{-0.051} \times 10^{-2}$, where the last error is from the scale uncertainty. Using a similar scale error for the CDHS result and averaging these two results, we obtain $(0.49 \pm 0.05) \times 10^{-2}$.

The semileptonic branching ratio β has been obtained using emulsion data on the mixture of charmed particle species [USH 88] and PDG values for their semileptonic branching fractions to give [BAZ 95]

$$\beta = 0.099 \pm 0.012. \tag{4.1.34}$$

These measurements then give a value

$$|V_{cd}| = 0.224 \pm 0.016. \tag{4.1.35}$$

The CHARM II collaboration reported a value [CHA 98] $|V_{cd}| = 0.219 \pm 0.017$.

Coupling V_{cs} In the past, this coupling was obtained from neutrino and antineutrino dimuon production data. Since the charm production from strange sea quarks is considered here, the quantity measured is the product $|V_{cs}|^2 \cdot 2S$, where $S = \int xs(x)\,dx$ is the integral of the strange sea structure function. In the absence of an independent determination of S, one obtains as an upper limit $2S \leq \bar{U} + \bar{D}$, where $\bar{U}$ and $\bar{D}$ are the momentum fractions of nonstrange sea antiquarks. We call $\alpha = 2S/(\bar{U} + \bar{D})$ the ratio of these momentum fractions, and $\alpha^* = 2S(|V_{us}|^2 + |V_{cs}|^2/r_s)/(\bar{U} + \bar{D})$ the same ratio modified by the threshold suppression factor r_s for the charm quark mass. Then from the x-distribution of neutrino dimuons [ABR 82c] one obtains

$$\frac{|V_{cs}|^2}{|V_{cd}|^2} = (6.26 \pm 0.73)\frac{1 + \alpha^*}{\alpha} \tag{4.1.36}$$

and from the cross-sectional ratios R^ν and $R^{\bar{\nu}}$ of dimuon production

$$\frac{|V_{cs}|^2}{|V_{cd}|^2} = (5.9 \pm 1.5 + (8.5 \pm 1.7)\alpha^*)/\alpha. \tag{4.1.37}$$

Making the most conservative assumption that the strange quark sea does not exceed the value corresponding to an $SU(3)$ symmetric sea, leads to the lower bound [ABR 82c] $|V_{cs}| > 0.59$. An alternative method of obtaining V_{cs} is based on the measurement and calculation of the decay rate for $D^+ \to \bar{K}^0 e^+ \nu_e$. This rate can be expressed as

$$(D^+ \to \bar{K}^0 e^+ \nu_e) = |f_+^D(0)|^2 |V_{cs}|^2 \ (1.54 \cdot 10^{11}\mathrm{s}^{-1})$$

where $|f_+^D(0)|$ is the form-factor for D_{13} decay for zero momentum transfer. Combining data on the branching ratios of the decays $D^+ \to \bar{K}^0 e^+ \nu$ and $D^0 \to K^- e^+ \nu$ and with the world average values of D^+ and D^0 lifetime [BAR 96], the experimental decay width is $\Gamma(D \to \bar{K}e^+\nu) = (0.818 \pm 0.041) \times 10^{11}\mathrm{s}^{-1}$. This leads to

$$|f_+^D(0)|^2 |V_{cs}|^2 = 0.531 \pm 0.027.$$

In three theoretical calculations using QCD sum rules [ALI 84; WIR 85; GRI 89], values were obtained for the transition form-factor $|f_+^D(0)|$. We use an intermediate value, namely,

$$|f_+^D(0)| = 0.70 \pm 0.10. \tag{4.1.38}$$

From this we obtain

$$|V_{cs}| = 1.04 \pm 0.16.$$

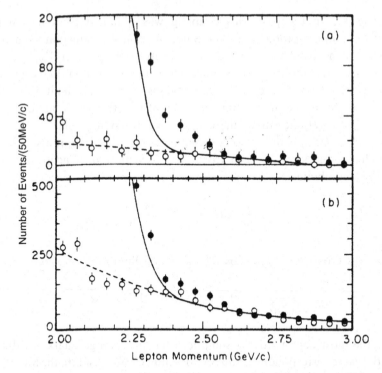

Fig. 4.1.14 Momentum distribution of electrons produced in e^+e^- collisions at the $\Upsilon(4S) = b\bar{b}$ resonance; (a) strict cuts, (b) loose cuts for continuum suppression. Filled points are ON resonance, open points OFF resonance. Dashed curves are fits to OFF data, solid curves show sum of predicted $b \to cl\nu$ and fitted OFF [BAR 93].

Bottom-quark couplings

$|V_{cb}|$ – The heavy quark effective theory [ISG 89] (HQET) provides a nearly model-independent treatment of B semileptonic decays to charmed mesons, assuming that both the b and c quarks are heavy enough for the theory to apply. From measurements of the exclusive decay $B \to \bar{D}^* l^+ \nu_l$, the value $|V_{cb}| = 0.0387 \pm 0.0021$ has been extracted [FEI 97] using corrections based on HQET. Exclusive $B \to \bar{D} l^+ \nu_l$ decays give a consistent, but less precise result. Analysis of inclusive decays, where the measured semileptonic bottom hadron partial width is assumed to be that of a b quark decaying through the usual $V-A$ interaction, depends on going from the quark to hadron level. This is also understood within the context of the HQET [NEU 97], and the results for $|V_{cb}|$ are again consistent with those from exclusive decays. Combining all these results [FEI 97]:

$$|V_{cb}| = 0.0395 \pm 0.0013, \tag{4.1.39}$$

which is now the third most accurately measured CKM matrix element.

$|V_{ub}|$ – The decay $b \to cl\bar{\nu}$ and its charge conjugate can be observed from the semileptonic decay of B mesons produced on the $\Upsilon(4S)$ $(b\bar{b})$ resonance by measuring the lepton energy spectrum above the endpoint of the $b \to cl\bar{\nu}_l$ spectrum. Fig. 4.1.14

shows an example for such an inclusive electron momentum spectrum [BAR 93]. There the $b \rightarrow ul\bar{\nu}_l$ decay rate can be obtained by subtracting the background from nonresonant e^+e^- reactions. The continuum background is determined from auxiliary measurements off the $\Upsilon(4S)$ [BAR 93, ALB 91]. The interpretation of the result in terms of $|V_{ub}/V_{cb}|$ depends fairly strongly on the theoretical model used to generate the lepton energy spectrum, especially for $b \rightarrow u$ transitions [WIR 85, GRI 86, ALT 82]. Combining the experimental and theoretical uncertainties, we quote

$$|V_{ub}/V_{cb}| = 0.08 \pm 0.02. \tag{4.1.40}$$

This result is supported by the first exclusive determinations of $|V_{ub}|$ from the decays $B \rightarrow \pi l \nu_l$ and $B \rightarrow \rho l \nu_l$ by the CLEO experiment [ALE 96] to obtain $|V_{ub}| = 3.3 \pm 0.4 \pm 0.7 \times 10^{-3}$, where the first error is experimental and the second reflects systematic uncertainty from different theoretical models of the exclusive decays. While this result is consistent with Eq. (4.1.40) and has a similar error bar, given the theoretical model dependence of both results we do not combine them, and retain the inclusive result for $|V_{ub}|$.

4.1.2.4 Determination of allowed ranges for coupling constants and angles

Three generations Using the constraints of Eqs. (4.1.28, 29, 35–40) we first do a fit using the six-quark parameterization of Table 4.1.2, along the lines of [KLE 83]. We obtain, for a $\chi^2 = 6.28$, the values for the angles $\sin\theta_{12} = 0.2196 \pm 0.0023$, $\sin\theta_{23} = 0.0395 \pm 0.0013$ and $\sin\theta_{13} = 0.0032 \pm 0.0008$, and $2S/(\bar{U} + \bar{D}) = 0.32 \pm 0.06$. The corresponding 90 percent C.L. allowed ranges for the mixing matrix elements are given in Table 4.1.3.

Four generations We use the standard parameterization. The allowed range of parameters was examined by a numerical scan of the parameter space. The free parameters of this scan were the six angles θ_{12}, θ_{13}, θ_{14}, θ_{23}, θ_{24}, and θ_{34}, the three phases and the values of α and $f^+(0)$. The parameters were varied in a region around the solution with the minimal $\chi^2 = \chi_0^2$. As experimental constraints, Eqs. (4.1.28–9, 35–7, 39–40) were used with the theoretical value for $f_+^D(0)$ as an optional constraint. All combinations of parameters leading to a χ^2 value below $\chi_0^2 + 1.6^2$ were accepted as allowed values. Table 4.1.4. gives the resulting bounds for the elements of the 4×4 unitary matrix with the constraint on $f_+^D(0)$. Also, Fig. 4.1.15 gives contour plots for allowed regions of each pair of angles θ_{ik} $(i < k)$.

Other constraints on elements of the KM matrix In this analysis, several experimental constraints that have implications on some of the elements of the mixing matrix were not used because their interpretation in terms of the KM elements needs more theoretical input. These experiments include observation of B-$\bar{B}$-mixing and of direct CP violation.

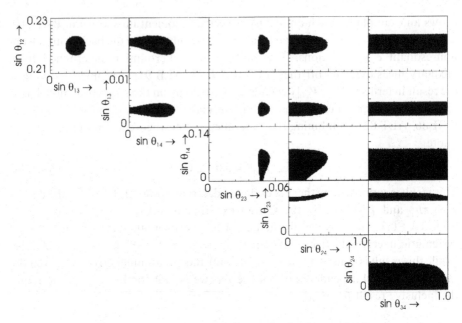

Fig. 4.1.15   Allowed ranges for the six mixing angles in the standard parameterization, new analysis along the line of [KLE 87].

Table 4.1.3.   *Elements of 3 × 3 quark mixing matrix V_{ik} from fit of experimental constraints (90 percent C.L. allowed ranges)*

	d	s	b
u	0.9750–0.9760	0.217–0.222	0.0018–0.0044
c	0.217–0.222	0.9743–0.9753	0.036–0.042
t	0.004–0.013	0.034–0.042	0.9991–0.9994

Table 4.1.4.   *Elements of 4 × 4 quark-mixing matrix V_{ik} from fit of experimental constraints (90 percent C.L. allowed ranges)*

	d	s	b	b'
u	0.9724–0.9755	0.217–0.223	0.0018–0.0044	0–0.08
c	0.199–0.231	0.847–0.975	0.036–0.042	0–0.48
t	0.000–0.100	0.000–0.360	0.050–0.9994	0–0.9999
t'	0.000–0.110	0.000–0.490	0.000–0.999	0.04–1.00

Note: Constraint $|f_+^D(0)| = 0.70 \pm 0.10$ used.
Allowed ranges for angles: $0.216 < s_{12} < 0.223$, $0.019 < s_{13} < 0.0044$, $0.000 < s_{14} < 0.080$, $0.037 \leq s_{23} \leq 0.046$, $0.000 \leq s_{24} < 0.500$, $0.000 \leq s_{34} < 1.000$.
Other parameters: $0.28 \leq 2S/(\bar{U} + \bar{D}) < 0.57$, $0.64 \leq |f_+^D(0)| \leq 0.82$.

1 B-$\bar{B}$ mixing, that is, the transition $B^0 \to \bar{B}^0$ has been observed by the ARGUS collaboration at DESY [ALB 87b] and confirmed by the CLEO collaboration at Cornell [FUL 88]. The mixing parameter $x_d = \Delta M/\Gamma$ is the ratio of the mass difference ΔM of the weak eigenstates (B_L and B_s) and the decay width Γ. Experimentally, $x_d = 0.70 \pm 0.13$ is very large. Theoretically, this mixing is due to a box graph similar to the one in the $K^0 - \bar{K}^0$ system. The dominant process is the one with an exchange of a top quark, and therefore x_d is proportional to $|m_t \times V_{td}|^2$. Using the top quark mass m_t ($m_t = 166 \pm 5\,\text{GeV}$) and the hadronic matrix element from QCD lattice calculations, one obtains

$$|V_{tb}^A V_{td}| = 0.0034 \pm 0.0018.$$

2 Direct CP violation Direct CP violation in the K^0 system has been seen via a small difference between K_L decay rates into $\pi^0\pi^0$ and $\pi^+\pi^-$ [BUR 88]. The amplitude for this effect, ϵ', relative to the amplitude ϵ of CP violation through $K^0\bar{K}^0$-mixing is found to be $\epsilon'/\epsilon = (3.3 \pm 1.1)10^{-3}$. The CP violating amplitudes vanish if any of the three angles θ_{12}, θ_{23}, or θ_{13} or the phase δ in the KM matrix are zero. Conversely, from this measurement, one can derive [KLE 88] that V_{ub} is nonzero:

$$|V_{ub} \times \sin\delta| > 1.7 \times 10^{-3}$$

and therefore also

$$|\sin\theta_{13}| \sim |V_{ub}| > 0.002.$$

Discussion It is evident from this analysis that for the first three generations of quarks, we observe a pattern of decreasing mixing angles: the angle connecting the first two generations, θ_{12}, is larger than the one connecting generations 2 and 3, θ_{23}, and this one again is larger than θ_{13}. This pattern remains to be explained by theories beyond the Standard Model, which should explain the pattern of quark masses as well. At the moment, only a few phenomenological models exist that connect the values of the mixing angles θ_{ik} to the ratio of quark masses in the generations i and k, m_k/m_i [FRI 79; STE 83].

For the angles connected to the fourth generation, only θ_{14} is constrained significantly. The upper limits obtained here follow the sequence $s_{34}^{\max} > s_{24}^{\max} > s_{14}^{\max}$. It is therefore possible that s_{34} and s_{24} are larger than s_{12}, contrary to the expectation of the simple models. The experimental situation can be improved significantly when the observation of W decays into heavy pairs at LEP 200 or at $\bar{p}p$ colliders allows a determination of their weak couplings.

Neutrino reactions, however, cannot contribute to the measurement of weak couplings of the heavy b or t quarks because their weak couplings to the light quarks are so small.

4.2 Neutrino reactions and the structure of the neutral weak current*

4.2.1 Introduction

In 1973 muonless neutrino reactions were discovered at CERN in the elastic scattering of antimuon neutrinos on electrons [HAS 73a],

$$\bar{\nu}_\mu e \rightarrow \bar{\nu}_\mu e$$

the neutral neutrino current $\bar{\nu}_\mu \nu_\mu$ and the neutral electron current $\bar{e}e$ couple by exchanging a neutral intermediate boson. In the reaction [HAS 73b]

$$\nu_\mu N \rightarrow \nu_\mu \pi^+ \pi^- \pi^0 N$$

muon neutrinos scatter inelastically on nucleons and transfer part of their energy to the neutral quark currents $\bar{u}u$ and $\bar{d}d$, which fragment into pions.

Observation of an asymmetry of the cross section of the scattering of left-handed and right-handed electrons on deuterons at SLAC in 1978 [PRE 78] demonstrated that the neutral weak $\bar{e}e$ current is violating parity. The discovery of parity violation of the weak $\bar{e}e$ current was confirmed by the observation that the polarization plane of a laser beam passing through bismuth vapor is rotated [BOU 84; NOE 88].

The existence of $\bar{\mu}\mu$ and $\bar{\tau}\tau$ neutral currents was deduced from the observation of a weak forward–backward charge asymmetry in the annihilation of electrons and positrons at the PETRA collider at DESY [SWU 87] in 1982. The existence of other neutral quark currents, for example, $\bar{s}s$ and $\bar{c}c$, was deduced from the analysis of the inelasticity distribution of deep inelastic neutral current neutrino scattering on nucleons [JON 81] and from J/ψ production [ABR 82a], respectively. Also the neutral current $\bar{\nu}_e \nu_e$ has been observed in neutrino experiments at CERN [DOR 86].

So far, only diagonal neutral currents have been observed that do not change the flavor of the particles involved. With three families of leptons and quarks, the standard theory [GSW 67] predicts the existence of six diagonal neutral lepton currents and six diagonal neutral quark currents.

The neutral current has a more complex helicity structure than the charged current. Experiments, for example, measurements of the inelasticity distribution of deep inelastic neutral current neutrino scattering [WIN 88; ALL 89a], have shown that coupling to both left-handed and right-handed quarks exists. This observation shows directly the existence of a unified, electroweak force.

The structure of the current as deduced from these experiments implies that left-handed fermions transform as doublets under a weak isospin rotation group and right-handed fermions transform as singlets. The Standard Model of the electroweak gauge theory predicted this structure of neutral currents and it successfully describes a large amount of experimental data [AMA 87; COS 88; FOG 88; KIM 81; LAN 95; SEG 81] which are all consistent with universal strength of the forces g_2 and g_1 associated with the $SU(2)$ and $U(1)$ symmetry groups, respectively.

---

* K. Winter, CERN, Geneva, Switzerland and Humboldt Universität, Berlin.

The fundamental quantities of the Standard Model, the coupling constants g_1 and g_2, and the masses of the weak bosons m_W and m_Z are related to the angle Θ_W that describes the mixing of the two local symmetries by the relations

$$e = \frac{g_1 g_2}{\sqrt{g_1^2 + g_2^2}} = g_2 \sin \Theta_W, \quad \sin^2 \Theta_W = 1 - \frac{m_W^2}{m_Z^2}. \tag{4.2.1}$$

The value of the mixing angle is not predicted by the Standard Model. Grand unified theories predict a value of $\sin^2 \Theta_W = \frac{3}{8}$ at the unification mass.

The occurrence of a nonzero mixing angle defines both the structure and the strength of the neutral currents. The left-handed "up" particles of the weak isospin doublets couple with a coefficient $(\frac{1}{2} - Q \sin^2 \Theta_W)$. Q is the electric charge of the particle. The coupling coefficient of the right-handed states is $-Q \sin^2 \Theta_W$. Hence the value of $\sin^2 \Theta_W$ can be deduced from all neutral current-induced processes.

The neutral-current interaction at low energies is given by the effective Lagrangian

$$L_{eff}^{NC} = \rho \frac{G_F}{\sqrt{2}} J_z^\mu J_{z\mu} \tag{4.2.2}$$

where $\rho = m_W^2 / m_Z^2 \cos^2 \Theta_W$ is unity in models with Higgs doublets. For a model independent analysis we shall write the terms in L_{eff}^{NC} for νe and ν quark processes in a form that is valid in arbitrary gauge theories with massless left-handed neutrinos

$$L^{\nu e} = -\rho \frac{G_F}{\sqrt{2}} \bar{\nu}_\mu \gamma^\mu (1 + \gamma_5) \nu_\mu \bar{e} \gamma^\mu (g_V^e - g_A^e \gamma_5) e \tag{4.2.3}$$

and

$$L^{\nu q} = -\rho \frac{G_F}{\sqrt{2}} \bar{\nu}_\mu \gamma^\mu (1 + \gamma_5) \nu_\mu$$

$$\left[\sum_i g_L^i \bar{q}_i \gamma_\mu (1 + \gamma_5) q_i + g_R^i \bar{q}_i \gamma_\mu (1 - \gamma_5) q_i \right] \tag{4.2.4}$$

where g_V^e and g_A^e are the vector and axial vector coupling constants of the electron, and g_L^i and g_R^i the chiral coupling constants of left-handed (L) and right-handed quarks (R) of generation i.

4.2.2 Neutrino identity

Since the discovery of the neutral weak current, it has generally been believed that in neutrino-induced processes

$$\nu_\mu N \rightarrow \nu' x; \quad \bar{\nu}_\mu N \rightarrow \bar{\nu}' x$$
$$\nu_\mu e \rightarrow \nu' e; \quad \bar{\nu}_\mu e \rightarrow \bar{\nu}' e$$

the neutral lepton in the final state (ν') is identical with the incident neutrino. This identity is suggested by the assumption that the neutral neutrino current $\bar{\nu}_\mu\nu_\mu$ and the neutral e^+e^- current couple by exchanging a neutral intermediate boson, in analogy with the charged-current coupling.

In general, however, one cannot establish whether the outgoing neutral lepton is completely or only partly identical with, or completely different from, the initial neutrino. The possibility of a complete nonidentity can be excluded by direct experimental evidence in two ways.

One source of information is the behavior of the differential cross sections of ν- and $\bar{\nu}$-induced processes in the limit of $Q^2 \to 0$ for exclusive reactions or

$$y = 1 - \frac{E_{\nu'}}{E_\nu} \to 0$$

for inclusive reactions. In these kinematical configurations, V amplitudes are required to vanish, and only A amplitudes contribute. As the only differences between ν and $\bar{\nu}$ cross sections must show up in the $V \times A$ interference term, we expect $d\sigma(\nu) = d\sigma(\bar{\nu})$ as $q^2 \to 0$ if $\nu' = \nu$ [WOL 75; SAK 75]. Any deviation from this equality would be evidence either that the incident and outgoing neutrino states are nonidentical, if V and A interactions are assumed, or that nondiagonal scalar or tensor interactions contribute, if one assumes neutrino identity ($\nu' = \nu$) [KIN 75; SEG 75].

To distinguish between these two interpretations of a nonvanishing difference $d\sigma(\nu) - d\sigma(\bar{\nu})$ at q^2 or $y \to 0$, it is necessary to look for direct evidence of scalar (S), pseudoscalar (P), and tensor (T) interactions (see Section 4.2.3, discussion of $\nu_e e \to \nu_e e$).

Under the assumption of V and A interactions, the results of measurements of neutral- and charged-current cross section ratios for ν and $\bar{\nu}$ on isoscalar targets

$$\left[\frac{d\sigma^{\bar{\nu}}(NC)/d\sigma^{\bar{\nu}}(CC)}{d\sigma^{\nu}(NC)/d\sigma^{\nu}(CC)}\right]_{y=0} = 0.95 \pm 0.15 \text{ (stat)} \pm 0.12 \text{ (syst)} \qquad (4.2.5)$$

by Holder *et al.* [HOL 77] and

$$\left[\frac{d\sigma^{\bar{\nu}}}{dy}(NC)/\frac{d\sigma^{\nu}}{dy}(NC)\right]_{y=0} = 1.16 \pm 0.14 \text{ (stat) and (syst)} \qquad (4.2.6)$$

by Jonker *et al.* [JON 81], and more recently of a new measurement by the CHARM collaboration [ALL 89a] (see also Section 4.2.4)

$$\left[\frac{d\sigma^{\bar{\nu}}}{dy}(NC)/\frac{d\sigma^{\nu}}{dy}(NC)\right]_{y=0} = 1.072 \pm 0.060 \qquad (4.2.7)$$

support the concept of neutrino identity. The average of these results, combining statistical and systematic errors quadratically is 1.075 ± 0.053.

The other source of information is electron–neutrino and antineutrino scattering on electrons

$$\nu_e e \rightarrow \nu_e e, \ \bar{\nu}_e e \rightarrow \bar{\nu}_e e$$

with contributions of amplitudes from both neutral-current and charged-current interactions. As we shall see in Section 4.2.3, quantum mechanical interference of these amplitudes demonstrates neutrino identity.

4.2.3 Lorentz structure

Observation of an asymmetry of the cross sections of the scattering of left-handed and right-handed electrons on deuterons at SLAC (Stanford) in 1978 [PRE 78] demonstrated that the neutral current interaction of the weak $e\bar{e}$ current is parity violating. In terms of the space-time structure of the current, because of the interference with the electromagnetic interaction this is direct evidence for vector (V) and axial vector (A) $e\bar{e}$ currents. This was confirmed by the observation that the plane of polarization of a beam of circularly polarized laser light passing through bismuth or cesium vapor is rotated (see Section 4.2.5).

Is the Lorentz structure of the neutrino neutral-current interaction of the helicity-conserving vector type (V, A) or of the helicity-changing scalar (S), pseudoscalar (P), or tensorial (T) type? Elucidating this question for the charged weak current interaction took nearly 25 years of experimental investigation [CWU 65]. In neutrino-induced reactions the answer is further obscured by the so-called confusion theorem, which was first derived by B. Kayser et al. [KAY 74]. These authors evaluated the angular (inelasticity) distribution $d\sigma/dy (y = E_\nu^{\text{out}}/E_\nu^{\text{in}})$ of inclusive deep inelastic neutrino scattering on isoscalar nuclear targets. For clarity we consider only valence u and d quarks with a distribution function (see Chapter 5) of $Q(x) = u(x) + d(x)$.

$$\frac{d\sigma^{\nu N}}{dy} = (a + by + cy^2) \int_0^1 \frac{G_F^2 M E_\nu}{16\pi} \times Q(x) \, dx$$

$$\frac{d\sigma^{\bar{\nu} N}}{dy} = (\bar{a} + \bar{b}y + \bar{c}y^2) \int_0^1 \frac{G_F^2 M E_\nu}{16\pi} \times Q(x) \, dx. \qquad (4.2.8)$$

The six parameters describing the y dependence are related to the coupling constants

$$a = \bar{a} = 2[(g_V + g_A)^2 + (g_V - g_A)^2 + 32g_T^2]$$
$$b = -4(g_V - g_A)^2 - 32g_T^2 - 8g_T(g_S - g_P)$$
$$\bar{b} = -4(g_V + g_A)^2 - 32g_T^2 + 8g_T(g_S - g_P)$$
$$c = 2(g_V - g_A)^2 + 8g_T^2 + (g_S^2 + g_P^2) + 4g_T(g_S - g_P)$$
$$\bar{c} = 2(g_V + g_A)^2 + 8g_T^2 + (g_S^2 + g_P^2) - 4g_T(g_S - g_P).$$

These equations establish three relations between the six parameters; only three independent equations remain for the determination of the neutral-current coupling constants g_V, g_A, g_S, g_P, and g_T,

$$a = 4(g_V^2 + g_A^2) + 32g_T^2$$
$$c + \bar{c} - a = 2(g_S^2 + g_P^2) - 16g_T^2$$
$$\bar{c} - c = 8g_V g_A - 8g_T(g_S - g_P).$$

Therefore the coupling constants for the five interactions V, A, S, P, and T cannot be determined from the y distribution.

Nevertheless, a component proportional to y^2 in the y distribution is expected for a linear combination of S and P covariant interactions

$$\frac{d\sigma^{\nu N}}{dy} = \frac{d\sigma^{\bar{\nu} N}}{dy} \propto (g_S^2 + g_P^2)y^2 \qquad (4.2.9)$$

and for a pure tensor interaction

$$\frac{d\sigma^{\nu N}}{dy} = \frac{d\sigma^{\bar{\nu} N}}{dy} \propto g_T^2 \left(1 - \frac{1}{2}y\right)^2. \qquad (4.2.10)$$

More generally, the equation

$$c + \bar{c} - a = 0 \qquad (4.2.11)$$

is a necessary but not a sufficient condition for pure V, A interaction, while a nonzero value would be indication for the presence of S, P, or T interactions. In Section 4.2.4 a limit of $(g_S^2 + g_P^2)/g_{V,A}^2 < 0.03$ (95 percent C.L.) is derived from measurements of $d\sigma/dy$ [JON 81; ALL 89a].

A more direct and elegant way of finding an answer to the question of the Lorentz structure is obtained by looking for quantum mechanical interference of a charged weak current reaction amplitude with that of a neutral-current process. Interference will be observed if, and only if, the two amplitudes have indistinguishably identical parts. As far as we know [CWU 65], the charged-current interaction at low energy is entirely of $V{-}A$ structure. Interference with a neutral-current amplitude would demonstrate directly that part of the neutral-current interaction is of $V{-}A$ structure as well.

Nature is providing the required interference laboratory by the reaction

$$\nu_e e \rightarrow \nu_e e \qquad (4.2.12)$$

which proceeds by charged- and neutral-current interaction amplitudes (see Fig. 4.2.4). To make these amplitudes indistinguishably identical requires also the identity of the outgoing and incoming neutrinos; hence, observation of interference answers also the question of neutrino identity in neutral-current neutrino reactions discussed in Section 4.2.2.

The interference term is proportional to the neutral-current coupling constant of left-handed electrons g_L^e. According to the Standard Model we expect an interference term I proportional to

$$I \propto g_L^e = g_V^e + g_A^e = 2\sin^2\Theta_W - 1 \qquad (4.2.13)$$

The electroweak mixing parameter $\sin^2\Theta_W$ has been determined from measurements of a large variety of neutral-current phenomena; its generally accepted best value is $\sin^2\Theta_W = 0.230$ [AMA 87]. Interference, if observed, should therefore be destructive ($I < 0$). Experimental evidence of destructive interference in reaction (4.2.12) has indeed been reported [ALL 93]. The observed interference coefficient is

$$\alpha = -0.9 \pm 0.17 \text{ (stat)} \pm 0.12 \text{ (syst)} \qquad (4.2.14)$$

confirming the expected value of $\alpha = -1$ and, hence neutrino identity and $V{-}A$ structure of part of the neutral-current $\nu\bar{\nu}$ interaction; this latter property demonstrates parity violation in this reaction.

Another direct demonstration of the helicity-conserving vector-nature of the neutral neutrino interaction can be derived from measurements of the process of coherent neutral pion production by neutral-current neutrino interactions on nuclei (A). Owing to the intrinsic quantum numbers (spin zero, negative parity, isospin one) of the neutral pion, the process

$$\nu_\mu A \to \nu_\mu \pi^0 A \qquad (4.2.15)$$

probes directly the Lorentz structure of the isovector neutral-current interaction. The nucleus A recoils without breaking up or excitation. In the limit of zero momentum transfer, the helicity of the outgoing neutrino (as depicted in Fig. 4.2.1) remains negative for V, A interactions and is flipped to a positive value for S, P, T interactions. In the latter case the differential cross section for π^0 emission in the forward direction will be suppressed because of angular momentum conservation, whereas in the former case (V, A) it will be peaked forward. Figure 4.2.2 shows the angular distribution of π^0 with energies in the range $3 < E_\pi < 6$ GeV for a neutrino energy of 25 GeV [LAC 79]. The expectation for V, A, and SPT interactions are shown by the two curves. For $\Theta(\pi^0) \sim 0°$ we expect $\Theta(\nu_\mu') \to 0$ if most of the energy of the incoming neutrino is transferred to the outgoing one (ν_μ'), and hence $Q^2 \to 0$. A nearly monoenergetic neutrino beam is, however, required to observe whether or not π^0 production at $\Theta(\pi^0) \sim 0°$ is suppressed. This experimental condition has not yet been realized. On the basis of the direct observation, we shall assume in the following discussion that the neutral-current interaction is entirely of the V, A type.

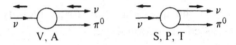

Fig. 4.2.1   Helicity of incident and outgoing neutrino (denoted by a black arrow) in the reaction $\nu_\mu A \to \nu_\mu \pi^0 A$ for V,A and for S,P,T interaction, in the limit $Q^2 = 0$.

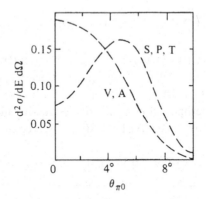

Fig. 4.2.2   Angular distribution of π^0 produced in the reaction $\nu_\mu A \to \nu_\mu \pi^0 A$, predicted for VA and for SPT interactions [LAC 79].

4.2.4 Chiral lepton and quark couplings

In the standard $SU(2) \times U(1)$ electroweak model, all left-handed fermions

$$\begin{pmatrix} \nu_e \\ e \end{pmatrix}_L \begin{pmatrix} \nu_\mu \\ \mu \end{pmatrix}_L \begin{pmatrix} \nu_\tau \\ \tau \end{pmatrix}_L \begin{pmatrix} u \\ d \end{pmatrix}_L \begin{pmatrix} c \\ s \end{pmatrix}_L \begin{pmatrix} t \\ b \end{pmatrix}_L$$

are assumed to transform as doublets with respect to weak $SU(2)$. One top particle, ν_τ has not yet been observed. The right-handed fermions

$$e_R^-, \mu_R^-, \tau_R^-, u_R, d_R, c_R, s_R, t_R, b_R$$

are assumed to transform as singlets. The motivations for these assignments were suggestive but not uniquely supported by experimental data.

Among these motivations was the well-established dominant left-handedness of fermions coupling to the charged weak currents. This established the doublet assignment of

$$\begin{pmatrix} \nu_e \\ e \end{pmatrix}_L \begin{pmatrix} \nu_\mu \\ \mu \end{pmatrix}_L \begin{pmatrix} \nu_\tau \\ \tau \end{pmatrix}_L \begin{pmatrix} u \\ d \end{pmatrix}_L .$$

The nonobservation of right-handed charged weak currents in muon decay [BUR 85; FET 86] and in deep inelastic neutrino scattering [ABR 82b] suggests that the right-handed fields are $SU(2)$ singlets. It should, however, be noted that this simplicity may be a low-energy phenomenon. A heavy quark U with charge $\frac{2}{3}$ that cannot be observed in the existing experiments could exist,

$$\begin{pmatrix} u \\ d \end{pmatrix}_L \quad \text{and} \quad \begin{pmatrix} U \\ d \end{pmatrix}_R,$$

would transform as doublets while u_R and U_L would be singlets.

The absence of flavor-changing neutral currents (see Section 4.2.6) between the observed fermions suggested that all fermions of the same charge, color, and chirality transform in the same way under $SU(2) \times U(1)$. Again, it should be noted that the argument would not be valid if there was strong mixing of doublet and singlet states [LAN 88a]. From the possible exceptions noted above it is clear that the suggested standard assignments of fermions are simpler than the alternatives.

There now exist high-precision measurements of charged- and neutral-current processes that establish directly the assignments of the known fermions and imply that a top quark and a ν_τ state must exist (see. e.g., [LAN 88b]). The remaining possible exceptions of placing e_R^-, μ_R^-, u_R, and d_R into doublets with heavy fermions can be eliminated by neutral-current data that directly measure the $SU(2) \times U(1)$ properties of the particle itself, independent of the mass of its partner. The neutral-current interaction of the fermions has been precisely measured in experiments [AMA 87; COS 88; FOG 88] including deep inelastic $(\bar{\nu})_\mu N$ scattering on isoscalar and proton targets, elastic $(\bar{\nu})_\mu p$ scattering, coherent π^0 production $\nu N \to \nu \pi^0 N$ elastic $(\bar{\nu})_e e$, and $(\bar{\nu})_\mu e$ scattering and $(\bar{\nu})_e d$ scattering.

In Section 4.2.5 we compare these results with those obtained in studies of scattering of polarized electrons on deuterium and of parity violation in atomic transitions.

The results obtained in neutrino experiments are described in Sections 4.2.4.1–7. They are used to determine the third component of weak isospin $I_L^3(i)$ and $I_R^3(i)$, $i = e, \mu, d, u$. In the Standard Model the neutral-current couplings i_L and i_R are related to the electric charge $Q(i)$ and the electroweak mixing parameter $\sin^2 \Theta_W$ by

$$i_L = I_L^3(i) - Q(i) \sin^2 \Theta_W$$
$$i_R = I_R^3(i) - Q(i) \sin^2 \Theta_W.$$

The results are summarized in Section 4.2.4.8.

4.2.4.1 Neutrino–electron scattering

There are four different reactions of elastic neutrino–electron scattering:

$$\begin{aligned} \nu_\mu e^- &\to \nu_\mu e^- \\ \bar{\nu}_\mu e^- &\to \bar{\nu}_\mu e^- \end{aligned} \tag{4.2.16}$$

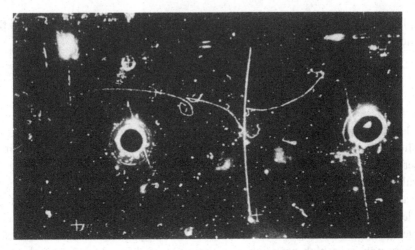

Fig. 4.2.3   First event of $\bar{\nu}_\mu e$ scattering observed in the Gargamelle bubble chamber [HAS 73a].

$$\nu_e e^- \rightarrow \nu_e e^-$$
$$\bar{\nu}_e e^- \rightarrow \bar{\nu}_e e^-. \tag{4.2.17}$$

The first experimental observation of a neutral weak current phenomenon was one event (reproduced in Fig. 4.2.3) in which a $\bar{\nu}_\mu$ scattered off an electron, uncovered in 1973 at CERN [HAS 73a] in the Gargamelle bubble chamber (see reprint in Section 1.11). Now, more than 25 years later, massive electronic detectors have achieved remarkable progress in this field. The CHARM collaboration [BER 84; DOR 89] has collected at CERN about 83 events of $\nu_\mu e$ and 112 events of $\bar{\nu}_\mu e$ scattering. A U.S.–Japan collaboration, working at the Brookhaven National Laboratory, has collected 160 $\nu_\mu e$ and 97$\bar{\nu}_\mu e$ events [ABE 87, 89]. The CHARM II collaboration has reported the largest samples of 2677 $\nu_\mu e$ and 2752 $\bar{\nu}_\mu e$ events [CHA 94].

The process ($\bar{\nu}_\mu e$) is described by the exchange of a neutral intermediate boson Z° (see Section 4.3) whereas for ($\bar{\nu}_e e$) elastic scattering additional charged-current amplitudes mediated by the exchange of a charged boson $W^\pm$ contribute (see Fig. 4.2.4) as well.

The main goal of studying these reactions has been to determine the coupling constants of the neutral $e\bar{e}$ current, the axial-vector coupling g_A^e, and the vector coupling constant g_V^e or their combinations, the chiral coupling constants $g_L^e = g_V^e + g_A^e$ and $g_R^e = g_V^e - g_A^e$. For comparison with the experiments, we are, however, attempting a phenomenological, model-independent analysis. We are assuming that leptons are pointlike and that the incoming and outgoing neutrinos are identical (Section 4.2.1); we are restricting the analysis to vector and axial-vector currents. 'tHooft [HOO 71] has derived the expressions for the differential cross section in the framework of the $SU(2) \times U(1)$ gauge theory.

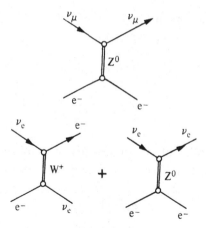

Fig. 4.2.4   Lowest-order Feynman diagrams for $\nu_\mu e^-$ and $\nu_e e^-$ elastic scattering.

Table 4.2.1.   *Standard Model predictions for the neutral-current coupling of leptons*

Lepton	g_A	g_V	g_L	g_R
ν	$\frac{1}{2}$	$\frac{1}{2}$	1	0
e	$-\frac{1}{2}$	$-\frac{1}{2}+2\sin^2\Theta_W$	$-1+2\sin^2\Theta_W$	$2\sin^2\Theta_W$

$$\frac{d\sigma^\nu}{dy} = \rho^2 \frac{\sigma_0}{2} E_\nu \left[g_L^2 + g_R^2(1-y)^2 + \frac{m_e}{E_\nu} g_L \cdot g_R \right]$$

$$\frac{d\sigma^{\bar\nu}}{dy} = \rho^2 \frac{\sigma_0}{2} E_\nu \left[g_L^2(1-y)^2 + g_R^2 - \frac{m_e}{E_\nu} g_L \cdot g_R \right] \qquad (4.2.18)$$

where $y = E_e/E_\nu$.

The nominal cross section slope is

$$\sigma_0 = \frac{G_F^2 m_e}{\pi} = 8.6 \times 10^{-42} \text{ cm}^2/\text{GeV} \qquad (4.2.19)$$

and $G_F = 1.105 \cdot 10^{-5} \text{ GeV}^{-2}$ is the Fermi coupling constant. The quantity $\rho = G_{NC}/G_F$ is equal to one in the minimal version of the Standard Model.

The predictions of the Glashow–Salam–Weinberg theory (Chapter 3) are summarized in Table 4.2.1. The fourfold ambiguity for exchanging g_V^e and g_A^e in Eq. (4.2.18) can be resolved, in principle, by the terms of order m_e/E_ν. Effects of this term may be detected in experiments using, for example, low-energy anti-neutrinos from fission reactors, but no result of the required accuracy has been reported. Two solutions can be eliminated by $\bar\nu_e e$ and $\nu_e e$ results, a third solution by $e^+e^- \to \mu^+\mu^-$ forward–backward asymmetries under the assumption that the neutral weak current is dominated by the exchange of single Z^0.

To extract g_V^e and g_A^e from the cross sections of neutrino electron scattering it is convenient to use the following expressions:

$$g_V^2 + g_A^2 = \frac{(3\pi/4G_F^2 m_e)}{E_\nu}(\sigma^\nu + \sigma^{\bar\nu})$$

$$g_V^e \times g_A^3 = \frac{(3\pi/4G_F^2 m_e)}{E_\nu}(\sigma^\nu - \sigma^{\bar\nu}). \qquad (4.2.20)$$

Assuming $\mu - e$ universality of the neutral-current coupling $(G_{\nu_\mu \bar\nu} = G_{\nu_e \bar\nu_e})$, we can describe $(\bar\nu_e e)$ scattering (4.2.17) by the same expressions if we replace g_V^e and g_A^e by

$$g_V^{e\prime} = 1 + g_V^e$$
$$g_A^{e\prime} = 1 + g_A^e \qquad (4.2.21)$$

to account for both charged- and neutral-current contributions.

Other measurable quantities can be derived from Eq. (4.2.18); they again apply also for $\nu_e e$ scattering with the substitution of Eq. (4.2.21). These are the total cross sections $(E_\nu \gg m_e)$

$$\sigma^\nu = \rho^2 \frac{\sigma_0}{2} E_\nu \left[g_L^2 + \frac{1}{3} g_R^2 \right]$$

$$\sigma^{\bar\nu} = \rho^2 \frac{\sigma_0}{2} E_\nu \left[\frac{1}{3} g_L^2 + g_R^2 \right] \qquad (4.2.22)$$

and the mean inelasticities, defined as

$$\langle y \rangle = \frac{\int_0^1 y(d\sigma/dy)\, dy}{\int_0^1 (d\sigma/dy)\, dy}. \qquad (4.2.23)$$

Expressing the mean inelasticity in term of the chiral coupling constants one finds

$$\langle y \rangle^\nu = \frac{6g_L^2 + g_R^2}{12g_L^2 + 4g_R^2}$$

$$\langle y \rangle^{\bar\nu} = \frac{6g_R^2 + g_L^2}{12g_R^2 + 4g_L^2}. \qquad (4.2.24)$$

In the framework of the Glashow–Salam–Weinberg Standard Model, all these quantities are determined by assuming a value for $\sin^2\Theta_W$ (Table 4.2.1) and $\rho = 1$.

The dependence on $\sin^2\Theta_W$ is shown in Fig. 4.2.5; predictions for $\sin^2\Theta_W = 0.23$ are summarized in Table 4.2.2.

The present experimental situation concerning $(\bar\nu_\mu e)$ scattering is summarized in Tables 4.2.3 and 4.2.4.

A discussion of the procedures used to extract these results can give a better appreciation of the discrepancies among the earlier experiments. The main selection

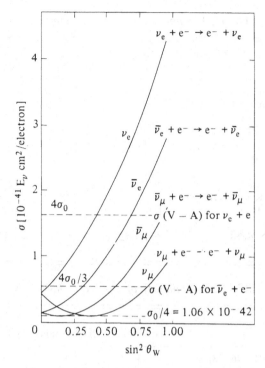

Fig. 4.2.5   Cross section of $\nu_\mu e$, $\bar\nu_\mu e$, $\nu_e e$ and $\bar\nu_e e$ scattering as a function of $\sin^2 \Theta_W$.

Table 4.2.2.   *Predictions of measurable quantities for $\nu e \rightarrow \nu e$ scattering for $\sin^2\Theta_W = 0.23$*

Reaction	g_L^e	g_R^e	σ/E_ν (cm^2/GeV)	$\langle y \rangle$
$\bar\nu_\mu e$	−0.54	0.46	$1.56 \cdot 10^{-42}$	0.451
$\bar\nu_\mu e$	−0.54	0.46	$1.32 \cdot 10^{-42}$	0.421
$\bar\nu_\mu e$	1.46	0.46	$9.47 \cdot 10^{-42}$	0.492
$\bar\nu_\mu e$	1.46	0.46	$3.96 \cdot 10^{-42}$	0.307

criterion is based on the kinematical consequences of the small mass of the electron target; this can be expressed by the invariant ($E_\nu \gg m_e$)

$$E_e \Theta^2 = 2m_e(1 - y) \leq 3m_e. \tag{4.2.25}$$

Electrons from $\nu_\mu e$ scattering are recoiling at small angles ($\Theta < 10\,\text{mrad}$ for $E_e = 10\,\text{GeV}$) whereas all background reactions resulting from semileptonic neutrino interactions on nucleons have much broader angular distributions because of the larger nucleon mass.

To select νe reactions it is thus necessary to identify reactions with an isolated electron shower and to measure its direction precisely. The backgrounds are composed of ($\bar\nu_e$) quasi-elastic scattering, giving an isolated high-energy electron in

Table 4.2.3. *Summary of $\nu_\mu e$ scattering results*

Experiment	$\nu_\mu e$ candidates	Background	σ/E_ν $(10^{-42}\ cm^2/GeV)$
GGM CERN-PS [BLI 76, 78]	1	0.3 ± 0.1	< 3 (90% C.L.)
Aachen-Padua counter exp. (CERN-PS) [FAI 78]	32	20.5 ± 2.0	1.1 ± 0.6
GGM CERN-SPS [ARM 79a]	9	0.5 ± 0.2	$2.4^{+1.2}_{-0.9}$
BNL-COL FNAL 15' [CNO 78]	11	0.5 ± 0.5	1.8 ± 0.8
VMNOP counter exp. (FNAL) [HEI 80]	46	12	1.4 ± 0.3
CHARM counter exp. CERN-SPS) [BER 84; DOR 89]	83 ± 16		2.2 ± 0.4 (stat) ± 0.4 (syst)
E734 counter exp. (BNL) [ABE 87, 89, 90]	160 ± 17		1.8 ± 0.20 (stat) ± 0.25 (syst)
CHARM II (CERN-SPS) [CHA 94]	2677 ± 82		1.53 ± 0.04 (stat) ± 0.12 (syst)

Table 4.2.4. *Summary of $\bar\nu_\mu e$ scattering results*

Experiment	$\bar\nu_\mu e$ candidates	Background	σ/E_ν $(10^{-42}\ cm^2/GeV)$
GGM CERN-PS [BLI 76, 78]	3	0.4 ± 0.1	$1.0^{+2.1}_{-0.9}$
Aachen-Padua counter exp. (CERN-PS) [FAI 78]	17	7.4 ± 1.0	2.2 ± 1.0
GGM CERN-SPS [BER 79a]	0	< 0.03	< 2.7 (90% C.L.)
FMMMS FNAL 15' [BER 79b]	0	0.2 ± 0.2	< 2.1 (90% C.L.)
BEBC-TST (CERN-SPS) [ARM 79b]	1	0.5 ± 0.2	< 3.4 (90% C.L.)
CHARM counter exp. (CERN-SPS) [BER 84; DOR 89]	112 ± 21		1.6 ± 0.3 (stat) ± 0.4 (syst)
E734 counter exp. (BNL) [ABE 87, 89, 90]	97 ± 14		1.17 ± 0.16 (stat) ± 0.13 (syst)
CHARM II (CERN-SPS) [CHA 94]	2752 ± 88		1.39 ± 0.04 (stat) ± 0.10 (syst)

the final state with $y \sim 1$, and $(\bar\nu_\mu)$ neutral-current interactions producing high energy neutral pions, predominantly by coherent production on nuclei (Section 4.2.4.5). To determine the angular distribution and the relative amount of these background components, it is important to discriminate electrons from photons in a sample of the data. This will now be described in more detail for two recent experiments.

The CHARM II experiment The CHARM collaboration [BER 84; DOR 89; CHA 94] have measured the cross sections for the reactions $\nu_\mu e^- \to \nu_\mu e^-$ and $\bar\nu_\mu e^- \to \bar\nu_\mu e^-$ using an electronic fine-grain calorimeter. The CHARM detector has been described in detail elsewhere [DID 80]. The CHARM II

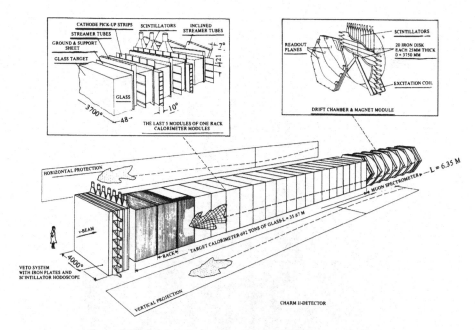

Fig. 4.2.6   Schematic layout of the CHARM II experiment.

detector [CHA 89b; CHA 93] is composed of a calorimeter with 420 identical modules of 500-t fiducial volume and of a magnetic muon spectrometer. Glass was chosen as a material for the calorimeter because of its low atomic number ($\bar{Z} \sim 11$). The accuracy of electron shower direction measurements is limited by the Z number of the material in which the shower propagates,

$$\sigma(\Theta) \sim \frac{\text{shower width}}{\text{shower length}} \sim \frac{R_{\text{Moliere}}}{X_0} \sim Z$$

where X_0 is the radiation length and R_{Moliere} the familiar Moliere radius of the shower. The accuracy depends further on the sampling frequency (plate thickness), the grain size of the detector, and the detection method.

The structure of this detector is shown schematically in Fig. 4.2.6. Figure 4.2.7 is a photograph of the CHARM II detector. It consists of 420 modules of $3.7 \times 3.7\,\text{m}^2$ surface area, each composed of a 4.8-cm thick glass plate ($\frac{1}{2}X_0$) and of a streamer tube plane with 1-cm wire spacing, read out by the wires and by crossed (90°) cathode strips 2 cm wide. Using pulse-height measurements of the charge on each strip, the centroid position of a track can be reconstructed with $\pm 3\,\text{mm}$ accuracy, whereas the wires are read out digitally to obtain unambiguous information about the track multiplicity near the vertex. A shower angular resolution of $\sigma(\Theta) \simeq 16\,\text{mrad}/\sqrt{E/\text{GeV}}$ has been achieved for electrons of a test beam, not far from the limiting resolution due to shower fluctuations.

Candidates for $(\bar{\nu}_\mu e)$ reactions have been searched for among events with an electron shower at a small angle ($\Theta < 100\,\text{mrad}$) between the shower axis and the

Fig. 4.2.7   Photograph of the CHARM II detector.

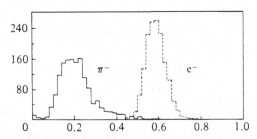

Fig. 4.2.8   Energy density distribution of 20 GeV electron and pion showers.

direction of the neutrino beam. Events with an isolated electron shower were selected by their characteristically large energy density, whereas the bulk of the data has broad hadronic showers of low-energy density. In the plane following the vertex a single hit was required. The results of measurements of the energy density of showers induced by electrons and pions is shown in Fig. 4.2.8. Selecting events as indicated in the figure reduces the background due to semileptonic neutrino interactions by a factor of ~ 100. Figure 4.2.9 shows two typical events: one (a) due to charged-current neutrino interaction, the other (b) a candidate for $\nu_\mu e \to \nu_\mu e$ scattering. Only events with a shower energy E_e between 3 GeV and 24 GeV have been retained in the final sample. The upper limit was applied to eliminate high-energy events dominantly due to quasi-elastic charged-current interactions induced

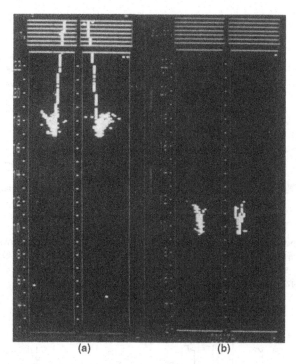

Fig. 4.2.9   Typical neutrino events in the CHARM II detector: (a) $\nu_\mu N \to \mu^- X$ charged-current interaction, (b) a candidate for $\nu_\mu e \to \nu_\mu e$ scattering.

by the small (~ 1 percent) $\nu_e(\bar{\nu}_e)$ component of the beam; the lower limit was applied because of increasing trigger inefficiency at lower electron energies. The $E\Theta^2$ distribution of the selected neutrino events is shown in Fig. 4.2.10. The narrow peak in the forward direction ($E\Theta^2 < 3\,\text{MeV}$) is due to νe scattering.

The background consists of (a) single π^0 production by neutral current interaction and of (b) quasi-elastic scattering of electron neutrinos on nucleons. A shower induced by a photon yields an even number of charged particles, while electron induced showers yield odd numbers. The characteristic difference between electron and photon induced showers is best observed in an early stage of shower development. Scintillation counters covered every fifth module of the target calorimeter. For events starting in the target plates in front of the scintillation counters [CHA 94] energy depositions such as shown in Fig. 4.2.11 were observed. Selecting events associated with a single electron by requiring an energy loss of less than 8 MeV, the distributions shown in Fig. 4.2.12 were obtained. Combining the background event numbers observed in the ν-beam and in the $\bar{\nu}$-beam in the reference region $5 < E\Theta^2 < 72\,\text{MeV}$ for all candidates (Fig. 4.2.10) and for those with $E_F < 8\,\text{MeV}$ (Fig. 4.2.12) and using the calculated efficiencies of the selection criteria, the ratio of the two background processes (a) and (b) were determined. Comparing them to the result of the simultaneous fit of the modelled distributions of events as a function of $E_e\Theta_e^2$ (see Fig. 4.2.10) and E_e the background ratios can be

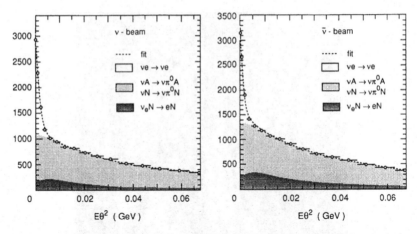

Fig. 4.2.10 Experimental data and the result of the best fit; data are shown as circles and the fit results are displayed as a dashed line. Only the projections in $E_e\Theta_e^2$ of the 2-dim. distributions are shown. The different background components are added on top of each other. The bin size varies with the experimental resolution [CHA 94].

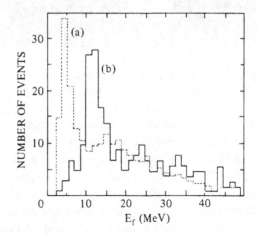

Fig. 4.2.11 Energy deposition in the scintillator plane following the shower vertex: (a) electrons and (b) neutral pions [DOR 89].

checked. The result of the fit is shown in Fig. 4.2.10 and in Fig. 4.2.12. The background ratios determined by the fit and from the analysis described above are in good agreement. 2677 ± 82 events in the ν-beam and 2752 ± 88 events in the $\bar{\nu}$-beam are attributed to neutrino–electron scattering. Turning now to the determination of the couplings g_V and g_A, they used the samples with and without energy deposition requirement, to reduce the statistical error. The absolute normalisation of neutrino fluxes and the presence of $\nu_e e$ and $\bar{\nu}_e e$ events detected in the same experiment allowed them to reduce the well known four-fold ambiguity of g_A and g_V to two solutions.

The ratio of ν_e/ν_μ and of $\bar{\nu}_e/\bar{\nu}_\mu$ fluxes were calculated using Monte Carlo methods and measured pion and kaon production. The agreement of the ratio of

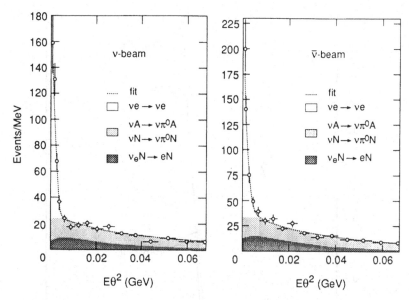

Fig. 4.2.12 Experimental data and the result of the best fit for the sample of events with energy loss information weighted with the electron probability. The signal to background ratio is improved by a factor 3.5 with respect to the total sample shown in Fig. 4.2.10. The bin size varies with the experimental resolution [CHA 94].

quasi-elastic ν_e scattering and single π^0 production determined from the energy loss in scintillation counters and from the fit of the modelled $E\Theta^2$ and E distributions is lending support to this flux evaluation.

The 67% confidence domains of g_A and g_V thus determined from the absolute differential cross sections of $\nu_\mu e$, $\bar{\nu}_\mu e$, $\nu_e e$ and $\bar{\nu}_e e$ scattering are shown in Fig. 4.2.13. They intersect each other in two regions of g_V and g_A. Also shown in Fig. 4.2.13 are two cones of values determined by the experiments on the forward–backward asymmetry in $e^+e^- \rightarrow e^+e^-$ annihilations at LEP [LEP 94].

One of the LEP cones is intersecting one of the two solutions of neutrino–electron scattering from the CHARM II experiment; this single solution is in agreement with $g_A^e = -1/2$:

$$g_A^{\nu e} = -0.503 \pm 0.006 \text{ (stat)} \pm 0.016 \text{ (syst)} \qquad (4.2.26)$$

$$g_V^{\nu e} = -0.035 \pm 0.012 \text{ (stat)} \pm 0.052 \text{ (syst)}. \qquad (4.2.27)$$

This had been demonstrated already in 1984 by combining results from the first CHARM experiment [BER 84], from $\bar{\nu}_e e$ scattering at a nuclear reactor [FRE 76] and from experiments on $e^+e^- \rightarrow \mu^+\mu^-$ annihilations at PETRA and PEP [SWU 87]. As a convention the sign and the value of $g^{\nu_\mu} = +1/2$ is used.

A value of $g_A^e = -1/2$ is predicted if the electron is assigned as the lower member of a doublet under $SU(2) \times U(1)$ transformation, in agreement with the experiment. The original guess is thus confirmed and established directly. As the Standard

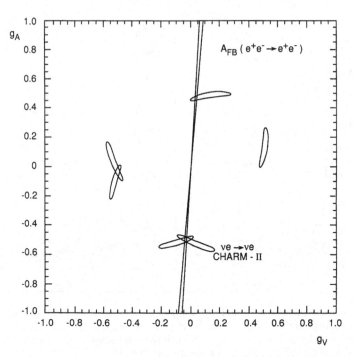

Fig. 4.2.13   90% confidence level contours in the $g_V - g_A$ plane, as obtained from the fit to the data from the ν-beam, the $\bar{\nu}$-beam and to both beams [CHA 94]. Only statistical errors are considered. Results from experiments on the forward–backward asymmetry for $e^+e^- \rightarrow e^+e^-$ at LEP are shown as well. Together they select a single solution in agreement with $g_A^e = -\frac{1}{2}$.

Model does not distinguish between families, we shall take the same assignment for muons and tauons as well (Section 4.2.6).

It has been shown by [NOV 93] that these coupling constants (4.2.26, 27) can be expressed by the following products

$$g_A^{\nu_\mu e} = 2g^{\nu_\mu} \cdot g_A^e$$
$$g_V^{\nu_\mu e} = 2g^{\nu_\mu} \cdot g_V^e.$$

Hence, the ν_μ coupling to the Z can be determined using the LEP results for g_A^e and g_V^e from $Z \rightarrow e^+e^-$ [LEP 94], with the result [CHA 94a]

$$2g^{\nu_\mu} = 1.004 \pm 0.033. \tag{4.2.28}$$

The agreement with the LEP result on the 3-flavor averaged value of neutrino–Z coupling derived from the 'invisible' decay width of the $Z \rightarrow \nu_i\bar{\nu}_i$, $i = e, \mu, \tau$,

$$2g^{\nu} = 0.9999 \pm 0.0043,$$

gives the first evidence for flavor universal neutrino–Z coupling (see also Table 4.2.20).

Table 4.2.5. *Summary of experiments reporting* $R = \sigma(\nu_\mu e)/\sigma(\bar{\nu}_\mu e)$

Experiment	$N(\nu_\mu e)$	$N(\bar{\nu}_\mu e)$	R^ν	$\sin^2\Theta_W$
E734, U.S.-Japan [ABE 87, 89, 90]	160	97	$1.56^{+0.31+0.18}_{-0.25-0.17}$	$0.198^{+0.020+0.014}_{-0.021-0.013}$
CHARM, CERN [DOR 89, BER 84]	80	112	1.20 ± 0.35	$0.210 \pm 0.035 \pm 0.011$
CHARM II, CERN [CHA 94][a]	2677	2752		$0.2324 \pm 0.0058 \pm 0.0059$
Average				0.2268 ± 0.0077

Note: First error is statistical, second error is systematic. Radiative corrections [BAR 92b] for $m_t = m_H = 100\,\text{GeV}$.
[a] from ratio of $d\sigma/dy$ for $\nu_\mu e$ and $\bar{\nu}_\mu e$.

Within the framework of the $SU(2) \times U(1)$ electroweak model the coupling constants can be expressed in terms of the electroweak mixing angle $\sin^2\Theta_W$ (Table 4.2.1) and ρ. It has been suggested [WIN 80] that from the ratio of $\nu_\mu e$ and $\bar{\nu}_\mu e$ differential cross sections the most direct and reliable determination of $\sin^2\Theta_W$ can be obtained without any hypothesis on the value of ρ. Constraining only the relative neutrino to antineutrino flux in the fit of the CHARM II experiment, one obtained

$$\sin^2\Theta_W = 0.2324 \pm 0.0058 \text{ (stat)} \pm 0.0059 \text{ (syst)}, \qquad (4.2.29)$$

in good agreement with the results obtained in e^+e^- annihilations at LEP and at SLC [LEP 94]. Results based on the ratio method are summarized in Table 4.2.5.

The helicity structure of the neutral current weak interaction can be investigated in a model independent way by studying the angular distribution of elastic neutrino–electron scattering. Because of the small mass of the electron, the scattering angles in the laboratory system are too small for a direct measurement. The angular distribution is, however, reflected in the energy distribution of the recoiling electrons which can be measured in a fine-grained target calorimeter. Therefore the y-distribution, where y is the fraction of the incident neutrino energy carried away by the recoiling electron, is equivalent to the angular distribution.

As the neutrino energies are not known on an event-by-event basis, only the convolution of the true y-distributions with the energy spectrum of incident neutrinos is accessible to experimental observation. The extraction of a statistically significant y-distribution by an unfolding procedure requires a large number of observed neutrino–electron scattering events as in the CHARM II experiment [CHA 93].

The unfolded differential distributions are shown in Fig 4.2.14. The y-distributions show deviations from isotropy in the $\nu_\mu e$ case and from a pure $(1-y)^2$ dependence in the $\bar{\nu}_\mu e$ case indicating contributions from scattering on right-handed electrons. A fit of the theoretical expressions Eq. (4.2.18) to the data gives a value

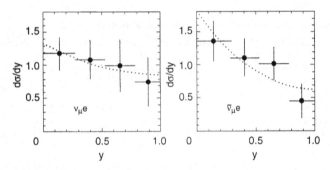

Fig. 4.2.14 Unfolded differential $d\sigma/dy$ cross sections for $\nu_\mu e$ scattering (left), and $\bar{\nu}_\mu e$ scattering (right) in arbitrary units [CHA 93]. The line overlaid corresponds to the prediction of the Standard Model for a value of the electroweak mixing angle of $\sin^2\Theta_W = 0.212$.

of the ratio of the squares of right-handed and left-handed electron–Z coupling,

$$g_R^2/g_L^2 = 0.60 \pm 0 \text{ (stat) } \pm 0.09 \text{ (syst)}.$$

It is confirming in a direct and model independent way non-maximal parity violation in neutral current interactions, in contrast to maximal parity violation ($g_R^{cc} = 0$) in charged current interactions (see also Eq. (4.2.38) for neutral current coupling to right-handed and left-handed quarks).

The U.S.–Japan experiment This experiment was performed at the 28 GeV proton synchrotron of Brookhaven National Laboratory. It followed a strategy very similar to the CHARM experiments [ABE 87, 89a; AHR 87b].

The detector used the fine-grain calorimeter technique pioneered by the CHARM collaboration [DID 80]. A schematic view is shown in Fig 4.2.15. Because of the lower average neutrino energy ($\bar{E}_\nu = 1.5$ GeV) the angular resolution ($\sigma(\Theta) \approx 16 \,\text{mrad}/\sqrt{E/\text{GeV}}$) for measuring the electron shower direction also allowed an analysis of the observed angular distribution of the recoil electrons in terms of three contributions:

1 a constant term in $d\sigma/dy$ [Eq. (4.2.18)]
2 a $(1-y)^2$ term
3 the background

The inelasticity $y = E_e/E_\nu$ can be expressed as

$$\Theta^2 = \frac{2m_e}{E_e}(1-y).$$

This method can be applied separately to the neutrino and antineutrino data and has the advantage over the ratio method that no neutrino flux monitoring is required. The statistical error is, however, larger; $\Delta\sin^2\Theta \sim (\frac{1}{2})N^{-\frac{1}{2}}$ is obtained from the analysis of $d\sigma/dy$ for N events and $\Delta\sin^2\Theta \sim (\frac{1}{4})N^{-\frac{1}{2}}$ from the ratio. The angular

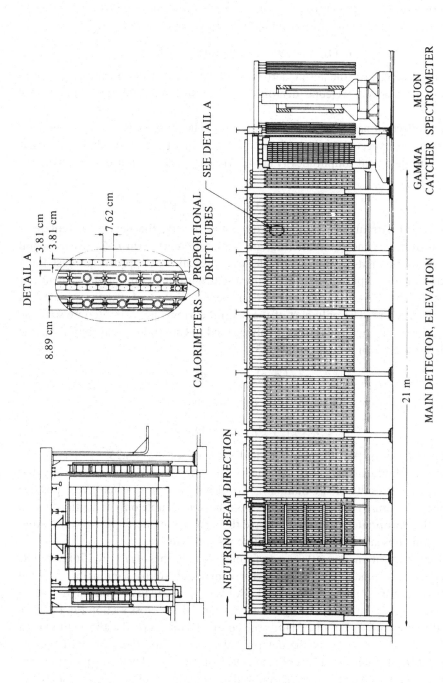

DETAIL A

3.81 cm

3.81 cm

7.62 cm

8.89 cm

CALORIMETERS

PROPORTIONAL
DRIFT TUBES

SEE DETAIL A

NEUTRINO BEAM DIRECTION

21 m

MAIN DETECTOR, ELEVATION

GAMMA   MUON
CATCHER SPECTROMETER

Fig. 4.2.15   Schematic view of the detector used by the U.S.–Japan collaboration [AHR 87b] to measure $\sigma(\nu_\mu e)$ and $\sigma(\bar{\nu}_\mu e)$.

resolution function has to be well-known and unfolded as it does not cancel in the ratio of term (2)/term (1) from which $\sin^2\Theta_W$ is determined. The result quoted is derived by combining the $d\sigma/dy$ analysis and the ratio method,

$$\sin^2\Theta_W = 0.195 \pm 0.18 \text{ (stat)} \pm 0.13 \text{ (syst)}. \qquad (4.2.30)$$

It is consistent with the result derived from the ratio alone (Table 4.2.5). The consistency is a very welcome demonstration of the validity of the methods used.

Electron-neutrino–electron scattering The cross section for $\nu_e e^-$ and $\bar{\nu}_e e^-$ scattering is expected to arise from diagrams involving both charged-current and neutral-current amplitudes (Fig. 4.2.4). The cross section is given by three terms:

$$\sigma(\nu_e e) = \sigma(CC) + \sigma(NC) + I$$

the charged-current term, known from muon decay, the neutral-current term known from $\nu_\mu e$ and $\bar{\nu}_\mu e$ scattering, and an interference term I. An experimental demonstration that an interference term is present would prove, at least for a fraction of the amplitude, that the neutral-current interaction preserves neutrino helicity, as the charged-current interaction does, implying V, A structure rather than S, P, T structure. The only part of the neutral current that can interfere with the charged-current amplitude is that involving the $V-A$ neutral-current coupling to the left-handed electron (see Section 4.2.3);

$$I \propto g_L^e = g_V^e + g_A^e$$

is therefore expected. In the $SU(2) \times U(1)$ Standard Model

$$I \propto g_V^e + g_A^e = 2\sin^2\Theta_W - 1 \qquad (4.2.31)$$

implying $I < 0$ for the currently accepted value of $\sin^2\Theta_W \sim 0.23$.

One can build models with two neutral intermediate bosons [KAY 79] that reproduce all results of the Standard Model with one boson but give a positive interference term. The second boson would have to couple only to leptons with a strength appropriately chosen to reverse the sign of the interference term. Nature has again chosen the simplest alternative, as we shall see.

An experiment on $\bar{\nu}_e e$ scattering performed by F. Reines *et al.* [FRE 76] from UC Irvine at the Savannah River fission reactor plant obtained cross sections for two recoil electron energy regions, at low (1.5 MeV $< E_e <$ 3.0 MeV) and high energy (3.0 MeV $< E_e <$ 4.5 MeV). Using the expression for the cross section Eq. (4.2.18), and the substitution Eq. (4.2.21) for the coupling constants, the allowed range of $(g_V + g_A)$ and $(g_V - g_A)$ can be determined as shown by the shadowed regions (Fig. 4.2.16) for the two ranges of recoil electron energies. The corresponding confidence region shown in Fig. 4.2.13 allowed us to eliminate two of the four solutions determined from $(\bar{\nu}_\mu e)$ scattering [KRE 82]. The presence of the predicted interference term (Section 4.2.3) cannot be inferred from these measurements owing to the large

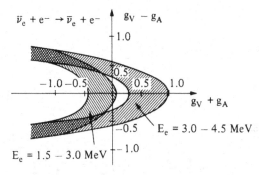

Fig. 4.2.16   Allowed range of $g_L^e = g_A^e + g_V^e$ and $g_R^e = g_V^e - g_A^e$ as determined from $\sigma(\bar{\nu}_e e)$ [FRE 76]. The shadowed regions give the 68 percent confidence intervals for two ranges of recoil electron energies.

Table 4.2.6.   *Cross sections for the $\bar{\nu}_e e^- \rightarrow \bar{\nu}_e e^-$ reactor experiment*

	Electron energy	
Case	$1.5\,\text{MeV} < E_e < 3\,\text{MeV}$	$3\,\text{MeV} < E_e < 4.5\,\text{MeV}$
Destructive interference $(\sin^2\Theta_W = \frac{1}{4})$	$0.85\sigma_{V-A}^{CC}$	$1.1\sigma_{V-A}^{CC}$
Constructive interference	2.2	2.7
No interference $(\nu^{\text{in}} \neq \nu^{\text{out}})$	1.5	1.9
Experiment	0.87 ± 0.25	1.70 ± 0.44

Note: In units of σ_{V-A} the charged-current cross section, compared with theoretical expectations.
Source: [FRE 76].

errors (Table 4.2.6). Using electron neutrinos from the decay chain

$$\pi^+ \rightarrow \mu^+ \, e^+ \nu_e \bar{\nu}_\mu \nu_\mu$$

at LAMPF (Los Alamos Proton Facility), yielding energies ranging from 0 to 53 MeV, a collaboration from UC Irvine–Los Alamos–Maryland [ALL 93] reported observation of 236 ± 35 events of $\nu_e e \rightarrow \nu_e e$ scattering, assuming a charged-current contribution, as expected from universality, destructive interference and $\sin^2\Theta_W = 0.23$ (see Fig. 4.2.17). In the absence of interference, 457 events should be observed. Hence, the relative interference coefficient is

$$\alpha = -0.99 \pm 0.17 \text{ (stat)} \pm 0.12 \text{ (syst)}$$

to be compared with a value of $\alpha = -1$, which is expected if the $\nu\bar{\nu}$ neutral current contains a $V{-}A$ part, as predicted by the Standard Model. This result therefore demonstrates directly that the $\nu\bar{\nu}$ neutral-current interaction is partially parity

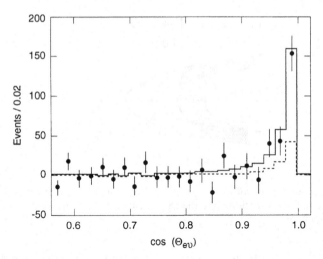

Fig. 4.2.17 Angular distribution of the measured elastic scattering signal in the ILM experiment [ALL 93]. The solid line is the result of the best fit, 295 ± 35 events. The dashed line is the background contribution from 59.2 $\nu_\mu e$ and $\bar{\nu}_\mu e$, scattering events.

violating and that $g^e_A < 0$. The cross section reported is

$$\sigma(\nu_e e \to \nu_e e) = [10.0 \pm 1.5 \, (\text{stat}) \pm 0.9 \, (\text{syst})] \times 10^{-42} \, \text{cm}^2/\text{GeV}. \qquad (4.2.32)$$

Neutrino trilepton production Coherent muon-neutrino scattering on nuclei with the production of a muon pair is another reaction that can reveal the Lorentz structure of the neutral weak current [FUJ 78]. As in $\nu_e e$ scattering, the reaction

$$(\bar{\nu})_\mu N \to (\bar{\nu})_\mu \mu^+ \mu^- N \qquad (4.2.33)$$

can receive contributions from both neutral-current and charged-current interactions and makes it possible to investigate the reactions (Fig. 4.2.18a)

$$(\bar{\nu})_\mu \mu \to (\bar{\nu})_\mu \mu \qquad (4.2.34)$$

which are related to $(\bar{\nu})_e e^- \to (\bar{\nu})_e e^-$ scattering by the assumption of $\mu - e$ universality in neutral weak current interactions. Experimental study of this reaction is therefore an important independent way of investigating both the Lorentz structure and the $\mu - e$ universality properties of the neutral current.

The CHARM II collaboration has reported the first experimental evidence for neutrino trident production [CHA 90]. They selected from samples of $4 \cdot 10^7$ neutrino-induced and $2 \cdot 10^7$ antineutrino-induced charged-current event candidates with a muon pair of opposite charge and momentum $p_\mu > 4 \, \text{GeV}/c$ and at most one or two additional tracks near the vertex. An enhancement is observed (Fig. 4.2.18b) if only two muon tracks are observed (20 hits in 10 planes) and a rising

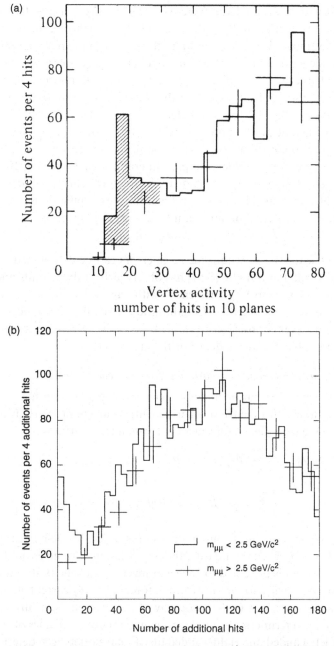

Fig. 4.2.18 (a) Neutrino trilepton production $\nu_\mu N \to \nu_\mu \mu^+ \mu^- N$ mediated by Z° and W^+ exchange. (b) Vertex activity determined by the number of additional hits in ten planes following the vertex for dimuon events of opposite charge [CHA 90].

background with additional vertex activity. This background is due to inclusive charm meson production in charged-current reactions in their semileptonic decay to a muon. To estimate their contribution to the background under the enhancement the hit dependence of events with a muon opening angle larger than $10°$ which cannot have contributions from trilepton production was determined. Normalizing this distribution to the total observed distribution with more than 30 hits the background (crosses in Fig. 4.2.18b) was determined and subtracted; 59 ± 15 events are remaining. Coherent production of single π^+ or π^- in ν_μ or $\bar{\nu}_\mu$ charged-current processes can also contribute to this sample through their decay into a muon. A sample of these events in which the pion interacts after traversing 10 planes has been identified. 10 ± 4 events contribute to the trilepton candidate sample, which is reduced to 49 ± 15 events. From this observation, the coupling constant of the diagonal four-fermion interaction and an interference coefficient

$$\alpha = -0.63 \pm 0.30 \qquad (4.2.35)$$

in agreement with $\alpha = -1$ within the large errors can be determined.

The CCFR collaboration [MIS 91] has observed 16 trident events and reported evidence for destructive interference with an interference coefficient of $\alpha = -1.25 \pm 0.42$. The CHORUS Collaboration [CHO 00b] has reported observation of 55 ± 15 events observed from a Pb-target in agreement with destructive interference implying z^2 dependence of the cross-section.

4.2.4.2 Deep inelastic neutrino scattering from isoscalar nuclear targets

Following the model-independent approach formulated in Section 4.2.1, we use an effective Lagrangian for neutrino–hadron neutral-current processes in a form that is valid in any gauge theory for massless left-handed neutrinos

$$L(\nu q) = -\frac{G_F}{\sqrt{2}} \bar{\nu}\gamma_\mu(1+\gamma_5)\nu$$

$$\left\{ \sum_i [g_L(i)\bar{q}_i\gamma_\mu(1+\gamma_5)q_i + g_R(i)\bar{q}_i\gamma_\mu(1-\gamma_5)q_i] \right\} \qquad (4.2.36)$$

$g_L(i)$ and $g_R(i)$ are the chiral coupling constants of left-handed and right-handed quarks of flavor i. Their Standard Model expressions are given in Table 4.2.1. The neutral current is predicted to have a more complex helicity structure than the charged current. Owing to the electroweak force, the $(\nu\bar{\nu})$ current should interact with left-handed fermions in the same way as the charged $(\nu_e e)$ current, and also with right-handed fermions, as the electromagnetic current. Values of the coupling strength of left-handed and right-handed quarks have often been determined from cross-sectional ratios [AMA 87]. For example,

$$\rho^2 g_R^2 = \frac{\bar{R} - R}{r^{-1} - r} + O(s, \bar{s})$$

where R and $\bar{R}$ are ratios of NC and CC cross sections for deep inelastic scattering of neutrinos and antineutrinos, respectively; $r = \sigma_\infty(\bar{\nu}N)/\sigma_{cc}(\nu N)$ and $O(s, \bar{s})$ is a correction for effects of strange sea quarks. However, it is evident that one cannot demonstrate the handedness of fermions participating in the NC reaction by measurements of scalar quantities. "You need a nut if you are looking for a screw," said Val Telegdi. For a direct demonstration of the handedness of the quarks participating in the neutral-current interaction we follow the approach of Ch. Llewellyn-Smith [LLE 83] relating the differential cross sections of neutrino (antineutrino) scattering on isoscalar targets (defined as nuclei with equal numbers of neutrons and protons) by the neutral-current and charged-current weak interactions

$$\frac{d\sigma_{NC}^{\nu}}{dy} = g_L^2 \frac{d\sigma_{CC}^{\nu}}{dy} + g_R^2 \frac{d\sigma_{CC}^{\bar{\nu}}}{dy}$$

$$\frac{d\sigma_{NC}^{\bar{\nu}}}{dy} = g_R^2 \frac{d\sigma_{CC}^{\nu}}{dy} + g_L^2 \frac{d\sigma_{CC}^{\bar{\nu}}}{dy}. \tag{4.2.37}$$

The validity of this fundamental relation depends on the assumption of weak isospin invariance and applies to u, $\bar{u}$ and d, $\bar{d}$ quarks only. Terms due to scattering on s and c quarks, quark mixing, and flavor-changing charged-current transitions are ignored and have to be corrected for.

For incoming neutrinos and antineutrinos of known helicity (ν_L, $\bar{\nu}_R$), the angular distributions $d\sigma/dy$ of the elementary $\nu(\bar{\nu})$ quark neutral-current scattering process has two components: an isotropic one for $\nu_L q_L$ and $\bar{\nu}_R q_R$ scattering and an anisotropic one proportional to $(1-y)^2$ for $\nu_L q_R$ and $\nu_R q_L$ scattering. From measurements of the four angular distributions, the coupling to left-handed (g_L^2) quarks and to right-handed (g_R^2) quarks has been directly determined by the CHARM collaboration [ALL 89a].

In neutral-current-induced neutrino reactions, the outgoing neutrino (ν') cannot be observed. Hence, the inelasticity y

$$y = \frac{E_h}{E_\nu} = \frac{E_h}{E_h + E_{\nu'}}$$

cannot be determined on an event-by-event basis. For this reason, previous attempts by the CHARM collaboration [JON 81] and by CDHS [HOL 77] to determine $d\sigma/dy$ of NC neutrino reactions have not achieved the required precision. One must resort to a sign and momentum selected parent beam, a so-called narrow-band neutrino beam. Owing to the dominant two-body decays $\pi \rightarrow \mu\nu_\mu$ and $K \rightarrow \mu\nu_\mu$ that produce the neutrinos, a measurement of the neutrino direction Θ_ν implies a determination of the neutrino energy, apart from the π/K ambiguity. The neutrino direction can be inferred from the distance R of the neutrino interaction point from the beam axis. The event density $d^2N/dE_h dR$ is related to $d\sigma/dy$ by the integral equation

$$d^2N/dE_h\, dR = \int F(E_\nu, R) \frac{d\sigma}{dy}\, dE_\nu.$$

It can be solved using an unfolding method developed by V. Blobel [BLO 84]. The data were obtained in an exposure to a 160 GeV narrow-band beam. Details

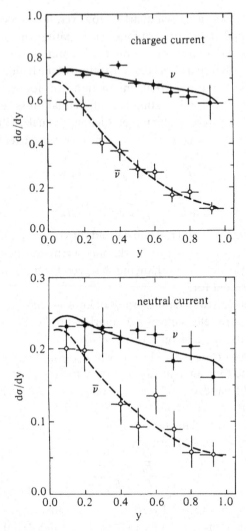

Fig. 4.2.19   Measurements of $d\sigma/dy$ of ν_μ and $\bar{\nu}_\mu$ induced CC and NC reactions by the CHARM collaboration [ALL 89a]. The curves show the result of a Monte Carlo simulation.

of the event selection will be described later [ALL 87]. The event samples consisted of 100 000 CC and 40 000 NC events in the neutrino beam and 6000 CC and 2000 NC events in the antineutrino beam. The CC and NC current y distributions were both determined in the way described above; they are shown in Fig. 4.2.19 together with the results of a Monte Carlo simulation. A maximum likelihood fit to the Llewellyn-Smith relations gives ($g_L^2 = u_L^2 + d_L^2$ etc.):

$$g_L^2 = 0.287 \pm 0.008$$

$$g_R^2 = 0.042 \pm 0.010$$

(4.2.38)

Table 4.2.7. *Results of simultaneous fits to CC and NC y distributions using an unfolding method*

Parameter	Fit A	Fit B	Fit C
α	0.129 ± 0.013	0.129 ± 0.011	0.129 ± 0.011
β	0.134 ± 0.034	0.118 ± 0.037	0.124 ± 0.030
g_L^2	0.315 ± 0.004		
g_R^2	0.056 ± 0.010		
g_S^2		0.204 ± 0.054	
$\sin^2\Theta_W$		0.241 ± 0.016	0.236 ± 0.008

Source: [ABT 86].

in good agreement with the expected values of $\sin^2\Theta_W = 0.23$ (Table 4.2.11). This is the first direct demonstration of a coupling of the $(\nu\bar{\nu})$ current to right-handed quarks ($g_R^2 > 0$ with 4σ). It confirms the prediction of the Standard Model.

Within the context of the quark model, and the assumption (see Chapter 3) that the weak currents contain only V and A terms, the differential cross sections for deep inelastic scattering on isoscalar nuclei can be written as [JON 81]

$$\frac{d\sigma}{dy}((\bar{\nu}) \to (\bar{\nu})) = A[g_{L(R)}^2(Q + \bar{Q}(1-y)) + g_{R(L)}^2(\bar{Q} + Q(1-y)^2)$$

$$+ g_S^2 Q_S(1 + (1-y)^2) - F_L]$$

(4.2.39)

where A is a normalization constant. The quark structure of the nucleon is described by Q, $\bar{Q}$, and Q_S; for example $Q = \int x(u(x) + d(x))\,dx$ is the momentum-weighted valence quark content of the nucleon; $\alpha = \bar{Q}/(Q + \bar{Q})$ is the fractional momentum-weighted sea quark content; and $\beta = \bar{Q}_S/(Q + \bar{Q})$ the fractional strange quark content. The constants g_L^2 and g_R^2 are the left-handed and right-handed couplings of the weak neutral current to "up" and "down" quarks, while g_S^2 is the sum of the right- and left-handed couplings to strange quarks. By simultaneously fitting these expressions to the four unfolded differential cross sections, the couplings of the weak neutral current have been determined; the resulting values are given in Table 4.2.7. The values of α, β, and g_L^2 are determined by neglecting the term in g_S (fit A). The form of the NC y-distribution in (4.2.39) precludes the possibility of varying all five parameters α, β, g_L^2, g_R^2, and g_S^2 simultaneously in a fit. In particular, g_S^2 is strongly correlated with the value of β. In fit B, the parameters α, β, $\sin^2\Theta_W$ and g_S^2 were determined, and $g_S/g_d = 1.06 \pm .014$ was found. Thus, one finds that the total coupling strength of the weak neutral current to the strange quark is consistent with being equal to that of the nonstrange down quark, an assumption implied by the GIM mechanism [GLA 70]. Motivated by this consistent result, one can go on to assume that the couplings of the u, d, and s quarks can all be described in terms of the Glashow–Salam–Weinberg Standard Model. The neutral-current sector is then described by a single parameter $\sin^2\Theta_W$. This fit gives $\sin^2\Theta_W = 0.2268 \pm 0.0031$ (see Eq (4.2.44)).

In the preceding sections, we first demonstrated the existence of a right-handed part in the weak neutral current, independent of the validity of the Standard Model. After showing that the neutral-current coupling of the strange quark is consistent with being equal to that of the "down" quark, we then described the neutral-current sector completely in terms of the Standard Model. This, of course, already assumes the presence of only V and A currents. However, if there were scalar (S) or pseudoscalar (P) parts of the neutral-current interaction, they would contribute equally to neutrino and antineutrino interactions, and would manifest themselves in $d\sigma/dy$, Eq. (4.2.39), as a term proportional to y^2 with a coefficient B. Then the ratio B/A gives the relative proportions of the S or P and V, A parts. A fit to the two NC distributions gives [JON 81] $B/A = -0.05 \pm 0.05$, implying

$$g_{SP}^2/g_{VA}^2 < 0.03 \ (95\% \ \text{C.L.}) \qquad (4.2.40)$$

where g_{SP}^2 and g_{VA}^2 are the S, P, and V, A coupling strengths, respectively. The analysis in this form disregards the possibility of a conspiracy of S, P, T terms mimicking a V, A structure in the neutral currents. This ambiguity has, however, already been resolved by the observation of coherent π^0 production and $W-Z$ interference in $\nu_e e$ scattering (Section 4.2.3).

The final step of the analysis will lead us to directly determine the value of $\sin^2\Theta_W$ from R and r alone,

$$R = \tfrac{1}{2} - \sin^2\Theta_W + \tfrac{5}{9}\sin^4\Theta_W + r(\tfrac{5}{9}\sin^4\Theta_W). \qquad (4.2.41)$$

As mentioned before, Eqs. (4.2.37) are valid if neutrino interactions with quarks and antiquarks other than u and d can be neglected and if the Cabibbo angle is set to zero. Higher twist effects (interactions with several quarks) have also been neglected; they have been estimated to give uncertainties smaller than $\Delta\sin^2\Theta_W \sim 0.005$ [LLE 83]. Of course, weak isospin symmetry is implied by Eq. (4.2.41). We know that it is broken by flavor-changing processes that have been observed to contribute to charged-current-induced reactions but not to neutral-current reactions. The energy threshold of the flavor transition $s(d) + W^+ \to c$ crosses the peak energy of the neutrino beam used for the measurement. To correct the measured cross sections we require knowledge of the mass of the charm (c) quark to describe the threshold behavior. The corrections were applied with the help of the quark model of the nucleon. Using the best knowledge of the charm quark mass, $m_c = (1.46 \pm 0.17)$ GeV, introduces an uncertainty of $\Delta(\sin^2\Theta_W) = \pm 0.0019$. Fixing the mass at $m_c = 1.46$ GeV, the remaining theoretical uncertainty is $\Delta(\sin^2\Theta_W) = \pm 0.003$ (see Table 4.2.10).

A high-precision measurement has recently been performed at CERN by the CHARM collaboration [ALL 87] and by the CDHS collaboration [ABR 86]. Ten years after the discovery of the neutral-current interaction by the Gargamelle team at CERN with a signal-to-background ratio of one to six, events have been classified as neutral-current (NC) or charged-current (CC) by direct recognition of the muon

Fig. 4.2.20   Photograph of CHARM detector.

in the fine-grain calorimeter of the CHARM detector with less than 0.2 percent ambiguity.

This progress is due to some important new features of electronic detectors:

> Fast timing is used and events occurring upstream are vetoed, thus eliminating the so-called associated neutron background that plagued the Gargamelle experiment [HAS 73b].
>
> The lateral and longitudinal dimensions of the target-calorimeters are more than 10 times larger than the interaction length of hadrons, thus giving clear signatures to neutrino interactions and muon tracks.
>
> Detector elements of small lateral dimensions (fine-grain) and frequent segmentation of the target plates allow detection of hadron showers with high efficiency and good energy resolution $(\sigma(E_H)/E_H = 0.47/\sqrt{E_H}/\mathrm{GeV})$ and the recognition of muons with momenta as low as $1\,\mathrm{GeV}/c$ [DID 80].
>
> Nearly equal response of the calorimeter $(E_e/E_H = 1.17)$ to electromagnetic and hadronic showers allows the definition of an effectively equal energy threshold in NC and CC events which have different π^0 content [DID 80].

The feasibility of a precision measurement was discussed by the author at the 1982 Javea Workshop on Weak Interaction [WIN 82] and was demonstrated in detail by the CHARM collaboration at a physics workshop at CERN [PAN 83].

Figure 4.2.20 is a photograph of the CHARM detector. It is composed of a fine-grained calorimeter and a muon spectrometer. The calorimeter consisted of 78 modules, each composed of a target plate of marble measuring 3 m × 3 m and 8 cm

Table 4.2.8. *Event numbers for neutrino exposure* ($E_h \geq 4\,GeV$)

	NC	CC
Uncorrected data sample	$39\,239 \pm 198$	$108\,472 \pm 329$
Trigger + filter efficiency	7 ± 4	0 ± 0
Scan correction	40 ± 0	60 ± 44
Corrected raw data sample	$39\,286 \pm 202$	$108\,532 \pm 332$
WB and cosmic correction:	-2310 ± 87	-4311 ± 119
Of which WB	-1998 ± 88	-4308 ± 119
Of which cosmic	-312 ± 8	-3 ± 1
	$36\,976 \pm 225$	$104\,220 \pm 361$
Possible difference in energy cut for NC and CC	—	0 ± 129
Lost muons	-3737 ± 50	-3735 ± 50
π and K decay	1893 ± 50	-1835 ± 50
K_{e3} CC	-1768 ± 68	-106 ± 6
K_{e3} NC	-532 ± 20	-33 ± 2
Corrected event numbers	$23\,831 \pm 283$	$105\,981 \pm 408$

Source: [ALL 87].

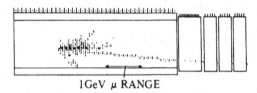

1 GeV μ RANGE

Fig. 4.2.21   Schematic view of a CC neutrino event recorded by the CHARM detector.

thick; a layer of 20 scintillation counters 15 cm wide, 3 cm long, and 3 cm thick; a plane of 128 proportional drift tubes (3 cm × 3 cm × 400 cm) oriented at 90° with respect to the scintillation counters, and a plane of digital wire chambers with 1-cm wire spacing oriented parallel to the scintillators. The calorimeter was surrounded by magnetized iron frames for the detection and measurement of large-angle muons. The orientation of the detector elements alternated from horizontal to vertical in successive modules. A detailed description can be found in [DID 80]. Figures 4.2.21 and 4.2.22 show schematic views of a CC and NC neutrino event, respectively. Scintillation counters and proportional drift tubes that are hit by the event are shown, and the range of a 1 GeV muon is indicated. Its track can also be recognized close to the hadron shower [ALL 87]. Charged current (CC) events for which the primary muon cannot be identified are classified as neutral current (NC) (Table 4.2.8). Some of these lost CC events have a muon with an energy less than 1 GeV, or a muon that left the detector at the sides before depositing 1 GeV, or a muon that was obscured by the hadronic shower. A correction is required for these CC losses. As this is the largest correction (11 percent of the NC events in the CHARM detector and 22 percent in the CDHS detector), the precision in measuring R^ν depends essentially on the reliability of estimating these losses. The uncertainty of the

Table 4.2.9. *Summary of experimental and theoretical errors on the value of* $sin^2\Theta_W$ *(CHARM)*

Error sources	CHARM
Statistical error	0.0040
Experimental systematical error from R_ν	0.0031
Experimental error from r	0.0006
Total experimental error	0.0051
Unitary CKM matrix $\|U_{ud}\|^2 = 0.9512 \pm 0.0012$	0.0000
Non-isoscalarity $D_V/U_V = 0.39 \pm 0.04$	0.0001
q_L QCD prediction $\pm 50\%$	0.0004
Quark sea $(\bar{U} + \bar{D})/(U + D) = 0.13 \pm 0.02$	0.0002
QCD evolution $\Lambda_{QCD} = 200 \pm 100$ MeV	0.0003
Higher twist effects	0.0003
Strange sea $\bar{S}/\bar{D} = 0.38 \pm 0.05$[47]	0.0004
Charm sea $C/S = 0.2 \pm 0.2$	0.0004
Strange sea asymmetry $S/\bar{S} = 1 \pm 0.1$	0.0001
Charmed sea asymmetry $C/\bar{C} = 1 \pm 0.3$	0.0002
Isospin breaking in the sea $\bar{U}/\bar{D} = 1 \pm 1$	0.0000
Radiative corrections	0.0020
Charm mass $m_c = 1.46 \pm 0.17$	0.0030
Total theoretical error	0.0037
Total error	0.0063

Source: [PER 95].

parent-beam momentum (± 3 percent) and of the muon-momentum measurement can affect the correction in a systematic way. A simple and beautiful method was used by the CHARM collaboration [ALL 87] to eliminate them. The correction was calculated relative to the number of events with muon momenta between 3 and 5 GeV/c. All scale errors cancel in this ratio, which was then applied by multiplying with the number of events observed in the muon-momentum interval. The remaining uncertainty affecting this correction contributed an error of $\Delta R^\nu/R^\nu = \pm 0.32$ percent. A summary of all experimental corrections is given in Table 4.2.8. A correction was applied for the small deviation from isoscalarity $(N-Z)$ of the target material. Selecting events induced by deep inelastic scattering $(E_{hadron} > 4$ GeV), the result of the ratio is

$$R^\nu = 0.3093 \pm 0.0031.$$

Table 4.2.9 shows the radiative and the various quark model corrections that have to be applied to determine $sin^2\Theta_W$ in the definition of Sirlin and Marciano. The final result is

$$sin^2 \Theta_W = 0.2330 + 0.111(m_c^{eff} - 1.46) \pm 0.0056. \qquad (4.2.42)$$

Table 4.2.10. *Values of R^0, $\bar{R}^0$ and r^0 corrected for non-isoscalarity, quark sea and $m_c^{(1)}$. $\sin^2\Theta_W$ is radiatively corrected for $m_t = 175\,GeV$ and $m_H = 150\,GeV$ [SHA 97]*

Experiment	R	$\bar{R}^0$	r^0	$\sin^2\Theta_W^{(1)}$
CCFR[b] [MCF 97]				0.2230 ± 0.0047
CDHS[b] [BLO 90]	0.3135 ± 0.0033	0.376 ± 0.016	0.409 ± 0.014	0.2250 ± 0.0059
CHARM[a] [ALL 87, CHA 97]	0.3093 ± 0.0031	0.390 ± 0.014	0.456 ± 0.011	0.2330 ± 0.0056
Average				0.2268 ± 0.0031

[a] Event-by-event method. [b] Event length method. [1] On-shell, corrected for $m_c = 1.46 \pm 0.17\,GeV$, the average value of the three experiments [CHA 97].

Table 4.2.11. *Values of model-independent coupling constants compared with the Standard Model predictions*

Quantity	Experimental value	Standard Model prediction $\sin^2\Theta_W = 0.230$
g_L^2	0.2987 ± 0.0039	0.301
g_R^2	0.0298 ± 0.0046	0.029

This result, obtained by the CHARM collaboration, is compared with other recent high-statistics results from semileptonic neutrino scattering experiments in Table 4.2.10, assuming a charm mass of $m_c = 1.46\,GeV/c^2$. The agreement between the experiments is good and significant in view of the fact that different experimental methods have been used, as indicated. From a model-independent fit [PER 95] one obtains the following values of the chiral coupling constants, assuming that both g_L and g_R contribute,

$$g_L^2 = 0.2987 \pm 0.0039$$
$$g_R^2 = 0.0298 \pm 0.0046 \qquad (4.2.43)$$

in good agreement with the values (4.2.38) determined directly. A combined value of $\sin^2\Theta_W$ from Table 4.2.10 ($m_c = 1.46\,GeV$) with radiative corrections [SHA 97] for $m_t = 175\,GeV$, $m_H = 150\,GeV$ is

$$\sin^2\Theta_W = 0.2268 \pm 0.0031. \qquad (4.2.44)$$

There is good agreement between the values of g_L^2 and g_R^2 determined by the model-independent analysis and those predicted by the Standard Model (Table 4.2.11) from the value of $\sin^2\Theta_W$ in Eq. (4.2.44). Thus, we have found strong support for the minimal Standard Model with a single Z^0.

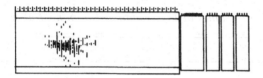

Fig. 4.2.22   Schematic view of an NC neutrino event recorded by the CHARM detector.

Using the most recent results for the Z boson mass [BLO 96]

$$m_Z = 91.1863 \pm 0.0020\,\text{GeV},$$

from LEP experiments we obtain from Eq. (4.2.44)

$$m_W = 80.16 \pm 0.16\,\text{GeV},$$

in good agreement with the direct measurements at $\bar{p}p$ colliders [BLO 96]

$$m_W = 80.36 \pm 0.13\,\text{GeV}.$$

We shall discuss radiative corrections in Section 4.2.7; if we assume their validity as calculated for $m_t = 175\,\text{GeV}$ and $m_H = 150\,\text{GeV}$, a precise value of the ρ parameter can be obtained by combining the measured value of R with the value of $\sin^2\Theta_W = 0.2234 \pm 0.0024$ determined from m_W and m_Z in the $\bar{p}p$ collider experiments (Section 4.2). From the CDHS [PER 95] and CHARM result of R [ALL 87], one obtains

$$\rho = 1.0007 - 0.022(m_c - 1.46) \pm 0.014 \qquad (4.2.45)$$

in good agreement with the minimal Standard Model. Another comparison recently made by Langacker [LAN 88b] begins by combining the results of R and $\bar{R}$ corrected for nonisoscalar target composition, different neutrino spectra, QCD, and radiative corrections. The resulting averages,

$$R = 0.311 \pm 0.002, \quad \bar{R} = 0.370 \pm 0.007 \qquad (4.2.46)$$

are in excellent agreement with the Standard Model predictions shown in Figure 4.2.23 for

$$\sin^2\Theta_W = 0.2234 \pm 0.0024. \qquad (4.2.47)$$

This value of $\sin^2\Theta_W$ is obtained from data [BLO 96] other than neutrino–hadron scattering. The same data can be used to obtain an important confirmation of the assignment of the right-handed strange quark (s_R) to an $SU(2)$ singlet. Although the strange (s) quark content of the nucleon is small,

$$\lambda = \frac{2s}{u+d} = 0.061 \pm 0.005$$

Table 4.2.12. *Values of $I_{3R}(i)$ obtained from neutral-current data*

i	$I_{3R}(i)$
u	0.003 ± 0.010
d	0.007 ± 0.012
e^-	-0.001 ± 0.022
μ^-	0.035 ± 0.038
τ^-	-0.039 ± 0.054
c	-0.15 ± 0.15
b	0.04 ± 0.15

Note: The values are predicted to vanish in the Standard Model, while for right-handed doublets one would have $I_{3R}(i) = \frac{1}{2}$, $(i = u, c)$, or $-\frac{1}{2}$, $(i = d, e, \mu, \tau,$ or $b)$. The values of $I_{3R}(i)$ are determined assuming canonical I_{3L} assignments (the results are insensitive to small deviations). This is justified independently by both charged and neutral-current data for the u, d, e, and μ, and by charged-current data for the c and τ. For the b, one needs the extra assumption that there are no exotic electric charges.
Source: [LAN 88b].

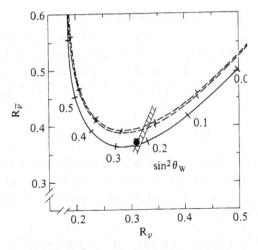

Fig. 4.2.23   Comparison of R and $\bar{R}$ with the prediction of the Standard Model (solid line) and with a model in which S_R is an $SU(2)$ doublet (dashed lines). Also shown is the range $\sin^2 \Theta_W = 0.2234 \pm 0.0024$ [BLO 96].

it is nevertheless large enough to have an important effect on R and $\bar{R}$ if the s_R neutral current couples as a doublet [LAN 88b] with $I^S_{3,R} = -\frac{1}{2}$. R and $\bar{R}$ would change by large amounts

$$\Delta R = 0.014, \quad \Delta \bar{R} = 0.031$$

compared to the experimental errors. The predictions are compared with the data in Fig. 4.2.23 and strongly favor the Standard Model assignment $(I^S_{3,R} = 0)$. The values of $I_{3,R}$ for all right-handed fermions as determined by a fit [LAN 88b] to neutral-current data are shown in Table 4.2.12. They are predicted to vanish in

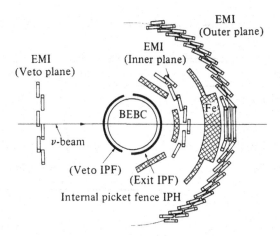

Fig. 4.2.24   The big European bubble chamber at CERN (BEBC) with external muon identifier (EMI), veto plant and internal picket fence (IPF).

the Standard Model, while for right-handed doublets one would have $I_{3R}^i = \frac{1}{2} (i = d, e, \mu, \tau,$ or $b)$ or $I_{3R}^i = \frac{1}{2} (i = u, c)$.

4.2.4.3 Deep inelastic neutrino scattering on proton and neutron targets

Simultaneous measurements of four ratios of NC and CC cross sections, of neutrinos and antineutrinos on protons and neutrons

$$R_{\nu p} = \frac{\sigma^\nu(\nu p \to \nu X)}{\sigma^\nu(\nu p \to \mu^- X)} \qquad R_{\bar\nu p} = \frac{\sigma^{\bar\nu}(\bar\nu p \to \bar\nu X)}{\sigma^{\bar\nu}(\bar\nu p \to \mu^+ X)}$$

$$R_{\nu n} = \frac{\sigma^\nu(\nu n \to \nu X)}{\sigma^\nu(\nu n \to \mu^- X)} \qquad R_{\bar\nu n} = \frac{\sigma^{\bar\nu}(\bar\nu n \to \bar\nu X)}{\sigma^{\bar\nu}(\bar\nu n \to \mu^+ X)} \qquad (4.2.48)$$

have been performed by a collaboration [DAL 88] using the big European bubble chamber (BEBC) filled with deuterium. They provided information about the chiral coupling constants of u and d quarks separately. Instead of reviewing the earlier set of combined results from different experiments [AMA 87], which may be subject to systematic uncertainties, we shall describe here this single experiment that gives the same statistical accuracy and minimizes systematical effects.

The bubble chamber was equipped (see Fig. 4.2.24) with an external muon identifier (EMI) consisting of two planes of proportional wire chambers separated by a hadron absorber, and an internal picket fence (IPF) detector consisting of two layers of proportional tube chambers surrounding the bubble chamber and covering 90 percent in azimuth. The EMI detected muons with $p_\mu \geq 4 \, \text{GeV}/c$ with an efficiency of (95 ± 0.4) percent. The picket fence provides timing information and was of crucial importance for the separation of neutral-current events from hadronic background produced upstream in the chamber walls.

Table 4.2.13. *Values of cross-sectional ratios of NC to CC on proton and neutron targets for* $E_h > 5\,\mathrm{GeV}/c$

Cross-sectional ratios[a]	Chiral coupling constants [DAL 88]
$R_{\nu p} = 0.405 \pm 0.024 \pm 0.021$	$u_L^2 = 0.099 \pm 0.018 \pm 0.008$
$R_{\nu n} = 0.243 \pm 0.013 \pm 0.016$	$d_L^2 = 0.202 \pm 0.020 \pm 0.019$
$R_{\bar{\nu} p} = 0.301 \pm 0.027 \pm 0.024$	$u_R^2 = 0.020 \pm 0.016 \pm 0.009$
$R_{\bar{\nu} n} = 0.490 \pm 0.050 \pm 0.037$	$d_R^2 = 0.002 \pm 0.017 \pm 0.010$
BEBC-TST [ARM 88]	Standard Model prediction for $\sin^2\Theta_W = 0.230$
$u_L^2 = 0.144^{+0.024}_{-0.029}$	0.119
$d_L^2 = 0.176^{+0.035}_{-0.040}$	0.182
$u_R^2 = 0.023 \pm 0.022$	0.023
$d_R^2 = 0.004^{+0.040}_{-0.004}$	0.0058

[a]First error is statistical, the second systematic. *Source:* [DAL 88].

As BEBC was filled with deuterium, the target particle could either be a proton or a neutron. The two cases were distinguished by the number of detected charged tracks; events with odd (even) numbers of tracks correspond to interactions of protons (neutrons). Two effects contribute that are not described by this simple scheme:

1 Spectator protons may have a momentum larger than $150\,\mathrm{MeV}/c$ due to Fermi motion changing the even number of tracks for a neutrino interaction with a neutron to an odd number.
2 Rescattering effects.

An estimate of the rescattering fraction f was made by a kinematical analysis of the neutrino and antineutrino data, yielding

$$f = 0.112 \pm 0.003 \ (\text{stat}) \pm 0.010 \ (\text{syst}).$$

Events were analyzed and classified according to the number of charged tracks (n or p targets) and the detection or absence of a muon with $p_\mu \geq 4\,\mathrm{GeV}/c$ (CC or NC). Corrections were then applied for efficiencies, for spectator protons and rescattering, for CC events with $p_\mu < 4\,\mathrm{GeV}/c$ (20 percent), for wrong helicity muon neutrinos, and for electron neutrinos that were classified as NC (muonless); 1305 NC νp, 1573 NC νn, 609 NC $\bar{\nu} p$, and 1207 $\bar{\nu} n$ events were obtained in this way. The values of the four cross-sectional ratios are summarized in Table 4.2.13 together with those of the chiral coupling constants determined from them. Also shown for comparison are the values predicted by the minimal Standard Model for $\sin^2\Theta_W = 0.230$; we note again good agreement. A summary of earlier results can be found in [AMA 87]. Figure 4.2.25 gives a comparison of the chiral coupling constants with the Standard Model as a function of $\sin^2\Theta_W$ [LAN 88]. The solid

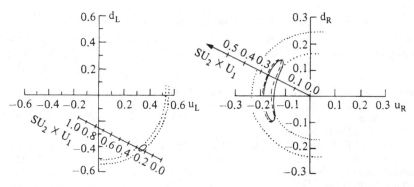

Fig. 4.2.25 Comparison of the chiral couplings of u and d quarks with the Standard Model predictions as a function of $\sin^2\Theta_W$. The solid lines are from a global analysis [AMA 87]; the dashed lines for u_R and d_R include the new BEBC WA25 data [DAL 88]. The new CHARM [WIN 88; ALL 89] determination of $g^2_{L,R}$ is shown by dotted lines.

points correspond to $\sin^2\Theta_W = 0.23$. The dashed line for d_R and u_R gives the combined results from the BEBC WA25 data [DAL 88] and from the previous global analysis [AMA 87], while the solid lines are from that global analysis alone. The dashed circular lines are from the direct values of cross-sectional ratios of NC to CC on proton and neutron targets for $E_h > 5\,\text{GeV}/c$ [DAL 88] and the direct determination of g^2_L and g^2_R by the CHARM collaboration [ALL 89a; WIN 88] is shown by dotted lines.

The analysis of semi-inclusive neutrino data can provide further information on u and d quark coupling constants. The primary neutrino–quark interaction produces a quark beam with a composition that reflects the structure of the neutral weak currents. For the reactions on isoscalar targets we expect

$$
\begin{aligned}
(\nu \to \mu^-) \quad & u : d = 1 : 0 \\
(\bar{\nu} \to \mu^+) \quad & u : d = 0 : 1 \\
(\nu \to \nu) \quad & u : d = u_L^2 + \tfrac{1}{3}u_R^2 : d_L^2 : \tfrac{1}{3}d_R^2 \\
(\bar{\nu} \to \bar{\nu}) \quad & u : d = u_R^2 + \tfrac{1}{3}u_L^2 : d_R^2 : \tfrac{1}{3}d_L^2.
\end{aligned}
\qquad (4.2.49)
$$

The fragmentation of the quarks produces the observed final state of hadrons. The ratio π^+/π^- is, in principle, sensitive to the quark composition produced in the primary neutrino–quark interaction. Following some early attempts using heavy liquid bubble chambers [AMA 87], a recent experiment used BEBC equipped with a track-sensitive liquid hydrogen target surrounded by Ne-H_2 mixture [ARM 88]. It was exposed to the neutrino and antineutrino wide-band beam from the CERN SPS. The external muon identifier (Fig. 4.2.24) was again used. Events were classified as NC or CC interactions using a multivariate discriminant analysis on an event-by-event basis. As a result, 456 neutrino-induced and 156 antineutrino-induced NC interactions on hydrogen were selected. Final state particles are then separated into forward- and backward-going ones (in the cm system) and a mass

Table 4.2.14. *Summary of neutrino-proton elastic scattering experiments*

Experiment	$R_\nu^{el} = \sigma_{\nu p}^{NC}/\sigma_{\nu p}^{CC}$	$R_{\bar\nu}^{el} = \sigma_{\bar\nu p}^{NC}/\sigma_{\bar\nu p}^{CC}$	Q^2 Region (in GeV^2)
Columbia–Illinois Rockefeller collaboration [LEE 76]	0.23 ± 0.09	—	$0.3 < Q^2 < 1.0$
Harvard–Pennsylvania–Wisconsin collaboration [HOR 82]	0.11 ± 0.015	0.19 ± 0.035	$0.4 \le Q^2 \le 0.9$
GARGAMELLE (Freon) [POH 78]	0.12 ± 0.06	—	$0.3 \le Q^2 \le 1.0$
Aachen–Padua [FAI 80]	0.10 ± 0.03	—	$0.2 \le Q^2 \le 1.0$
Columbia–Illinois Brookhaven collaboration [COT 81]	0.11 ± 0.03	—	$0.3 \le Q^2 \le 0.9$
U.S.–Japan E 734 (BNL) [AHR 87a]	0.153 ± 0.018	0.218 ± 0.024	$0.4 \le Q^2 \le 1.1$

attribution is made for positive particles to select pions. The charged pion ratios for forward-going particles were found

$$R_{\nu p\,NC}^{\pm} = 1.15 \pm 0.10$$
$$R_{\bar\nu p\,NC}^{\pm} = 1.22 \pm 0.19. \qquad (4.2.50)$$

From these ratios and the cross-sectional ratios $R_{\nu p}$ and $R_{\bar\nu p}$, they determined the chiral coupling constants given in Table 4.2.13. They are in good agreement with those of [DAL 88]. At first glance, it may seem surprising to find very similar errors for the results of [DAL 88] and [ARM 88], despite the large differences in statistics. However, this may be traced back to the method of π^+/π^- ratios, which, although somewhat model dependent, gives better precision. A summary and the results of a global fit for all data is given in Section 4.2.4.8.

4.2.4.4 Elastic neutrino proton scattering

The relative signs of the chiral coupling constants cannot be determined from inclusive experiments. Choosing one sign (e.g., $u_L > 0$) and supposing $d_R = 0$, there remain two sign ambiguities

$$sign(u_L d_L)$$
$$sign(u_L u_R)$$

and hence four possible solutions that are not shown in Fig. 4.2.25. They correspond to ambiguities between dominant V or A coupling and between dominant isoscalar ($I = 0$) or isovector ($I = 1$) coupling.

Analysis [HUN 77] of data from the elastic neutrino–proton scattering reactions (see Table 4.2.14)

$$\nu p \to \nu p$$
$$\bar\nu p \to \bar\nu p \qquad (4.2.51)$$

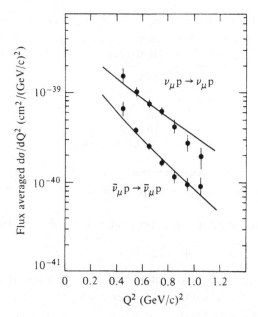

Fig.   4.2.26 $d\sigma/dQ^2$ distribution of elastic $\nu_\mu p$ and $\bar{\nu}_\mu p$ scattering [AHR 87a].

favors the solution implying

$$sign(u_L\, u_R) < 0 \qquad\qquad (4.2.52)$$

indicating that the neutral-current coupling of u quarks is dominantly axial vector. From a recent high-statistics study by Abe *et al.* [ABE 86; AHR 87a] of the differential cross section of reactions (4.2.51) at eight values of Q^2 (see Fig. 4.2.26), a value of $\sin^2\Theta_W = 0.205 \pm 0.041$ has been deduced.

4.2.4.5 Coherent π^0 production

Coherent π^0 production on nuclei by the neutral weak current is characterized by a constructive interference of the neutrino interactions on neutron and protons within the same nucleus [LAC 79; REI 83]. The nucleus does not break up in this reaction and therefore a negligible amount of recoil energy is transferred to it. Owing to the helicity-conserving Lorentz structure (V, A) of the neutral-current interaction (see Section 4.2.3), coherently produced neutral pions (π^0) are emitted at small angles compared to those produced in the incoherent and resonant π^0 production. These events are the main background in experiments studying elastic muon-neutrino scattering on electrons (see Section 4.2.4.1). Owing to the intrinsic quantum numbers of the neutral pion, coherent π^0 production is probing directly the axial-vector–isovector neutral-current coupling. The cross section can be calculated

Table 4.2.15. Summary of data on coherent π production on nuclei

Group	Reference	$|\beta|$
Aachen–Padova	[FAI 83]	0.93 ± 0.16
Gargamelle	[ISI 84]	
CHARM	[BER 85]	1.08 ± 0.24
Skat	[GRA 86]	0.99 ± 0.20
FNAL 15'BCH	[BAL 86]	0.98 ± 0.24
BEBC	[MAR 84a, 86b]	
Average		0.99 ± 0.10

Note: $|\beta|$ is the isovector axial-vector coupling constant.

using the PCAC theorem; it is proportional to the factors [LAC 79]

$$\sigma(\nu A \rightarrow \nu \pi^0 A) \propto (u_a - d_a)^2 \rho^2 \qquad (4.2.53)$$

where u_a and d_a are the axial-vector coupling constants of u and d quarks, respectively, and ρ the ratio of the neutral- and charged-current coupling strengths. The Standard Model predicts $u_a = -d_a = \frac{1}{2}$ and $\rho = 1$ and, hence, a value of one is expected for expression (4.2.53).

From the experiment one therefore determines in the most direct and model-independent way the difference and the relative sign of u_a and d_a, namely, the axial-vector–isovector coupling

$$|\beta| = |u_a - d_a|. \qquad (4.2.54)$$

Because of the pure axial-vector nature of the process, the cross sections of the neutrino- and antineutrino-induced reactions are expected to be equal.

The reaction has been extensively studied both at low ($\bar{E}_\nu \sim 1.8\,\text{GeV}$) and high ($\bar{E}_\nu \sim 30\,\text{GeV}$) neutrino energies, using bubble chamber and electronic calorimeter techniques. A summary of the results on $|\beta|$ is given in Table 4.2.15. The world average is

$$|\beta| = 0.99 \pm 0.10 \qquad (4.2.55)$$

in close agreement with the predicted value of one, and hence with a negative relative sign of the axial-vector coupling constants of the u and d quarks.

The energy dependence of the cross section is compared with theoretical models in Fig. 4.2.27. Also shown are results of coherent $\pi^+(\pi^-)$ production by $\nu_\mu(\bar{\nu}_\mu)$ charged-current interactions. References for Fig. 4.2.27 are given in Table 4.2.15. The cross-sectional ratio of ν- and $\bar{\nu}$-induced coherent π^0 production was determined in one experiment [BER 85] with the result

$$\sigma(\nu_\mu A \rightarrow \nu_\mu \pi^0 A)/\sigma(\bar{\nu}_\mu A \rightarrow \bar{\nu}_\mu \pi^0 A) = 1.22 \pm 0.33 \qquad (4.2.56)$$

in agreement with the expected equality.

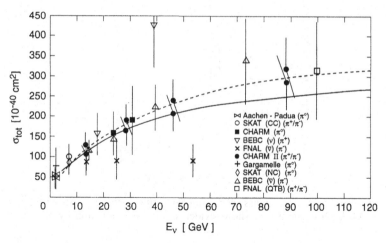

Fig. 4.2.27 Compilation of experiments on coherent single pion production. For this experiment the results from a scaling according to the Rein–Sehgal model (upper points) and the Bel'kov–Kopeliovich approach (lower points) are shown, averaged over ν_μ and $\bar{\nu}_\mu$. The predictions of the Rein–Sehgal model for $m_A = 1.3\,\text{GeV}/c^2$ (full line) and the Bel'kov–Kopeliovich approach (dashed line) are indicated. Data have been duly scaled to allow comparison.

4.2.4.6 Neutrino disintegration of the deuteron

Low-energy $\bar{\nu}_e$ from a fission reactor can induce the disintegration of the deuteron by neutral-current interaction

$$\bar{\nu}_e + d \rightarrow \bar{\nu}_e + p + n. \qquad (4.2.57)$$

Close to threshold ($E_\nu \geq 2.225\,\text{MeV}$) only a Gamov–Teller transition can contribute. In reaction (4.2.57) this is due to the isovector–axial-vector current giving rise to the spin-flip and isospin-flip transition

$$d(^3S_1, I = 0) \rightarrow pn(^1S_0, I = 1).$$

Therefore the cross section is only sensitive to the coupling parameter β (4.2.54) and independent of the electroweak mixing angle.

An experimental result has been reported by an Irvine group [PAS 79] working at the Savannah River fission reactor. Using an instrumented target of 268 kg of D_2O they detected the neutron from the deuteron disintegration by capture in ^{3}He. The measured cross section of $(3.8 \pm 0.9) \cdot 10^{-45}\,\text{cm}^2$ corresponds to

$$|\beta| = 0.9 \pm 0.1 \qquad (4.2.58)$$

in excellent agreement with results determined from the reaction $\nu_\mu A \rightarrow \nu_\mu \pi^0 A$ and with the predicted value of one.

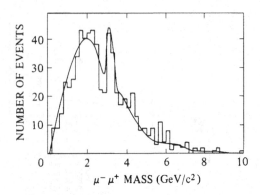

Fig. 4.2.28   Invariant $\mu^+\mu^-$ mass spectrum of neutrino induced events with small hadronic recoil [ABR 82b]. The fit line shows evidence for J/ψ production by NC reactions.

4.2.4.7 Neutral-current production of J/ψ

A study of the invariant mass spectrum of neutrino-produced opposite-sign muon pairs (Fig. 4.2.28) by the CDHS collaboration [ABR 82a] has shown evidence of neutral-current production of J/ψ

$$\nu_\mu N \rightarrow \nu_\mu J/\psi N$$
$$\rightarrow \mu^+\mu^- \qquad\qquad (4.2.59)$$

demonstrating directly the existence of a neutral $c\bar{c}$ current interaction with the $\nu\bar{\nu}$ current. Recent results of the CHORUS Collaboration [CHO 00] confirm these data but invalidate the charm coupling strength determination.

To determine the coupling strength, the cross section of reaction (4.2.59) has been compared to the related muon-induced reaction

$$\mu N \rightarrow \mu J/\psi N. \qquad\qquad (4.2.60)$$

Selecting events with small hadronic recoil energy and assuming that the observed samples of reactions (4.2.59) and (4.2.60) are due to the elastic channel without excitation of the nucleon N, it has been inferred that the NC coupling strength of c quarks is comparable to that of u quarks

$$(c_L^2 + c_R^2)/(u_L^2 + u_R^2) = 2.1 \pm 1.0 \qquad\qquad (4.2.61)$$

as expected from generation symmetry.

4.2.4.8 Summary of neutrino–quark coupling

Following the direct demonstrations of the V, A Lorentz structure of the neutral-current interaction and of the fundamentally important existence of right-handed

Table 4.2.16. *Values of neutral-current quark coupling constants determined from neutrino experiments*

Coupling constant	Value	Standard Model prediction
u_L	0.339 ± 0.017	0.345
d_L	-0.429 ± 0.014	-0.427
u_R	-0.172 ± 0.014	-0.152
d_R	$-0.011 + 0.081$	-0.076
	$\quad\;\; -0.057$	

Note: The Standard Model prediction is given from $\sin^2\Theta_W = 0.230$.
Source: [AMA 87; LAN 88b].

quark coupling, we can conclude this section by summarizing the overall picture in Fig. 4.2.25.

The annular domains are determined from deep inelastic scattering on isoscalar targets. The separations of u and d quark couplings are based on measurements of deep inelastic scattering on proton and neutron targets and on studies of the ratio π^+/π^- in the final state. The fourfold ambiguity of solutions is resolved by information from exclusive channels (elastic $(\bar{\nu})_\mu p$ scattering, $\nu_\mu A \to \nu_\mu \pi^0 A$, $\bar{\nu}_e d \to \bar{\nu}_e pn$), which determine the relative signs of the couplings to be

$$\text{sign}(u_L d_L) < 0, \; \text{sign}(u_L u_R) < 0 \qquad (4.2.62)$$

the sign of $u_L d_R$ remains undetermined, but d_R is close to zero. The overall sign of the chiral coupling constants is determined from interference experiments with the electromagnetic current (ed scattering and atomic parity violation) giving

$$u_L > 0 \qquad (4.2.63)$$

in agreement with the Standard Model.

Table 4.2.16 summarizes best fits to the coupling constants obtained by [AMA 87] and [LAN 88b]. Very similar results were obtained in other analyses [COS 88; FOG 88]. The values predicted by the Standard Model for the best fit of $\sin^2\Theta_W = 0.230$ are given as well.

4.2.5 Comparison of neutral-current phenomena induced by neutrinos and electrons

The processes to which neutral-current interactions may give rise can be depicted by a tetragon (Fig. 4.2.29), in analogy to the famous Puppi triangle of charged-current reactions [PUP 48].

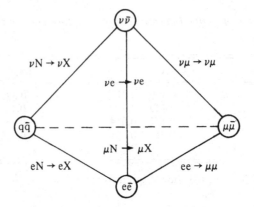

Fig. 4.2.29   The tetragon of neutral currents.

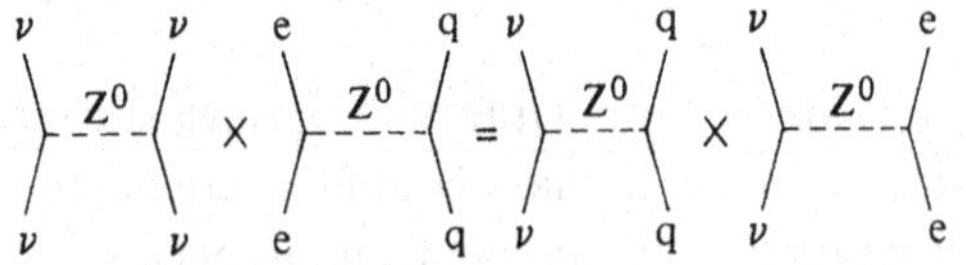

Fig. 4.2.30   Factorization relation for single Z^0 exchange.

The reactions

$$\nu N \rightarrow \nu X$$
$$\nu e \rightarrow \nu e$$
$$e N \rightarrow e X$$
$$\nu \mu \rightarrow \nu \mu$$

have been observed to be parity violating, except for the latter where the existence of an interference term with the $V{-}A$ interaction of the charged current has not yet been demonstrated. Other reactions have been observed as well

$$e\bar{e} \rightarrow \mu\bar{\mu}, \tau\bar{\tau}, q\bar{q}$$
$$\mu N \rightarrow \mu X$$

with the exception of $\nu_\mu \mu \rightarrow \nu_\mu \mu$, which remains to be uncovered. In the latter reactions, parity violation has not been directly observed.

  If these phenomena can be described by the exchange of a single intermediate Z^0 boson, there must exist relations between the different sectors. These relations have already been used for solving the fourfold ambiguity of g_A^e and g_V^e determined from $\nu_\mu e \rightarrow \nu_\mu e$ scattering (Section 4.2.4.) Here we shall relate ν–hadron scattering, electron–deuteron scattering, and atomic physics experiments. The corresponding factorization relations can be diagrammatically described as shown in Fig. 4.2.30. From each process a product of two vertex functions can be determined. These products satisfy relations such as the one sketched in Fig. 4.2.30.

Table 4.2.17. *Value of the weak charge Q_W extracted from experiments on atomic parity violation*

Group (reference)	Atom	$-Q_W$ measured	Predicted for $\sin^2\Theta_W = 0.230$
Bouchiat, Paris [BOU 82, 84]	Cesium	$68 \pm 9 \pm 3$	71.8
Gilbert, Boulder	Cesium	$74 \pm 6 \pm 3$	71.8
[GIL 85, 86a; NOE 88]		$69.4 \pm 1.5 \pm 3.8$	
Commins, Berkeley [DRE 84]	Thallium	$164 \pm 31 \pm 50$	113.7
Emmons, Seattle [EMM 84]	Lead	$84 \pm 21 \pm 13$	115.6
Hollister, Seattle [HOL 81]	Bismuth	$116 \pm 19 \pm 29$	116.5
Birich, Moscow [BIR 84]	Bismuth	$70 \pm 60 \pm 17$	116.5

Note: First error experimental, second theoretical (wave function). *Source:* [PIK 86].

There are 13 phenomenological parameters describing all processes allowed by the neutral-current tetragon (Fig. 4.2.29) and six factorization relations between them. Hence, seven independent parameters will completely specify the couplings of the fermions involved ($\nu_L, u_L, d_L, e_L, u_R, d_R, e_R$). These relations impose constraints that can be tested.

Scattering of polarized electrons on deuterons and parity violation in atomic transitions are probing the isospin structure of the electron–quark neutral-current interaction. From the parity-violating cross-sectional asymmetry

$$A(x, y, Q^2) = \frac{d\sigma(e_R^- d) - d\sigma(e_L^- d)}{d\sigma(e_R^- d) + d\sigma(e_L^- d)}$$

mainly the $A_{\text{electron}} \times V_{\text{quark}}$ coupling is determined in a linear combination:

$$\alpha + \gamma/3 = 0.60 \pm 0.16. \tag{4.2.64}$$

We have chosen the notation of Sakurai [SAK 81], which emphasizes the isospin structure; α, β, γ, and δ denote the couplings of the isovector vector, isovector axial-vector, isoscalar vector, and isoscalar axial-vector currents, respectively.

From measurements of parity violation in atomic transitions the so-called weak charge of the nucleus is determined. This quantity is given by the sum of the vector couplings ($\alpha + \gamma$) of all u and d quarks.

$$\begin{aligned} Q_W &= -[(\alpha + \gamma)N_u + (\gamma - \alpha)N_d] \\ &= -[\alpha(Z - N) + 3\gamma(Z + N)] \end{aligned} \tag{4.2.65}$$

where Z and N are the numbers of protons and neutrons. The results of recent measurements are summarized in Table 4.2.17. Earlier measurements that are inconsistent with those in the table have been omitted. The regions of α and γ determined by these two experiments are shown in Fig. 4.2.31. The signs are fixed by

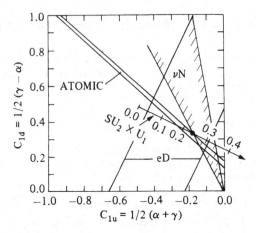

Fig. 4.2.31   Regions of α (isovector V) and γ (isoscalar V) allowed (68 percent confidence intervals) by the SLAC ed experiment [PRE 78], the Seattle [HOL 81] and Paris [BOU 84] atomic parity violation experiments, and ν–hadron scattering [ALL 87] assuming factorization (single Z^0).

Table 4.2.18. *Determination of ρ and $\sin^2\Theta_W$ from various reactions*

Reaction	$\sin^2\Theta_W$	ρ
$\nu_\mu N \rightarrow \nu_\mu X$	0.2268 ± 0.0031	1.0007 ± 0.014
$\bar{\nu}_\mu e \rightarrow \bar{\nu}_\mu e$	0.2324 ± 0.0083	1.006 ± 0.036
$M_W, M_Z(\mathrm{LEP}, p\bar{p})$	0.2234 ± 0.0024	1.0004 ± 0.0025
All data	0.2251 ± 0.0019	1.0004 ± 0.0025

Note: Experimental and systematic errors are combined quadratically. Radiative corrections for $m_t = 175\, m_H = 150\,\mathrm{GeV}$.

interference with the electromagnetic current:

$$\alpha = -0.65 \pm 0.17$$
$$\gamma = +0.14 \pm 0.05. \qquad (4.2.66)$$

The region constrained by ν–hadron data through the factorization relation is shown in Fig. 4.2.31 as well. The results are consistent with the prediction of the Standard Model ($\sin^2\Theta_W = 0.230$) indicated by a solid point.

   A comparison of $\sin^2\Theta_W$ and ρ determined from all neutral-current phenomena [CHA 94, BLO 96, SHA 97] is given in Table 4.2.18 and shown in Fig. 4.2.32. The allowed domains are 90 percent confidence intervals. A single set of parameters

$$\rho = 1.0004 \pm 0.0025$$
$$\sin^2\Theta_W = 0.2251 \pm 0.0019 \qquad (4.2.67)$$

gives the best fit in good agreement with earlier studies [LAN 95b].

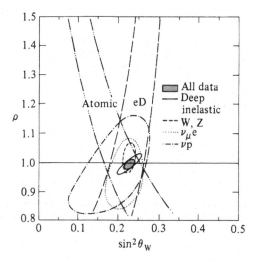

Fig. 4.2.32 Domains of ρ, $\sin^2\Theta$ for various neutral-current phenomena (90 percent confidence intervals) [AMA 87].

4.2.6 Generation universality and flavor conservation

So far only flavor-conserving neutral currents have been observed. In the GSW model this is due to the so-called GIM mechanism (Chapter 3), which requires that the weak interactions of all fermions are completely determined by their weak isospin and electric charge without distinguishing between leptons and quarks or between generations. The universality of the neutral-current interaction can therefore be examined experimentally in two ways: in a model-independent way by comparing the values of coupling constants of the different fermions; or in a model-dependent way by searching for flavor-changing neutral-current processes.

Lepton-quark universality can be tested in neutrino reactions with electrons and quarks, and lepton-generation universality in e^+e^- annihilations, ν_μ and ν_e interactions and Z^0 decays. A test of the relations

$$g_A^e = g_A^d = -g_A^u \qquad (4.2.68)$$

determined from neutrino reactions is given in Table 4.2.19 together with tests of the relations.

Further tests of generation universality of neutral-current coupling constants from neutrino reactions at typical values of $Q^2 \sim 4$–$10\,\text{GeV}^2$ and from branching ratios of Z^0 decays are summarized in Table 4.2.20. There is no evidence for a deviation from universality. Flavor-changing neutral currents have been searched for in kaon, D, and B decays as well as in neutrino reactions. A flavor-changing neutral current neutrino reaction

$$\nu_\mu u \to \nu_\mu c, \quad c \to s l^+ \nu_\mu \qquad (4.2.69)$$

Table 4.2.19. *Values of the effective axial vector coupling constants*

Fermion	$g_A^{(a)}$	Reaction	Reference
e	-0.503 ± 0.017	$\nu_\mu(\bar{\nu}_\mu)e \to \nu_\mu(\bar{\nu}_\mu)e$	[CHA 94]
	-0.50119 ± 0.00045	$e^+e^- \to e^+e^-$	[BLO 96]
μ	-0.50086 ± 0.00068	$e^+e^- \to \mu^+\mu^-$	[BLO 96]
τ	-0.50117 ± 0.00079	$e^+e^- \to \tau^+\tau^-$	[BLO 96]
u	0.511 ± 0.025	$\nu_\mu u \to \nu_\mu X$	[WIN 91]
d	-0.44 ± 0.030	$\nu_\mu d \to \nu_\mu X$	[WIN 91]
c	0.502 ± 0.010	$e^+e^- \to c\bar{c}$	[BLO 96]
b	-0.527 ± 0.007	$e^+e^- \to b\bar{b}$	[BLO 96]

Note: [a] For μ, τ, c, b, d sign $(g_A) = -1$ was assumed, for e it has been established from neutrino electron scattering and LEP $e^+e^- \to e^+e^-$ *forward–backward asymmetry* [CHA 94].

Table 4.2.20. *Test of generation universality of neutral current coupling*

Quantity	Value	Experiment	Reference
g_s/g_d	1.05 ± 0.15	$d\sigma/dy(\nu_\mu N \to \nu_\mu X)$	[JON 81], [WIN 87]
g_c/g_u	0.982 ± 0.057	LEP	[BLO 96]
g_b/g_d	1.20 ± 0.08	LEP/$\nu_\mu N \to \nu_\mu X$	[BLO 96; WIN 87]
$2g^{\nu_\mu}$	1.004 ± 0.033	$\bar{\nu}_\mu e \to \bar{\nu}_\mu e, Z \to e^+e^-$	[CHA 93]
$2g^\nu$	0.9999 ± 0.0043	Γ_{inv}^Z	[LEP 94]
$2g^{\nu_e}$	1.019 ± 0.142	$\nu_e N \to \nu_e X$	[DOR 86], [ALL 93]
$2g^{\nu_\tau}$	0.98 ± 0.15	Assuming $N_\nu = 3$	[CHA 93], [LEP 94]
$g_{\mu\mu}/g_{ee}$	1.0006 ± 0.0061	$Z \to \mu^+\mu^-, e^+e^-$	[LEP 94]
$g_{\tau\tau}/g_{ee}$	1.0024 ± 0.0076	$Z \to \tau^+\tau^-, e^+e^-$	[LEP 94]

Note: as example, $g_s^2 = g_{s,L}^2 + g_{s,R}^2$.

will give rise to a lepton of the wrong charge. A search for events with a wrong-charge muon has been performed by the CDHS [HOL 78] collaboration at CERN using a sign and momentum selected (so-called narrow-band) neutrino beam and a massive magnetized Fe calorimeter. A background due to antineutrinos from decays of negative pions before the sign and momentum selection has been eliminated by selecting events with $E_\nu > 60$ GeV. Efremenko *et al.* [EFR 79] have searched for events with a positron in an exposure of the 15′ FNAL bubble chamber filled with neon–hydrogen in a beam of $\bar{\nu}_\mu$. The limits obtained are summarized in Table 4.2.21.

Flavor-changing neutral currents have also been searched for in decays of $K_L^0 \to \mu^+\mu^-$, $K^+ \to \pi^+\nu\bar{\nu}$, $D^0 \to \mu^+\mu^-$, and in the decays $B^0 \to \mu^+\mu^-$. The limits are summarized in Table 4.2.21 as well. These limits imply, in the context of the GSW model, very stringent constraints on the generation universality of the neutral-current interaction.

Table 4.2.21. *Experimental limits (90% C.L.) on flavor-changing neutral-current processes*

Process	Upper limit 90% C.L.	Reference
$(\sigma(\nu_\mu u \to \nu_\mu c \to s\mu^+ \nu_\mu)/(\sigma(\nu_\mu N \to \nu_\mu X))$	0.026	[HOL 78]
$(\sigma(\bar{\nu}_\mu u \to \bar{\nu}_\mu c \to se^+ \nu_\mu))/(\sigma(\bar{\nu}_\mu N \to \bar{\nu}_\mu X))$	0.04	[EFR 79]
$(\Gamma(C \to e^+ e^- X)/(\Gamma(C \to e^+ \nu_e X))$	0.02	[BAL 77]
$BR(K_L^0 \to \mu^+ \mu^-)$	$(7.2 \pm 0.5) \times 10^{-9}$	[PDG 96]
$BR(K^+ \to \pi^+ \nu \bar{\mu})$	$< 24 \times 10^{-9}$	[PDG 96]
$BR(D^0 \to \mu^+ \mu^-)$	$< 8 \times 10^{-6}$	[PDG 96]
$BR(B^0 \to \mu^+ \mu^-)$	$< 5.9 \times 10^{-6}$	[PDG 96]

Table 4.2.22. *Values of* $\sin^2 \Theta_W^{\text{eff}}$ *(on-shell) from recent measurements. Radiaive corrections for* $m_t = 175\,GeV$, $m_H = 150\,GeV$

Process	Reference	$\sin^2 \Theta_W^{\text{eff}}$
$\nu_\mu(\bar{\nu}_\mu)e$ scattering	[CHA 94]	0.2324 ± 0.0083
$\nu_\mu N \to \nu_\mu X$	[SHA 97]	0.2268 ± 0.0031
$m_W/m_Z(p\bar{p}, LEP)$	[BLO 96]	0.2234 ± 0.0024
m_Z	[BLO 96]	0.2319 ± 0.0002

4.2.7 $\sin^2 \Theta_W$ and radiative corrections

Values of $\sin^2 \Theta_W$ determined from different neutral-current phenomena are summarized in Table 4.2.22. Radiative corrections have been applied for $m_t = 175\,GeV$ and $m_H = 150\,GeV$ (see Chapter 3 and Section 4.4). The theoretical uncertainties for deep inelastic neutrino scattering have been given in some detail in Table 4.2.9 (Section 4.2.4). We draw attention to the remarkable agreement of the radiatively corrected values determined over a very large range of Q^2 and for different sectors of the neutral-current tetragon. This quantitative agreement is a major success of the theory. In contrast to this the measured, uncorrected values differ significantly from each other.

The best overall fit is obtained for

$$\sin^2 \Theta_W = 0.23607 \pm 0.00020. \tag{4.2.70}$$

A comparison of these neutral-current data with the measured W and Z masses (Section 4.3) provides a fundamental test of the theory at the quantum level.

The Fermi interaction mediated by the weak bosons gives a relation between the boson masses, the fine structure constant α, the Fermi coupling constant G_F and

Table 4.2.23. *Values of* m_W, m_W/m_Z, m_Z *from* $\bar{p}p$ *[BLO 96], LEP [BLO 96] and* νN *[BLO 96; CHA 97], of the radiative correction* Δr *and of the corresponding predictions of the Standard Model (on-shell)*

Quantity	Experimental result	Reference	Standard Model
m_W	80.339 ± 0.098 GeV	[BLO 96]	80.310 ± 0.02
m_W^2/m_Z^2	0.2268 ± 0.0031	[BLO 96; CHA 97]	0.23165 ± 0.00024
m_Z	91.1863 ± 0.0020 GeV	[BLO 96]	91.1861
Δr	0.044 ± 0.004	[PER 95]	0.040 ± 0.004

the weak mixing parameter $\sin^2 \Theta_W$

$$m_W = m_Z \cos \Theta_W = \frac{A_0}{\sin^2 \Theta_W (1 - \Delta r)^{\frac{1}{2}}} \qquad (4.2.71)$$

where $A_0 = (\pi \alpha / \sqrt{2} G_F)^{\frac{1}{2}} = 37.281$ GeV using $\alpha^{-1} = 137.035963(15)$ from measurements of the Josephson effect, and $G_F = 1.16637(2) \cdot 10^{-5}$ GeV^{-2} from measurements of the muon lifetime; $\sin^2 \Theta_W = 1 - m_W^2/m_Z^2 = 0.2268 \pm 0.0031$ is the renormalized weak mixing parameter determined from deep inelastic neutrino scattering [SHA 97]. The radiative corrections Δr relating muon decay and the boson mass scale are very large, in comparison, for example, with the Lamb shift or $(g-2)$ of the muon and electron in QED; assuming $m_t = 175$ GeV, $m_H = 150$ GeV, Δr is predicted [LAN 95]

$$\Delta r = 0.040 \pm 0.004. \qquad (4.2.72)$$

The predicted and measured mass values m_W, m_Z are summarized in Table 4.2.23. The quantitative success of predicting the Fermi mass scale has been one of the triumphs of the GSW theory.

The value of Δr in Table 4.2.23 determined from the W and Z masses and the neutrino data is in agreement with the calculated radiative corrections and demonstrates their existence. It should, however, be noted that these corrections are to a large extent electromagnetic owing to the running of α, and not electroweak. Nevertheless, compared to the divergent electromagnetic corrections in the Fermi theory, the progress is very significant.

Novikov *et al.* [NOV 93] have noted that there is a large cancellation between electroweak fermionic and bosonic loop terms to the W and Z self-energies; only recently data have become sufficiently precise to require [BLO 96] these loop terms.

The relative strengths of the neutral- and charged-current couplings are exactly equal in $SU(2) \times U(1)$ models with spontaneous symmetry breaking by weak isospin doublets of scalar Higgs mesons ($\rho = G^{NC}/G_F$),

$$\rho = m_W^2/m_Z^2 \cos^2 \Theta_W. \tag{4.2.73}$$

Values of $\rho \neq 1$ can occur if Higgs triplets with weak isospin $I = 1$ and $I_{i3} = -1, 0, +1$ exist and have a vacuum expectation value $\langle \Phi \rangle$,

$$\rho = \frac{\sum_i (I_i^2 - I_{i3}^2 + I_i) \langle \Phi_i \rangle^2}{\sum_i 2 I_{i3}^2 \langle \Phi_i \rangle^2}. \tag{4.2.74}$$

A summary of ρ and $\sin^2 \Theta_W$ values determined from various neutral-current phenomena was given in Table 4.2.18 (Section 4.2.5). A global fit [AMA 87] determined $\rho = 0.998 \pm 0.0086$ in remarkable agreement with $\rho = 1$, implying the following (90 percent C.L.) constraints on the relative vacuum expectation values of Higgs triplets [AMA 87] with $I = 1$, $I_3 = 0$ and with $I = 1$, $I_3 = \pm 1$

$$\frac{|\langle \Phi_{10} \rangle|}{|\langle \Phi_{\frac{1}{2}\frac{1}{2}} \rangle|} < 0.047$$

$$\frac{|\langle \Phi_{1\pm 1} \rangle|}{|\langle \Phi_{\frac{1}{2}\frac{1}{2}} \rangle|} < 0.081. \tag{4.2.75}$$

One of the limitations of the GSW theory is the absence of a relation between the weak isospin ($SU(2)$) charge and the hypercharge ($U(1)$). This is reflected in the occurrence of two independent coupling constants Ig_1 ($U(1)$) and g_2 ($SU(2)$). The hypercharge Y is defined as

$$Q = I_3 + Y$$

to give the correct electric charge. To obtain an algebraic relation between I and Y, the group $SU(2) \times U(1)$ has to be embedded into a higher symmetry. In such a scheme g_1/g_2 and $\sin^2 \Theta_W$ would be predictable if that group G has a representation containing all 15 fermions of given helicity belonging to one generation (e.g., ν_{eL}, e_L^-, $e_L^+, u_{iL}, d_{iL}, \bar{u}_{iL}, \bar{d}_{iL}$, where i is the color index of quarks). Georgi, Quinn, and Weinberg [GEO 74a] obtained the relation

$$\sin^2 \Theta_W = \sum_i I_{3i}^2 \bigg/ \sum_i Q_i^2 = \frac{8(\frac{1}{4})}{2[0 + 1 + 3(\frac{4}{9}) + 3(\frac{1}{9})]} = \frac{3}{8}. \tag{4.2.76}$$

One example of a group G is the "grand unifying" group $SU(5)$ [GEO 74b], which is embedding the symmetry group $SU(3)_c \times SU(2) \times U(1)$ of the strong and the weak interactions ($SU(3)_c$ is the color group of QCD). Evaluating $\sin^2 \Theta_W$ at $Q^2 = m_W^2$ in the modified minimal subtraction scheme gives a predicted value

$$\sin^2 \hat{\Theta}_W(m_Z^2) = 0.2100 \pm 0.0026 \tag{4.2.77}$$

Table 4.2.24. *Outlook on future determinations of* $\sin^2\Theta_W$

Quantity	$\Delta\sin^2\Theta_W$	Date, installation
m_Z ($\pm 2\,\text{MeV}$)	0.0002	LEP 1995 (1996)
m_W ($\pm 125\,\text{MeV}$)	0.0008	Tevatron (1998)
m_W ($\pm 100\,\text{MeV}$)	0.0006	LEP 200 (1998)
R_ν (± 0.001)	0.002	$\nu_\mu N$ (2002)
A_{LR} ($e^+e^- \to \mu^+\mu^-$)	0.00025	SLD (2000)

in disagreement by 5 standard deviations with the experimental value from νN data [SHA 97]

$$\sin^2 \hat{\Theta}_W(m_Z^2) = 0.2316 \pm 0.0031 \text{ (experiment)} \tag{4.2.78}$$

while the prediction derived from supersymmetry

$$\sin^2 \hat{\Theta}_W(m_Z^2) = 0.2334 \pm 0.0035 \tag{4.2.79}$$

is consistent with the measured value.

4.2.8 Status of the Glashow–Salam–Weinberg Standard Model

The GSW model seems to describe all neutral-current phenomena in a quantitatively correct way. Radiative corrections at the quantum level are calculable to all orders and have been successfully confronted with experimental data. To some extent this confrontation is testing the gauge character of the theory.

Concerning the assignment of the known fermions to doublets for left-handed and to singlets for right-handed states, no exception has been found, although in some cases the conclusion depends on the assumption that large singlet–doublet mixing is excluded. Data from recent experiments have completed the verification of the assignments. In particular the τ_L is found to require a doublet partner; hence the fermion ν_τ must exist.

The top quark has been discovered at Fermilab with a mass of $m_t = (175 \pm 6)\,\text{GeV}$ [BLO 96]; the mass value derived from radiative corrections is $m_t^{rad} = (157 \pm 9)$ GeV (see Section 4.4). The experimental value of m_t and the constraints from neutrino experiments and from LEP give now the possibility of estimating the Higgs boson mass, with the result $m_H = 56^{+101}_{-31}$ GeV. It is remarkable that this value falls into the narrow window between the lower limit of direct searches, $m_H > 85$ GeV, and the theoretical upper limit in the Standard Model, $m_H < 600$–$800\,\text{GeV}$ (see Section 4.4). An outlook on more precise measurements of $\sin^2\Theta_W$ is given in Table 4.2.24. If electron beams can be polarized in LEP and SLC, another order of magnitude in precision can be gained.

It is of great interest to improve the precision of measuring neutral weak current phenomena. Further progress in our understanding of the electroweak sector is expected to come from an observation of small deviations from the predictions of the Minimal Standard Model. It is hoped that such deviations would help to fix the loose ends of the model and to determine into which grand unified theory it could be successfully embedded or which gauge groups have to be appended.

Appending an additional group $U(1)_X$ introduces an additional neutral boson Z_X. In the simplest case [ROS 85] with one extra Z_X and no mixing of Z^0 and Z_X, the couplings are modified, for example, for $\nu_\mu e$ scattering, as follows:

$$\Delta g_A^e = (3/10)(m_{Z_x}^2/m_Z^2)(g/g_x)^2$$
$$\Delta g_V^e = (3/5)(m_{Z_x}^2/m_Z^2)(g/g_x)^2. \tag{4.2.80}$$

Comparing the values of g_A^e and g_V^e determined from $(\bar{\nu})_\mu e$ scattering with those determined independently from $\sin^2 \Theta_W = 1 - m_W^2/m_Z^2$ the CHARM collaboration [CHA 94] has derived an upper limit on the quantity

$$(g_x/g)^2(m_Z/m_{Z_x})^2 < 0.11 \quad \text{(95 percent C.L.)}. \tag{4.2.81}$$

Assuming equal coupling strength (g_x) to Z_x and (g) to Z, a lower limit

$$m_{Z_x} > 398 \,\text{GeV} \quad \text{(95 percent C.L.)} \tag{4.2.82}$$

was obtained. Direct search by the CDF collaboration [ABE 97] gives $m_{Z_x} \geq 690 \,\text{GeV}$ (95% C.L.).

The other essential phenomena in support of the theory, the gauge field interactions of the W and Z bosons, which manifest themselves through their self-coupling, are discussed in Sections 4.3 and 4.4.

4.3 The weak bosons W and Z*

The weak bosons were discovered in 1983 at the CERN $p\bar{p}$ collider [UA1 83a,b; UA2 83a,b] and the comparison of the values of their masses [CDF 89; UA2 90] with the cross section of the deep inelastic neutrino scattering on isoscalar target [CDH 86; CHA 87] was the first stringent and successful test of the electroweak theory.

Since 1985 the Fermilab Tevatron Collider has provided the highest energies available to accelerator-based experiments today. This machine operates at 1.8 TeV in the $p\bar{p}$ center of mass, and provides parton–parton collisions at energies of up to several hundred GeV, allowing direct production of not only W and Z bosons, but also pairs of the recently discovered top quarks. Although the production of heavy objects at hadron colliders is accompanied by many additional soft particles,

* K. Einsweiler, Lawrence Berkeley National Laboratory, Berkeley, USA, L. Rolandi, Experimental Physics Division, CERN, Geneva, Switzerland.

making the analysis of the data significantly more challenging than in e^+e^- colliders, contemporary experiments using sophisticated tracking, calorimetry, and trigger systems have mastered this environment to provide precise measurements of the weak boson properties.

The ideal machine to measure the properties of the weak bosons in the most convenient experimental conditions is an e^+e^- collider where the Z boson is directly produced by e^+e^- annihilation ($e^+e^- \rightarrow Z$) when the energy of each beam is equal to $M_Z/2$ and W bosons are produced in pairs ($e^+e^- \rightarrow W^+W^-$) at beam energies in excess of M_W. Two e^+e^- colliders came into operation in 1989 to study the reaction ($e^+e^- \rightarrow Z$), the SLAC linear collider (SLC) and the CERN large electron positron ring (LEP). The latter machine has been subsequently upgraded in energy with superconducting cavities and in 1996 crossed the threshold for W pair production.

The measurements of the properties of the Z bosons obtained with the large data samples collected at LEP and SLC, the measurement of the W mass performed at LEP and measurements of the W mass and of the top mass performed at the Tevatron challenge the Standard Model of the electroweak interaction with unprecedented precision.

4.3.1 LEP and SLC

In a circular e^+e^- collider the energy loss due to synchrotron radiation in the bending magnets is compensated by the acceleration in the radio frequency cavities. The maximum beam energy is limited by the available RF power. Since the energy loss is inversely proportional to the radius of the machine, the maximum beam energy at fixed power is proportional to radius, resulting in very large accelerators. Linear e^+e^- colliders do not suffer this limitation and can be smaller in size. However, their luminosity is limited because the bunches are lost after a single collision.

LEP, CERN's large electron positron collider, is a circular machine. It is located in a 26.7 kilometers long, 3.8 meters wide underground tunnel situated 50 to 170 meters below the surface. The accelerator consists of eight arcs 2.8 kilometers long linked by eight straight sections. The particles are kept on their track in the arcs by 3400 bending magnets and are focused by 800 quadrupoles and 500 sextupoles.

In the first phase (LEP1) electrons and positrons were accelerated by copper cavities located in two diametrically opposite positions in straight sections on either side of the underground experimental halls. The radio frequency system has been upgraded for the second phase (LEP2) with the installation of superconducting cavities that started in fall 1995 and that will be completed by 1999 when the beam energy will eventually reach almost 100 GeV.

During the years 1989–1995 LEP delivered an integrated luminosity of about $200\,\text{pb}^{-1}$ to each of the its four experiments. About 80% of this integrated luminosity was delivered at a center-of-mass energy within 100 MeV of the Z mass. The remaining 20% was used to scan the resonance. A total of 15 million

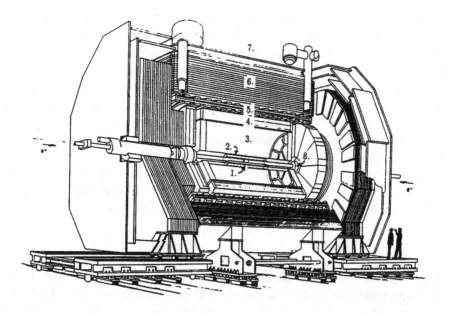

Fig. 4.3.1. Cut-view of the ALEPH detector showing the main detector elements. 1-silicon vertex detector, 2- inner trigger chamber, 3- time projection chamber, 4- electromagnetic calorimeter, 5- superconducting coil, 6- hadron calorimeter, 7- muon chambers, 8- luminosity monitors.

hadronic Z decays and 1.6 million leptonic Z decays have been recorded by the four experiments.

SLC, the Stanford Linear Accelerator, is a 3.2 kilometer long linear electron positron collider. It started its operation in 1989 at a center-of-mass energy close to the Z mass. In 1993 a new source of polarized electrons was commissioned, based on an electron gun with a GaAs strained cathode providing electrons with about 90% polarization at a rate of 120 Hz with the possibility of reversing the spin at each pulse. SLC operates at a luminosity smaller than LEP but takes advantage of the polarized electron beam and of the very small dimensions of the luminous region in the directions perpendicular to the beams. SLC has delivered to the SLD experiment an integrated luminosity of about $8 \, \text{pb}^{-1}$ corresponding to more than 200 000 Z decays.

4.3.2 The LEP and SLC detectors

A typical experimental apparatus at a colliding beam facility consists, in general, of many successive layers of different kinds of detectors inside a solenoidal magnet. An example of such an apparatus is given in Fig. 4.3.1. Moving outward from the region where the beams collide, one finds first the tracking detectors, which serve to reconstruct charged particle tracks to measure their momenta and impact parameters with respect to the collision point. Following the tracking detectors, one finds an electromagnetic calorimeter with large angular coverage, which is used to measure the energy of the electrons and photons, and the hadron calorimeter, which absorbs the hadronic showers associated with strongly interacting particles and measures their energies. If the calorimeters are thick enough, no long-lived particles can escape, except for muons and neutrinos. Muon detectors, consisting in general of large area tracking chambers, can therefore be used as the outermost shell of the apparatus. The luminosity monitors are small angle calorimeters that measure the electrons and positrons scattered in the t-channel reaction $e^+e^- \to e^+e^-$.

The four LEP detectors ALEPH [ALE 90], DELPHI [DEL 91] , L3 [L3 90] and OPAL [OPA 91] and the SLC detector SLD [SLD 84] follow this basic pattern with a different choice of the main detector elements.

4.3.3 The Z lineshape

The precise determination of the Z parameters (mass, total and partial widths) is the most important test of the Minimal Standard Model which can be done at LEP. This study has been done measuring the leptonic and hadronic cross sections at different center-of-mass energies around the nominal value of the Z mass.

The measured cross sections are fitted to a formula that depends on the Z parameters and takes into account the important effects caused by initial state radiation. The absolute scale of the Z mass and widths is given by the precise calibration of the center-of-mass energy at the collision points. The luminosity was collected at three scan points (named peak -2, peak and peak $+2$) separated by roughly 1790 MeV, almost symmetrically placed around the energy corresponding to the maximum of the cross section. In this configuration the errors on M_Z and Γ_Z depend approximately only on the errors in the sum and on the difference of center-of-mass energies at the two off-peak points.

$$\Delta M_Z \approx 0.5\Delta(E_{+2} + E_{-2})$$
$$\Delta\Gamma_Z \approx \frac{\Gamma_Z}{(E_{+2} - E_{-2})} \Delta(E_{+2} - E_{-2}) = 0.71\Delta(E_{+2} - E_{-2}),$$

where E_{-2} and E_{+2} are the luminosity-weighted center-of-mass energies at the two off-peak points.

Energy calibration   The average energy of the circulating beam, E_{beam}, can be measured with a precision of 1 MeV [ARN 94] using the resonant depolarization method. The emission of the synchrotron radiation in the vertical bending field polarizes the LEP beams in the vertical direction. The spin vector of each electron precesses on average $a_e\gamma$ times during one turn around the ring, where γ is its average Lorentz factor and a_e is the electron magnetic moment anomaly. The *spin tune* is defined as $a_e\gamma$ and the time-averaged spin tune, ν_0, of each electron is proportional to the average beam energy, E_{beam}:

$$\nu_0 = a_e \gamma = \frac{a_e E}{m_e c^2} = \frac{E_{\text{beam}}}{440.6486(1)[\text{MeV}]},$$

where m_e is the mass of the electron and c is the speed of light.

The precession frequency of the polarization vector is precisely measured by inducing a resonant depolarization of the beam with a radial oscillating field from a coil. If the perturbation from the radial field is in phase with the spin precession, then the spin rotations about the radial direction add up coherently from turn to turn and the beam is depolarized.

The average beam energy cannot be measured continuously since in standard LEP running conditions the beams are not polarized. A model based on a large set of monitored quantities (currents in the magnets, temperatures, measurement of magnetic fields, status of RF units, etc...) is used to follow the evolution of the energy as a function of time [ASS 98]. The effects of changing the beam energy have been studied with dedicated experiments in order to provide an assessment of the systematic errors. The model that describes E_{beam} as a function of time is precisely calibrated using the measurements with resonant depolarization. This calibrated model is eventually used to compute the luminosity-weighted energies with a typical precision of 2 parts in 10^5 at each collision point. The resulting errors on M_Z and Γ_Z are about 1.9 MeV and 1.2 MeV.

Cross section measurements   About 70% of the Z bosons produced decay to a $q\bar{q}$ pair which fragments, producing a multihadronic final state. These events carry most of the weight in the lineshape analysis. They can be easily selected with acceptances larger than 97% and very small background, exploiting their multiparticle structure and their high visible energy [ALE 94; DEL 94; L3 94; OPA 94]. Figure 4.3.2 shows the hadronic cross sections measured by the ALEPH collaboration.

Charged lepton pair decays of the Z account for only 10% of the Z decays. Their selections are typically based on criteria of low-multiplicity or high visible energy or momentum. Electron pairs are further separated thanks to the energy deposited in the electromagnetic calorimeters, and muon pairs thanks to their penetration through the dense hadron calorimeters. Typical efficiencies inside the detector acceptance are from around 85% for taus to above 95% for electrons and muons. The main systematic errors concern the reliability of detector simulations (for electrons and muons) and background contamination (for taus).

The cross sections are measured by selecting and counting the number of events, correcting them for the small contribution of background events and the selection efficiency and normalizing them to the rate of a process with well known cross section. At LEP the normalization is given by the Bhabha scattering $e^+e^- \rightarrow e^+e^-$.

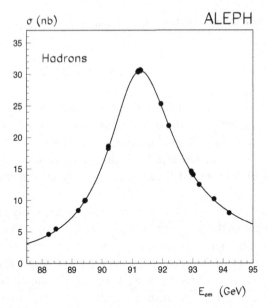

Fig. 4.3.2. Hadronic cross section as a function of center-of-mass energy as measured by the ALEPH collaboration. The solid line represents the MSM fit to the data.

The systematic error on the cross section is dominated by the error on the normalization which is limited by the knowledge of the internal geometry of the detector (0.05%) and by the theoretical error [JAD 96] on the Bhabha cross sections (0.11%).

Results from the Z lineshape fit   The measured cross sections are fitted using a formula that convolutes a reduced cross section $\hat{\sigma}$, function of the Z parameters, with a radiator function $H(s, s')$ which takes into account the corrections for initial state radiation:

$$\sigma_{f\bar{f}}(s) = \int_{4m_f^2}^{s} ds' H(s, s') \hat{\sigma}_{f\bar{f}}(s').$$

The radiator function includes the information on the energy spectrum of radiated photons and is peaked at $s' = s$.

The reduced cross section $\hat{\sigma}$ is

$$\hat{\sigma}_{f\bar{f}}(s) = \sigma_{f\bar{f}}^{peak} \cdot \frac{s\Gamma_Z^2}{\left(s - M_Z^2\right)^2 + \left(s\Gamma_Z/M_Z\right)^2} + (\gamma - Z) + |\gamma|^2.$$

The first term includes the relativistic Breit–Wigner distribution corresponding to the Z exchange. The photon term $|\gamma|^2$ is only a few percent of the Z term and is taken from theory in the fit. The interference term $(\gamma - Z)$ is even smaller and is zero when

Table 4.3.1. *Average line shape parameters from the results of the four LEP experiments*

Parameter	Average value
M_Z (GeV)	91.1867 ± 0.0020
Γ_Z (GeV)	2.4948 ± 0.0025
σ^0_{had} (nb)	41.486 ± 0.053
R_ℓ	20.775 ± 0.027

$s = M_Z^2$. It is taken from theory assuming the validity of the Minimal Standard Model.

The cross section at the peak can be written in terms of the Z mass and width and the Z partial widths to the initial state, Γ_e, and the final state, Γ_f, as:

$$\sigma_{f\bar{f}}^{peak} = \sigma_{f\bar{f}}^0 \left(\frac{1}{1 + \delta_{QED}} \right) = \frac{12\pi}{M_Z^2} \cdot \frac{\Gamma_e \Gamma_f}{\Gamma_Z^2} \cdot \frac{1}{1 + (3\alpha/4\pi)}.$$

Here Γ_f represents the physical partial width of the Z into the fermion pair $f\bar{f}$, and therefore includes by definition all radiative corrections. Since the initial state radiation is taken into account by the convolution procedure, the contribution of the QED final state radiative corrections δ_{QED} is removed from the initial state width Γ_e, thus avoiding a double counting.

Assuming lepton universality, four parameters are needed to describe the s dependence of the hadronic and leptonic cross sections. The set of parameters used is the Z mass (M_Z) and total width (Γ_Z), the ratio of hadronic to leptonic partial widths ($R_\ell = \Gamma_h / \Gamma_\ell$) and the hadronic peak cross section (σ_h^0). These parameters have small correlations. The lineshape parameters, fitted from the data collected by the four LEP collaborations [LWG 97], are shown in Table 4.3.1.

The invisible width Γ_{inv} is obtained from the equation:

$$\Gamma_Z = \Gamma_h + 3\Gamma_\ell + \Gamma_{inv}.$$

The number of fermion generations with a light neutrino, N_ν, is obtained from the ratio of the invisible width to the leptonic width, assuming that the invisible width is only due to neutrino final states:

$$\frac{\Gamma_{inv}}{\Gamma_\ell} = N_\nu \cdot \frac{\Gamma_\nu}{\Gamma_\ell}.$$

The ratio Γ_ν / Γ_ℓ is taken from the MSM: $\Gamma_\nu / \Gamma_\ell = 1.991 \pm 0.001$. The small error in the Minimal Standard Model prediction for this ratio results from the large cancellations of the top and Higgs mass dependences. The result is:

$$N_\nu = 2.993 \pm 0.011,$$

showing that only three fermion generations exist with light neutrinos.

4.3.4 Asymmetries at the Z pole

Parity violation of the weak neutral current is caused by the difference of the couplings of the Z to right-handed and left-handed fermions and anti-fermions. The asymmetries in the measured cross sections are proportional to the quantity:

$$A_f = \frac{2\,g_V^f g_A^f}{(g_V^f)^2 + (g_A^f)^2} = \frac{2\,g_V^f/g_A^f}{1 + (g_V^f/g_A^f)^2},$$

where f is the flavor of the initial or final state fermions.

Beam polarization provides the most natural way to separate initial from final state couplings. The simplest way to access initial state couplings with polarized beams is the measurement of the left–right asymmetry A_{LR}. The cross sections for Z production σ_l and σ_r are measured at SLD with the electron beam having left-handed(l) or right-handed(r) polarization while the positron beam is unpolarized. The left–right asymmetry is defined as:

$$A_{LR} = \frac{1}{\mathcal{P}}\frac{\sigma_l - \sigma_r}{\sigma_l + \sigma_r},$$

where $\mathcal{P}$ is the average beam polarization. Up to very small corrections A_{LR} is equal to $\mathcal{A}_e$, thus providing a direct measurement of the ratio of the couplings of the neutral current to the electron.

The sign of the polarization is randomly chosen at the frequency of the SLAC machine pulse rate; in this way the measurement is not affected by time variations of the apparatus efficiency. The longitudinal beam polarization is measured with an error smaller than 1% [SLD 97a] giving the main systematic error on the asymmetry determination. The measured value of the asymmetry A_{LR} is corrected by +0.0029 to take into account the effects of the photon exchange, the $Z - \gamma$ interference and initial and final state radiation. SLD has recently obtained [SLD 97a]:

$$A_{LR}^0 = 0.1525 \pm 0.0029.$$

The ratio between the couplings of the leptons is usually expressed in terms of the effective electroweak mixing parameter $\sin^2\theta_{eff}$:

$$\sin^2\theta_{eff} = \frac{1}{4}\left(1 - \frac{g_V^\ell}{g_A^\ell}\right).$$

This SLD measurement corresponds to $\sin^2\theta_{eff} = 0.23084 \pm 0.00035$.

The effective mixing parameter is measured at LEP in many different reactions, notably the forward–backward asymmetry in the Z decays into charged leptons, the tau polarization asymmetries, the charge asymmetry in hadronic Z decays and the forward–backward asymmetry of the Z decays into $b\bar{b}$ and $c\bar{c}$. A compilation of

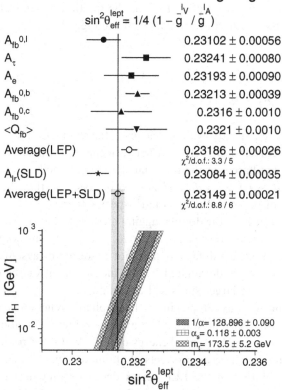

Fig. 4.3.3. The most precise determinations of $\sin^2\theta_{eff}$ and their average are compared with the Standard Model expection as function of the Higgs Boson mass. The error on the prediction is shown as a hatched zone.

the various measurements is shown in Fig. 4.3.3. The average of the different determinations [LWG 97], including the SLD measurement reported above, is $\sin^2\theta_{eff} = 0.23149 \pm 0.00021$ with a χ^2 of eight for six degrees of freedom.

4.3.5 The measurement of R_b

The $Z \rightarrow b\bar{b}$ vertex receives additional radiative corrections from diagrams involving the top quark which are suppressed for other flavors. The size of these effects is about 2% and a precise measurement of the ratio

$$R_b \equiv \Gamma_{b\bar{b}}/\Gamma_{had}$$

is sensitive to them.

The most precise measurements of R_b take advantage of the fact that the b hadrons produced in $Z \rightarrow b\bar{b}$ are typically boosted in opposite directions. Therefore it is useful to divide the events into two hemispheres according to the thrust axis, and to apply a *b* tag on both sides of the event. This allows us to measure the *b* tagging efficiency directly on data. If $\mathcal{F}_1$ is the fraction of hemispheres which are tagged and $\mathcal{F}_2$ the fraction of events which are tagged in both hemispheres, one can write

$$\mathcal{F}_1 = R_b \cdot (\epsilon_b - \epsilon_{uds}) + R_c \cdot (\epsilon_c - \epsilon_{uds}) + \epsilon_{uds}$$
$$\mathcal{F}_2 = R_b \cdot (C_b \cdot \epsilon_b^2 - \epsilon_{uds}^2) + R_c \cdot (\epsilon_c^2 - \epsilon_{uds}^2) + \epsilon_{uds}^2,$$

where ϵ_b, ϵ_c, ϵ_{uds} are respectively the efficiency of the tag on *b*, *c* and *uds* events and C_b is the efficiency correlation coefficient between the two hemispheres for *b* events. These relations are based on the fact that the tagging methods have very similar efficiencies for $Z \to u\bar{u}$, $Z \to d\bar{d}$ and $Z \to s\bar{s}$ decays, and that the sum of partial decay fractions of the *Z* to the five quark species is one.

The main advantage of the double tagging method is that the two equations can be used simultaneously to measure both R_b and ϵ_b, in this way the *b* tagging efficiency is measured on data and it is not a source of systematic error. The statistical error on R_b is dominated by the measurement of the double tagging fraction which has the larger statistical uncertainty.

The most efficient way of tagging *b* hemispheres relies on the large impact parameter of the *b* decay products. The average *b* lifetime is about 1.5 ps and *b* hadrons are produced with a typical energy of 30 GeV at the *Z* peak, resulting in an impact parameter of the decay products of about 300 μm to be compared with an experimental resolution of the LEP vertex detectors ranging from 20 to 70 μm depending on the momentum of the track. By using these variables very high purity samples (more than 96% purity) can be selected while keeping efficiencies in excess of 25%.

The measurement of R_b is directly affected by the knowledge of C_b, i.e. by the uncertainty on the correlation coefficient. Efficiency correlations originate from detector effects, from the physics or from the algorithm itself. Great care is taken by the most recent and precise measurements [ALE 96; DEL 97; L3 94; OPA 97; SLD 97b] to identify the cause of the correlations and make sure that the Monte Carlo simulation is reliable in predicting the size of the effect. In these measurements the value of the total correlation coefficient C_b is close to one, any deviation due to individual components is of the order of 1%, or less.

Lighter quark efficiencies are another source of systematic errors. In particular, the uncertainty in ϵ_c is typically the greatest source of systematic error, this quantity is evaluated with the help of Monte Carlo simulations tuned to reproduce charm hadron properties (lifetimes, decay multiplicities) measured in lower energy experiments.

The average [LWG 97] of LEP and SLD measurements of R_b is:

$$R_b = 0.2177 \pm 0.0011.$$

4.3.6 The W mass measurement at LEP

Pairs of W bosons have been produced at LEP near the production threshold (161 GeV) and at higher center-of-mass energies (172 and 183 GeV). The measurement of the cross section at threshold ($\sqrt{s} \sim 2m_W + 0.5$ GeV) provides a sensitive measurement of the W mass because the dependence of the cross section on the mass is mainly of kinematical origin. However, since the cross section is measured only at one point, the determination of the mass can only be done within the framework of the Standard Model, though with very small dependence on its input parameters. At higher energies the W mass is measured from direct reconstruction of the final state.

The event selection is simple for the final states where at least one W decays into lepton plus neutrino. The final state $\ell\ell\nu\nu$ (11%) has two acoplanar leptons and missing energy, while the final state $q'\bar{q}\ell\nu$ (43%) has an isolated lepton, two hadronic jets and isolated missing energy. These channels can be selected with high efficiency and very low background. The totally hadronic channel (46%) has a large QCD background that is less relevant at 172 GeV where the WW cross section is larger. The events are selected using many topological properties that are combined in a single distribution (neural networks, likelihood function, weights) which is compared with the predicted distributions for the signal and the background obtained with Monte Carlo simulation.

In the 161 GeV data sample each experiment has selected typically 5 events in the $\ell\ell\nu\nu$ channel and 15 events in the $q'\bar{q}\ell\nu$ channel. The cross sections measured in each channel are combined using the SM branching ratios for the W decays. The results of the four experiments are averaged giving a cross section for the production of two *real* Ws in the final state of 3.69 ± 0.45 pb that corresponds to $M_W = 80.40 \pm 0.22$ GeV.

At higher energies the W mass is measured by direct reconstruction using the $q'\bar{q}\ell\nu$ and the fully hadronic channels. A value of the W mass is extracted for each event using the energies and the directions of the reconstructed jets and leptons, applying the constraints of energy and momentum conservation and imposing, in some cases, the equality of the two W masses. With this procedure, the absolute energy scale of the W mass is constrained by the beam energy. The measured distribution is compared with the Monte Carlo expectations for many W masses to fit M_W. Each Collaboration selected about 800 WW pairs for this measurement that is statistically limited. The main systematic errors come from the simulation of the jets and, in the fully hadronic events, from final state effects involving quarks or hadrons from the decays of the two Ws.

The two values of the W mass measured at LEP with two different techniques are comparable in precision. Their average is [LWG 97]:

$$M_W = 80.35 \pm 0.09 \text{ GeV}.$$

4.3.7 *The Tevatron $p\bar{p}$ collider*

The first truly high-energy hadron collider was the CERN proton-antiproton collider built and operated in the SPS tunnel at CERN (known as the $Sp\bar{p}S$ Collider). This machine was conceived to discover the W and Z bosons. The $p\bar{p}$ collisions required the capability to create and accelerate large numbers of antiprotons. This was a great technical challenge, but the investment was justified both by the simplification of using a single ring for both beams, and by the need to provide the highest possible energy in the parton–parton center-of-mass, which could only be achieved by colliding valence quarks and antiquarks. The antiproton source was developed in the late 70s, and required initial production of antiprotons on a target using a proton beam. The antiprotons were then accumulated in a special storage ring and 'cooled' to reduce their momentum and angular spread [VDM 72, 81]. The accumulation of a large enough number of antiprotons for colliding beam operation typically requires almost 24 hours. Once there was an adequate store of antiprotons, bunches of protons and antiprotons were injected into the SPS accelerator ring in opposite directions. Since the two beams have the same mass and energy, they follow identical, but counter-rotating, paths in the accelerator. These bunches of protons and antiprotons were brought into collision in the experimental areas providing a center-of-mass energy of 630 GeV, equal to twice the individual beam energy.

A parallel development effort was carried out at Fermilab, including the installation of the new superconducting proton accelerator ring capable of providing operation at 900 GeV [EDW 85], a significant improvement over the existing conventional magnet 400 GeV Main Ring accelerator, and an antiproton source capable of providing a luminosity of at least $10^{30}\,\mathrm{cm}^{-2}\,\mathrm{s}^{-1}$ [PEO 83]. First collisions in this new complex were obtained in 1985, but real physics data was first available during what is now called Run 0 in the 1988–1989 period. The era which provided the data samples whose results are summarized here began with Run 1 in 1992. This run extended over a period of roughly three years, and the total integrated luminosity acquired was roughly 100 pb^{-1} in each experiment. The first part of this Run, known as Run 1a, provided an integrated luminosity of about 20 pb^{-1}, and the reminder was delivered in a later period referred to as Run 1b.

During Run 1, six bunches each of protons and antiprotons were collided in the Tevatron ring, with a separation between collisions of about 3.5 µs. The peak luminosity during Run 1a reached $0.8 \times 10^{31}\,\mathrm{cm}^{-2}\,\mathrm{s}^{-1}$, and after further improvements for Run 1b, a peak luminosity of $2.5 \times 10^{31}\,\mathrm{cm}^{-2}\,\mathrm{s}^{-1}$ was achieved. At this bunch spacing interval, this corresponds to almost four inelastic interactions per bunch crossing in each experiment, noticeably increasing the soft particle backgrounds. The bunch lengths in the Tevatron are relatively large, leading to a luminous region with an RMS size of approximately 25 cm. This allows relatively easy separation of different inelastic collisions along the beam direction, but places a premium on coverage to allow interactions that are significantly displaced along the beam direction to be reconstructed without bias.

4.3.8 The Tevatron detectors: CDF and D0

The CDF detector [CDF 88] is a magnetic spectrometer based on a large super-conducting solenoid. The tracking system is contained inside the uniform 1.5 T field created by this magnet, and includes a silicon strip microvertex detector, followed by a small Time Projections Chamber which provides overall vertex information, followed by a large open-cell drift chamber which provides 84 samples along the track length to give excellent momentum resolution. The coil is surrounded by scintillation calorimetry in the central region, and gas proportional chamber calorimetry in the forward region, extending down to an angle of several degrees from the beam axis to provide the complete coverage needed for missing energy analyses. The calorimeters are surrounded in the central region by a muon system consisting of gas chambers behind the calorimeters, and additional chambers after layers of steel for improved muon identification.

The D0 detector [D0 94] is a non-magnetic detector with a small radius gas-based tracking system to reconstruct charged particle tracks. This is followed by a large and highly segmented liquid argon calorimeter which uses uranium absorber plates. This system provides a hermetic energy measurement which is segmented into approximately 5000 projective towers. The calorimeters are surrounded by a large iron-toroid muon spectrometer which provides muon coverage over the full solid angle. Gas drift tubes are placed in front of and behind the toroids to measure the direction of the muon before and after bending in the magnetic field.

4.3.9 Production of W and Z in hadron colliders

The production of W and Z bosons in a hadron collider occurs predominantly via the Drell–Yan mechanism [DRE 70], whereby a quark from an incoming proton annihilates with an antiquark from an incoming antiproton to form the gauge boson via a process such as $u\bar{d} \rightarrow W^+$ or $u\bar{u} \rightarrow Z$. The longitudinal momentum distribution is related to the difference between the x values of the quark and anti-quark:

$$P_L(W) = \sqrt{s}(x_q - x_{\bar{q}})/2,$$

whereas the transverse momentum distribution is approximately zero. Higher order QCD corrections modify this picture somewhat, leading to observed W and Z transverse momenta of 5–10 GeV. The W is detected via its decay to a lepton ($\ell = e, \mu$), and a neutrino. In the Standard Model, BR($W \rightarrow \ell\nu$) = 0.108. The Z is detected by its decay to two leptons. In the Standard Model, BR($Z \rightarrow \ell^+\ell^-$) = 0.033.

The identification of electrons and muons in a hadron collider environment has now become a well-established art, with rejections against QCD jets of 10^5 for efficiencies of 90% and lepton $P_T > 15$ GeV. For the $W \to \ell\nu$ decay, it is also vital to reconstruct the missing neutrino via energy conservation. It is essentially impossible to reconstruct the total energy of an event in a hadron collider, as typically only about 5% of the total center-of-mass energy for a typical 'minimum bias' event[1] appears in the central rapidity plateau covered by active calorimetry in the collider detectors. Thus, one is forced to rely on transverse energy balance to reconstruct $P_T(\nu)$ in W events:

$$\vec{P}_T(W) = \vec{P}_T(\ell) + \vec{P}_T(\nu) \simeq -\vec{P}_T(\text{hadrons}),$$

where

$$\vec{P}_T(\text{hadrons}) = \left\{ \sum E_{\text{tower}} \hat{v}_{\text{tower}} \right\}_T,$$

with E_{tower} being the energy in a given calorimeter tower, and $\hat{v}_{\text{tower}}$ being a unit vector from the event vertex to the centre of the given tower. Note that it is important to exclude any energy deposited in the calorimeter by the lepton in order to avoid double-counting.

The W mass measurement at the Tevatron   The single most important property of the W and Z bosons from the perspective of electroweak theory is their mass. The $Z \to \ell\ell$ decay, with the observation and accurate reconstruction of both leptons, provides a direct measurement of the Z mass. The $W \to \ell\nu$ decay, with its unobserved neutrino, requires an indirect measurement of the mass using variables which are strongly correlated with M_W.

The basic variables are the 3-vectors of the lepton and the neutrino. Since the longitudinal momentum of the neutrino is not measurable, that of the lepton is also of little use, and we must rely on $\vec{P}_T(\ell)$ and $\vec{P}_T(\nu)$. Since the W decay is a 2-body decay, the distributions of these momenta carry substantial information about the W mass. Unfortunately, they are also very sensitive to assumptions about the transverse momentum of the W. A very useful combination, with greatly reduced sensitivity to the transverse momentum of the W, is the transverse mass:

$$M_T^2 = 2P_T(\ell)P_T(\nu)[1 - \cos\phi_{\ell\nu}],$$

where $\phi_{\ell\nu}$ is the angle between the two vectors in the transverse plane. This variable is not a mass (it is not Lorentz invariant), but it contains very useful information about the W mass.

---

[1] So-called 'minimum bias' events are selected by making very minimal requirements, such as the presence of at least two charged particles in a large region of the detector. Such triggers generally see $\geq 95\%$ of the total inelastic cross section in a hadron collider, and thus provide an unbiased sample of 'typical' events.

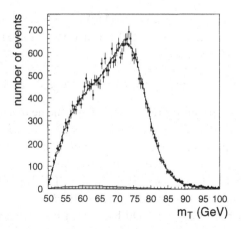

Fig. 4.3.4. The fit to the transverse mass for the Run 1b $W \rightarrow e\nu$ data in the D0 experiment. The shaded area represents the background contribution.

The D0 experiment at the Tevatron uses a large sample of the electron decays of the W and Z to measure the W mass in Run 1b. The momentum resolution of the muon toroid system is not good enough to use the muon decays for this analysis. The $Z \rightarrow e^+e^-$ events are used as a calibration sample in order to fix the calorimeter energy scale by forcing the Z mass to agree with the precise measurements from LEP. A total sample of 3.5K events provides a measurement of the energy scale to 0.09%, while also constraining the energy resolution by imposing the measured $\Gamma(Z)$ from LEP. A much larger sample of 28.3 K $W \rightarrow e\nu$ decays is used to measure the W mass. The resulting fit, including the region in transverse mass from 60 GeV to 90 GeV, is shown in Fig. 4.3.4, and leads to a mass of:

$$M_W = 80.44 \pm 0.10 \text{ (stat)} \pm 0.07 \text{ (syst) GeV}.$$

Here, the statistical uncertainty from the limited Z statistics used for the energy scale calibration has also been included in the statistical error. The dominant systematic errors arise from the resolutions for both the electron and the neutrino, as well as uncertainties in the W production details [D0 97]. When combined with the earlier published mass from Run 1a [D0 96], this leads to the final Run 1 D0 result of $M_W = 80.43 \pm 0.11$ GeV.

The CDF experiment uses both the electron and muon final states to measure the W mass. The muon measurement uses the momentum scale, calibrated using the $\psi(3096)$, and checked using the $\Upsilon(1S)$ and Z masses, as the foundation for the measurement. A sample of about 250 K $\psi(3096) \rightarrow \mu^+\mu^-$ events is used to fit for $M(\psi)$, including radiative effects and B backgrounds. This result can then be normalized to the extremely precise measurement of this quantity from e^+e^- colliders. The fitted mass has an error of 1.5 MeV, which is dominated by uncertainties in energy loss corrections for the low P_T muons, and by possible nonlinearities in the extrapolation to the W mass scale. The total uncertainty on the

W mass from the tracking scale is about 40 MeV. A sample of about 21 K $W \to \mu\nu$ events from Run 1b are fitted, including the transverse mass region from 65 GeV to 100 GeV to produce the result of:

$$M_W = 80.43 \pm 0.10 \text{ (stat)} \pm 0.12 \text{ (syst) GeV}.$$

The dominant uncertainty in this result arises from the neutrino resolution and the knowledge of the W transverse momentum distribution. This result has been combined with the previous CDF results from Run 1a [CDF 95a] correctly propagating the systematic errors, to give a Run 1 CDF result of $M_W = 80.38 \pm 0.12$ GeV. Upon completion of the CDF Run 1b electron analysis, the total error is expected to be reduced below 100 MeV.

The results from both CDF and D0 have been combined to give the present Tevatron result of:

$$M_W = 80.41 \pm 0.09 \text{ GeV}.$$

The average of the Tevatron measurement of the W mass and the LEP measurement of the W mass gives:

$$M_W = 80.375 \pm 0.064 \text{ GeV}.$$

4.3.10 *The properties of the top quark*

When the top quark is heavier than the W, the Standard Model predicts that it is produced at the Tevatron predominantly by the pair-production process $q\bar{q} \to t\bar{t}$, and it should decay essentially only to $t \to Wb$. This results in three types of final states: $t\bar{t} \to WWb\bar{b} \to \ell\nu\ell\nu b\bar{b}$ (dileptons), $t\bar{t} \to \ell\nu jjb\bar{b}$ (lepton plus jets), and $t\bar{t} \to jjjjb\bar{b}$ (all-hadronic).

The first evidence for the observation of the top quark at the Fermilab Tevatron appeared during Run 1a, and was published in 1994 by CDF [CDF 94]. With the increased statistics of Run 1b, this evidence was clearly confirmed by both collaborations [CDF 95b, D0 95]. In the case of CDF, it is easier to isolate a clean signal for $t\bar{t}$, through the use of the micro-vertex detector to tag b-quarks via the displaced secondary vertices in their decays. In the D0 experiment, sophisticated kinematic criteria are used to produce an enriched sample of $t\bar{t}$ events.

The top quark mass   The best sample of $t\bar{t}$ events to use for measuring the top quark mass is the 'lepton + jets' sample, in which one W has decayed hadronically, and the other decays leptonically, leading to a $W + 4$ jet signature. These samples are selected by requiring an isolated high-P_T lepton, missing transverse energy from the neutrino in the W decay, and at least four jets. Jets arising from decaying b-quarks are tagged either by the presence of a secondary vertex, or by the presence of a high-P_T lepton in the jet from the semi-leptonic decay

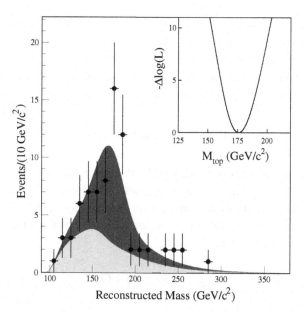

Fig. 4.3.5. The reconstructed top mass for the Run 1b CDF $W+4$ jet analysis. The data (points) are compared with the fit result. The light shading indicates the backround contribution.

of the b-quark. The top quark mass is reconstructed using kinematic fitting to improve the resolution. The jets are treated as elementary particles, and in addition to energy and momentum conservation, the constraints that the t and $\bar{t}$ masses are identical, and that there are two W masses in the decay chain, are imposed. The result is a two-constraint (2C) kinematic fit. All permutations of b assignments for the jets are considered, unless there is b-tagging information available.

The CDF mass measurement begins with a sample of 76 $W+4$ jet candidate events which satisfy event selection criteria and a χ^2 cut on the kinematic fit hypothesis. These events are then partitioned into four sub-samples: events with a single vertex b-tag, events with two vertex b-tags, events with a lepton b-tag and no vertex b-tag, and events with no b-tag which satisfy tight kinematic requirements. A separate background analysis is done for each sub-sample. Each sample is fitted separately, and the results are combined to give:

$$m_{\text{top}} = 175.9 \pm 4.8 \text{ (stat)} \pm 4.9 \text{ (syst) GeV}.$$

The fit results, superimposed on the combined data sample, are shown in Fig. 4.3.5. The systematic error is dominated by the jet energy scale uncertainty which contributes ± 4.4 GeV. Additional uncertainties arise from initial and final state radiation effects and backgrounds.

In the D0 experiment, an initial sample of 77 $W+4$ jet candidate events is selected, all of which pass a χ^2 cut on the kinematic fit as well as event selection criteria. A further subset of 31 events are selected which satisfy a likelihood requirement based on a set of four kinematic criteria designed to separate signal from background without introducing a bias in the top mass fit. In order to use all of the available events to extract a mass, the full sample of 77 events is fitted to the signal plus background hypothesis. This leads to the result:

$$m_{top} = 173.3 \pm 5.6 \text{ (stat)} \pm 5.5 \text{ (syst) GeV}.$$

The systematic uncertainty is dominated by the uncertainty on the jet energy scale, which is ± 4.0 GeV. Additional uncertainties arise from the variations between different models for $t\bar{t}$ production and for the fragmentation of the decay products.

The combined CDF and D0 top mass, including results from fits to the dilepton and all-hadronic final states, is:

$$m_{top} = 174.1 \pm 5.4 \text{ GeV}.$$

4.4 Precision tests of the electroweak theory*

4.4.1 The data

In recent years new powerful tests of the Standard Model (SM) have been performed mainly at LEP but also at SLC, at the Tevatron and in fixed target experiments. The running of LEP1 was terminated in 1995 and close-to-final results of the data analysis are now available [ABB 98]. The experiments at the Z_0 resonance have enormously improved the accuracy in the electroweak neutral-current sector. The LEP2 programme is in progress. Here we report on the results of these tests. The validity of the SM has been confirmed to a level that we can say was unexpected at the beginning. In the present data there is no significant evidence for departures from the SM, no convincing hint of new physics (also including the first results from LEP2). The impressive success of the SM poses strong limitations on the possible forms of new physics. Favoured are models of the Higgs sector and of new physics that preserve the SM structure and only very delicately improve it, as is the case for fundamental Higgs(es) and Supersymmetry. Disfavoured are models with a nearby strong non-perturbative regime that almost inevitably would affect the radiative corrections, as for composite Higgs(es) or technicolour and its variants.

The relevant electroweak data together with their SM values are presented in Table 4.4.1 [ABB 98]. The SM predictions correspond to a fit of all the available data (including the directly measured values of m_t and m_W) in terms of m_t, m_H and $\alpha_s(m_Z)$, described later in Section 4.4.2, Table 4.4.4

---

* G. Altarelli, Theoretical Physics Division, CERN, Switzerland and Università di Roma Tre, Rome, Italy. F. Caravaglios, Theoretical Physics Division, CERN, Switzerland. For a review of the theoretical background see Chapter 3 by L. Maiani.

Table 4.4.1 *The electroweak data*

Quantity	Data (May 98)	Standard model fit	Pull
m_Z (GeV)	91.1867(20)	91.1866	0.1
Γ_Z (GeV)	2.4948(25)	2.4965	−0.7
σ_h (nb)	41.486(53)	41.466	0.4
R_h	20.775(27)	20.757	0.7
R_b	0.2173(9)	0.2159	1.6
R_c	0.1731(44)	0.17225	0.2
A_l^{FB}	0.0171(10)	0.01621	0.9
A_τ	0.1400(63)	0.1470	−1.1
A_e	0.1438(71)	0.1470	−0.5
A_b^{FB}	0.0998(22)	0.10305	−1.5
A_c^{FB}	0.0735(45)	0.07363	−0.0
A_b (SLD direct)	0.899(49)	0.9347	−0.7
A_c (SLD direct)	0.660(45)	0.668	−0.2
$\sin^2\theta_{\text{eff}}$ (LEP-combined)	0.23185(26)	0.23152	1.5
$A_{LR} \rightarrow \sin^2\theta_{\text{eff}}$	0.23084(35)	0.23152	−2.0
m_W (GeV) (LEP2 + $p\bar{p}$)	80.375(64)	80.3707	0.3
$1 - (m_W^2/m_Z^2)$ (νN)	0.2253(21)	0.22316	1.0
Q_W (Atomic PV in Cs)	−72.11(93)	−73.20	1.2
m_t (GeV)	174.1(5.4)	170.4	0.7

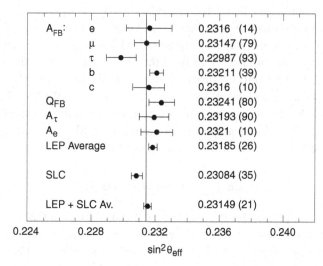

Fig. 4.4.1.   The collected measurements of $\sin^2\theta_{eff}$.

Other important derived quantities are, for example, N_ν, the number of light neutrinos, obtained from the invisible width: $N_\nu = 2.993(11)$, which shows that only three fermion generations exist with $m_\nu < 45$ GeV, or the leptonic width Γ_l, averaged over e, μ and τ: $\Gamma_l = 83.91(10)$ MeV, or the hadronic width: $\Gamma_h = 1743.2(2.3)$ MeV.

For indicative purposes, in Table 4.4.1 the 'pulls' are also shown, defined as: pull = (data point − fit value)/(error on data point). At a glance we see that the agreement with the SM is quite good. The distribution of the pulls is statistically normal. The presence of few ∼ 2σ deviations is what is to be expected. However, it is maybe worthwhile to give a closer look at these small discrepancies.

Perhaps the most annoying feature of the data is the persistent difference between the values of $\sin^2\theta_{eff}$ measured at LEP and at SLC (although the discrepancy went down in the latest data). The value of $\sin^2\theta_{eff}$ is obtained from a set of combined asymmetries. From asymmetries one derives the ratio $x = g_V^l/g_A^l$ of the vector and axial vector couplings of the Z_0, averaged over the charged leptons. In turn $\sin^2\theta_{eff}$ is defined by $x = 1 - 4 \sin^2\theta_{eff}$. SLD obtains x from the single measurement of A_{LR}, the left–right asymmetry, which requires longitudinally polarized beams. The distribution of the present measurements of $\sin^2\theta_{eff}$ is shown in Fig. 4.4.1. The LEP average, $\sin^2\theta_{eff}$=0.23185(26), differs by 2.3σ from the SLD value $\sin^2\theta_{eff}$=0.23084(35). The most precise individual measurement at LEP is from A_b^{FB}: the combined LEP error on this quantity is about the same as the SLD error, but the two values are 2.4σ away. One might attribute this to the fact that the b measurement is more delicate and affected by a complicated systematics. In fact one notices from Fig. 4.4.1 that the value obtained at LEP from A_l^{FB}, the average for $l = e, \mu$ and τ, is somewhat low (indeed quite in agreement with the SLD value). However, the statement that LEP and SLD agree on leptons while they only disagree when the b quark is considered is not quite right. First, the value of A_e, a quantity essentially identical to A_{LR}, measured at LEP from the angular distribution of the τ polarization, differs by 1.1σ from the SLD value. Second, the low value of $\sin^2\theta_{eff}$ found at LEP from A_l^{FB} turns out to be entirely due to the τ lepton channel which leads to a central value above that of e and μ [ABB 98]. The e and μ asymmetries, which are experimentally simpler, are perfectly on top of the SM fit. Thus it is difficult to find a simple explanation for the SLD–LEP discrepancy on $\sin^2\theta_{eff}$. In the following we will tentatively use the official average [ABB 98]

$$\sin^2\theta_{eff} = 0.23149 \pm 0.00021 \tag{4.41}$$

obtained by a simple combination of the LEP and SLD data. One could be more conservative and somewhat enlarge the error because of the larger dispersion of the data (as we did until recently [ALT 98]), but the difference would not be large. In fact this dispersion has decreased in the last run of data, which is encouraging. The data-taking by the SLD experiment is still in progress and also at LEP sizeable improvements on A_τ and A_b^{FB} are foreseen as soon as the corresponding analyses are completed. We hope to see the difference further reduced in the end.

From the above discussion one may wonder if there is evidence for something special in the τ channel, or equivalently if lepton universality is really supported by the data. Indeed this is the case: the hint of a difference in A_τ^{FB} with respect to the corresponding e and μ asymmetries is not confirmed by the measurements of A_τ

and Γ_τ which appear normal [ABB 98]. In principle the fact that an anomaly shows up in A_τ^{FB} and not in A_τ and Γ_τ is not unconceivable because the FB lepton asymmetries are very small and very precisely measured. For example, the extraction of A_τ^{FB} from the data on the angular distribution of τs could be biased if the imaginary part of the continuum was altered by some non-universal new physics effect [CAR 97]. But a more trivial experimental problem is at the moment quite plausible.

A similar question can be asked for the b couplings. We have seen that the measured value of A_b^{FB} is 1.6σ below the SM fit. At the same time R_b, which used to show a major discrepancy, is now only about 1.6σ away from the SM fit (as a result of the more sophisticated second generation experimental procedures). It is often stated that there is a -2.1σ deviation on the measured value of A_b vs the SM expectation [ABB 98]. But in fact that depends on how the data are combined. In our opinion one should rather talk of a -1.5σ effect. We recall that A_b can be measured directly at SLC by taking advantage of the beam longitudinal polarization. At LEP one measures $A_b^{FB} = 3/4\, A_e A_b$. One can then derive A_b by inserting a value for A_e. The question is what to use for A_e: the LEP value obtained, using lepton universality, from the measurements of A_l^{FB}, A_τ, A_e which is: $A_e = 0.1465(33)$, or the combination of LEP and SLD etc. The LEP electroweak working group adopts for A_e the SLD + LEP average value which also includes A_{LR} from SLD: $A_e = 0.1499(21)$. This procedure leads to a -2.1σ deviation. However, in this case, the well known $\sim 2\sigma$ discrepancy of A_{LR} with respect to A_e measured at LEP (and also to the SM fit), which is not related to the b couplings, further contributes to inflate the number of σs. Since we are here concerned with the b couplings it is perhaps wiser to obtain A_b from LEP by using the SM value for A_e (that is the pull-zero value of Table 4.4.1): $A_e^{SM} = 0.1475(16)$. With the value of A_b derived in this way from LEP we finally obtain $A_b = 0.902 \pm 0.022$(LEP + SLD, $A_e = A_e^{SM}$: -1.5σ). In the SM A_b is so close to 1 because the b quark is almost purely left-handed. A_b only depends on the ratio $r = (g_R/g_L)^2$ which in the SM is small: $r \sim 0.033$. To adequately decrease A_b from its SM value one must increase r by a factor of about 1.6, which appears large for a new physics effect. Also such a large change in r must be compensated by decreasing g_L^2 by a small but fine-tuned amount in order to counterbalance the corresponding large positive shift in R_b. In view of this, the most likely way out is that A_b^{FB} and A_b have been a bit underestimated at LEP and actually there is no anomaly in the b couplings. Then the LEP value of $\sin^2\theta_{eff}$ would slightly move down, in the direction of decreasing the SLD–LEP discrepancy.

4.4.2 Precision electroweak data and the Standard Model

For the analysis of electroweak data in the SM one starts from the input parameters: some of them, α, G_F and m_Z, are very well measured, some other ones, $m_{f_{light}}$, m_t and $\alpha_s(m_Z)$ are only approximately determined, while m_H is largely unknown.

With respect to m_t the situation has much improved since the CDF/D0 direct measurement of the top quark mass [GIR 97]. From the input parameters one computes the radiative corrections [ALT 89] to a sufficient precision to match the experimental capabilities. Then one compares the theoretical predictions and the data for the numerous observables which have been measured, checks the consistency of the theory and derives constraints on m_t, $\alpha_s(m_Z)$ and hopefully also on m_H.

Some comments on the least known of the input parameters are now in order. The only practically relevant terms where precise values of the light quark masses, $m_{f_{\text{light}}}$, are needed are those related to the hadronic contribution to the photon vacuum polarization diagrams, that determine $\alpha(m_Z)$. This correction is of order 6%, much larger than the accuracy of a few per mille of the precision tests. Fortunately the imaginary part of the hadronic contribution to the photon vacuum polarization diagram is proportional to the $e^+e^- \rightarrow hadrons$ cross section. Thus one can insert the actual data in a dispersive integral to solve the related ambiguity. But the leftover uncertainty is still one of the main sources of theoretical error. In recent years there has been a lot of activity on this subject and a number of independent estimates of $\alpha(m_Z)$ have appeared in the literature [PIE 97]. A consensus has been established in the sense that from the data alone the result is

$$\alpha(m_Z)^{-1} = 128.90 \pm 0.09. \tag{4.4.2}$$

For the derivation of this result the QCD theoretical prediction is actually used for large values of s where the data do not exist. But the sensitivity of the dispersive integral to this region is strongly suppressed, so that no important model dependence is introduced. More recently some analyses have appeared where one studied by how much the error on $\alpha(m_Z)$ is reduced by using the QCD prediction down to $\sqrt{s} = m_\tau$, with the possible exclusion of the regions around the charm and beauty thresholds [GRO 98]. These attempts were motivated by the apparent success of QCD predictions in τ decays, despite the low τ mass value (note however that the relevant currents are different: e.g. $V-A$ in τ decay but V in the present case and so on). One finds that the error is reduced from 0.09 down to something like 0.03–0.04, but of course at the price of more model dependence. For this reason, in the following we shall keep the conservative value in Eq. (4.4.2). As for the strong coupling $\alpha_s(m_Z)$, the world average central value is by now quite stable. The error is going down because the dispersion among the different measurements is much smaller in the most recent set of data. The most important determinations of $\alpha_s(m_Z)$ are summarized in Table 4.4.2 [CAT 97]. For all entries, the main sources of error are the theoretical ambiguities which are larger than the experimental errors. The only exception is the measurement from the electroweak precision tests (see later, Table 4.4.4), but only if one assumes that the SM electroweak sector is correct. Our personal views on the theoretical errors are reflected in Table 4.4.2. The error on the final average is taken by all authors between ±0.003 and ±0.005 depending on how

Table 4.4.2 *Measurements of $\alpha_s(m_Z)$. In parenthesis we indicate if the dominant source of errors is theoretical or experimental. For theoretical ambiguities our personal figure of merit is given*

Measurements	$\alpha_s(m_Z)$
R_τ	0.122 ± 0.006 (Th)
Deep inelastic scattering	0.116 ± 0.005 (Th)
Y_{decay}	0.112 ± 0.010 (Th)
Lattice QCD	0.117 ± 0.007 (Th)
Re^+e^- ($\sqrt{s} < 62\,\text{GeV}$)	0.124 ± 0.021 (Exp)
Fragmentation functions in e^+e^-	0.124 ± 0.012 (Th)
Jets in e^+e^- at and below the Z	0.121 ± 0.008 (Th)
Z line shape (assuming SM)	0.120 ± 0.004 (Exp)

Table 4.4.3 *Errors from different sources: Δ^{exp}_{now} is the present experimental error; $\Delta\alpha^{-1}$ is the impact of $\Delta\alpha^{-1} = \pm 0.09$; Δ_{th} is the estimated theoretical error from higher orders, Δm_t is from $\Delta m_t = \pm 6\,GeV$; Δm_H is from $\Delta m_H = 60-1000\,GeV$; $\Delta\alpha_s$ corresponds to $\Delta\alpha_s = \pm 0.003$. The epsilon parameters are defined in [ALT 98]*

Parameter	Δ^{exp}_{now}	$\Delta\alpha^{-1}$	Δ_{th}	Δm_t	Δm_H	$\Delta\alpha_s$
Γ_Z (MeV)	± 2.5	± 0.7	± 0.8	± 1.4	± 4.6	± 1.7
σ_h (pb)	53	1	4.3	3.3	4	17
$R_h \cdot 10^3$	27	4.3	5	2	13.5	20
Γ_1 (keV)	100	11	15	55	120	3.5
$A^l_{FB} \cdot 10^4$	10	4.2	1.3	3.3	13	0.18
$\sin^2\theta \cdot 10^4$	2.1	2.3	0.8	1.9	7.5	0.1
m_W (MeV)	64	12	9	37	100	2.2
$R_b \cdot 10^4$	9	0.1	1	2.1	0.25	0
$\epsilon_1 \cdot 10^3$	1.2		~ 0.1			0.2
$\epsilon_3 \cdot 10^3$	1.2	0.6	~ 0.1			0.12
$\epsilon_b \cdot 10^3$	2.1		~ 0.1			1

conservative one is. Thus, in the following our reference value will be

$$\alpha_s(m_Z) = 0.119 \pm 0.004. \tag{4.4.3}$$

In order to appreciate the relative importance of the different sources of theoretical errors for precision tests of the SM, we report in Table 4.4.3 a comparison for the most relevant observables, evaluated using [ALT 89] and [BAR 92a]. What is important to stress is that the ambiguity from m_t, once by far the largest one, is by now smaller than the error from m_H. We also see from Table 4.4.3 that the error from $\Delta\alpha(m_Z)$ is especially important for $\sin^2\theta_{eff}$ and, to a lesser extent, is also sizeable for Γ_Z and ϵ_3.

Table 4.4.4 *Standard Model fits of electroweak data*

Parameter	LEP (incl. m_W)	All but m_W, m_t	All data
m_t(GeV)	$157 + 12 - 10$	$156.7 + 8.7 - 8.5$	170.4 ± 5.3
m_H(GeV)	$56 + 101 - 31$	$30 + 36 - 14$	$74 + 89 - 47$
$log[m_H(\text{GeV})]$	$1.75 + 0.45 - 0.35$	$1.48 + 0.34 - 0.27$	$1.87 + 0.34 - 0.43$
α_s (m_Z)	0.122 ± 0.003	0.121 ± 0.003	0.121 ± 0.003
χ^2/dof	6/9	11.6/12	14.4/15

The most important recent advance in the theory of radiative corrections is the calculation of the $o(g^4 m_t^2/m_W^2)$ terms in $\sin^2\theta_{eff}$, m_W and, more recently on δ_ρ [DEG 97]. The result implies a small but visible correction to the predicted values, but especially a sizeable decrease of the ambiguity from scheme dependence (a typical effect of truncation). These calculations are now implemented in the fitting codes used in the analysis of LEP data. The overall effect of the new corrective terms on the fitted central value of the Higgs mass is a decrease by about 30 GeV: a quite noticeable difference.

We now discuss fitting the data in the SM. Similar studies based on older sets of data are found in [ALT 98] and [ELL 96]. As the mass of the top quark is finally rather precisely known from CDF and D0, one must distinguish two different types of fit. In one type one wants to answer the question: is m_t from radiative corrections in agreement with the direct measurement at the Tevatron? Similarly how does m_W inferred from radiative corrections compare with the direct measurements at the Tevatron and LEP2? For answering these interesting but somewhat limited questions, one must clearly exclude the direct measurements of m_t and m_W from the input set of data. Fitting all other data in terms of m_t, m_H and α_s (m_Z) one finds the results shown in the second column of Table 4.4.4 [ABB 98]. The extracted value of m_t is typically a bit too low. For example, as shown in Table 4.4.4, from all the electroweak data except the direct production results on m_t and m_W, one finds $m_t = 157 \pm 9$ GeV. There is a strong correlation between m_t and $m_H \sin^2\theta_{eff}$, and m_W drives the fit to small values of m_H. Then, at small m_H the widths, in particular the leptonic width (whose prediction is nearly independent of α_s) drive the fit to small m_t. In a more general type of fit, e.g. for determining the overall consistency of the SM or the best present estimate for some quantity, say m_W, one should of course not ignore the existing direct determinations of m_t and m_W. Then, from all the available data, by fitting m_t, m_H and α_s (m_Z) one finds the values shown in the last coloumn of Table 4.4.4. This is the fit also referred to in Table 4.4.1. The corresponding fitted values of $\sin^2\theta_{eff}$ and m_W are:

$$\sin^2\theta_{eff} = 0.23152 \pm 0.00022; \quad m_W = 80.371 \pm 0.028 \text{ GeV}. \qquad (4.4.4)$$

The fitted value of $\sin^2\theta_{eff}$ is practically identical to the LEP + SLD average. The error of 27 MeV on m_W clearly sets up a goal for the direct measurement of m_W at LEP2 and the Tevatron.

As a final comment we want to recall that the radiative corrections are functions of $\log(m_H)$. It is truly remarkable that the fitted value of $\log(m_H)$ is found to fall right into the very narrow allowed window around the value 2 specified by the lower limit from direct searches, $m_H > \sim 85$ GeV, and the theoretical upper limit in the SM $m_H < 600$–800 GeV. The fulfilment of this very stringent consistency check is a beautiful argumeut in favour of a fundamental Higgs (or one with a compositeness scale much above the weak scale).

4.4.3 A more model independent analysis

We now discuss an update of the epsilon analysis [ALT 92; 98] which is a method to look at the data in a more general context than the SM. The starting point is to isolate from the data that part which is due to the purely weak radiative corrections. In fact the epsilon variables are defined in such a way that they are zero in the approximation when only effects from the SM at tree level plus pure QED aud pure QCD corrections are taken into account. This very simple version of improved Born approximation is a good first approximation according to the data and is independent of m_t and m_H. In fact the whole m_t and m_H dependence of the electroweak measured quantities arises from weak loop corrections and therefore is only contained in the epsilon variables. Thus the epsilons are extracted from the data without need of specifying m_t and m_H. But their predicted value in the SM or in any extension of it depend on m_t and m_H. This is to be compared with the competitor method based on the S, T, U variables [PES 90; ALT 90]. The latter cannot be obtained from the data without specifying m_t and m_H because they are defined as deviations from the complete SM prediction for specified m_t and m_H. Of course there are very many variables that vanish if pure weak loop corrections are neglected, at least one for each relevant observable. Thus for a useful definition we choose a set of representative observables that are used to parameterize those hot spots of the radiative corrections where new physics effects are most likely to show up. These sensitive weak correction terms include vacuum polarization diagrams which, being potentially quadratically divergent, are likely to contain all possible non-decoupling effects (like the quadratic top quark mass dependence in the SM). There are three independent vacuum polarization contributions. In the same spirit, one must add the $Z \rightarrow b\bar{b}$ vertex which also includes a large top mass dependence. Thus altogether we consider four defining observables: one asymmetry, for example A_l^{FB}, (as representative of the set of measurements that lead to the determination of $\sin^2\theta_{eff}$), one width (the leptonic width Γ_l is particularly suitable because it is practically independent of α_s), m_W and R_b. Here lepton universality has been taken for granted, because the data show that it is verified within the present accuracy. The four variables, ϵ_1, ϵ_2, ϵ_3 and ϵ_b, are defined in [ALT 92] in one to one correspondence with the set of observables A_l^{FB}, Γ_l, m_W, and R_b. The definition is so chosen that the quadratic top mass dependence is only present in ϵ_1 and ϵ_b, while the m_t dependence of ϵ_2 and ϵ_3 is only logarithmic. That definition of ϵ_1 and ϵ_3 is specified in terms of A_l^{FB}

Table 4.4.5 *Experimental values of the epsilons in the SM from different sets of data. These values (in 10^{-3} units) are obtained for $\alpha_s(m_Z) = 0.119 \pm 0.003$, $\alpha(m_z) = 1/128.90 \pm 0.09$, the corresponding uncertainties being included in the quoted errors [ALT 98]*

$\epsilon\ 10^3$	Only def. quantities	All asymmetries	All high energy	All data
$\epsilon_1\ 10^3$	4.0 ± 1.2	4.3 ± 1.2	3.9 ± 1.2	3.6 ± 1.2
$\epsilon_2\ 10^3$	-7.5 ± 2.1	-8.3 ± 1.9	-8.6 ± 2.0	-8.8 ± 2.0
$\epsilon_3\ 10^3$	2.9 ± 1.9	4.2 ± 1.2	4.0 ± 1.2	3.7 ± 1.2
$\epsilon_b\ 10^3$	-2.4 ± 2.3	-2.5 ± 2.3	-3.6 ± 2.1	-3.3 ± 2.1

and Γ_l only. Then adding m_W or R_b one obtains ϵ_2 or ϵ_b. The values of the epsilons as obtained from the defining variables (we update here [ALT 98]), following the specifications of [ALT 92], are shown in the first column of Table 4.4.5. To proceed further and include other measured observables in the analysis we need to make some dynamical assumptions. The minimum amount of model dependence is introduced by including other purely leptonic quantities at the Z pole such as A_τ, A_e (measured from the angular dependence of the τ polarization) and A_{LR} (measured by SLD). For this step, one is simply assuming that the different leptonic asymmetries are equivalent measurements of $\sin^2\theta_{eff}$ (for an example of a peculiar model where this is not true, see [CAR 95]). We add, as usual, the measure of A_b^{FB} because this observable is dominantly sensitive to the leptonic vertex. We then use the combined value of $\sin^2\theta_{eff}$ according to Eq. (4.4.1). At this stage the best values of the epsilons are shown in the second column of Table 4.4.5.

All observables measured on the Z peak at LEP can be included in the analysis provided that we asssume that all deviations from the SM are only contained in vacuum polarization diagrams (without demanding a truncation of the q^2 dependence of the corresponding functions) and/or the $Z \rightarrow b\bar{b}$ vertex. From a global fit of the data on m_W, Γ_T, R_h, σ_h, R_b and $\sin^2\theta_{eff}$ (for LEP data, we have taken the correlation matrix Γ_T, R_h, and σ_h given by the LEP experiments [ABB 98], while we have considered the additional information on R_b and $\sin^2\theta_{eff}$ as independent) we obtain the values shown in the third column of Table 4.4.5.

To include in our analysis lower energy observables as well, a stronger hypothesis needs to be made: vacuum polarization diagrams are allowed to vary from the SM only in their constant and first derivative terms in a q^2 expansion [PES 90; ALT 90]. In such a case, one can, for example, add to the analysis the ratio R_ν of neutral to charged-current processes in deep inelastic neutrino scattering on nuclei [ALL 86; MCF 98], the weak charge Q_W measured in atomic parity violation experiments on Cs [WOO 97] and the measurement of g_V/g_A from $\nu_\mu e$ scattering [VIL 97]. In this way one obtains the global fit given in the fourth column of Table 4.4.5. With the progress of LEP the low energy data, while important as a check that no deviations from the expected q^2 dependence arise, play a lesser role in the global fit. Note that the present ambiguity on the value of $\delta\alpha^{-1}(m_Z) = \pm 0.09$ [PIE 97] corresponds to an

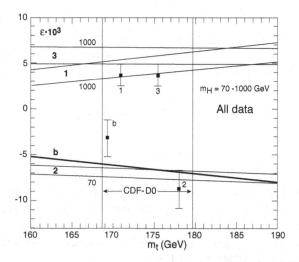

Fig. 4.4.2. The bands (labelled by the ϵ index) are the predicted values of the epsilons in the SM as functions of m_t for $m_H = 70$–$1000\,\text{GeV}$ (the m_H value corresponding to one edge of the band is indicated). The CDF/D0 experimental 1-σ range of m_t is shown. The experimental results for the epsilons from all data are displayed (from the last column of Table 4.4.5). The position of the data on the m_t axis has been arbitrarily chosen and has no particular meaning.

uncertainty on ϵ_3 (the other epsilons are not much affected) given by $\Delta\epsilon_3 10^3 = \pm 0.6$ [ALT 92] (see Table 4.4.3). Thus the theoretical error is still comfortably less than the experimental error. In Fig. 4.4.2 we present a summary of the fitted values for the epsilon parameters corresponding to all available data (Table 4.4.5, fourth column). Also shown are the SM predictions, as a function of m_t, for $m_H = 70$–$1000\,\text{GeV}$. The data points are plotted at an indicative m_t value in the CDF/D0 range. The overall agreement is quite good. We see that ϵ_1 and ϵ_b prefer a low m_t value. We also confirm that the ϵ_3 central value is on the low side by about 1σ. This is a not significant but persistent feature of the data. We also see that the ϵ_b point is still somewhat high due to the remaining excess of R_b in comparison with the predicted value. Finally, ϵ_2 is in very good agreement with the SM prediction. Together ϵ_2 and ϵ_3 point to a light m_H. There is remarkable evidence for weak corrections, measured by the distance of the data from the improved Born approximation point (based on tree level SM plus pure QED corrections). In other words a strong evidence for the pure weak radiative corrections has been obtained, and LEP/SLC are measuring the various components of these radiative corrections. For example, some authors [GAM 94] have studied the sensitivity of the data to a particularly interesting subset of the weak radiative corrections, i.e. the purely bosonic part. These terms arise from virtual exchange of gauge bosons and Higgs. The result is that indeed the measurements are sufficiently precise to require the presence of these contributions in order to fit the data. The good agreement of the fitted epsilon values with the SM imposes strong constraints on possible forms of new physics. Consider, for example, new quarks or leptons. Mass-split multiplets contribute to $\Delta\epsilon_1$, in analogy to the $t-b$ quark doublet. Recall that $\Delta\epsilon_1 \sim +9.5 \times 10^{-3}$ for the $t-b$ doublet, which is about eight σ in terms of the present error [ALV 83]. Even mass degenerate multiplets are strongly constrained. They contribute to $\Delta\epsilon_3$ according to [VEL 77; ALT 92]

$$\Delta\varepsilon_3 \sim N_C \frac{G_F m_W^2}{8\pi^2 \sqrt{2}} \frac{4}{3} (T_{3L} - T_{3R})^2. \tag{4.4.5}$$

For example a new left-handed quark doublet, degenerate in mass, would contribute $\Delta\epsilon_3 \sim +1.3 \times 10^{-3}$, that is about one σ, but in the wrong direction, in the sense that the experimental value of ϵ_3 favours a displacement, if any, with negative sign. Only vector fermions ($T_{3L} = T_{3R}$) are not constrained. In particular, naive technicolour models, that introduce several new technifermions, are strongly disfavoured because they tend to produce large corrections with the wrong sign to ϵ_1, ϵ_3 and also to ϵ_b [CHI 98; ELL 95].

The situation is different for the Minimal Supersymmetric Standard Model (MSSM). The MSSM [NIL 84] is a completely specified, consistent and computable theory. In the 'heavy' limit where all particles are sufficiently massive, still within the limits of a natural explanation of the weak scale of mass, a very important result holds [BAR 92b]: for what concerns the precision electroweak tests, the MSSM predictions tend to reproduce the results of the SM with a light Higgs, say $m_H \sim 100$ GeV. Given that this value for the Higgs mass is compatible with the data, it follows that if the masses of SUSY partners are pushed at sufficiently large values the same quality of fit as for the SM is guaranteed. In the 'light' MSSM option, where some of the superpartners have a relatively small mass, close to their experimental lower bounds, the pattern of radiative corrections could in principle sizeably deviate from that of the SM [BAR 92b]. But in practice, given the present limits on charginos, sleptons and stops, it turns out that it is quite natural that the effects from the MSSM can go unnoticed [ALT 98; CHA 98].

It is interesting that some phenomenological evidence in favour of a super-symmetric extension of the SM is obtained from imposing coupling unification in Grand Unified Theories (GUT) [ROS 85a]. The idea of Grand Unification is so attractive that by now it is difficult to imagine particle physics and cosmology without some realization of it. GUTs have great conceptual appeal. The observed low energy fragmentation of gauge interactions is recomposed. The quark and lepton quantum numbers in each generation are explained in terms of (possibly) one single irreducible representation of the unifying group (the 16 of $SO(10)$ being the best candidate at present). The small breakings of baryon and lepton numbers which are required for baryogenesis in the Universe are naturally introduced, with rates compatible with proton stability. Neutrino masses, which are by now indicated by the data on solar and atmospheric neutrinos explained in terms of neutrino oscillations [SUZ 98], occur naturally in GUTs, and their smallness can be understood as a reflection of the large scale of mass where lepton number is violated [GEL 79]. While the general framework is widely accepted, the detailed structure of the GUT model is still quite open. In this respect it is important to note that given the measured values of $\alpha(m_Z)$ and $\sin^2\theta_{eff}$ (equivalent to the knowledge of the $SU(2) \otimes U(1)$ gauge couplings) one can predict the value of $\alpha_s(m_Z)$ required for the unification of all three gauge couplings at a single value of M_{GUT}. The predicted value depends on the input spectrum at the weak scale and on the assumption of no new physics between the weak scale and M_{GUT} (in particular one assumes one stage only of breaking for the GUT group down to the weak scale symmetry). With these assumptions the result is [LAN 95a]

$$\alpha_s(m_Z) \sim 0.073 \pm 0.002 \quad Standard\ Model. \qquad (4.4.6)$$

The error is small because the SM spectrum is well known. This result is completely excluded in comparison with the world average given in Eq. (4.4.3). Instead in the MSSM one obtains

$$\alpha_s(m_Z) \sim 0.130 \pm 0.010 \quad MSSM, \qquad (4.4.7)$$

which is compatible with the measured value.

References

[ABB 98]   D. Abbaneo *et al.*, The LEP Electroweak Working Group, *CERN report LEPEWWG/CERN-EP* 98–01.
[ABE 83]   R. Abela *et al.*, *Nucl. Phys.* A395 (1983) 413.
[ABE 86]   K. Abe *et al.*, *Phys. Rev. Lett.* 56 (1986) 1107 and 1183 E.
[ABE 87]   K. Abe *et al.*, *Phys. Rev. Lett.* 58 (1987) 636.
[ABE 89]   K. Abe *et al.*, *Phys. Rev. Lett.* 62 (1989) 1709; and M.V. Diwan, Ph.D. thesis, Brown Univ. R.I. (1988).
[ABE 97]   CDF Collaboration, F. Abe *et al.*, *Phys. Rev. Lett.* 79 (1997) 2191.
[ABR 82a] H. Abramowicz *et al.*, *Phys. Lett.* B109 (1982) 115.
[ABR 82b] H. Abramowicz *et al.*, (CDHS), *Z. Phys.* C12 (1982) 225.
[ABR 82c] H. Abramowicz *et al.*, *Z. Phys.* C15 (1982) 19.
[ABT 86]   I. Abt, Ph.D. thesis, Univ. Hamburg (1986).
[AHR 87a] L.A. Ahrens *et al.*, *Phys. Rev.* D35 (1987) 785.
[AHR 87b] L.A. Ahrens *et al.*, *Nucl. Instrum. Methods* A254 (1987) 515.
[ALB 86]   UA1 Collab., C. Albajar *et al.*, *Europhys. Lett.* 1 (1986) 327.
[ALB 87b] H. Albrecht *et al.*, *Phys. Lett.* B192 (1987) 245.
[ALB 87c] UA1 Collab., C. Albajar *et al.*, *Phys. Lett.* B198 (1987) 271 and *Z. Phys.* C371 (1988) 505.
[ALB 91]   H. Albrecht *et al.*, *Phys. Lett.* B255 (1991) 297.
[ALE 89]   ALEPH Collab., D. Decamp *et al.*, *Phys. Lett.* B235 (1990) 399.
[ALE 90]   ALEPH Collaboration, D. Decamp *et al.*, *Nucl. Instrum. Methods* A294 (1990) 121.
[ALE 94]   ALEPH Collaboration, D. Buskulic *et al.*, *Z. Phys* C62 (1994) 539.
[ALE 96]   J.P. Alexander *et al.*, *Phys. Rev. Lett.* 77 (1996) 5000.
[ALE 97a] ALEPH Collaboration, D. Buskulic *et al.*, *Phys. Lett.* B401 (1997) 150.
[ALE 97b] ALEPH Collaboration, D. Buskulic *et al.*, *Phys. Lett.* B401 (1997) 163.
[ALI 84]   T.M. Aliev *et al.*, *Sov. J. Nucl. Phys.* 40 (1984) 527 [*Yad. Fiz.* 40 (1984) 823.]
[ALI 92]   S. Alitti *et al.*, *Phys. Lett.* B276 (1992) 354.
[ALL 86]   CHARM Collaboration, J.V. Allaby *et al.*, *Phys. Lett.* B177 (1986) 446; *Z. Phys.* C36 (1987) 611; CDHS Collaboration, H. Abramowicz *et al.*, *Phys. Rev. Lett.* 57 (1986) 298;

              A. Blondel *et al.*, *Z. Phys.* C45 (1990) 361; CCFR Collaboration, K. McFarland, *Eur. Phys. J.* C1 (1998) 509–513.

[ALL 87]   CHARM Collab., J.V. Allaby *et al.*, *Z. Phys.* C36 (1987) 611.

[ALL 88a]   CHARM Collab., J.V. Allaby *et al.*, *Z. Phys.* C38 (1988) 403.

[ALL 89]   CHARM Collab., J.V. Allaby *et al.*, *Phys. Lett.* B231 (1989) 317.

[ALL 93]   R.C. Allen *et al.*, ILM Collaboration, *Phys. Rev.* D47 (1993) 47.

[ALT 82]   G. Altarelli *et al.*, *Nucl. Phys.* B208 (1982) 365.

[ALT 89]   G. Altarelli, R. Kleiss and C. Verzegnassi (eds.), *Z Physics at LEP1* (CERN 89-08, Geneva, 1989), Vols. 1–3; Precision Calculations for the Z Resonance, ed. D. Bardin, W. Hollik and G. Passarino, *CERN Rep. 95–03* (1995); M.I. Vysotskii, V.A. Novikov, L.B. Okun and A.N. Rozanov, *Phys. Usp.* 39 (1996) 503–538; *Usp. Fiz. Nauk* 166 (1996) 539–574.

[ALT 90]   G. Altarelli and R. Barbieri, *Phys. Lett.* B253 (1990) 161; B.W. Lynn, M.E. Peskin and R.G. Stuart, *SLAC-PUB-3725* (1985); in *Physics at LEP, Yellow Book CERN 86-02*, Vol. I, p. 90; B. Holdom and J. Terning, *Phys. Lett.* B247 (1990) 88; D.C. Kennedy and P. Langacker, *Phys. Rev. Lett.* 65 (1990) 2967.

[ALT 92]   G. Altarelli, R. Barbieri and S. Jadach, *Nucl. Phys.* B369 (1992) 3; G. Altarelli, R. Barbieri and F. Caravaglios, *Nucl. Phys.* B405 (1993) 3; *Phys. Lett.* B349 (1995) 145.

[ALT 98]   G. Altarelli, R. Barbieri and F. Caravaglios, *Int. J. Mod. Phys.* A13 (1998) 1031.

[ALV 83]   L. Alvarez-Gaume, J. Polchinski and M. Wise, *Nucl. Phys.* B221 (1983) 495; R. Barbieri and L. Maiani, *Nucl. Phys.* B224 (1983) 32.

[AMA 87]   U. Amaldi *et al.*, *Phys. Rev.* D36 (1987) 1385.

[APP 86]   UA2 Collab., J.A. Appel *et al.*, *Z. Phys.* C30 (1986) 1.

[ARM 79a]   N. Armenise *et al.*, *Phys. Lett.* B86 (1979) 225.

[ARM 79b]   N. Armenise *et al.*, *Phys. Lett.* B81 (1979) 385.

[ARM 79c]   N. Armenise *et al.*, *Phys. Lett.* B84 (1979) 137.

[ARM 88]   N. Armenise *et al.*, BEBC TST, submitted to 13th Int. Conf. on Neutrinos and Astrophysics, Boston, 1988, CO 23.

[ARN 85]   UA1 Collab., G. Arnison *et al.*, *Nuovo Cimento* 44 (1985) 1.

[ARN 94]   L. Arnaudon *et al.*, *Z. Phys* C66 (1994) 45.

[ASA 81]   Y. Asano *et al.*, *Phys. Lett.* B107 (1981) 159.

[ASS 98]   R. Assman *et al.*, The LEP energy working group, 1998, *Calibration of centre of mass energies at LEP1 for precise mesurement of Z properties*, CERN-EP/98-40.

[BAC 61]   G. Backenstoss *et al.*, *Phys. Rev. Lett.* 6 (1961) 415.

[BAL 77]   Columbia–BNL Collab., C. Baltay *et al.*, *Phys. Rev. Lett.* 39 (1977) 62.

[BAL 86]   C. Baltay *et al.*, *Phys. Rev. Lett.* 57 (1986) 2629.

[BAR 92a]   ZFITTER: D. Bardin *et al.*, CERN-TH. 6443/92 and refs. therein; TOPAZ0: G. Montagna *et al.*, *Nucl. Phys.* B401 (1993) 3, *Comp. Phys. Comm.* 76 (1993) 328; BHM: G. Burgers *et al.*, LEPTOP: A.V. Novikov, L.B. Okun and M.I. Vysotsky, *Mod. Phys. Lett.* A8 (1993) 2529; WOH, W. Hollik: see ref. [ALT 89]; G. Montagna *et al.*, hep-ph/9804211, D. Bardin and G. Passarino, hep-ph/9803425.

[BAR 92b]   R. Barbieri, F. Caravaglios and M. Frigeni, *Phys. Lett.* B279 (1992) 169.

[BAR 93]   F. Bartelt *et al.*, *Phys. Rev. Lett.* 71 (1993) 4111.

[BAR 96]   R.M. Barnett *et al.*, *Rev. Part. Prop.*, *Phys. Rev.* D54 (1996) 1.

[BAZ 95]   A.O. Bazarko *et al.*, *Phys.* C65 (1995) 189; S.A. Rabinowitz *et al.*, *Phys. Rev. Lett.* 70 (1993) 134.

[BEB 88]   C. Bebek *et al.*, *Proc. 24th Int. Conf. on High-Energy Physics*, Munich (1988).

[BEG 77]   M.A. Beg *et al.*, *Phys. Rev. Lett.* 38 (1977) 1252.

[BER 79a]   J.P. Berge *et al.*, *Phys. Lett.* B84 (1979) 357.

[BER 79b]   D. Bertrand *et al.*, *Phys. Lett.* B84 (1979) 354.

[BER 83]   CHARM Collab., F. Bergsma *et al.*, *Phys. Lett.* B122 (1983) 185, 465.

[BER 84]   CHARM Collab., F. Bergsma *et al.*, *Phys. Lett.* B147 (1984) 481.

[BER 85]   CHARM Collab., F. Bergsma *et al.*, *Phys. Lett.* B157 (1985) 469.

[BIR 84]   G.N. Birich *et al.*, *Zh. Eksp. Teor. Fiz.* 87 (1984) 776. [Transl. *JETP* 60 (1985) 441.]

[BLI 76]   J. Blietschau *et al.*, *Nucl. Phys.* B114 (1976) 189.

[BLI 78]	J. Blietschau *et al.*, *Phys. Lett.* B73 (1978) 232.
[BLO 84]	V. Blobel, DESY 84/118, Hamburg (1984) and CERN 85–09, Geneva (1985) 88.
[BLO 90]	A. Blondel *et al.*, CDHS Coll., *Z. Phys.* C45 (1990) 361.
[BLO 96]	A. Blondel, *Proc. XXVIII Int. Conf. HEP* Warsaw, Poland 1996, Vol. 1, p. 205, World Scientific, Singapore.
[BOG 85]	D. Bogart *et al.*, *Phys. Rev. Lett.* 55 (1985) 1969.
[BOS 78]	P.C. Bosetti *et al.*, *Nucl. Phys.* B142 (1978) 1.
[BOS 80]	S.K. Bose and E.A. Paschos, *Nucl. Phys.* B169 (1980) 384.
[BOU 82]	M.A. Bouchiat *et al.*, *Phys. Lett.* B117 (1982) 358.
[BOU 83]	M. Bourquin *et al.*, *Z. Phys.* C21 (1983) 27.
[BUR 84]	A. Buras, W. Slominski and H. Steger, *Nucl. Phys.* B238 (1984) 529.
[BUR 85]	H. Burkard *et al.*, *Phys. Lett.* B160 (1985) 343.
[BUR 88]	H. Burkhardt *et al.*, *Phys. Lett.* 206B (1988) 169.
[CAR 83]	J. Carr *et al.*, *Phys. Lett.* 51 (1983) 627.
[CAR 95]	F. Caravaglios and G.G. Ross, *Phys. Lett.* B346 (1995) 159.
[CAR 97]	F. Caravaglios, *Phys. Lett.* B394 (1997) 359.
[CAT 97]	S. Catani, *Proc. LP'97*, Hamburg, 1997.
[CDF 88]	CDF Collaboration, F. Abe *et al.*, *Nucl. Instr. Meth* A271 (1988) 387.
[CDF 89]	CDF Collaboration, F. Abe *et al.*, *Phys. Rev. Lett.* 63 (1989) 720.
[CDF 94]	CDF Collaboration, F. Abe *et al.*, *PRL* 73 (1994) 225 and *PRD* 50 (1994) 2966.
[CDF 95a]	CDF Collaboration, F. Abe *et al.*, *PRL* 75 (1995) 11 and *PRD* 52 (1995) 4784.
[CDF 95b]	CDF Collaboration, F. Abe *et al.*, *PRL* 74 (1995) 2626.
[CDH 86]	CDHS Collaboration, H. Abramowicz *et al.*, *Phys. Rev. Lett.* 57 (1986) 298.
[CHA 83]	L.L. Chau *et al.*, *Phys. Rev.* D27 (1983) 2145; L.L. Chau, *Phys. Rep.* 95 (1983) 3; L.L. Chau and K. Keung, *Phys. Rev. Lett.* 53 (1984) 1802.
[CHA 87]	CHARM Collaboration, J.V. Allaby *et al.*, *Z. Phys.* C36 (1987) 611.
[CHA 89b]	CHARM II Collab., K. De Winter *et al.*, *Nucl. Instrum. Methods*, A278 (1989) 670.
[CHA 90]	CHARM Collab., D. Geiregat *et al.*, *Phys. Lett.* B245 (1990) 271.
[CHA 93]	CHARM II Collab., P. Vilain *et al.*, *Nucl. Phys.* B31 (*cont. Suppl.*) (1993) 287.
[CHA 93b]	CHARM II Collab., P. Vilain *et al.*, *Phys. Lett.* B313 (1993) 267.
[CHA 94]	CHARM II Collab., P. Vilain *et al.*, *Phys. Lett.* B335 (1994) 246.
[CHA 97]	CHARM II Collab., V. Lemaître *et al.*, *Evr. Phys. J.* C11 (1999) 19.
[CHA 98]	P.H. Chankowski, J. Ellis and S. Pokorski, *Phys. Lett.* B423 (1998) 327.
[CHE 71]	T.P. Cheng and W.K. Tung, *Phys. Rev.* D3 (1971) 733.
[CHE 86]	H.H. Chen *et al.*, *Proc. 12th Int. Conf. on Neutrino Physics and Astrophysics*, Sendai, eds. T. Kitagaki and H. Yuta, Singapore (1986).
[CHI 98]	For a review see R. Chivukula, hep-ph/9803219.
[CLE 88]	CLEO Collab., A. Jawakery, *Proc. 24th Int. Conf. on High-Energy Physics*, Munich, eds. R. Kotthaus and J.H. Kúhn, Springer-Verlag (1988) 545.
[CNO 78]	A.M. Cnops *et al.*, *Phys. Rev. Lett.* 41 (1978) 357.
[COT 81]	P. Cotens *et al.*, *Phys. Rev.* D24 (1981) 1420.
[CWU 65]	C.S. Wu and T.D. Lee, 'Weak Interactions', *Ann. Rev. Nucl. Sci.* 15 (1965) 381.
[D0 94]	D0 Collaboration, S. Abachi *et al.*, *Nucl. Instr. Meth.* A338 (1994) 185.
[D0 95b]	D0 Collaboration, S. Abachi *et al.*, *PRL* 74 (1995) 2632.
[D0 96]	D0 Collaboration, S. Abachi *et al.*, *PRL* 77 (1996). 3309.
[D0 97]	D0 Collaboration, B. Abbott *et al.*, submitted to *PRL* and *PRD*, hep-ex/9712028 and hep-ex/9712029.
[DAL 88]	WA 25 Collab., D. Allasia *et al.*, *Nucl. Phys.* B307 (1988) 1.
[DAN 62]	G. Danby *et al.*, *Phys. Rev. Lett.* 9 (1962) 36.
[DEG 97]	G.Degrassi, P. Gambino and A. Vicini, *Phys. Lett.* B383 (1996) 219; G. Degrassi, P. Gambino and A. Sirlin, *Phys. Lett.* B394 (1997) 188; G. Degrassi, P. Gambino, M. Passera and A. Sirlin, *Phys. Lett.* B418 (1998) 209–213.
[DEL 91]	DELPHI Collaboration, P. Aarnio *et al.*, *Nucl. Instr. Meth.* A303 (1991) 233.
[DEL 94]	DELPHI Collaboration, P. Abreu *et al.*, *Nucl. Phys.* B418 (1994) 403.

[DEL 97]   DELPHI Collaboration, Measurement of the partial decay width $R_b^0 = \Gamma_{barb}/\Gamma_{had}$ with the DELPHI detector at LEP. Contributed paper to *EPS EP 97*, Jerusalem, EPS-419.

[DID 80]   CHARM Collab., A.N. Diddens *et al.*, *Nucl. Instr. Meth.* 178 (1980) 27; 200 (1982) 183; 215 (1983) 361; A253 (1987) 203.

[DOM 88]   C.A. Dominguez and N. Paver, *Phys. Lett.* B207 (1988) 499.

[DOR 86]   CHARM Collab., J. Dorenbosch *et al.*, *Phys. Lett.* B180 (1986) 303.

[DOR 89]   CHARM Collab., J. Dorenbosch *et al.*, *Z. Phys.* C41 (1989) 567.

[DRE 70]   S.D. Drell and T.M. Yan, *Phys. Rev. Lett.* 25 (1970) 316.

[DRE 84]   P.S. Drell and E.D. Commins, *Phys. Lett.* 53 (1984) 968.

[EDW 85]   H. Edwards, *Ann. Rev. Nucl. Part. Sci.* 35 (1985) 605.

[EFR 79]   V. Efremenko *et al.*, *Phys. Lett.* B88 (1979) 181.

[ELL 95]   J. Ellis, G.L. Fogli and E. Lisi, *Phys. Lett.* B343 (1995) 282.

[ELL 96]   J. Ellis, G.L. Fogli and E. Lisi, *Phys. Lett.* B389 (1996) 321; A. Gurtu, *Phys. Lett.* B385 (1996) 415; P. Langacker and J. Erler, hep-ph/9703428; J.L. Rosner, *Comm. Nucl. Part. Phys.* 22 (1998) 205–219; K. Hagiwara, D. Haidt and S. Matsumoto, *Eur. Phys. J.* C2 (1998) 95–122.

[EMM 84]   T.P. Emmons *et al.*, *Phys. Rev. Lett.* 51 (1983) 2089.

[FAI 78]   H. Faissner *et al.*, *Phys. Rev. Lett.* 41 (1978) 213.

[FAI 80]   H. Faissner *et al.*, *Phys. Rev.* D21 (1980) 555.

[FAI 83]   H. Faissner *et al.*, *Phys. Lett.* B125 (1983) 230.

[FEI 97]   M. Feindt, plenary talk at EPS Conf. On High Energy Physics, Jerusalem, August 1997.

[FEL 78]   F.M. Feller *et al.*, *Phys. Rev. Lett.* 40 (1978) 274; W. Bacino *et al.*, *Phys. Rev. Lett.* 43 (1979) 1073; R.H. Schindler *et al.*, *Phys. Rev.* D24 (1981) 78; R.M. Baltrusaitis *et al.*, *Phys. Rev. Lett.* 54 (1985) 1976.

[FET 86]   W. Fetscher, H.-J. Gerber and K.F. Johnson, *Phys. Lett.* B173 (1986) 102.

[FEY 58]   R.P. Feynman and M. Gell-Mann, *Phys. Rev.* 109 (1958) 193.

[FRE 76]   F. Reines, H. Gurr and H. Sobel, *Phys. Rev. Lett.* 37 (1976) 315.

[FRI 87]   H. Fritzsch and J. Plankl, *Phys. Rev.* D35 (1987) 1732.

[FUL 88]   R. Fulton *et al.*, *Proc. 24th Int. Conf. on High-Energy Physics*, Munich (1988).

[GAB 84]   K. Gabathuler *et al.*, *Phys. Lett.* B138 (1984) 449.

[GAM 94]   P. Gambino and A. Sirlin, *Phys. Rev. Lett.* 73 (1994) 621; S. Dittmaier *et al.*, *Nucl. Phys.* B426 (1994) 249; S. Dittmaier, D. Schildknecht and G. Weiglein, *Nucl. Phys.* B465 (1996) 3.

[GEE 87]   S. Geer, *Proc. Europhysics Conf. on High-Energy Physics*, Upsalla, ed. O. Botner, Vol. 1 (1987) 219.

[GEI 90]   D. Geiregat *et al.*, *Phys. Lett.* B247 (1990) 131.

[GEI 93]   D. Geiregat *et al.*, *Nucl. Instr. Meth.* A325 (1993) 92.

[GEL 78]   M. Gell-Mann *et al.*, *Rev. Mod. Phys.* 50 (1978) 721.

[GEL 79]   M. Gell-Mann, P. Ramond and R. Slansky, in *Supergravity*, eds. P. van Nieuwenhuizen and D. Z. Freedman, North Holland, 1979; T. Yanagida, *Proc. Workshop on Unified Theory and Baryon Number in the Universe*, ed. O. Sawada *et al.*, KEK, 1979.

[GEO 74a]   H. Georgi, H. Quinn and S. Weinberg, *Phys. Rev. Lett.* 33 (1974) 451.

[GIL 85]   S.L. Gilbert *et al.*, *Phys. Rev. Lett.* 55 (1985) 2680.

[GIL 86a]   S.L. Gilbert *et al.*, *Phys. Rev.* A34 (1986) 792.

[GIL 86b]   F.J. Gilman and K. Kleinknecht, *Phys. Lett.* B170 (1986) 74.

[GIL 88]   F.J. Gilman, K. Kleinknecht and B. Renk, *Phys. Lett.* B204 (1988) 1.

[GIR 97]   P. Giromini, *Proc. of LP'97*, Hamburg, 1997.

[GLA 70]   S.L. Glashow, J. Iliopoulos and L. Maiani, *Phys. Rev.* D2 (1970) 1285.

[GRA 86]   H.J. Grabosch *et al.*, *Z. Phys.* C31 (1986) 203.

[GRI 86]   B. Grinstein *et al.*, *Phys. Rev. Lett.* 56 (1986) 289.

[GRI 89]   B. Grinstein *et al.*, *Phys. Rev.* D39 (1989) 799.

[GRO 79]   H. de Groot *et al.*, *Z. Phys.* C1 (1979) 143.

[GRO 85]   M. Gronau and J. Schechter, *Phys. Rev.* D31 (1985) 1668.

[GRO 98]	S. Groote *et al.*, hep-ph/9802374; M. Davier and A. Hocker, *Phys. Lett.* B419 (1998) 419, J.H. Kuhn and M. Steinhauser, hep-ph/9802241; J. Erler, hep-ph/9803453; M. Davier and A. Hocker, hep-ph/9805470.
[GSW 67]	S.L. Glashow, *Nucl. Phys.* 22 (1961) 579; A. Salam and J. Ward, *Phys. Lett.* 13 (1963) 168; S. Weinberg, *Phys. Rev. Lett.* 19 (1967) 1264.
[HAR 86]	H. Harari and M. Leurer, *Phys. Lett.* B181 (1986) 123.
[HEI 80]	R.H. Heisterberg *et al.*, *Phys. Rev. Lett.* 4 (1980) 635.
[HOL 77]	M. Holder *et al.*, *Phys. Lett.* B72 (1977) 254.
[HOL 78a]	M. Holder *et al.*, *Nucl. Instr. Meth.* 148 (1978) 235.
[HOL 78b]	M. Holder *et al.*, *Phys. Lett.* B74 (1979) 277.
[HOL 81]	H. Hollister *et al.*, *Phys. Rev. Lett.* 46 (1981) 643.
[HOO 71]	G. 'tHooft, *Phys. Lett.* 37 (1971) 195.
[HOR 82]	J. Horstkoffe *et al.*, *Phys. Rev.* D25 (1982) 2743.
[HUN 77]	P.Q. Hung and J. Sukarai, *Phys. Lett.* B72 (1977) 20.
[HUN 78]	P.Q. Hung and J. Sukarai, *Nucl. Phys.* B143 (1978) 81.
[ISG 89]	N. Isgur and M.B. Wise, *Phys. Lett.* B232 (1989) 113 and *Phys. Lett.* B237 (1990) 527; E. Eichten and B. Hill, *Phys. Lett.* B234 (1990) 511; M.E. Luke, *Phys. Lett.* B252 (1990) 447.
[ISI 84]	E. Isiksal, D. Rein and J.G. Morfin, *Phys. Rev. Lett.* 52 (1984) 1096.
[JAD 96]	S. Jadach and B.F.L. Ward, *Phys. Lett.* B 389 (1996) 129.
[JAR 70]	C. Jarlskog, *Nuovo Cimento* 4 (1970) 377.
[JEG 86]	F. Jegerlehner, *Z. Phys.* C32 (1986) 425.
[JON 81]	CHARM Collab., M. Jonker *et al.*, *Phys. Lett.* B102 (1981) 67.
[JON 81b]	M. Jonker *et al.*, *Phys. Lett.* B107 (1981) 241.
[JON 82]	M. Jonker *et al.*, *Phys. Lett.* B109 (1982) 133.
[JON 83]	M. Jonker *et al.*, *Z. Phys.* C17 (1983) 211.
[KAM 88]	T. Kamae, *Proc. 24th Int. Conf. on High-Energy Physics*, Munich, eds. R. Kotthaus and J.H. Kühl, Springer-Verlag (1988) 165.
[KAY 74]	B. Kayser *et al.*, *Phys. Lett.* B52 (1974) 385.
[KAY 79]	B. Kayser, E. Fishbach, S.P. Rosen and H. Spivack, *Phys. Rev.* D20 (1979) 87.
[KIM 81]	J.E. Kim *et al.*, *Rev. Mod. Phys.* 53 (1981) 211.
[KIN 74]	R.L. Kingsley *et al.*, *Phys. Rev.* D10 (1974) 2216.
[KIN 75]	R.L. Kingsley *et al.*, *Phys. Rev.* D11 (1975) 1043.
[KLE 82a]	K. Kleinknecht, *Proc. 10th Int. Neutrino Conference*, Balatonfured, Central Res. Inst. Physics, Budapest, Vol. 1 (1982) 115.
[KLE 82b]	K. Kleinknecht and B. Renk, *Z. Phys.* C16 (1982) 7; *Z. Phys.* C20 (1983) 67.
[KLE 83]	K. Kleinknecht and B. Renk, *Phys. Lett.* B130 (1983) 459; *Comm. Nucl. Part. Phys.* 13 (1984) 219.
[KLE 87]	K. Kleinknecht and B. Renk, *Z. Phys.* C34 (1987) 209.
[KLE 88]	K. Kleinknecht, Plenary talk, 'Weak Mixing, CP Violation and Rare Decays.' *Proc. 24th Int. Conf. on High-Energy Physics*, Munich, Springer-Verlag (1988).
[KOB 73]	M. Kobayashi and T. Maskawa, *Prog. Theor. Phys.* 49 (1973) 652.
[KRE 82]	W. Krenz, preprint Aachen Phys. Inst. Techn. Hochschule 82/26 (1982).
[L3 90]	L3 Collaboration, B. Adeva *et al.*, *Nucl. Instr. Meth.* A289 (1990) 35.
[L3 94]	L3 Collaboration, M. Acciarri *et al.*, *Z. Phys.* C62 (1994) 551.
[LAC 79]	K. Lackner, *Nucl. Phys.* B153 (1979) 526.
[LAN 57]	L. Landau, *Nucl. Phys.* 3 (1957) 127.
[LAN 88a]	P. Langacker and D. London, *Phys. Rev.* D38 (1988) 886, 907.
[LAN 88b]	P. Langacker, *Proc. 24th Int. Conf. on High-Energy Physics*, Munich, Springer Verlag (1988) 190.
[LAN 95a]	P. Langacker and N. Polonsky, *Phys. Rev.* D52 (1995) 3081; P. Langacker and J. Erler, hep-ph/9703428.
[LAN 95b]	P. Langacker, in *Precision tests of the Standard electroweak model*, ed. P. Langacker, World Scientific, Singapore, 1995.
[LEE 57]	T.D. Lee and C.N. Yang, *Phys. Rev.* 105 (1957) 1671.

[LEE 76]   W. Lee *et al.*, *Phys. Rev. Lett.* B228 (1983) 205.

[LEP 90]   Combined results of [ALE 89 and L3 90].

[LEP 94]   D. Schaile, Rapporteur talk, *Proc. 27th Int. Conf. HEP*, Glasgow 1994, Vol. 1, p. 27.

[LEP 95]   P. Renton, *Proc. Int. Lepton-Photon Symposium*, Beijing (China) 1995, p. 35, World Scientific, Singapore (1996).

[LEU 84]   H. Leutwyler and M. Roos, *Z. Phys.* C25 (1984) 91.

[LLE 83]   Ch. Llewellyn-Smith, *Nucl. Phys.* B228 (1983) 205.

[LOU 86]   W.C. Louis *et al.*, *Phys. Rev. Lett.* 56 (1986) 1027.

[LWG 97]   LEPEWG, The LEP Collaborations ALEPH, DELPHI, L3, OPAL, The LEP Electroweak Working Group and the SLD Heavy Flavour Group, 1997, *A Combination of Preliminary Electroweak Measurements and Constraints on the Standard Model*, CERN-PPE/97–154.

[MAI 77]   L. Maiani, *Proc. Int. Symposium on Lepton and Photon Interactions at High Energies*, Hamburg, ed. F. Gutbrod, DESY, Hamburg (1977) 867.

[MAN 83]   UA2 Collab., B. Mansoulie, *Proc. Moriond Workshop on Antiproton-Proton Physics and the W Discovery*, La Plagne, Savoie, France, ed. Frontière (1983) 609.

[MAR 84a]   P. Marage *et al.*, *Phys. Lett.* B140 (1984) 137.

[MAR 86a]   W.J. Marciano and A. Sirlin, *Phys. Rev. Lett.* 56 (1986) 22.

[MAR 86b]   P. Marage *et al.*, *Z. Phys.* C31 (1986) 191.

[MAR 88]   R. Marshall, *Z. Phys.* C43 (1989) 607.

[MCF 97]   McFarland *et al.*, CCFR/NUTEV Coll., *Phys. Rev. Lett.* 75 (1997) 3993, *Evr. Phys. J.* C1 (1998) 509.

[MCF 98]   NuTeV Collaboration, K.S. McFarland *et al.*, *Proc. XXXIII Rencontres de Moriond*, France, 1998.

[MIS 90]   S.R. Mishra *et al.*, *Phys. Lett.* B252 (1990) 170.

[MIS 91]   S.R. Mishra *et al.*, CCFR Collab., *Phys. Rev. Lett.* 66 (1991) 3117.

[NAR 83]   B. Naroska, *Proc. Int. Symp. Lepton and Photon Interactions at High Energies*, eds. D.G. Cassel and L. Kreinick, Cornell Univ., Ithaca, N.Y. (1983) 96.

[NEU 97]   M. Neubert, in *Heavy Flavours*, Second edition, eds. A.J. Buras and M. Lindner, World Scientific, Singapore, 1997.

[NEZ 66]   F. Nezrick and F. Reines, *Phys. Rev.* 142 (1966) 852.

[NIL 84]   H.P. Nilles, *Phys. Rep.* C110 (1984) 1; H.E. Haber and G.L. Kane, *Phys. Rep.* C117 (1985) 75; R. Barbieri, *Riv. Nuovo Cim.* 11 (1988) 1.

[NOV 93]   V.A. Novikov, L.B. Okun and M.I. Vysotzky, *Nucl. Phys.* B397 (1993) 35.

[OPA 91]   OPAL Collaboration, K. Ahmet *et al.*, *Nucl. Instr. Meth.* A305 (1991) 34.

[OPA 94]   OPAL Collaboration, R. Akers *et al.*, *Z. Phys.* C61 (1994) 19.

[OPA 97]   OPAL Collaboration, K. Ackerstaff *et al.*, *Z. Phys.* C74 (1997) 1.

[ORM 95]   W.E. Ormand *et al.*, Preprint nucl-th/9504017 (1995) unpublished.

[PAK 82]   S. Pakvasa, *Proc. 12th Int. Conf. on High-Energy Physics*, Paris (1982), *J. Phys.* 43, Suppl. 12 (1982) C 3-234.

[PAN 83]   J. Panman, *Workshop on SPS Fixed-Target Physics*, CERN 83–02, Vol. 2 (1983) 146.

[PAS 79]   E. Pasierb *et al.*, *Phys. Rev. Lett.* 43 (1979) 96.

[PAS 82]   E.A. Paschos and U. Türke, *Phys. Lett.* B116 (1982) 360.

[PAT 74]   J.C. Pati and A. Salam, *Phys. Rev.* D10 (1974) 275.

[PDG 96]   Particle DATA Group, Review of Particles Properties, *Phys. Rev.* D54 (1996) 1.

[PDG 98]   Particle Data Group 1998, *Evr. Phys. J.* C3 (1998) 3.

[PEO 83]   J. Peoples, *IEEE Trans on Nucl. Sci.*, NS30 (1983) 1970.

[PER 95]   F. Perrier *et al.*, in *Precision tests of the Standard electroweak model*, ed. P. Langacker, World Scientific, Singapore, 1995.

[PES 90]   M.E. Peskin and T. Takeuchi, *Phys. Rev. Lett.* 65 (1990) 964 and *Phys. Rev.* D46 (1991) 381.

[PIE 97]   For a review see B. Pietrzyk, *Acta Phys. Polon.* B28 (1997) 673–678.

[PIK 86]   C.A. Picketty, *Proc. Int. Conf. on Weak and Electromagnetic Interactions in Nuclei*, Heidelberg, Springer-Verlag (1986) 603.

[POH 78]   M. Pohl *et al.*, *Phys. Lett.* B72 (1978) 489.

[PRE 78] C.Y. Prescott *et al.*, *Phys. Lett.* B (1978) 347; B84 (1979) 524.

[PUP 48] G. Puppi, *Nuovo Cimento* 5 (1948) 505. See also O. Klein, *Nature* 161 (1948) 897, and B. Pontecorvo, *Phys. Rev.* 72 (1947) 246.

[REI 59] F. Reines and C.L. Cowan, *Phys. Rev.* 113 (1959) 273.

[REI 83] D. Rein and L.M. Seghal, *Nucl. Phys.* B223 (1983) 29.

[REU 85] P.G. Reutens *et al.*, *Phys. Lett.* B152 (1985) 404.

[ROS 75] D. Ross and M. Veltman, *Nucl. Phys.* B95 (1975) 135.

[ROS 85a] G.G. Ross, *Grand Unified Theories*, Benjamin, 1985; R.N. Mohapatra, *Unification and Supersymmetry*, Springer-Verlag, 1986; C. Kounnas *et al.*, *Grand Unification with and without Supersymmetry and Cosmological Implications*, World Scientific, Singapore (1996).

[ROS 85] J.L. Rosner, *Comm. Nucl. Part. Phys.* 14 (1985) 229.

[SAI 95] K.P. Saito *et al.*, *Phys. Lett.* B363 (1995) 157.

[SAK 75] J.J. Sakurai, *Int. Summer Institute on Theor. Phys. in Hamburg, Physics* 56 (1975) 258.

[SAK 81] P.Q. Hung and J.J. Sakurai, *Ann. Rev. Nucl. Part. Sci.* 31 (1981) 375.

[SAL 57] A. Salam, *Nuovo Cimento* 5 (1957) 299.

[SAV 95] G. Savard *et al.*, *Phys. Rev. Lett.* 74 (1995) 1521.

[SCH 87] W. Schmidt-Parzefall, *Proc. Int. Symp. on Lepton and Photon Interactions*, Hamburg, eds. W. Bartel and R. Rückl, *Nucl. Phys. B Suppl.* 3 (1988) 257.

[SEG 75] L.M. Seghal, *Phys. Lett.* B55 (1975) 205.

[SEG 81] L.M. Seghal, *Progress in Particle and Nuclear Physics*, ed. A. Faeesler, Vol. 14, Pergamon Press (1981) 1.

[SHA 97] M. Shaevitz *et al.*, *Rev. Mod. Phys.* (1998) 1.

[SHO 79] M.J. Shochet *et al.*, *Phys. Rev.* D19 (1979) 1965.

[SIR 86] A. Sirlin and R. Zucchini, *Phys. Rev. Lett.* 57 (1986) 1994.

[SLD 84] SLD Collaboration, 1984, *The SLD design report*, SLAC Report 273.

[SLD 97a] SLD Collaboration, K. Abe *et al.*, *Phys. Rev. Lett.* 78 (1997) 2075; B. Schumm, *Electroweak Results from the SLD*, SLAC-PUB-7697 (1997).

[SLD 97b] SLD Collaboration, 1997, *A Measurement of R_b using a Vertex Mass Tag*, Contributed paper to EPS EP 97, Jerusalem, EPS-118.

[STE 83] B. Stech, *Phys. Lett.* B130 (1983) 189.

[STR 84] M. Strovink, *Proc. 11th Int. Conf. on Neutrino Physics and Astrophysics*, Nordkirchen, eds. K. Kleinknecht and E.A. Paschos, Singapore (1984).

[SUD 58] E.C.G. Sudarshan and R.E. Marshak, *Phys. Rev.* 109 (1958) 1860.

[SUZ 98] Superkamiokande Collaboration, Y. Suzuki and T. Kajita, *Proc. Neutrino '98 Conference*, 1998.

[SWU 87] S.L. Wu, Rapporteur Talk, *Proc. Int. Symp. on Lepton and Photon Interactions at High Energies*, Hamburg, ed. W. Bartel (1987) 39.

[THO 86] E.H. Thorndike, *Proc. Int. Symp. on Lepton and Photon Interactions at High Energies*, Kyoto, eds. M. Komuna and K. Takahashi (1986) 406.

[TUR 85] U. Türke *et al.*, *Nucl. Phys.* B285 (1985) 313.

[UA1 83a] UA1 Collaboration, G. Arnison *et al.*, *Phys. Lett.* B122 (1983) 103.

[UA1 83b] UA1 Collaboration, G. Arnison *et al.*, *Phys. Lett.* B126 (1983) 398.

[UA2 83a] UA2 Collaboration, M. Banner *et al.*, *Phys. Lett.* B122 (1983) 476.

[UA2 83b] UA2 Collaboration, P. Bagnaia *et al.*, *Phys. Lett.* B129 (1983) 130.

[UA2 90] UA2 Collaboration, J. Alitti *et al.*, *Phys. Lett.* B241 (1990) 150.

[USH 83] N. Ushida *et al.*, *Phys. Lett.* B121 (1983) 292.

[USH 88] N. Ushida *et al.*, *Phys. Lett.* B206 (1988) 375.

[VDM 72] S. van der Meer, *Stochastic damping of betatron oscillations in the ISR*, CERN/ISR-PO/72-31 (1972).

[VDM 81] S. van der Meer, Stochastic cooling in the Antiproton Accumulator, *IEEE Trans. Nucl. Sci.* NS28 (1981) 1994.

[VEL 77] M. Veltman, *Nucl. Phys.* B123 (1977) 589; S. Bertolini and A. Sirlin, *Nucl. Phys.* B248 (1984) 589.

[VIL 95] P. Vilain *et al.*, CHARM II Collab., *Phys. Lett.* B364 (1995) 121.

[VIL 97]   P. Vilain *et al.*, CHARM II Collab., *Phys. Lett.* B335 (1997) 246.

[VIL 98]   P. Vilain *et al.*, CHARM II Collab., CERN-EP/98-73, *Phys. Lett.* B434 (1998) 205.

[WEY 29]   H. Weyl, *Z. Phys.* 56 (1929) 330.

[WIN 80]   K. Winter, *Neutrino Physics: The New Aspects of Subnuclear Physics*, Plenum Press, N.Y. (1980) 205.

[WIN 82]   K. Winter, *Weak Interaction Workshop*, Javea (Spain) 1982.

[WIN 83]   K. Winter, *Electroweak Effects at High Energy*, Europhysics Study Conference, Erice, Sicily, Plenum Press, N.Y. (1983) 41.

[WIN 87]   K. Winter, *Proc. Int. Workshop on Neutrino Physics*, Heildelberg, eds. H.V. Klapdor and B. Povh, Springer, Berlin (1987) 68.

[WIN 88]   K. Winter, *Proc. 13th Int. Conf. on Neutrino and Astrophysics*, World Scientific, Singapore (1988) 403.

[WIN 91]   K.Winter, *Neutrino Physics*, Cambridge University Press, 1991.

[WIR 85]   M. Wirbel, B. Stech and M. Bauer, *Z. Phys.* C29 (1985) 637.

[WOL 75]   L. Wolfenstein, *Nucl. Phys.* B91 (1975) 95.

[WOO 97]   C.S.Wood *et al.*, *Science* 275 (1997) 1759.

5

Study of nucleon structure by neutrinos*

Deep inelastic scattering of neutrinos and antineutrinos on nucleons has provided valuable information on the nucleon structure and it has allowed us to test extensively the validity of the parton model and of Quantum Chromodynamics (QCD).

The study of the nucleon structure using neutrinos and antineutrinos as probes is complementary to deep inelastic electron and muon experiments. In fact, in spite of poorer statistics with respect to muon experiments, neutrinos and antineutrinos have the unique feature, because of parity violation, of distinguishing quarks from antiquarks. Thus it is possible to isolate flavor nonsinglet combinations of structure functions. Moreover, because of the different couplings to the virtual vector bosons, in both charged- and neutral-current reactions, one can study combinations of quarks, antiquarks, and gluons different from those of electro-production experiments.

This chapter describes the physics of neutrino deep inelastic scattering and the study of the nucleon structure functions in the framework of the QCD improved parton model.

The relevant kinematics and notation are introduced in Section 5.1. The structure functions and their expressions in terms of parton distributions are discussed in Section 5.2. The experimental results on total cross sections and inelasticity distributions are given in Section 5.3. The experimental methods used to extract the structure functions from the differential cross sections and the results are given in Sections 5.4–5.5. The QCD analysis of the data at the leading and beyond the leading logarithmic approximation and the determination of the gluon distribution are given in Sections 5.6–5.8. In Section 5.9, the determination of the structure functions in neutral-current reactions is briefly reported. Section 5.10 contains a summary of the present situation and perspective in this field.

* M. Diemoz, F. Ferroni, E. Longo and G. Martinelli, Dipartimento di Fisica, Università "La Sapienza," Roma and Istituto Nazionale di Fisica Nucleare, Sezione di Roma, Italy; and M. Mangano, CERN, Geneva, Switerland.

5.1 Foundations of neutrino–nucleon interactions

5.1.1 *Kinematics and experimental requirements*

The relevant diagram describing charged-current neutrino–nucleon interactions $\nu_\mu N \to \mu X$ is shown in Fig. 5.1.1, where k and k' are the four-momenta of the incoming and outgoing leptons, respectively, and p and p' are those of the nucleon and of the final hadronic state. The energies E of the neutrino, E' of the muon, and E_h of the hadronic system refer to the laboratory reference frame; M indicates the nucleon rest mass; and θ is the laboratory scattering angle of the muon. In the kinematical description of the inclusive process the following Lorentz-invariant variables are currently introduced:

The center-of-mass squared energy

$$s = (p+k)^2. \tag{5.1.1}$$

The four-momentum and the energy transferred from the leptonic to the hadronic system

$$Q^2 \equiv -q^2 = -(k-k')^2 \tag{5.1.2}$$

$$\nu = \frac{(p \cdot q)}{M}. \tag{5.1.3}$$

The invariant mass squared of the hadronic system

$$W^2 = (p+q)^2 = M^2 - Q^2 + 2M\nu. \tag{5.1.4}$$

And finally the two dimensionless Bjorken variables

$$x = \frac{Q^2}{2p \cdot q} \tag{5.1.5}$$

$$y = \frac{p \cdot q}{p \cdot k}. \tag{5.1.6}$$

In the laboratory system, considering the nucleon at rest and neglecting the muon mass:

$$\nu = E - E' = E_h - M \tag{5.1.7}$$

$$Q^2 = 4EE' \sin^2 \frac{\theta}{2}. \tag{5.1.8}$$

In many experimental papers E_h indicates ν; hereafter we shall follow this convention.

The kinematical region of the process is bounded by the conditions

$$Q^2 > 0, \quad W^2 \geq M^2 \tag{5.1.9}$$

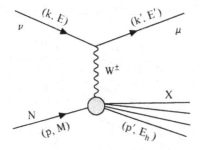

Fig. 5.1.1   Neutrino–hadron charged-current scattering.

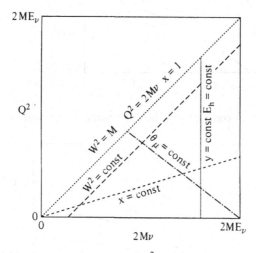

Fig. 5.1.2   Allowed kinematical region in the Q^2, ν plane. Only the region $Q^2 \le 2M\nu$ is accessible.

which corresponds in terms of x and y to

$$0 \le x \le 1, \quad 0 \le y \le \frac{1}{1 + Mx/2E}. \tag{5.1.10}$$

The allowed region and the curves of constant x and W^2 are given in Fig. 5.1.2. At constant Q^2 the kinematical x-range is not completely accessible to measurements, given that $x > Q^2/2ME$. On the contrary, at constant ν (i.e., constant hadron energy) the whole x-range is accessible.

The study of the nucleon structure through neutrino interactions requires a complete determination of the relevant kinematical variables x, y, and E. For charged currents, this requirement implies a detector able to measure with good resolution at the same time the muon vector momentum and the hadron energy over an extended range in x and Q^2. Curves of constant y and θ in the (x, Q^2) plane are shown in Figs. 5.1.3 and 5.1.4 for a neutrino energy of 20 GeV.

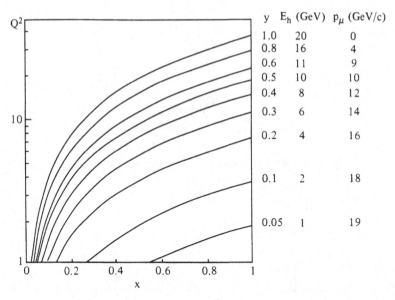

y	E_h (GeV)	p_μ (GeV/c)
1.0	20	0
0.8	16	4
0.6	11	9
0.5	10	10
0.4	8	12
0.3	6	14
0.2	4	16
0.1	2	18
0.05	1	19

Fig. 5.1.3   Curves of constant y in the x, Q^2 plane for a neutrino energy of 20 GeV.

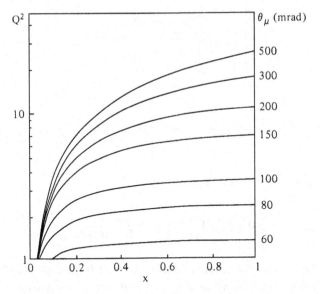

Fig. 5.1.4   Curves of constant θ_μ in the x, Q^2 plane for a neutrino energy of 20 GeV.

In the case of neutral currents, the complete kinematical reconstruction of the events must rely on the knowledge of the neutrino energy and on the measurement of the total vector momentum of the hadronic fragments. This is achieved by using narrow-band neutrino beams and fine-grain calorimetric detectors able to measure, in addition to the hadronic energy, the direction of the hadronic shower.

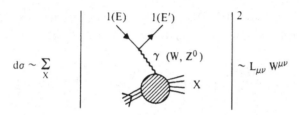

$$\mathrm{d}\sigma \sim \sum_X$$

Fig. 5.1.5   Lepton–nucleon inclusive scattering.

5.1.2 The general form of the neutrino–nucleon cross section

The cross section for the deep inelastic scattering of a neutrino on a nucleon is proportional to

$$L_{\mu\nu} W^{\mu\nu} \tag{5.1.11}$$

where $L_{\mu\nu}$ is the leptonic tensor and $W^{\mu\nu}$ the hadronic one (Fig. 5.1.5). $W^{\mu\nu}$ contains the strong interaction dynamics probed by the virtual vector boson:

$$W_{\mu\nu} = \frac{1}{4\pi} \sum (2\pi)^4 \delta^4(q + p - p')\langle p \,|\, J_\mu^+ | X \rangle\langle X | J_\nu | p \rangle \tag{5.1.12}$$

where X are the hadronic states with momentum p'. The constraints due to Lorentz and CP invariance and to currrent conservation imply

$$W_{\mu\nu} = \left(-g_{\mu\nu} + \frac{q_\mu q_\nu}{q^2}\right) M W_1 + \frac{1}{M}\left(p_\mu - \frac{p \cdot q}{q^2} q_\mu\right)\left(p_\nu - \frac{p \cdot q}{q^2} q_\nu\right) W_2$$

$$+ i\epsilon_{\mu\nu\alpha\beta} \frac{p^\alpha q^\beta}{2M} W_3 \tag{5.1.13}$$

where $W_{1,2,3} = W_{1,2,3}(\nu, Q^2)$. The inclusive differential cross section is then expressed in terms of these structure functions $W_{1,2,3}$:

$$\frac{d^2\sigma^{\nu,\bar\nu}}{dQ^2 d\nu} = \frac{G^2}{2\pi}\left(\frac{M_{W,Z}^2}{Q^2 + M_{W,Z}^2}\right)^2 \frac{E'}{E}\left(2W_1^{\nu,\bar\nu}(Q^2, \nu)\sin^2\frac{\theta}{2} + W_2^{\nu,\bar\nu}(Q^2, \nu)\cos^2\frac{\theta}{2}\right.$$

$$\left. \pm \left(\frac{E + E'}{M}\right) W_3^{\nu,\bar\nu}(Q^2, \nu)\sin^2\frac{\theta}{2}\right). \tag{5.1.14}$$

Typically, $Q^2 \ll M_{W,Z}^2$. For this reason, hereafter the term coming from the W, Z propagator will be set to one. The experimental study of the structure functions $W_{1,2,3}$ gives direct information on the nucleon structure. W_3 comes from the interference between the vector and the axial-vector part of the weak current, so that its sign is opposite in the ν and $\bar\nu$ cases.

The structure functions W_1, W_2, and W_3 are related to the cross sections for the absorption of transversally and longitudinally polarized bosons by nucleons:

$$\sigma_\pm = \frac{G\pi\sqrt{2}}{\Phi}\left(W_1(\nu, Q^2) \pm \frac{\sqrt{\nu^2 + Q^2}}{2M}W_3(\nu, Q^2)\right) \qquad (5.1.15)$$

$$\sigma_L = \frac{G\pi\sqrt{2}}{\Phi}\left(W_2(\nu, Q^2)\left(1 + \frac{\nu^2}{Q^2}\right) - W_1(\nu, Q^2)\right) \qquad (5.1.16)$$

where $\Phi = (W^2 - M^2)/2M$ is an overall flux factor. It is convenient to introduce the following dimensionless structure functions

$$MW_1(Q^2, \nu) = F_1(x, Q^2) \qquad (5.1.17)$$

$$\nu W_2(Q^2, \nu) = F_2(x, Q^2) \qquad (5.1.18)$$

$$\nu W_3(Q^2, \nu) = F_3(x, Q^2) \qquad (5.1.19)$$

and to express the differential, transverse, and longitudinal cross sections in terms of F_i, x, and y:

$$\frac{d^2\sigma^{\nu,\bar{\nu}}}{dx\,dy} = \frac{G^2 s}{2\pi}\left(xy^2 F_1^{\nu,\bar{\nu}} + \left(1 - y - \frac{Mxy}{2E}\right)F_2^{\nu,\bar{\nu}} \pm \left(y - \frac{y^2}{2}\right)xF_3^{\nu,\bar{\nu}}\right) \qquad (5.1.20)$$

$$\sigma_\pm^{\nu,\bar{\nu}} = \frac{G\pi\sqrt{2}}{\Phi M}\left(F_1^{\nu,\bar{\nu}}(x, Q^2) \pm \frac{1}{2}\left(1 + \frac{Q^2}{2\nu^2}\right)F_3^{\nu,\bar{\nu}}(x, Q^2)\right) \qquad (5.1.21)$$

$$\sigma_L^{\nu,\bar{\nu}} = \frac{G\pi\sqrt{2}}{\Phi}\left(F_2^{\nu,\bar{\nu}}(x, Q^2)\left(\frac{2Mx}{Q^2} + \frac{1}{2Mx}\right) - \frac{1}{M}F_1^{\nu,\bar{\nu}}(x, Q^2)\right)$$

$$= \frac{G\pi\sqrt{2}}{\Phi}\frac{F_L^{\nu,\bar{\nu}}(x, Q^2)}{2Mx} \qquad (5.1.22)$$

where $F_L = (1 + Q^2/\nu^2)F_2 - 2xF_1$ is the longitudinal structure function. Finally, the average transverse cross section σ_T is defined as

$$\sigma_T = \frac{1}{2}(\sigma_+ + \sigma_-). \qquad (5.1.23)$$

The previous formulas are derived without any hypothesis on the internal structure of the nucleons. Within the parton model the structure functions F_i can be expressed in terms of simple combinations of the partons contained in the nucleon, as discussed in the next sections.

5.2 The parton model in QCD

5.2.1 Generalities

The idea of composite nature of the nucleons, and particularly of the presence of a granular structure inside them [BJO 68], can be experimentally tested by probing the nucleon at increasing values of Q^2.

Consider the scattering of a lepton on a hadron as was shown in Fig. 5.1.5. When the spacelike squared four-momentum $Q^2 = -q^2$ is much smaller than the typical hadronic scale, $Q^2 \ll M^2 \simeq 1 \text{ GeV}^2$, the target will appear to be a pointlike source carrying a certain charge (magnetic moment). At larger values of Q^2, the virtual probe (photon, W or Z^0) will start to explore the distribution of charge due to the presence of a pion cloud and the hadrons will appear as extended objects. At even larger momentum transfer, $Q^2 \gg M^2$, the probe will be able to resolve the elementary constituents inside the hadron. Neglecting terms down by powers of $1/Q^2$, the cross section will then essentially be given by the scattering of the virtual boson on the constituents (partons) in the target, Fig. 5.2.1.

Defining $q(z)$ to be the density of partons q with a fraction z of the longitudinal momentum p of the hadron, one then can write:

$$\sigma \propto \int \frac{dz}{z} q(z) \sigma_{\text{parton}}(zp + q) \tag{5.2.1}$$

where σ_{parton} is the pointlike elementary cross section and the factor $1/z$ on the r.h.s. of Eq. (5.2.1) comes from the four-dimensional invariant phase space of an incoming massless particle. In order for Eq. (5.2.1) to be valid, the effective virtuality k^2, the transverse momentum k_T^2, and the mass of the parton inside the hadron must be so small that they indeed produce corrections of order $1/Q^2$, which can be neglected for $Q^2 \gg M^2$. On the other hand, $\sigma_{\text{parton}} \propto \delta(1 - x/z)$, where x must satisfy the relation $x = Q^2/2p \cdot q$, to ensure the vanishing of the squared invariant mass of the final state parton $(zp + q)^2 = 0$. Thus, up to kinematical factors the cross section is a measure of the density of partons with a fraction x of the hadron momentum:

$$\sigma \propto q(x). \tag{5.2.2}$$

This means that the structure functions introduced in the previous section must be identified with suitable combinations of the parton densities $q(x)$ (see Section 5.2.3).

The simple parton model described above may be spoiled by strong interactions, which can provide the partons with a transverse momentum (virtuality) of the order of Q^2. Only in a weakly interacting theory, as QCD at short distances, can the parton model survive. In QCD in fact, because of asymptotic freedom [POL 73; GRO 73a,b], the effect of strong interactions leads, at large Q^2 and retaining only the leading corrections, to replacement of the original "native" parton model formulas with expressions where the densities of partons with a fraction x of the hadron

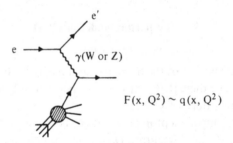

Fig. 5.2.1   Partonic diagram for deep inelastic scattering.

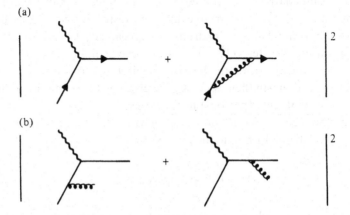

Fig. 5.2.2   Feynman diagrams for the first order QCD corrections to the parton scattering of Fig. 5.2.1. (a) Virtual gluon exchange, (b) real gluon emission.

momentum $q(x)$ are now given by parton densities $q(x, Q^2)$, which vary as a function of Q^2. The violations of the simple scaling formula, Eq. (5.2.2), are at most logarithmic because of the fact that QCD is a vectorlike renormalizable gauge theory [t'HO 71].

At lowest order in QCD perturbation theory, one has to consider the virtual and real Feynman diagrams shown in Fig. 5.2.2a and b, respectively. Strictly speaking, the following simplified discussion is valid only for a nonsinglet structure function like F_3. Otherwise other contributions should be included. Taking only the leading logarithmic corrections, that is, the terms of order $\alpha_s \ln(Q^2)$, where α_s is the strong coupling constant, Eq. (5.2.2) becomes [ALT 77]

$$\sigma(x, Q^2) \propto \int_x^1 \frac{dz}{z} q(z) \left[\delta\left(1 - \frac{x}{z}\right) + \frac{\alpha_s}{2\pi} \ln\left(\frac{Q^2}{\mu^2}\right) P\left(\frac{x}{z}\right) \right]$$

$$+ \text{(subleading corrections)}. \qquad (5.2.3)$$

where $P(x/z)$ is some calculable kernel. The $\ln(Q^2/\mu^2)$ term in the above equation arises from the collinear singularity present in the emission of a gluon by a quark, analogously to what happens for the emission of a photon by an electron in electrodynamics, and μ^2 is a reference scale where $\sigma(x, \mu^2) = q(x)$.

In analogy with Eq. (5.2.2), Eq. (5.2.3) can be written as

$$\sigma(x, Q^2) \propto q(x, Q^2) \tag{5.2.4}$$

by including the first-order correction in the definition of the parton distribution $q(x, Q^2)$. Eq. (5.2.3) is valid at lowest order in perturbation theory. The leading terms $\sim (\alpha_s t)^n$, where $t = \ln(Q^2/\mu^2)$, can be resummed to all orders, with the result that the cross section still has the same expression as in Eq. (5.2.4) in terms of $q(x, Q^2)$, and $q(x, Q^2)$ obeys the evolution equation

$$\frac{dq(x, Q^2)}{d \ln(Q^2/\mu^2)} = \frac{\alpha_s(Q^2)}{2\pi} \int_x^1 \frac{dz}{z} q(z, Q^2) P\left(\frac{x}{z}\right). \tag{5.2.5}$$

Equation (5.2.5) is the generalization to all orders in perturbation theory of the result obtained in Eq. (5.2.3). At order α_s, in fact, by deriving Eq. (5.2.3) w.r.t. Q^2 one gets

$$\frac{d\sigma(x, Q^2)}{dQ^2} \propto \frac{dq(x, Q^2)}{dQ^2} = \frac{\alpha_s}{2\pi} \frac{dz}{z} q(z) P\left(\frac{x}{z}\right). \tag{5.2.6}$$

In the leading logarithmic approximation, higher-order terms change $q(z)$ into $q(z, Q^2)$ on the r.h.s. of Eq. (5.2.3) and replace α_s with the "running" coupling constant $\alpha_s(Q^2)$. The dependence of the coupling constant on the scale of the reaction shows up only at the second order in perturbation theory. Equation (5.2.6) shows that when the external lepton current probes the hadron at different Q^2's the number of partons changes with the scale explored.

The quantity appearing in Eq. (5.2.3)

$$\delta\left(1 - \frac{x}{z}\right) + \frac{\alpha_s}{2\pi} \ln\left(\frac{Q^2}{\mu^2}\right) P\left(\frac{x}{z}\right) \tag{5.2.7}$$

can be interpreted as the probability density of finding a parton, with a fraction of momentum x/z, in a parent parton of momentum z and transverse momentum (virtuality) $k_T^2 \ll Q^2$. Thus in the QCD improved parton model (in the leading logarithmic approximation), the structure functions are still written in terms of the parton densities, which, however, now depend on x and Q^2. This dependence is calculable in perturbation theory through the parton evolution equations as schematically indicated in Eq. (5.2.5).

The validity of the QCD improved parton model is of fundamental importance in our understanding of hadronic processes. In fact, it allows us to express the results of deep inelastic scattering in terms of the electroweak charges of the elementary constituents. It is then possible, on the basis of the so-called factorization theorem

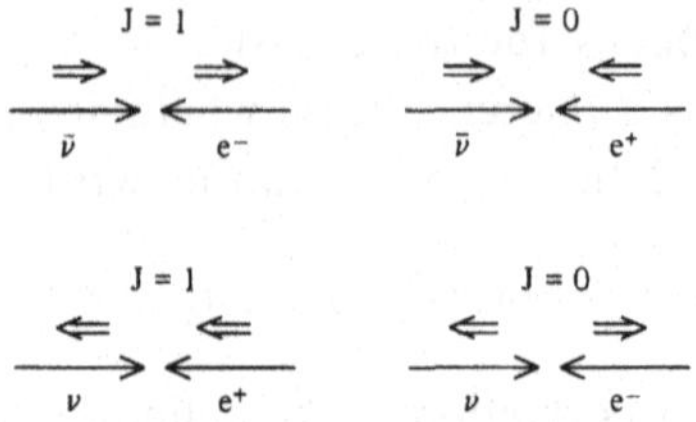

Fig. 5.2.3 Spin alignment in the neutrino–electron scattering.

[AMA 78a,b; ELL 78a,b; MUL 78] and of the distributions measured in deep inelastic scattering, to formulate predictions on other hadronic reactions such as Drell–Yan processes, direct photon production, and jet physics.

5.2.2 *Elementary quark charged-current cross sections*

Assuming that the partons in the nucleon are pointlike free quarks, their elementary charged-current interactions with neutrinos are described by formulas similar to those that are valid for neutrino–electron scattering. It is then very easy to derive the partonic elementary cross sections appearing in Eq. (5.2.1).

The high-energy neutrino–electron cross sections increase linearly with the square of the center-of-mass energy s and are characterized by the s-wave angular distributions resulting from the alignment of the lepton spins (see Fig. 5.2.3):

$$\frac{d\sigma}{dy} = \frac{G^2 s}{\pi} \qquad \text{for } \nu e^- \text{ or } \bar{\nu} e^+$$

and

$$\frac{d\sigma}{dy} = \frac{G^2 s}{\pi}(1-y)^2 \quad \text{for } \bar{\nu} e^- \text{ or } \nu e^+. \tag{5.2.8}$$

For a quark (or an antiquark) carrying a fraction x of the total momentum of the nucleon, s must be replaced by the νq center-of-mass energy $s' = (xp+k)^2 \simeq 2pxk = xs$. Then, using Eqs. (5.2.8), one has[1]

$$
\begin{aligned}
&\nu d \rightarrow l^- u + l^- c: \\
&\nu s \rightarrow l^- u + l^- c: &&\frac{d\sigma}{dy} = \frac{G^2 s x}{\pi} \\
&\bar{\nu} \bar{d} \rightarrow l^+ \bar{u} + l^+ \bar{c}: \\
&\bar{\nu} \bar{s} \rightarrow l^+ \bar{u} + l^+ \bar{c}: \\
&\nu \bar{u} \rightarrow l^- \bar{d} + l^- \bar{s}: \\
&\nu \bar{c} \rightarrow l^- \bar{d} + l^- \bar{s}: &&\frac{d\sigma}{dy} = \frac{G^2 s x}{\pi}(1-y)^2. \\
&\bar{\nu} u \rightarrow l^+ d + l^+ s: \\
&\bar{\nu} c \rightarrow l^+ d + l^+ s:
\end{aligned}
\tag{5.2.9}
$$

---

[1] We assume, on the basis of the available E_ν range of deep inelastic neutrino scattering, four active quark flavors: u, d, s, c. We also assume (though this is not always the case) that the reactions occur well above the charm threshold.

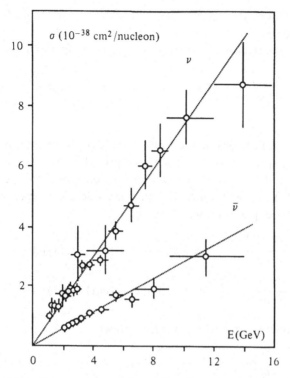

Fig. 5.2.4 Low-energy neutrino and antineutrino cross sections.

Note that the Cabibbo angle does not appear in the above formulas since we have neglected quark masses. Integration over y gives the following relation:

$$\sigma_{\nu q} : \sigma_{\bar{\nu}\bar{q}} : \sigma_{\nu\bar{q}} : \sigma_{\bar{\nu}q} = 1 : 1 : \frac{1}{3} : \frac{1}{3}. \tag{5.2.10}$$

If neutrinos actually interact on quarks, on the basis of Eqs. (5.2.9) neutrino– and antineutrino–nucleon cross sections are expected to increase linearly with the neutrino energy ($s \simeq 2ME_\nu$). Under the hypothesis that the nucleon is mainly composed of (valence) up and down quarks, one also predicts $\sigma_{\nu N}/\sigma_{\bar{\nu}N} \simeq 3$. The experimental confirmation of these predictions (Fig. 5.2.4 [PER 75]) was the most spectacular success of the parton model, a few years after its first evidence in deep inelastic electron scattering [PAN 68; BRE 69; BLO 69; TAY 69]. The observed deviation from 3 of the cross-sectional ratio, rather than creating a problem, represents a new insight into the nucleons, revealing their antiquark content, as will be described in the next sections.

5.2.3 Neutrino–nucleon charged-current cross section

As discussed above, the general form of neutrino–nucleon cross section involves six independent functions, $F_i^{\bar{\nu}\nu}$ for $i = 1,2,3$. Given that spin one-half massless partons can only absorb transversally polarized W bosons, σ_L vanishes, leading (up to terms order M^2/Q^2) to the Callan–Gross relation [CAL 75], $2xF_1 = F_2$. Neglecting terms of order M/E_ν, the neutrino cross sections can then be written as

$$\frac{d^2\sigma^{\nu,\bar{\nu}}}{dx\,dy} = \frac{G^2 s}{2\pi}\left(\frac{1}{2}F_2^{\nu,\bar{\nu}}(x,Q^2)(1+(1-y)^2) \pm \frac{x}{2}F_3^{\nu,\bar{\nu}}(x,Q^2)(1-(1-y)^2)\right).$$

(5.2.11)

To express $\nu(\bar{\nu})$–nucleon cross sections in terms of constituents, one has to weigh each elementary quark cross section by the probabilities $q_i(x)$ of finding in the nucleon a quark of the given flavor with a fraction x of the nucleon momentum $(q_i(x) = u(x), d(x), \ldots, \text{etc.})$ which in the parton model are independent from Q^2. In terms of the above quantities we obtain

$$\frac{d^2\sigma^{\nu p \to \mu^- X}}{dx\,dy} = \frac{G^2 s}{\pi}x(d(x) + s(x) + (\bar{u}(x) + \bar{c}(x))(1-y)^2)$$

(5.2.12)

$$\frac{d^2\sigma^{\bar{\nu} p \to \mu^+ X}}{dx\,dy} = \frac{G^2 s}{\pi}x(\bar{d}(x) + \bar{s}(x) + (u(x) + c(x))(1-y)^2).$$

(5.2.13)

By isospin symmetry we get for a neutron target

$$\frac{d^2\sigma^{\nu n \to \mu^- X}}{dx\,dy} = \frac{G^2 s}{\pi}x(u(x) + s(x) + (\bar{d}(x) + \bar{c}(x))(1-y)^2)$$

(5.2.14)

$$\frac{d^2\sigma^{\bar{\nu} n \to \mu^+ X}}{dx\,dy} = \frac{G^2 s}{\pi}x(\bar{u}(x) + \bar{s}(x) + (d(x) + c(x))(1-y)^2).$$

(5.2.15)

By defining $q(x) = u(x) + d(x) + c(x) + s(x)$ and $\bar{q}(x) = \bar{u}(x) + \bar{d}(x) + \bar{c}(x) + \bar{s}(x)$ one can write:

$$\frac{d^2\sigma^{\nu N}}{dx\,dy} = \frac{G^2 s}{2\pi}x(q(x) + (s(x) - c(x)) + (\bar{q}(x) - (\bar{s}(x) - \bar{c}(x)))(1-y^2))$$ (5.2.16)

$$\frac{d^2\sigma^{\bar{\nu} N}}{dx\,dy} = \frac{G^2 s}{2\pi}x(\bar{q}(x) + (\bar{s}(x) - \bar{c}(x)) + (q(x) - (s(x) - c(x)))(1-y^2))$$ (5.2.17)

for an isoscalar target N.

Comparing these formulas with the general expression for the cross sections in Eq. (5.1.20) we are led to the identification

$$F_2^{\nu\bar{\nu}} = x(q + \bar{q}) \quad \text{and} \quad xF_3^{\nu\bar{\nu}} = x(q - \bar{q} \pm 2(s - c))$$

(5.2.18)

or, defining a neutrino average $F_3 = (F_3^\nu + F_3^{\bar{\nu}})/2$,

$$xF_3 = x(q - \bar{q}).$$

(5.2.19)

For isoscalar targets, $F_2^{\nu\bar{\nu}}$ can be directly related to the analogous structure function F_2^{lN} that appears in charged lepton–nucleon scattering. Taking into account the electric quark charges the following relation holds: $F_2^{lN} = \frac{5}{18} F_2^{\nu\bar{\nu}} + \frac{1}{6} x(c - s + \bar{c} - \bar{s})$. From $F_3^{\nu\bar{\nu}}$ in Eq. (5.2.18) the Gross–Llewellyn Smith sum rule can be derived:

$$\int_0^1 dx\, F_3(x) = \frac{1}{2} \int_0^1 dx\, (F_3^\nu + F_3^{\bar{\nu}}) = \sum_f v_f \qquad (5.2.20)$$

where v_f is the valence value for a given flavor f in the hadron (e.g., in a proton $v_f = 2$ or $v_f = 1$ for up and down valence quarks, respectively).

By integrating Eqs. (5.2.16–17) over x and neglecting the s and c contributions, we can write:

$$\frac{d\sigma^{\nu N}}{dy} = \frac{G^2 s}{2\pi} (Q + \bar{Q}(1 - y)^2) \qquad (5.2.21)$$

$$\frac{d\sigma^{\bar{\nu} N}}{dy} = \frac{G^2 s}{2\pi} (\bar{Q} + Q(1 - y)^2) \qquad (5.2.22)$$

where $Q = \int xq(x)dx$ and $\bar{Q} = \int x\bar{q}(x)\, dx$. Equations (5.2.21–22) display the unique feature of neutrino processes to tell quarks from antiquarks. A plot of the distributions in the inelasticity y (Fig. 5.2.5) visualizes immediately the quark and antiquark content of the nucleon.

Integrating Eqs. (5.2.21–22) over y, we get

$$\sigma^{\nu N} \propto Q + \frac{1}{3} \bar{Q} \qquad (5.2.23)$$

$$\sigma^{\bar{\nu} N} \propto \frac{1}{3} Q + \bar{Q}. \qquad (5.2.24)$$

Thus, in the naive parton model any deviation from 3 of the ratio $\sigma_{\nu N}/\sigma_{\bar{\nu} N}$ is an evidence for the presence of antiquarks.

5.2.4 Neutral-current cross sections

For neutral currents, in order to relate F_2 and xF_3 to the quark densities we can proceed as before, with the only difference that quarks and antiquarks have now both right- and left-handed couplings. In the minimal standard Weinberg–Salam model [SAL 64, 68; WEI 67, 71] these couplings depend only on the charge and the weak isospin assignment of quarks and leptons all expressed in terms of a single parameter, $\sin^2\theta_W$. Their expressions are given in Table 5.2.1.

Summing up the contribution from all the quarks in a proton we get

$$\frac{d^2\sigma_{NC}^{\nu p}}{dx\, dy} = \frac{G^2 s}{\pi} x((u_L^2 + u_R^2(1 - y)^2)(u(x) + c(x)) + (u_R^2 + u_L^2(1 - y)^2)$$
$$\times (\bar{u}(x) + \bar{c}(x)) + (d_L^2 + d_R^2(1 - y)^2)(d(x) + s(x))$$
$$+ (d_R^2 + d_L^2(1 - y)^2)(\bar{d}(x) + \bar{s}(x))). \qquad (5.2.25)$$

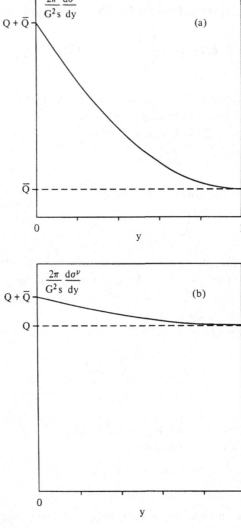

Fig. 5.2.5 Expected quark and antiquark contribution to the inelasticity distributions for antineutrino (a) and neutrino (b) scattering on nucleon.

For the neutron one only has to exchange $u(x)$ with $d(x)$ in Eq. (5.2.25), leaving $s(x)$ and $c(x)$ unchanged. For an isoscalar target one thus obtains

$$\frac{d^2\sigma^{\nu p}_{NC}}{dx\,dy} = \frac{G^2 s}{2\pi} x((u_L^2 + d_L^2 + (u_R^2 + d_R^2)(1-y)^2)q(x)$$
$$+ (u_R^2 + d_R^2 + (u_L^2 + d_L^2)(1-y)^2)\bar{q}(x)$$
$$+ (u_L^2 - d_L^2 + (u_R^2 - d_R^2)(1-y)^2)(c(x) - s(x))$$
$$+ (u_R^2 - d_R^2 + (u_L^2 - d_L^2)(1-y)^2)(\bar{c}(x) - \bar{s}(x))). \qquad (5.2.26)$$

Table 5.2.1 *Standard Model neutral-current quark couplings*

u_L	$\frac{1}{2} - \frac{2}{3}\sin^2(\theta_W)$
d_L	$-\frac{1}{2} + \frac{1}{3}\sin^2(\theta_W)$
u_R	$-\frac{2}{3}\sin^2(\theta_W)$
d_R	$\frac{1}{3}\sin^2(\theta_W)$

The antineutrino cross section can be immediately derived by interchanging the left- and the right-handed couplings in Eq. (5.2.26).

The comparison of Eq. (5.2.26) with the general form of Eq. (5.1.20) gives

$$F_2^{NC} = x((u_L^2 + d_L^2 + u_R^2 + d_R^2)(q + \bar{q}) - 2(u_L^2 - d_L^2 + u_R^2 - d_R^2)(s - c)) \quad (5.2.27)$$

and

$$xF_3^{NC} = x((u_L^2 - u_R^2 + d_L^2 - d_R^2)(q - \bar{q})). \quad (5.2.28)$$

5.2.5 QCD effects and scaling violations

In QCD, for deep inelastic scattering at large Q^2, the main effects of strong interactions can be summarized as follows:

1 A natural explanation to the observation that only ~ 50 percent of the total nucleon momentum is carried by the quark is found by introducing a new type of parton, the gluon, which carries momentum without participating directly in the electroweak reactions. It is the quark–gluon interaction that confines the quarks into the hadrons.

2 The approximate validity of scaling supports (indirectly) the idea that an asymptotically free theory like QCD is the basis of strong interactions. On the other hand QCD predicts scaling violations (Eq. (5.2.5)), originally observed at low energy by the SLAC–MIT experiments [CHA 75; RIO 75] and then confirmed in $\nu(\bar{\nu})$ and muon deep inelastic scattering. As a consequence, the structure functions, which in the naive parton model depend only on x, become functions of both x and Q^2.

3 In the naive parton model helicity suppresses the longitudinal cross section, which is expected to be of order m^2/Q^2. The emission of a gluon allows a nonzero cross section also in the case of a massless parton. Hence one expects the longitudinal structure function, $F_L = F_2 - 2xF_1$, to be of order α_s.

4 The emission of gluons provides the quark with a hard tail in the transverse momentum (k_T) distribution, which is expected to increase with Q^2:

$$\langle k_T^2 \rangle_{QCD} \sim \alpha_s \int^{Q^2} \frac{dk_T^2}{k_T^2} k_T^2 \sim \alpha_s Q^2. \quad (5.2.29)$$

The factor dk_T^2/k_T^2 in Eq. (5.2.29) is the same as that at the origin of the leading logarithmic scaling violation $\sim \ln Q^2$.

As briefly sketched above, in the QCD improved parton model the structure functions can be written as suitable combinations (convolutions) of quark, antiquark, and gluon densities. At the leading logarithmic approximation (LLA), the results of the QCD corrections can be interpreted by saying that the $F_i(x, Q^2)$ are given by the naive parton model formulas expressed in terms of Q^2 dependent effective parton densities obeying the first-order Altarelli–Parisi equations [ALT 77].

Beyond the LLA, the expressions of the $F_i(x, Q^2)$ deviate from the parton model formulas by terms of order α_s (coefficient functions) and the evolution equations must include the next-to-leading terms of the second-order Altarelli–Parisi kernels, the so-called anomalous dimensions. As discussed below, the explicit form of the corrections to the parton model expressions depends on the definition of the effective parton densities. Different definitions will give the same physical results.

The variation of the parton densities with Q^2 is governed by the Altarelli–Parisi evolution equations. For quark, antiquark, and gluon densities, these equations can be written as

$$\frac{dq_t}{d(\ln Q^2)} = \frac{\alpha_s}{2\pi}\left(\sum_j (P_{q_i q_j} \otimes q_j + P_{q_i \bar{q}_j} \otimes \bar{q}_j) + P_{q_i G} \otimes G\right)$$

$$\frac{d\bar{q}_t}{d(\ln Q^2)} = \frac{\alpha_s}{2\pi}\left(\sum_j (P_{\bar{q}_i q_j} \otimes q_j + P_{\bar{q}_i \bar{q}_j} \otimes \bar{q}_j) + P_{\bar{q}_i G} \otimes G\right) \qquad (5.2.30)$$

$$\frac{dG}{d(\ln Q^2)} = \frac{\alpha_s}{2\pi}\left(\sum_j (P_{G q_j} \otimes q_j + P_{G \bar{q}_j}) + P_{GG} \otimes G\right)$$

where $P \otimes f$ denotes

$$P \otimes f(x) = \int_x^1 \frac{dz}{z} P(z) f\left(\frac{x}{z}\right). \qquad (5.2.31)$$

The solution of Eqs. (5.2.30), together with the renormalization group equation for the running coupling constant,

$$\frac{d\alpha_s}{d(\ln Q^2)} = -\frac{\beta_0}{4\pi}\alpha_s^2 - \frac{\beta_1}{16\pi^2}\alpha_s^3 + O(\alpha_s^4) \qquad (5.2.32)$$

where

$$\beta_0 = 11 - \frac{2}{3}N_f \qquad (5.2.33)$$

$$\beta_1 = 102 - \frac{38}{3}N_f \qquad (5.2.34)$$

allows the computation of the parton densities at any scale Q^2 once initial conditions are known.

The solution of the renormalization group equation for the running coupling constant entails the definition of an integration constant that is usually expressed through the scale Λ:[2]

$$\alpha_s(Q^2) = \frac{1}{b \cdot \ln(Q^2/\Lambda^2)} \left[1 - \frac{\ln(\ln(Q^2/\Lambda^2))}{b' \ln(Q^2/\Lambda^2)} \right] \tag{5.2.35}$$

where

$$b = \frac{33 - 2N_f}{12\pi}$$

$$b' = \frac{(33 - 2N_f)^2}{6(153 - 19N_f)}. \tag{5.2.36}$$

A more convenient way [FUR 82] of writing Eqs. (5.2.30) is the following. Let us introduce

$$V_i(x,Q^2) = q_i(x,Q^2) - \bar{q}_i(x,Q^2) \tag{5.2.37}$$

and given $q_i^+ = q_i + \bar{q}_i$, define

$$T_8(x,Q^2) = u^+(x,Q^2) + d^+(x,Q^2) - 2s^+(x,Q^2)$$
$$T_{15}(x,Q^2) = u^+(x,Q^2) + d^+(x,Q^2) + s^+(x,Q^2) - 3c^+(x,Q^2) \tag{5.2.38}$$

u, d, s, and c are the up, down, strange, and charm quark densities, respectively. Defining

$$P_{q_i q_j} = \delta_{ij} P_{qq}^V + P_{qq}^S \tag{5.2.39}$$

$$P_{q_i \bar{q}_j} = \delta_{ij} P_{q\bar{q}}^V + P_{q\bar{q}}^S \tag{5.2.40}$$

and using the following relations (valid at the next-to-leading order),

$$P_{qq}^S = P_{q\bar{q}}^S$$

$$P_{q_i q_j} = P_{\bar{q}_i \bar{q}_j}; \quad P_{q_i \bar{q}_j} = P_{\bar{q}_i q_j} \tag{5.2.41}$$

$$P_{q_i G} = P_{\bar{q}_i G} \equiv P_{qG}; \quad P_{Gq_i} = P_{G\bar{q}_i} \equiv P_{Gq}.$$

---

[2] In the following sections, if not explicitly stated, Λ refers to the case of four flavors $N_f = 4$.

Eqs. (5.2.30) become

$$\frac{dV_i(x,Q^2)}{d(\ln Q^2)} = \frac{\alpha_s}{2\pi}[P_- \otimes V_i(x,Q^2)]$$

$$\frac{dT_i(x,Q^2)}{d(\ln Q^2)} = \frac{\alpha_s}{2\pi}[P_+ \otimes T_i(x,Q^2)]$$

$$\frac{d}{d(\ln Q^2)}\begin{pmatrix} \Sigma(x,Q^2) \\ G(x,Q^2) \end{pmatrix} = \frac{\alpha_s}{2\pi}\begin{pmatrix} P_{FF} & P_{FG} \\ P_{GF} & P_{GG} \end{pmatrix} \otimes \begin{pmatrix} \Sigma(x,Q^2) \\ G(x,Q^2) \end{pmatrix}$$

$$\equiv \frac{\alpha_s}{2\pi}\hat{P} \otimes \begin{pmatrix} \Sigma(x,Q^2) \\ G(x,Q^2) \end{pmatrix} \tag{5.2.42}$$

where

$$P_\pm = P_{qq}^V \pm P_{q\bar{q}}^V$$

$$P_{FF} = P_+ + 2N_f P_{qq}^S$$

$$P_{FG} = 2N_f P_{qG} \tag{5.2.43}$$

$$P_{GF} = P_{Gq}.$$

The T_8 and T_{15} can be used to evaluate the c quark density separately. The V_i and G distributions are usually called the valence quark and gluon density, respectively. The nonvalence quarks present in the nucleon are usually referred to as sea quarks. The valence distribution is also quoted as nonsinglet, and the gluon and Σ distribution as singlet components of the structure functions.

Following [FUR 82], we change the evolution variable $\ln(Q^2)$ into

$$t = \frac{2}{\beta_0}\ln\left(\frac{\alpha_s(Q_0^2)}{\alpha_s(Q^2)}\right) \tag{5.2.44}$$

and we expand the Altarelli–Parisi kernels P in powers of α_s:

$$P = P^{(0)} + \frac{\alpha_s}{2\pi}P^{(1)} + O(\alpha_s^2). \tag{5.2.45}$$

Equations (5.2.42) then become

$$\frac{dV_i(x,t)}{dt} = \left[P^{(0)} + \frac{\alpha_s}{2\pi}\left(P_-^{(1)} - \frac{\beta_1}{2\beta_0}P^{(0)}\right)\right] \otimes V_i(x,t)$$

$$\equiv \left(P^{(0)} + \frac{\alpha_s}{2\pi}R_-\right) \otimes V_i(x,t)$$

$$\frac{dT_i(x,t)}{dt} = \left[P^{(0)} + \frac{\alpha_s}{2\pi} \left(P_+^{(1)} - \frac{\beta_1}{2\beta_0} P^{(0)} \right) \right] \otimes T_i(x,t)$$

$$\equiv \left(P^{(0)} + \frac{\alpha_s}{2\pi} R_+ \right) \otimes T_i(x,t)$$

$$\frac{d}{dt} \begin{pmatrix} \Sigma(x,t) \\ G(x,t) \end{pmatrix} = \left[\hat{P}^{(0)} + \frac{\alpha_s}{2\pi} \left(\hat{P}^{(1)} - \frac{\beta_1}{2\beta_0} \hat{P}^{(0)} \right) \right] \otimes \begin{pmatrix} \Sigma(x,t) \\ G(x,t) \end{pmatrix} \qquad (5.2.46)$$

$$\equiv \left(\hat{P}^{(0)} + \frac{\alpha_s}{2\pi} \hat{R} \right) \otimes \begin{pmatrix} \Sigma(x,t) \\ G(x,t) \end{pmatrix}$$

where we have used $P^{(0)} = P_+^{(0)} = P_-^{(0)}$.

At the leading order we can still use the expressions of the structure functions in terms of partons given in Eq. (5.2.18), provided we replace the scaling parton densities $q(x)$ with the Q^2-dependent ones, $q(x, Q^2)$, which are obtained by solving the lowest-order evolution equations. Consequently the longitudinal structure function $F_L = F_2 - 2xF_1$ is still zero (up to terms of order α_s).

In the leading logarithmic approximation, the explicit expressions of the kernels appearing on the r.h.s. of Eqs. (5.2.46) are [ALT 77]

$$P_{qq}^V(x) = P_{qq}^{(0)}(x) = \frac{4}{3} \left(\frac{1 + x^2}{(1 - x)_+} + \frac{3}{2} \delta(1 - x) \right) \qquad (5.2.47)$$

$$P_{Gq}^{(0)}(x) = \frac{4}{3} \frac{1 + (1 - x)^2}{x} \qquad (5.2.48)$$

$$P_{qG}^{(0)}(x) = \frac{1}{2} (x^2 + (1 - x)^2) \qquad (5.2.49)$$

$$P_{GG}^{(0)}(x) = 6 \left(\frac{x}{(1 - x)_+} + \frac{1 - x}{x} + x(1 - x) \right)$$

$$+ \frac{33 - 2N_f}{6} \delta(1 - x) \qquad (5.2.50)$$

where the distribution

$$\frac{1}{(1 - x)_+} \qquad (5.2.51)$$

is defined as

$$\int_0^1 \frac{f(x)}{(1 - x)_+} = \int_0^1 \frac{f(x) - f(1)}{(1 - x)}. \qquad (5.2.52)$$

All the kernels that do not appear in Eqs. (5.2.47–50) vanish at the lowest order.

The evolution equations and their solution take a simpler form under a Mellin transformation. The Mellin transform or moments of a given function $g(x)$ are defined as

$$g_n = \int_0^1 x^{n-1} g(x)\, dx. \tag{5.2.53}$$

Under this transformation, for example, the nonsinglet evolution equation for V_i becomes:

$$\frac{dV_i^n(Q^2)}{d\ln Q^2} = \frac{\alpha_s}{2\pi} P_-^n V_i^n(Q^2) \tag{5.2.54}$$

and the convolution $P_- \otimes V_i$ is replaced by the product of

$$P_-^n = \int_0^1 x^{n-1} P_-(x)\, dx \text{ with } V_i^n = \int_0^1 x^{n-1} V_i(x)\, dx.$$

In the leading logarithmic approximation the solution of Eq. (5.2.54) is immediately found:

$$V_i^n(Q^2) = C_n \left(\ln \frac{Q^2}{\Lambda^2} \right)^{d_n} \tag{5.2.55}$$

where

$$d_n = \frac{P_-^n}{2\pi b} \tag{5.2.56}$$

is the anomalous dimension and b is given in Eq. (5.2.36).

5.2.6 *Parton densities beyond the leading order*

The next-to-leading order must include the $O(\alpha_s^3)$ term of the β-function for the running coupling constant β_1 in Eq. (5.2.34), the two-loop kernels $\hat{P}^{(1)}$ [FLO 77a,b, 79; GON 79; CUR 80; FUR 80], and the coefficient functions [ALT 79a; FLO 79], that is, the terms of order α_s, which are left over after the leading logarithmic corrections have been included in the definition of the parton densities.

As discussed in [ALT 78a,b] one can partly reabsorb the coefficient functions by a suitable redefinition of the parton densities. A popular choice for the definition of the quark densities is to demand that $F_2(x, Q^2)$ maintains the same form as in the naive parton model [ALT 78a,b]. More explicitly:

$$F_2(x,Q^2) = x \int_x^1 \frac{dy}{y} (q_f(y,Q^2) + \bar{q}_f(y,Q^2)) \left(\delta\left(1 - \frac{x}{y}\right) + \frac{\alpha_s(Q^2)}{2\pi} C_{qq}^2\left(\frac{x}{y}\right) \right)$$

$$+ \frac{\alpha_s(Q^2)}{2\pi} 2G(y,Q^2) C_{qG}^2\left(\frac{x}{y}\right) \to x[q_f(x,Q^2) + \bar{q}_f(x,Q^2)] \tag{5.2.57}$$

C_{qq}^2 and C_{qG}^2 are the coefficient functions for F_2, and Eq. (5.2.57) refers to one flavor, V–A coupling. This choice guarantees that the quark densities obey the conservation of charge because of the Adler sum rule, which is valid to all orders in perturbation theory

$$\int\limits_0^1 dx[q_f(x,Q^2) - \bar{q}_f(x,Q^2)] = v_f \tag{5.2.58}$$

where v_f has been defined previously (Eq. (5.2.20)).

We introduce the notation

$$\Sigma(x,Q^2) = \sum_f [q_f(x,Q^2) + \bar{q}_f(x,Q^2)] \tag{5.2.59}$$

and with the definition of Eq. (5.2.57) we find:

$$2F_1(x,Q^2) = \int\limits_x^1 \frac{dy}{y} \left(\Sigma(y,Q^2) \left(\delta\left(1 - \frac{x}{y}\right) - \frac{\alpha_s(Q^2)}{2\pi} C_{qq}^L\left(\frac{x}{y}\right) \right) \right.$$
$$\left. - 2N_f G(y,Q^2) \frac{\alpha_s(Q^2)}{2\pi} C_{qG}^L\left(\frac{x}{y}\right) \right)$$

$$F_3(x,Q^2) = \int\limits_x^1 \frac{dy}{y} \sum_f (q_f(y,Q^2) - \bar{q}_f(y,Q^2)) \left(\delta\left(1 - \frac{x}{y}\right) - \frac{\alpha_s(Q^2)}{2\pi} \tilde{C}_{qq}^3\left(\frac{x}{y}\right) \right)$$
$$= F_3^0(x,Q^2) - \Delta F_3(x,Q^2)$$
$$\tag{5.2.60}$$

where the coefficient functions C_{pp}^i, computed in [RUJ 77; ALT 78a, 79a; KUB 79], are

$$C_{qq}^L(z) = 8z/3$$
$$C_{qG}^L(z) = 2z(1 - z)$$
$$\tilde{C}_{qq}^3(z) = C_{qq}^3 - C_{qq}^2 = 4(1 + z)/3. \tag{5.2.61}$$

From Eqs. (5.2.57) and (5.2.60) one gets [ALT 78c]:

$$F_L(x,Q^2) = F_2(x,Q^2) - 2xF_1(x,Q^2)$$

$$= \frac{\alpha_s(Q^2)}{2\pi} x \int\limits_x^1 \frac{dy}{y} \left(\Sigma(y,Q^2) C_{qq}^L\left(\frac{x}{y}\right) + 2N_f G(y,Q^2) C_{qG}^L\left(\frac{x}{y}\right) \right) \tag{5.2.62}$$

and the Gross–Llewellyn Smith relation becomes

$$\int_0^1 dx\, F_3 = \left(1 - \frac{\alpha_s}{\pi}\right) \sum_f v_f. \tag{5.2.63}$$

Besides the definition of the quark densities beyond the LLA, we also need a definition of the gluon density. However, it will probably be impossible to measure the gluon in deep inelastic scattering at such a level of accuracy as to detect next-to-leading effects due to the coefficient functions and the two-loop anomalous dimensions, since the gluon is never directly probed by the external current. For this reason a more appropriate definition of the gluon can be derived by considering the next-to-leading corrections to direct photon production [AUR 84] or two-jet production in proton–proton (proton–antiproton) collisions (ELL 85, 88a,b; SOP 88; AVE 88], where the gluon plays an important role.

Hereafter we assume, for simplicity, the following redefinition of the gluon density [DIE 88]:

$$G(x, Q^2) \to G(x, Q^2) - \frac{\alpha_s(Q^2)}{2\pi} \int_x^1 \frac{dy}{y} \left(\Sigma(y, Q^2) C_{qq}^2 \left(\frac{x}{y}\right)\right.$$

$$\left. + 2N_f G(y, Q^2) C_{qG}^2 \left(\frac{x}{y}\right)\right) + O(\alpha_s^2) \tag{5.2.64}$$

which guarantees, at this order in α_s the total momentum conservation

$$\int_0^1 dx\, x \left[\sum_f (q_f(x, Q^2) + \bar{q}_f(x, Q^2)) + G(x, Q^2)\right] = 1. \tag{5.2.65}$$

Once a definition of the parton densities at the next-to-leading order has been given, one has to modify the second-order Altarelli–Parisi kernels $P^{(1)}$ to include the coefficient functions. Usually the kernels $P^{(1)}$ are given in the so-called $\overline{MS}$ scheme (see, e.g., [FLO 81]). In this case, the correct procedure is as follows. Let us write the matrix of the coefficient function as

$$\hat{C} = \hat{C}^{(0)} + \frac{\alpha_s}{2\pi} \hat{C}^{(1)} \tag{5.2.66}$$

then

$$P_{\pm}^{(1)} \to P_{\pm}^{(1)} - \frac{\beta_0}{2} C_{qq}^{(1)}$$

$$\hat{P}^{(1)} \to \hat{P}^{(1)} + [\hat{C}^{(1)}, \hat{P}^{(0)}] - \frac{\beta_0}{2} \hat{C}^{(1)}. \tag{5.2.67}$$

In the present case (see Eqs. (5.2.62) and (5.2.64)), where we define the quark and antiquark densities by incorporating the next-to-leading corrections to F_2 in q, $\bar{q}$

and we impose the total momentum conservation, the matrix $C^{(1)}$ is given by

$$\hat{C}^{(1)} = \begin{pmatrix} C_{qq}^2 & C_{qG}^2 \\ -C_{qq}^2 & -2N_f C_{qG}^2 \end{pmatrix}. \tag{5.2.68}$$

We close this section by noting again that the particular choice of the definition of the parton densities is completely arbitrary and that the physical results do not depend on this choice. It is a trivial change to define the parton densities with a different set of coefficient functions.

5.3 Experimental results on cross sections

Experimental results from deep inelastic neutrino scattering can be ordered in four classes:

1 absolute cross sections
2 ratios of cross sections
3 y distributions
4 x distributions as a function of Q^2.

The last point will be the main subject of the rest of this chapter, as it is directly connected with our understanding of quark interactions and of the intrinsic nature of nucleon structure. In this section we shall review the other points, which represent a kind of account of the amount of various quark components (see Sections 5.2.3–4).

Total cross sections are by far the most delicate measurement, as they imply precise absolute normalization of the incoming neutrino fluxes. Ratios of cross sections only require a reproducible monitor of the fluxes, to compare neutrino and antineutrino cross sections, while the comparison of neutral- to charged-current cross section suffers only from possible misidentification of muon and muonless events; y distributions, which carry a really important information on the nucleon content, are the easiest to study. In fact, the limitations in the experimental acceptance and the region of maximum confusion between charged- and neutral-current events (corresponding to very low muon momenta) are directly correlated to y and can be easily corrected in the analysis of the results.

5.3.1 Total cross sections

The measurement of absolute neutrino cross sections depends on precise monitoring of the incoming neutrino fluxes. A big benefit, particularly in the study of the dependence upon E_ν, comes from the knowledge of the neutrino energy on an event-by-event basis. Both requirements call for the use of narrow-band neutrino beams (NBB), where π's and K's from primary interactions are selected immediately after their production in a small momentum bite. By virtue of the correlation between the neutrino decay angle and energy, the latter can be deduced from the knowledge of

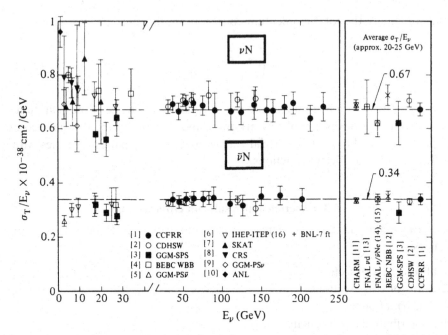

Fig. 5.3.1 Neutrino and antineutrino charged-current cross sections averaged (right) and as a function of neutrino energy (left). References: (1) [MAC 84], (2) [BER 87], (3) [MOR 81], (4) [COL 79], (5) [ERR 79], (6) [VOV 79], (7) [BAR 79], (8) [BAL 80], (9) [CIA 79], (10) [BAR 79], (11) [ALL 88a], (12) [ADE 86], (13) [KIT 82], (14) [BAK 83], (15) [TAY 83], (16) [BAK 82].

the meson momentum and from the measurement of the neutrino impact radius in a downstream detector. The ambiguity on the nature of the parent meson results in a dichromatic beam. The neutrino fluxes can be calculated measuring the properties of the primary and secondary parent beams and of the flux of the decay muons. High-energy (up to $\simeq 200\,\text{GeV}$ and more) NBBs were available at CERN and at Fermilab and they were exploited by several detectors.

A comprehensive plot of measured neutrino and antineutrino charged-current cross sections is given in Fig. 5.3.1, taken from [PDG 88]. A comparison of the average values with the earlier experiments at $E_\nu < 15\,\text{GeV}$ (Fig. 5.2.4) shows that $\sigma^\nu/\sigma^{\bar\nu}$ is decreasing with increasing energy and that, at the maximum explored energies, it is very close to 2. This indicates an increase in the antiquark content, as predicted by quantum chromodynamics. Notice the overall agreement of all the high-energy, high-statistics results fiom CERN and from FNAL [BLA 83]; subsequently, more reliable measurements of neutrino fluxes at CERN were achieved [BER 87; ADE 86; ALL 88a]. The world average calculated in [PDG 88] gives

$$\sigma^\nu/E_\nu = 0.67 \cdot 10^{-38}\,\text{cm}^2/\text{GeV}. \tag{5.3.1}$$

$$\sigma^{\bar\nu}/E_{\bar\nu} = 0.34 \cdot 10^{-38}\,\text{cm}^2/\text{GeV}. \tag{5.3.2}$$

The measurements of the total neutrino and antineutrino cross section provides a direct way of counting the total momentum carried by all the quarks,

$$Q + \bar{Q} = \frac{3\pi}{2G^2 s}(\sigma^\nu + \sigma^{\bar{\nu}}) \tag{5.3.3}$$

and by antiquarks alone

$$\frac{\bar{Q}}{Q + \bar{Q}} = \frac{1}{2}\frac{3r - 1}{1 + r} \tag{5.3.4}$$

where $r = \sigma^\nu / \sigma^{\bar{\nu}}$. From the values given in (5.3.1–2) it follows that charged partons carry only one-half of the nucleon momentum and that among them the antiquark component, at the higher explored energies, is close to 15 percent. A precise determination required the subtraction of elastic and quasi-elastic components and correction for $(s-c)$ contribution. A recently published analysis [ALL 88a] gives for instance,

$$Q + \bar{Q} = 0.492 \pm 0.006 \text{ (stat)} \pm 0.019 \text{ (syst)} \tag{5.3.5}$$

$$\frac{\bar{Q}}{Q + \bar{Q}} = 0.154 \pm 0.005 \text{ (stat)} \pm 0.011 \text{ (syst)}. \tag{5.3.6}$$

A further interesting test of the parton model is the comparison of measurements of cross sections on heavy nuclei with those on hydrogen, which are dominated by the valence down quarks. On this subject, in addition to several bubble chamber results [EFR 79; HAN 80; ARM 81; ALL 81a, 84], the CDHS collaboration [ABR 84] has simultaneously measured neutrino events originating in the iron calorimeter and in a tank of liquid hydrogen located in front of the detector, with the obvious advantage of a direct rate comparison without any normalization problem. These results, which agree with those from bubble chamber experiments, give $\sigma^{\nu p}/\sigma^{\nu\text{Fe}} = 0.63 \pm 0.02$ and $\sigma^{\bar{\nu} p}/\sigma^{\bar{\nu}\text{Fe}} = 1.31 \pm 0.08$. The quark parton model predicts $\sigma^{\nu p}/\sigma^{\nu\text{Fe}} = 0.73$ and $\sigma^{\bar{\nu} p}/\sigma^{\bar{\nu}\text{Fe}} = 1.2$ under the assumption of equal up and down quark momentum distributions. Taking into account the difference between the up and down distributions (see Section 5.5.5) one predicts $\sigma^{\nu p}/\sigma^{\nu\text{Fe}} = 0.61$ and $\sigma^{\bar{\nu} p}/\sigma^{\bar{\nu}\text{Fe}} = 1.3$, in very good agreement with the experimental result.

5.3.2 Comparison of neutral- and charged-current cross section

In the naive parton model, the ratio of the neutral- to charged-current cross section depends upon the right- and left-handed weak couplings of the quarks, which are all expressed in the minimal Standard Model in terms of a single parameter, $\sin^2 \theta_W$ (for $\rho = 1$). The measurement of this ratio is a fundamental test of the Standard Model, and is discussed in detail in Chapter 3. Here, for completeness, we recall the results coming from high-statistics experiments [ABR 86; ALL 87a; ARR 94].

Experimental uncertainties come from the separation of the two classes of events (charged and neutral). This can be done by two methods:

- The first, suited for fine-grained calorimeters, is based on the direct recognition of the muon in the final state, on an event by event basis.
- The second, used in heavy calorimeters, is statistically based on the different length of the neutral- and charged-current events, caused by the presence of long penetrating muon tracks.

The two methods require different corrections, resulting in independent experimental systematic errors. Most of the uncertainties coming from neutrino spectra cancel in the ratio. Given the differences in the experimental methods, all the results are in significant agreement. The total experimental uncertainty on $\sin^2\theta_W$ is 0.003.

From a theoretical point of view, the ratio has to be corrected for radiative corrections and strong interaction contributions. The largest theoretical uncertainty is associated with the c-quark threshold effect. Using the slow rescaling prescription [GEO 76; BRN 76] the correction can be parameterized as 0.013 (m_c (GeV)−1.3), where m_c is the effective c-mass. For $m_c = 1.31 \pm 0.24$ GeV [ARR 94] this contributes ± 0.003 to the total theoretical uncertainty $\Delta\sin^2\theta_W = \pm 0.004$.

The deep-inelastic neutrino world average is then [PDG 96]

$$\sin^2\theta_W^{\text{average}} = 0.2259 \pm 0.0043, \tag{5.3.7}$$

in impressive agreement with the SM prediction of 0.2237, corresponding to the LEP determination of $M_Z = 91.1884$ GeV [PDG 96].

5.3.3 Inelasticity distributions

Inelasticity distribution in charged-current events is the most direct measurement of the relative contribution of quarks and antiquarks in the nucleons. In terms of statistics, it carries the same information of total cross-sectional measurements, with the additional advantage coming from the knowledge of the shape of the distributions. Corrections for detector acceptance and for the elastic contribution are easily accounted for, as they appear in well-defined y regions.

A classical paper on the subject is the first report from CDHS [GRO 79] where a detailed analysis on y distributions is presented. Data were collected at the CERN NBB for $30 < E_\nu < 200$ GeV. Shower energy and muon momentum were both measured by the magnetized calorimeter, allowing a direct reconstruction of y for each event. The detector acceptance is shown in Fig. 5.3.2 for different neutrino energies. The data corrected for quasi-elastic contribution are presented in Fig. 5.3.3. After the application of radiative corrections, a global fit to these distributions gives

$$\frac{\bar{Q}}{Q + \bar{Q}} = 0.15 \pm 0.03. \tag{5.3.8}$$

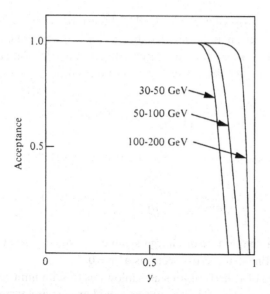

Fig. 5.3.2   Acceptance of the CDHS detector for charged-current events as a function of y for three ranges of incident neutrino energies.

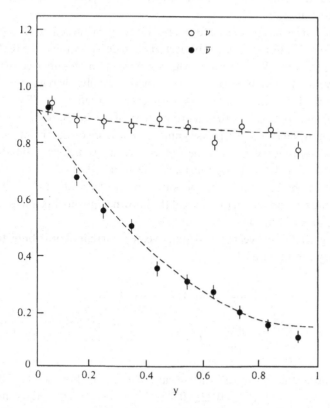

Fig. 5.3.3   Charged-current inelasticity (y) distributions as measured by CDHS collaboration.

Even more precise results can be obtained looking at the high y antineutrino region, where the cross section is dominated by the $(\bar{q} + \bar{s})$ component and the small contribution $\propto (1 - y)^2 q$ can be subtracted using the neutrino data at the same value of y. This method gives the extremely accurate result:

$$\frac{\bar{Q}}{Q + \bar{Q}} = 0.15 \pm 0.01. \tag{5.3.9}$$

The study of the cross sections at small y is also reported, giving

$$\left(\frac{d\sigma^{\bar{\nu}}}{dy}\right)_{y=0} \bigg/ \left(\frac{d\sigma^{\nu}}{dy}\right)_{y=0} = 1.01 \pm 0.07 \tag{5.3.10}$$

thus providing a simple test of charge symmetry, which predicts the equality of neutrino and antineutrino cross sections at $y = 0$.

The knowledge of y distributions also allows us to set a limit to other contributions to the cross section. A scalar current would appear as a term proportional to $(1 - y)$ in the cross section and it would violate the Callan–Gross relation, giving a large longitudinal component to the cross section. Negative evidence is reported for such a contribution.

Additional information can be found by comparing charged- and neutral-current inelasticity distributions. This has been studied in detail by the CHARM collaboration [JON 81a]. In order to perform the comparison in an unbiased way, both CC and NC events are analyzed in the same way, namely, disregarding in CC any measurement of the muon solely used for event classification. The main experimental problem comes from the dichromaticity of the NBB, resulting in a twofold ambiguity in the neutrino energy assignment for a given detector radius. To solve it, the form of $d\sigma/dy$ is assumed to be represented as a linear combination of bell-shaped spline functions (*B*-splines) (see Section 5.4) and the coefficients are determined by the best fit to the measured event distribution $d^2N/dE_h\,dr$ (where r is the interaction radius) on the basis of the known neutrino flux and of the detector acceptance and resolution.

Assuming that only vector and axial-vector currents contribute to the cross section, they are fitted as

$$\frac{d\sigma^{\nu}}{dy} = A((1 - \alpha) + \alpha(1 - y)^2) \tag{5.3.11}$$

$$\frac{d\sigma^{\bar{\nu}}}{dy} = A(\alpha + (1 - \alpha)(1 - y)^2). \tag{5.3.12}$$

The data and the result of the fit are shown in Fig. 5.3.4. The fit gives $\alpha^{CC} = 0.16 \pm 0.02$ and $\alpha^{NC} = 0.22 \pm 0.02$; α^{CC} is easily identified as the antiquark content of the nucleon, corrected for the small $(s-c)$ contribution, and it is in nice

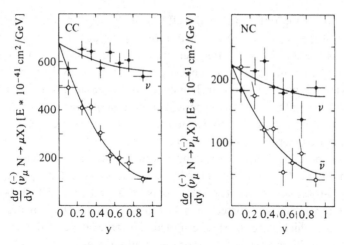

Fig. 5.3.4 Charged- and neutral-current inelasticity distributions as measured by CHARM collaboration.

agreement with the completely different determination of CDHS. The value of α^{NC}, close indeed to α^{CC}, indicates that neutral currents are also predominantly $V-A$.

From the complete quark-parton model cross sections given in Section 5.2.4, the couplings of the weak neutral currents can be extracted fitting the data with suitable combinations of free parameters. The best results give $g_L^2 = u_L^2 + d_L^2 = 0.32 \pm 0.02$ and $g_R^2 = u_R^2 + d_R^2 = 0.05 \pm 0.02$, while for the sum of the right- and left-handed couplings to strange quarks one gets $g_s^2 = 0.26 \pm 0.06$. However, all the couplings can be calculated in the Standard Model using as input the measured value of $\sin^2\theta_W$. Fixing these parameters, the fit can be repeated to determine at best the fractional momentum carried by the strange quark-antiquark, found to be (0.06 ± 0.04).

The above analysis can be repeated, abandoning the hypothesis of pure $V-A$ interactions and searching for the presence of a term proportional to y^2 in the neutral-current cross sections. The negative result puts a limit of 3 percent on the ratio of the scalar to vector squared couplings.

Recently the CHARM collaboration presented a new analysis of the inelasticity distributions, based on high-statistics data taken at the 160 GeV CERN NBB [ALL 89]. The results for the neutral-current couplings are $g_L^2 = 0.287 \pm 0.008$ and $g_R^2 = 0.042 \pm 0.010$, in good agreement with the predictions of the Standard Model and demonstrating a nonvanishing coupling of the neutral current to right-handed quarks with a significance of more than four standard deviations.

5.4 Methods to extract structure functions from differential cross sections

To unravel the nucleon picture, one must know something about the dependence of quark distributions in terms of the two variables x and Q^2. The structure function analysis requires the unfolding of the effects of the finite resolution and limited acceptance of the detector on the distributions of the data in these two variables. Nevertheless the unfolding is often performed on the distributions in x and E_h, rather than Q^2, because the hadron energy is a variable directly measured and its resolution is experimentally known. This choice offers the additional advantage that for fixed E_h values the x variable spans its whole range (see Section 5.1.1).

The method commonly used to get rid of the resolution effects assumes a model for the structure functions to be used as input to a Monte Carlo simulation based on the known experimental acceptance and resolution functions. The observed event number in any (x, E_h) bin is corrected by an unsmearing factor equal to the ratio of generated-to-accepted Monte Carlo events in the same bin. This procedure has the disadvantage that the correction relies just on the (x, Q^2) dependence of the cross sections, which have to be determined. The true shape of the cross sections must be iteratively approached by repeated comparisons of the simulated results and the data.

All the experiments following this recipe restrict the analysis to the region where the unsmearing correction is sufficiently close to unity. This criterion usually results in the elimination of bins at large x, which are essentially populated by events shifted from smaller values of x by the effect of the resolution and of those bins affected by the inefficiency in the low-momentum muon detection.

More rigorous approaches can be followed to solve the problem. Let $f(x)$ be the unknown distribution function of a measured variable x. Owing to finite resolution and limited acceptance effects, the observed event distribution in bins of x_m will be

$$G_i = \int dx_m \int dx f(x) r(x, x_m) \qquad i = 1, \ldots, n \qquad (5.4.1)$$

where $r(x, x_m)$ represents the distribution of measured values x_m for a given value of x. The method to estimate $f(x)$ consists in assuming a representation

$$f(x) = \sum_{j=1,m} a_j B_j(x) \qquad (5.4.2)$$

as being B_j a set of suited functions, so that

$$G_i = \sum_{j=1,m} a_j \int dx_m \int dx B_j(x(r), x, r_m) = \sum_j a_j R_{ij} \qquad (5.4.3)$$

and performing (for $n > m$) a least-square fit to solve the system. The R_{ij} integrals can be calculated by numerical integration. Once the coefficients a_j are determined, an approximate analytical form of the distribution $f(x)$ is known. In this way, however, a correlation between neighboring bins of x is introduced, making the statistical analysis of the results less transparent. The extension of the method to two-dimensional distributions as $d^2\sigma/dx\,dE_h$ is straightforward. This method does not involve any theoretical prejudice on the form of the distributions to be unfolded and allows, as a further advantage, an analytical transformation from the (x, E_h) to (x, Q^2) plane.

The accuracy of the approximation of $f(x)$ by the linear combination $\sum_j a_j B_j(x)$ is obviously strongly related to the choice of the functions B_j. The B-splines [SCH 46; BOO 78] are widely used for that. They are bell-shaped smooth functions defined in a limited interval and made up of kth order polynomial pieces, connected in a definite number of points called knots. In the CHARM analysis [BER 83], for example, cubic splines and equidistant knots in $\sqrt{E_h}$, according to the resolution, were used for E_h, while quartic splines and not-equidistant knots were used for x, matching the good resolution for small x and the coarse resolution at large x.

5.4.1 The structure functions xF_3, F_2

The differential cross sections of ν and $\bar{\nu}$ scattering on isoscalar targets can be written in terms of xF_3, F_2, and F_L in the following way:

$$
\frac{d^2\sigma^{\nu,\bar{\nu}}}{dx\,dy} = \frac{G^2 s}{4\pi} \left(\left(1 + (1-y)^2 + \frac{Mxy}{E}\right) F_2 - y^2 F_L \right.
$$
$$
\left. + (1 - (1-y)^2)(2x(s-c) \pm xF_3) \right) \tag{5.4.4}
$$

where F_3 is defined as $q - \bar{q}$ (Eq. (5.2.19)), contrary to $F_3^{\nu,\bar{\nu}}$, whose expression is given in Eq. (5.2.18). It is straightforward to extract xF_3 and F_2 by a suitable combination of the above expressions:

$$
xF_3 = \frac{2\pi}{G^2 s} \left(\frac{d^2\sigma^\nu}{dx\,dy} + \frac{d^2\sigma^{\bar{\nu}}}{dx\,dy} \right) \frac{1}{1 - (1-y)^2} \tag{5.4.5}
$$

$$
F_2 = \frac{\frac{2\pi}{G^2 s} \left(\frac{d^2\sigma^\nu}{dx\,dy} + \frac{d^2\sigma^{\bar{\nu}}}{dx\,dy} \right) + y^2 F_L - 2x(s-c)(1 - (1-y)^2)}{1 + (1-y)^2 + \frac{Mxy}{E}}. \tag{5.4.6}
$$

The determination of xF_3 does not entail any assumption. Conversely, F_2 is affected by the small but poorly known contributions due to F_L (mainly for $y \sim 1$) and to the difference $(s-c)$ (mainly for $x \sim 0$). Neglecting these contributions and the kinematical term $Mxy/E \sim M^2/Q^2$, the expression for F_2 simplifies to

$$
F_2 = \frac{2\pi}{G^2 s} \left(\frac{d^2\sigma^\nu}{dx\,dy} + \frac{d^2\sigma^{\bar{\nu}}}{dx\,dy} \right) \frac{1}{1 + (1-y)^2}. \tag{5.4.7}
$$

Let us write Eq. (5.4.6) as a function of $R = F_L/2xF_1$ rather than F_L:

$$F_2 = \frac{\frac{2\pi}{G^2 s}\left(\frac{d^2\sigma^\nu}{dx\,dy} + \frac{d^2\sigma^{\bar\nu}}{dx\,dy}\right) - 2x(s-c)(1-(1-y)^2)}{1+(1-y)^2 - y^2\left(R\left(1+\frac{Q^2}{\nu^2}\right)\right)\Big/\left((1+R) - \frac{Q^2}{2\nu^2}\right)}. \tag{5.4.8}$$

The simplest choice is to assume for R the form predicted by QCD. However, this assumption is not fully justified, as in the low x region the nonperturbative part of R is expected to be at least comparable with the perturbative one and its x dependence is a phenomenological guess subject to a large uncertainty.

A reasonable choice for the correction $(s(x)-c(x))$ is to neglect the charmed sea contribution and to assume

$$s(x) \simeq 0.2(\bar{u}(x) + \bar{d}(x)) \tag{5.4.9}$$

as suggested by the ν-, $\bar\nu$-induced dimuon events [ABR 82b].

5.4.2 The antiquark density $\bar{q}$

At high values of the variable y, the main contribution to the antineutrino cross section comes from the scattering off antiquarks. The distribution $\bar{q}(x, Q^2)$ can therefore be directly measured by high-statistics experiments using the approximate relation

$$\bar{q} = \frac{2\pi}{G^2 xs}\left(\frac{d^2\sigma^{\bar\nu}}{dx\,dy} - (1-y)^2 \frac{d^2\sigma^\nu}{dx\,dy}\right) \tag{5.4.10}$$

where terms $\sim F_L$, $(s-c)$, and Q^2/ν^2 are neglected. Equation (5.4.10) has been obtained by replacing F_2 and xF_3 in Eq. (5.4.4) with their partonic expressions. In this evaluation, only values of y larger than a suitable y_0 are used, where y_0 is chosen in order to minimize the statistical uncertainty. For instance, the CDHS [ABR 83] and CHARM [BER 83] collaborations chose a value $y_0 = 0.5$ and $y_0 = 0.6$, respectively.

The determination of $\bar{q}$ depends somewhat on the assumptions on F_L and $(s-c)$, as in the case of F_2.

5.4.3 The longitudinal structure function F_L

The longitudinal structure function is related to the sum of the differential cross sections

$$\frac{d^2\sigma^\nu}{dx\,dy} + \frac{d^2\sigma^{\bar\nu}}{dx\,dy} = \frac{G^2 s}{2\pi}\left(\left(1+(1-y)^2 + \frac{Mxy}{E}\right)F_2 - y^2 F_L\right.$$

$$\left. + 2x(s-c)(1-(1-y)^2)\right). \tag{5.4.11}$$

Its small contribution, limited to a restricted kinematical region, can be disentangled from the contribution of F_2 by exploiting their different y dependence. It is possible to write two equations to extract F_L and F_2 by considering the relation (5.4.1) in two different regions of y: the region $y \sim 1$ where the term proportional to F_L is appreciable and the region $y \sim 0$ where F_2 completely dominates.

5.4.4 Extraction from data

To extract the structure functions from the observed event rates, the neutrino and antineutrino fluxes $\Phi^\nu(E)$ and $\Phi^{\bar\nu}(E)$ have to be known, together with the total charged-current cross sections $\sigma^\nu(E)$ and $\sigma^{\bar\nu}(E)$. There is a substantial difference between NBB- and WBB-like exposures. Whereas in the former fluxes are measured at the level of percent, in the latter they are not directly measured but have to be deduced from the observed charged-current event rates and the total cross sections. For this reason, problems may emerge because of the absolute value and the energy dependence of the cross section.

Let the number of neutrino-induced events observed in a given bin $(\Delta x_i, \Delta Q_j^2)$ be

$$N_{ij}(x, Q^2) = A \int dE \Phi(E) \int_{\Delta x_i} dx \int_{\Delta Q_j^2} \frac{d^2\sigma}{dx\, dQ^2} \qquad (5.4.12)$$

where A is a normalization factor related to the total number of events. Let us choose the bins in x and Q^2 such that $2xF_1$, F_2, xF_3 can be considered almost constant and define

$$N_{1ij}(x, Q^2) = A \frac{G^2}{2\pi} \int dE\, \Phi(E) \int_{\Delta x_i} dx \int_{\Delta Q_j^2} dQ^2 \frac{y^2(x, Q^2)}{2x}$$

$$N_{2ij}(x, Q^2) = A \frac{G^2}{2\pi} \int dE\, \Phi(E) \int_{\Delta x_i} dx \int_{\Delta Q_j^2} dQ^2 \left(1 - y(x, Q^2) - \frac{Mxy(x, Q^2)}{2E}\right) \frac{1}{x}$$

$$N_{3ij}(x, Q^2) = A \frac{G^2}{2\pi} \int dE\, \Phi(E) \int_{\Delta x_i} dx \int_{\Delta Q_j^2} dQ^2 \frac{(1 - (1 - y(x, Q^2))^2)}{2x}$$

$$(5.4.13)$$

where the inelasticity $y = y(x, Q^2)$ is expressed as a function of x and Q^2. Then N_{ij} can be written as

$$N_{ij}(x, Q^2) = N_{1ij}(x, Q^2)\langle 2xF_1 \rangle_{ij} - N_{2ij}(x, Q^2)\langle F_2 \rangle_{ij} + N_{3ij}(x, Q^2)\langle xF_3 \rangle_{ij} \qquad (5.4.14)$$

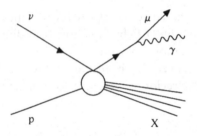

Fig. 5.4.1   Real photon radiation by the muon.

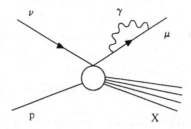

Fig. 5.4.2   Virtual photon radiation by the muon.

where $\langle F \rangle_{ij}$ represents the average over the bin ij. For the antineutrino one has

$$\bar{N}_{ij}(x, Q^2) = \bar{N}_1(x, Q^2)\langle 2xF_1 \rangle_{ij} + \bar{N}_2(x, Q^2)\langle F_2 \rangle_{ij} - \bar{N}_3(x, Q^2)\langle xF_3 \rangle_{ij}. \quad (5.4.15)$$

Relations (5.4.14) and (5.4.15) are deduced from (5.1.20) for isoscalar targets under the assumption $(s-c) = 0$. The structure functions F_2 and xF_3 can be obtained from the above relations when a relationship between $2xF_1$ and F_2 is given.

     Similar procedures can be applied to determine the $\bar{q}(x, Q^2)$ distribution. As for F_L, the same method leads to one equation for two unknowns, $\langle F_2 \rangle$ and $\langle F_L \rangle$ and one has to integrate on two different regions of y to obtain $\langle F_L \rangle$.

5.4.5 Radiative corrections

To derive the structure functions from the measured event distributions, the effect of electromagnetic radiation from charged particles involved in the scattering must be taken into account.

     Radiative corrections to neutrino data are usually applied following the prescription of [RUJ 79]. In experiments where the hadron energy is inclusively measured, the main contribution to the corrections is due to the collinear emission of photons by the lepton involved in the scattering (Figs. 5.4.1 and 5.4.2). The collinear photon emitted by the muon will not change the muon production angle but will make events migrate from large to smaller E', that is, from small to larger y. Then the observed y distribution at fixed E will be underestimated at small y and overestimated at large y with respect to the radiatively corrected distribution. Just

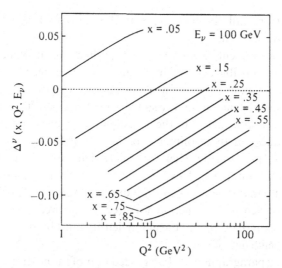

Fig. 5.4.3 Radiative corrections for charged-current neutrino scattering as a function of Q^2 at $E_\nu = 100\,\mathrm{GeV}$. The correction Δ defined as:

$$\Delta^\nu(x, Q^2, E_\nu) = \frac{(d\sigma_{obs}(E_\nu)/dx\,dQ^2 - d\sigma_{bare}(E_\nu)/dx\,dQ^2)}{d\sigma_{bare}(E_\nu)/dx\,dQ^2}$$

is reported at several values of x.

the contrary will happen in the case of the distributions in x and Q^2, as is readily seen by looking at them in terms of E' (Fig. 5.4.3).

5.5 Results on structure functions

5.5.1 Introduction

Neutrino experiments allow a complete determination of parton densities through the study of structure functions.

Results are available for

- $F_2, xF_3, \bar{q}$
- $u_v, d_v,$ *and their ratio*
- *Strange sea*
- *Gluon*
- F_L
- *Nuclear effects*

None of the existing detectors was able to span over the whole set of possible measurements. However, the presence of detectors possessing complementary features, like the electronics detector and the bubble chambers, permitted us not only to perform all the measurements but also to have reasonably large intersection between them.

Bubble chambers present the unique feature of changing target material and more specifically make it possible to compare deuterium to hydrogen cross sections. This enables us to tell the interactions of neutrinos on protons from those on neutrons. Therefore this kind of detector is more suited for studying the flavor composition of the nucleon. It also does better in the investigation of nuclear effects.

There is no question that counter experiments are definitely superior where statistics are a vital prerequisite. They are preferred in the determination of the total valence quark and antiquark densities, the longitudinal structure function, the strange sea density that is determined from dimuon events, and above all the gluon density. The gluon density is a function not directly measured since direct neutrino–gluon scattering cannot occur and the gluon density must be extracted from a QCD fit to the Q^2 dependence of the data.

In the following paragraphs we take a guided tour of the neutrino results, without pretending any completeness in what is shown, but rather to see where the level of knowledge stands at present. At the end of this section we describe the attempt to recombine most of the information known from data and theory to give a quantitative description of the proton in terms of its partonic constituents.

5.5.2 Results on F_2, xF_3, $\bar{q}$

The distribution of the momentum carried by all quarks (F_2) and valence quarks alone (xF_3) has been measured by many experiments [BOS 78a, 82; GRO 79; HEA 81; MOR 81; JON 82; ABR 83; BER 83; MAC 84; SEL 97]. A compilation of most of the results is found in [DIE 86].

As already mentioned, more precise data have come from the experiments that have exploited higher-statistics samples and so we present in Fig. 5.5.1 as a typical result for $F_2(x, Q^2)$ and $xF_3(x, Q^2)$ data from large counter experiments that have been undertaken in this field.

One would like to compare at a glance in a quantitative way data from different experiments. It must be recalled here that although structure functions are extracted under the same general rules, slightly different assumptions are made in the analysis and also corrections are applied in a different fashion from one experiment to another. In Table 5.5.1 we summarize the different choices made by some of the experiments.

Another factor that complicates the comparison of data is the value used by each experiment for the total neutrino and antineutrino cross sections and their ratio (see Section 5.3.1).

In order to visualize these effects, we plot in Figs. 5.5.2a and 5.5.2b the results from several experiments for F_2 and xF_3 as a function of Q^2 at a fixed value of x ($x = 0.25$). At this value of the quark momentum the functions are expected to be nearly Q^2 independent and the spread of the data represents a measure of the relative

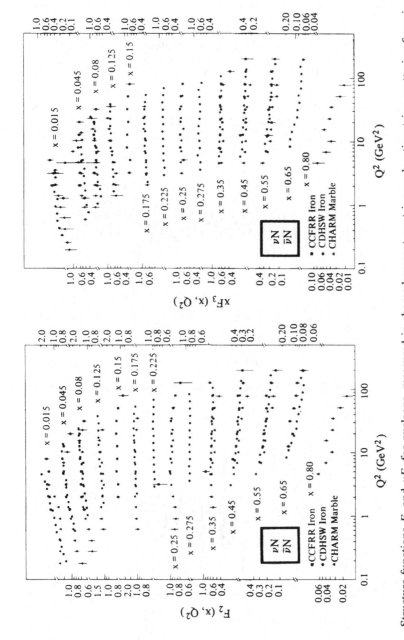

Fig. 5.5.1 Structure functions F_2 and xF_3 for nucleons, measured in charged-current neutrino and antineutrino scattering from iron (CCFR, CDHS) and marble (CHARM) targets plotted versus Q^2 for fixed x bins.

Table 5.5.1. *Assumptions applied in the extraction of structure functions*

	CDHS	CCFR	CHARM	BEBC
Sea content of the nucleon	$\bar{s} = 0.25(\bar{u} + \bar{d})$ $\bar{c} = c = 0$	$\bar{s} = 0.25(\bar{u} + \bar{d})$ $\bar{c} = c = 0$ slow rescaling $m_c = 1.7$ Gev	$s - c = 0$	$s - c = 0$
$R = \dfrac{\sigma_L}{\sigma_T}$	0.1	QCD	0.0	0.0
Fermi motion	No	No	Yes	Yes
Radiative corrections	Yes	Yes	Yes	No

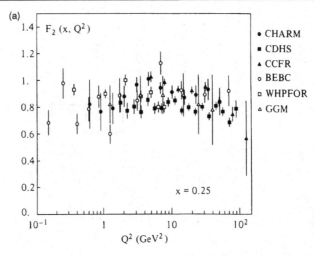

Fig. 5.5.2(a)   Comparison of several results on F_2 as a function of Q^2 at a value of $x = 0.25$.

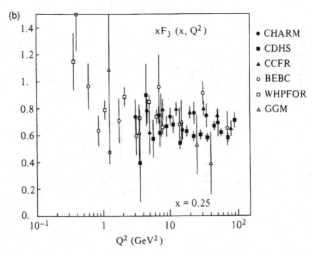

Fig. 5.5.2(b)   Comparison of several results on xF_3 as a function of Q^2 at a value of $x = 0.25$.

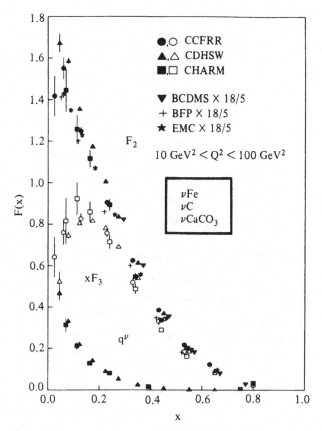

Fig. 5.5.3 Comparison of neutrino results on $F_2(x)$ and $xF_3(x)$ with those from muon production properly rescaled by the factor $\frac{18}{5}$ for a Q^2 ranging between 10 and 100 GeV^2.

systematic uncertainties among different sets. The data lie inside a ± 20 percent band, as one could expect from the previous discussion.

Despite systematic effects, it is possible to compare the nucleon structure functions extracted from neutrino scattering data with those measured in charged leptoproduction. Figure 5.5.3 shows the neutrino results on $F_2(x)$ and $xF_3(x)$ together with those from several μN experiments properly scaled by the factor $\frac{18}{5}$, which approximately takes into account the different charges seen in the scattering by the incident lepton. The agreement is fairly good and this can be considered an undisputable success of the quark parton model.

The picture of the nucleon structure derived from deep inelastic neutrino interactions has been recently complemented by the data coming from electromagnetic scattering of electrons on protons at HERA. These data cover a completely new Q^2 domain, up to 10 000 GeV^2, with a particular sensitivity for low x values. HERA data for F_2 are reported in Fig. 5.5.4 [PDG 96 and references therein] compared to muon scattering results at $Q^2 < 100$ GeV^2, showing an impressive agreement.

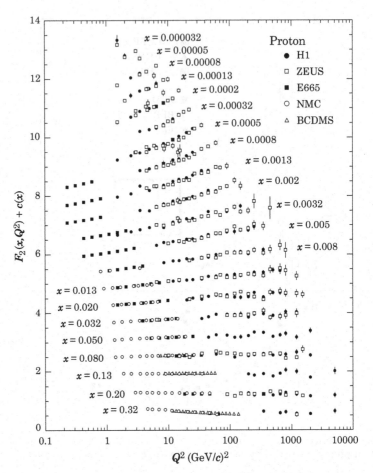

Fig. 5.5.4 F_2^p measured in electromagnetic scattering of electrons and muons. Only statistical errors are shown. For purpose of plotting, a constant $c(x) = 0.6(i_x - 0.4)$ is added to F_2^p, where i_x is the number of the x bin ranging from $i_x = 1$ ($x = 0.32$) to $i_x = 21$ ($x = 0.000032$). From [PDG 96] and references therein.

CCFR collaboration [LEU 93] presented a precise determination of the Gross–Llewellyn Smith sum rule (see Section 5.2.1), extracted from an interpolation of the measured values of xF_3 at $Q^2 = 3 \, \text{GeV}^2$:

$$\int_0^1 F_3(x, Q^2 = 3 \, \text{GeV}^2) dx = 2.50 \pm 0.018(\text{stat}) \pm 0.078 \, (\text{syst}),$$

to be compared to 3, as expected in the naive parton model, or to 2.66 ± 0.04, as predicted by a QCD evolution of the nonsinglet structure function.

5.5.3 Strange sea

A favorable process for digging out the strange quark component from the rest of the sea in the nucleon is the production of charm quarks by antineutrino.

The relevant reactions are

$$\bar{\nu}_\mu + \bar{s} \rightarrow \mu^+ + \bar{c} \tag{5.5.1}$$

$$\bar{\nu}_\mu + \bar{d} \rightarrow \mu^+ + \bar{c} \tag{5.5.2}$$

the second of which is Cabibbo-suppressed. The final state should in fact be constituted mostly (~ 90 percent) of events in which a ($\bar{s} \rightarrow \bar{c}$) transition occurs.

The events containing a charmed particle in the final state are easily detectable from the semileptonic decay of the charm quark

$$\bar{c} \rightarrow \bar{s} + \mu^- + \bar{\nu}_\mu \tag{5.5.3}$$

whose branching ratio is of order 10 percent.

The events are therefore characterized by having two muons of opposite charge in the final state (so-called dimuon events). The study of the x distribution of these events gives almost directly the x shape of the strange sea together with its integral. In addition, the cross section for dimuon events depends on the mass of the c quark and the ratio of dimuon to single muon production provides a direct test of the slow rescaling hypothesis [GEO 76; BRN 76].

Many experiments have measured the s distribution [BEN 75a,b, 78; BLI 75, 76; BAR 76, 77; KRO 76; HOL 77; BOS 77, 78b; BAL 77; JON 81b; ABR 82b]. The most recent and statistically significant results are from CCFR [BAZ 95] and include a next-to-leading order QCD analysis of their data. The neutrino energy dependence of the dimuon cross section clearly shows the presence of the heavy charm quark threshold. An NLO fit indicates a value of the charm quark mass of $m_c = 1.70 \pm 0.19$. The strange sea content of the nucleon, defined as $(S + \bar{S})/(\bar{U} + \bar{D})$ is found to be $0.477^{+0.063}_{-0.053}$. The x distribution is similar to that of the light antiquark sea. The Q^2 dependence of the strange sea is shown in Fig. 5.5.5 for several values of x: scaling violations are clearly observed.

5.5.4 Longitudinal structure function

The determination of the longitudinal structure function is one of the tasks that have not yet been accomplished by neutrino experiments.

Its importance is indeed crucial, as this function is of order α_s, and therefore it allows in principle a direct estimate of the strong coupling constant and of the gluon distribution.

The data available so far [ABR 83; BER 84; AUB 85; BEN 89; BER 91] are summarized in Fig. 5.5.6, where the ratio $R = \sigma_L/\sigma_T$ is reported for all the measurements performed by lepton–hadron scattering. Deviations from the

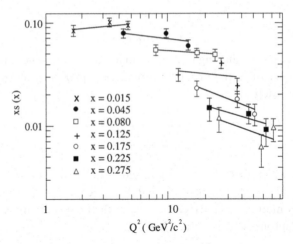

Fig. 5.5.5   Strange sea structure functions $xs(x)$ versus Q^2 for several values of x, as determined by CCFR [SMI 93]. The lines are power-law fits to the data. Errors are statistical.

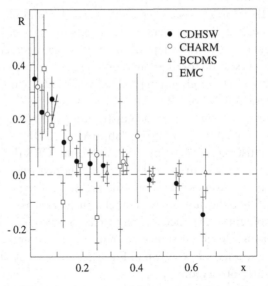

Fig. 5.5.6   The ratio $R = \sigma_L/\sigma_T$ as measured by several collaborations in lepton–hadron scattering (black dots from [BER 91], open circles from [BER 84], open squares from [AUB 85] and open triangles from [BEN 89]).

Callan–Gross relation are observed; still missing is a measurement of the longitudinal cross section as a function of (x, Q^2).

5.5.5 *Separation of valence quark of different flavors*

Bubble chamber experiments, as already mentioned, have a good chance of performing these measurements. The field was first explored by several experiments

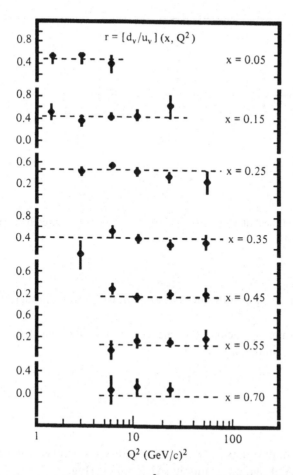

Fig. 5.5.7 The ratio $r = d_v/u_v$ versus x and Q^2 as measured in BEBC. The dashed lines are the values averaged over all Q^2 at the corresponding values of x.

of this kind [EFR 79; HAN 80; ARM 81; ALL 81a,b]. The best results on the flavor composition of the proton come from an experiment performed in BEBC, filled by deuterium, where each neutrino (antineutrino) interaction could be almost unambiguously attributed to scattering on proton or neutron by a prong-counting technique [ALL 84, 85]. This experiment yielded a measurement of the distributions $xu_v(x, Q^2)$, $xd_v(x, Q^2)$ and of their ratio $r = xd_v/xu_v$, at different values of Q^2.

The experimental results are summarized in Fig. 5.5.7, where the ratio r is plotted versus x and Q^2. The ratio appears to be Q^2 independent and fairly well described by a relation

$$r \sim 0.57(1 - x). \tag{5.5.4}$$

The d_v quark distribution is therefore substantially softer than that of the u_v. The ratio is consistent with the QPM expectation $u = 2d$ at $x = 0$ and it falls

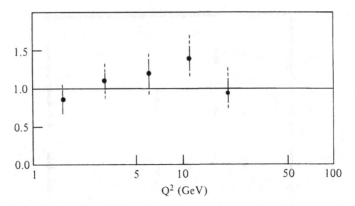

Fig. 5.5.8   Test of the Adler sum rule. $\int (F_2^{\nu n} - F_2^{\nu p}/2x)\,dx$ measured by BEBC is plotted at different Q^2 values.

significantly for $x \to 1$ in substantial agreement with several models [CLO 73; FAR 75; FIE 77].

The separate measurement of the valence component of the proton also allows us to test the Adler sum rule, namely,

$$\int_0^1 (u_v - d_v)\,dx = 1 \tag{5.5.5}$$

which is valid at all orders in QCD. The validity was proven with good accuracy both at fixed [ALL 84] and running [ALL 85] Q^2 (Fig. 5.5.8).

Other results in good agreement with those just mentioned were obtained by the CDHS detector supplied with an external hydrogen target [ABR 84].

5.5.6 Gluon density

The gluon density is not directly measured in neutrino interactions. Gluons in fact do not undergo weak interactions so that their fractional momentum distribution inside the nucleon, whose integral amounts to ~ 50 percent, has to be derived by looking at the variation in Q^2 of quark densities that are related to the gluon density by the Altarelli–Parisi equations (see Section 5.2).

The exercise is therefore delicate and the result not only depends on the available statistics but is also correlated to a specific QCD analysis. A more detailed discussion is presented in Section 5.7 in junction with the QCD analysis. Here, for completeness, we just report on the two existing determinations on the shape of the gluon density and its evolution [ABR 82a, 83; BER 83]. The results (Fig. 5.5.9) show significant differences: One of the two determinations (CHARM) presents a gluon considerably softer than the other.

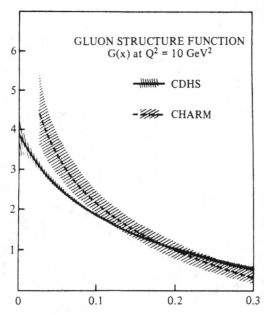

Fig. 5.5.9 Comparison of the gluon densities and their uncertainty at $Q^2 = 10 \, \text{GeV}^2$ as determined by the CDHS and CHARM collaborations.

5.5.7 Nuclear effects

Results from charged leptoproduction experiments [STE 75; AUB 83, 87; ARN 84a; BAR 85; BEN 87a; ALP 89; AMA 91, 92, 92a, 95; ADA 95] indicate that the nucleon structure functions measured on heavy nuclei differ from those measured in hydrogen. Differences have been observed over the whole x-spectrum and, given the difficulty of comparing data from different experiments, mostly due to application of different radiative correction model prescriptions, the conclusion is that the general pattern is consistent. The most recent results can be seen in Fig. 5.5.10 where NMC [AMA 95] data are compared to a re-analysis of SLAC-E139 [ARN 84a] data [GOM 94]. For a comprehensive review on the subject see [ARN 94b]. The contribution of neutrino physics to this subject is limited by the difficulty of changing different target materials in the same experiment, in order to avoid the systematic errors involved in the comparison of the results of different detectors and by the need of collecting sufficient statistics on the low-Z target. Heavy electronics detectors [ABR 84] had to add a low atomic number target in front so that the number of the events eventually collected is comparable to those of bubble chambers [COO 84]. Both experiment results agree on the possibility that a softening of the valence quark distribution at intermediate x ($0.3 < x < 0.6$) is present in heavy nuclei. The most interesting result from neutrino scattering comes from the BEBC-WA59 [ALL 89] experiment where ν–Ne cross sections were

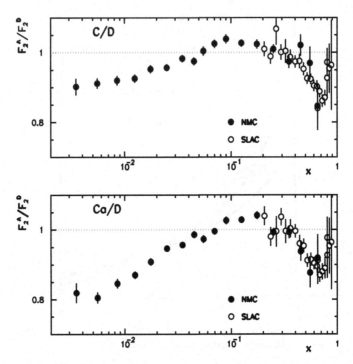

Fig. 5.5.10   The ratios of structure functions determined on C and D (top) and Ca and D (bottom) for NMC and for a reanalysis of SLAC-E139 results.

compared to ν–D ones. The deviation of the ratio from the simple ratio of effective atomic numbers was studied as a function of both x and Q^2. The main conclusion was that the results could be explained by an old model [BEL 64] which predicts a shadowing effect coming from a PCAC-induced term certainly present at least at low Q^2.

5.5.8 The proton composition

The calculation of the cross sections of hard processes occurring in the present and future colliders requires a knowledge of the parton densities of the proton.

Deep inelastic neutrino scattering, as already seen, mainly produces results on structure functions determined in heavy nuclei, at least for the high-statistics measurements. It is possible to combine the results, however, and to call for theoretical help, in such a way that a coherent and quantitatively correct picture is extracted.

This attempt is described in detail in [ALL 87b]. Its main features consist in choosing a reference value of Q_0^2, say $10\,\text{GeV}^2$, taking the experimental results for xF_3, $\bar{q}$, G, decomposing the valence distribution into its components d_v and u_v correcting the antiquark distribution for the strange component, applying corrections to take into account the nuclear effects at intermediate x values for the valence

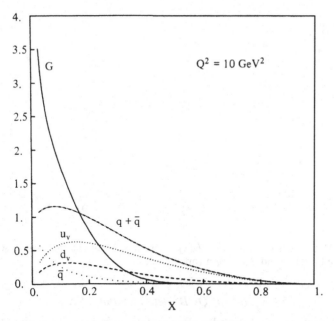

Fig. 5.5.11 Momentum distribution of the different partons at fixed $Q^2 = 10\,\text{GeV}^2$. The contribution of gluon (xG), quarks and antiquarks ($x(q+\bar{q})$), antiquarks alone ($x\bar{q}$), valence up component (xu_v), and valence down component (xd_v) is shown as a function of x.

quarks, including next-to-leading QCD corrections to the measured densities in order to resolve the ambiguity present in their definition at the next order, and finally imposing the available sum rules. The procedure results in a realistic picture of the proton. Different momentum distributions of quarks, antiquarks, and gluons present inside the proton are shown in Fig. 5.5.11.

5.6 QCD analysis of the data

The QCD analysis of the data coming from deep inelastic neutrino scattering can be divided in three main parts:

1 the observation of scaling violations, their interpretation in the framework of QCD, and the subsequent determination of the parameter Λ,
2 the extraction of the gluon distribution function,
3 the comparison of the measured longitudinal structure function with the expected contribution of order $\alpha_s(Q^2)$.

Points 1 and 3 are discussed here, while Section 5.7 is devoted to the gluon density.

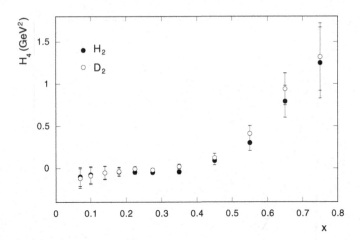

Fig. 5.6.1   Higher twist coefficients $H_4(x)$ determined from charged-lepton scattering on hydrogen (full circles) and deuterium (open circles).

5.6.1 *Tests of QCD; determination of* Λ_{QCD}

In order to analyze the observed scaling violations in terms of leading twist QCD effect, one has to exclude the kinematical region where target mass effects and higher twist contributions are expected to be important, namely, the region where Q^2 and the invariant hadronic mass squared W^2 are small.

Insidious terms are the so-called higher twist effects whose $(1/Q^2)^n$ behavior might mimic the $\ln Q^2$ dependence predicted by QCD at the leading twist level. Fundamental contributions to the theoretical understanding of this point come from the work of [POL 80; JAF 81; ELL 83b]. The structure functions $F_i(x, Q^2)$ can be expanded as a power series in $1/Q^2$:

$$F_i(x, Q^2) = F_i^{LT}(x, Q^2) \left[1 + \frac{H_4(x, Q^2)}{Q^2} + \frac{H_6(x, Q^2)}{Q^4} + \cdots \right] \qquad (5.6.1)$$

where the leading twist terms F_i^{LT} obey the Altarelli–Parisi equations. The coefficients H_n can only be determined in QCD as a result of a nonperturbative calculation.

A recent evaluation of higher twist effects is presented in [VIR 92] based on the data from deep inelastic scattering of charged leptons on hydrogen and deuterium targets, collected by electron scattering experiments at SLAC and by BCDMS at CERN. H_4 as a function of x has been determined as a correction to the NLO perturbative QCD fit to the data and is shown in Fig. 5.6.1. The fit includes target mass corrections from [GEO 76]. The introduction of an H_6 term does not induce appreciable improvements in the fit.

The possibility that the contribution to the scaling violation of the higher twist coefficient is different in sign and in size for neutrino and charged lepton scattering has been pointed out [LUT 81; SHU 82a,b].

Assuming that the scaling violations observed at large Q^2 are due to leading twist perturbative QCD, one is left with the problem of their interpretation, since a perturbation theory can only predict the Q^2 evolution of structure functions (through Eqs. (5.2.42)), but not the distributions themselves. In other terms, the initial conditions concerning the wave function of the nucleon are needed to solve the set of evolution equations. This additional piece of knowledge has to be inferred by the data. Several numerical methods have been developed, differing in the way to prescribe the functional form of the solution in the variable x. They will be discussed in Section 5.6.4.

The determination of the parameter Λ defined in Eq. (5.2.35) is obtained from two different kinds of analysis. The nonsinglet analysis profits from the decoupled evolution of nonsinglet quark combinations such as xF_3, or F_2 in the high x region, where $\bar{q}$ can be neglected. The singlet analysis entails the comparison of $F_2(x, Q^2)$ and $\bar{q}(x, Q^2)$ with QCD predictions. This is more complicated from the theoretical point of view since it involves the handling of the gluon density, which is either determined together with Λ or must be frozen through some assumption.

As a general comment on the comparative merits of different structure functions, the xF_3 structure function turns out to be the cleanest distribution for QCD tests. Its determination does not suffer any theoretical uncertainty (see Section 5.4.1) and its evolution can be easily interpreted. Unfortunately, it suffers a poor experimental determination, resulting from the difference of ν and $\bar{\nu}$ cross sections. The values of Λ derived from nonsinglet analysis are then usually characterized by a large statistical uncertainty, although being theoretically more reliable. The F_2 analysis produces smaller statistical uncertainties but requires more theoretical assumptions. This limitation is particularly severe in the next-to-leading order analysis. Nevertheless, the singlet analysis is mandatory for determining other crucial implications of the theory, such as the gluon distribution function.

5.6.2 Moments analysis

From the theoretical point of view, the simplest way to interpret scaling violations would be to look at the Q^2 dependence of the Mellin transform of the structure functions.

The leading logarithmic expression for the nth moment of the nonsinglet structure function has been given in Eq. (5.2.55). The simplest test of the theory can be done by observing that the relation between different moments $\ln V_n(Q^2)$ *vs* $\ln V_m(Q^2)$ is linear with a slope given by the ratio of the corresponding anomalous dimensions (Eq. (5.2.56)):

$$\frac{d \ln V_n(Q^2)}{d \ln V_m(Q^2)} = \frac{d_n}{d_m}. \tag{5.6.2}$$

At the leading order, the above ratio is a constant independent of Λ as well as the number of colors and flavors. These properties do not hold if the next-to-leading terms are included. In this case, one has [PAR 79]

$$\frac{d \ln V_n(Q^2)}{d \ln V_m(Q^2)} = \frac{d_n}{d_m(1 + \alpha_s C_{nm})} \tag{5.6.3}$$

so that the slopes depend on Q^2 and Λ through α_s.

The $\ln Q^2$ dependence predicted by QCD can be directly tested by looking at the plot of a single moment raised to the power $-1/d_n$ versus Q^2. This plot should in principle test QCD allowing the determination of Λ.

A serious drawback of the moments method is the need of data in the whole x range: high-order moments enhance the high x region, where data are scarce or absent. In addition, the determination of the Q^2 dependence of moments in a single experiment suffers from the kinematical limit $x < Q^2/2ME$. Then the experimental results are poor precisely in the region where the scaling violation effects are expected to be the strongest. Another problem comes from the unavoidable correlation between different moments. Finally, the region where nonperturbative effects can dominate the Q^2 dependence of structure functions should be eliminated from the data with a selection in W^2, which is impossible.

High-statistics data, covering a Q^2 range $5.6 \div 56.6 \, \text{GeV}^2$, were published by the CDHS collaboration [ABR 82c]. In order to increase the statistical accuracy, the F_2 measurements were included for $x > 0.4$, since the antiquark contribution is negligible in the high x region. The large x, low Q^2 range was complemented by SLAC data [BOD 79]. Figure 5.6.2 shows the log-log plot of V_3 vs V_5, together with predictions from second-order QCD. Figure 5.6.2 also shows the predictions obtained by assuming that the gluons are scalar particles. Vector gluons are clearly favored, and that the inclusion of next-to-leading corrections improves the agreement.

The same collaboration has also performed a QCD analysis in order to determine Λ. The data were fitted for $5 < Q^2 < 60 \, \text{GeV}^2$ using $n = 3,4,5$ moments. QCD turns out to be in good agreement with the data, with $\Lambda = (0.30 \pm 0.04) \, \text{GeV}$, but alternative hypotheses such as Abelian vector gluon or scalar gluon theories cannot be excluded.

In summary, the moment analysis shows that the vector gluon hypothesis is favored over scalar gluons, but the tests of the predictions of QCD are not conclusive at all.

5.6.3 *Slopes of structure functions*

The comparison between the slopes of structure functions $d \ln F/d \ln Q^2$ with the QCD predictions for the Q^2 evolution of the parton distributions allows a direct determination of Λ.

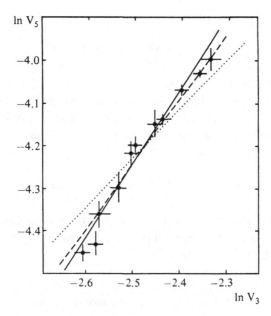

Fig. 5.6.2. Log-log plot of the fifth versus the third moment measured by CDHS collaboration. Also shown is the best straight-line fit to the data (solid line) and the predicted lines for second order QCD (dashed line) and scalar gluons (dotted line).

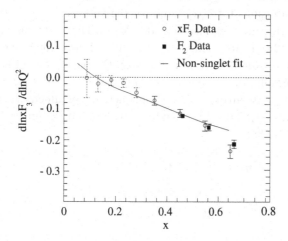

Fig. 5.6.3 The slopes of the structure functions as measured by CCFR experiment. Errors are statistical only. The curve is a prediction from perturbative QCD with target mass correction.

As the data are not precise enough to measure the slopes locally, they can be evaluated for each x bin from a power law fit to the Q^2 dependence of the data. Such an analysis has been performed for neutrino data by CCFR experiment [QUI 93]. The result is shown in Fig. 5.6.3: the logarithmic slopes of the data agree well with the QCD prediction throughout the entire x range. The value of Λ_{QCD} resulting from a fit to these data is

$$\Lambda_{QCD} = 179 \pm 36 \text{ (stat.)} \pm 54 \text{ (syst.) MeV} \quad \text{for } Q^2 > 15 \text{ GeV}^2.$$

A more precise determination of Λ_{QCD} is obtained by substituting F_2 for xF_3 for $x > 0.5$, where the effects of antiquarks, gluons and the longitudinal structure function are negligible and the evolution of F_2 should conform to that of a nonsinglet structure function (the slopes for F_2 are also shown in the same figure). With this substitution, the result of the fit is

$$\Lambda_{QCD} = 210 \pm 28 \text{ (stat.)} \pm 41 \text{ (syst.) MeV} \quad \text{for } Q^2 > 15 \text{ GeV}^2.$$

Most of the systematic error comes from a possible miscalibration of E_{had} with respect to E_μ. The contributions from cross section uncertainties are negligible, in particular after the F_2 substitution.

5.6.4 *Direct fits of structure functions*

The comparison of the data with QCD, available for a limited set of (x, Q^2) points, entails the numerical integration of the Altarelli–Parisi equations (5.2.42). The analysis can be performed at the leading or next-to-leading order using the appropriate expressions for $\alpha_s(Q^2)$ and for the kernels P.

Several numerical methods have been developed. They differ mainly on the choice of the initial conditions to be imposed on the distribution functions. Most of them assume a suitable parameterization at a given Q_0^2, partly suggested by theoretical prejudices (especially for $x \to 0$ and $x \to 1$) partly empirically dictated in order to reproduce the data. As an example we quote here the parameterization used by the CCFR collaboration [SEL 97] as input to an evolution program developed by Duke and Owens [DUK 84]:

$$xq_{NS}(x, Q_0^2) = A_{NS} x^{\eta_1} (1 - x)^{\eta_2}$$
$$xq_S(x, Q_0^2) = xq_{NS}(x, Q_0^2) + A_S (1 - x)^{\eta_S}$$
$$xG(x, Q_0^2) = A_G (1 - x)^{\eta_G},$$

where Q_0^2 is fixed to 5 GeV^2.

A quite different approach is given by the Furmanski–Petronzio method [FUR 82] adopted by the CHARM and BEBC collaborations and based on the expansion of the parton densities in a series of Laguerre polynomials in the variable $y = \ln(1/x)$. The choice of this variable is suggested by the simple rules governing the convolution of the polynomials in y. The choice of Laguerre polynomials offers other additional advantages: The series converges quickly and uniformly owing to the orthogonality of the polynomials and can then be truncated after a few terms with quite a high accuracy.

Table 5.6.1. *Summary of QCD results*

Experiment	Q^2 range (GeV^2)		Fit		Fitting program	Λ(MeV)
CDHS	1–180	N–S	xF_3	$\overline{\text{MS}}$	Abbot-Barnett	200^{+200}_{-100}
	1–180	N–S	F_2	LO	Abbot-Barnett	275 ± 80
	1–180	N–S	F_2	$\overline{\text{MS}}$	Abbot-Barnett	300 ± 80
	1–180	S	$F_2, \bar{q}$	LO	Abbot-Bernett	(a) 180 ± 80
	1–180	S	$F_2, \bar{q}$	LO	Abbot-Bernett	(b) 290 ± 30
CHARM	3–78	N–S	xF_3	LO	Furmanski-Petronzio	187^{+130}_{-110}
	3–78	N–S	xF_3	$\overline{\text{MS}}$	Furmanski-Petronzio	310 ± 150
	3–78	S	$F_2, \bar{q}$	LO	Furmanski-Petronzio	190^{+70}_{-40}
CCFR	5–125	N–S	xF_3	$\overline{\text{MS}}$	Duke-Owens	381 ± 53
		N–S	xF_3, F_2	$\overline{\text{MS}}$	Duke-Owens	337 ± 28
BEBC	1.5–55	N–S	xF_3, F_2	LO	Furmanski-Petronzio	180 ± 60
		N–S	xF_3, F_2	LO	Odorico	220 ± 70

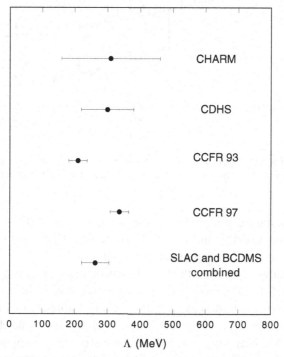

Fig. 5.6.4 Statistically most significant measurements of $\Lambda_{\overline{\text{MS}}}$ assuming $N_f=4$ for neutrino experiments. At the bottom is shown for comparison a combined analysis [VIR 92] of SLAC and BCDMS data. All the errors are statistical only.

A summary of the published analysis is given in Table 5.6.1 [ABR 83; BER 83, 85; SEL 97; ALL 85], and in Fig. 5.6.4 the statistically more significant results on Λ measurement at next-to-leading order for $N_f=4$ are shown.

A few explanatory comments have to be added for each experiment.

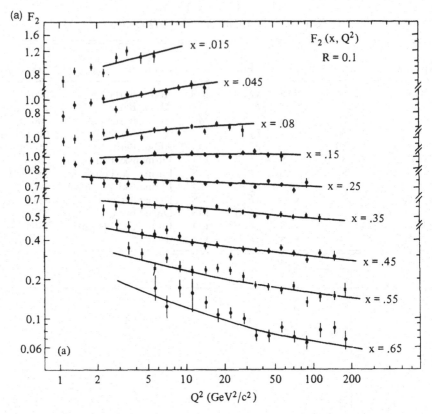

Fig. 5.6.5(a)   Structure functions versus Q^2 for different bins of x, as measured by the CDHS collaboration: F_2. The solid lines are the result of a leading-order QCD fit to the data.

Besides the standard nonsinglet xF_3 analysis, CDHS [ABR 83] performed a nonsinglet analysis to determine Λ from Q^2 evolution of F_2 at $x > 0.3$, by subtracting the sea-quark contribution and assuming $R \sim 0$.

The singlet analysis combined the F_2 and $\bar{q}$ data in order to extract simultaneously Λ and the gluon distribution whose shape is strongly constrained by the fact that the sea density is very small at large x. Their standard analysis ((a) in Table 5.6.1) assumes for the sea $(\bar{s} - \bar{c}) = 0.2(\bar{u} + \bar{d})$ (and no charm threshold effects), $R = 0.1$, $M_W = 80$ GeV. The normalization of the gluon distribution is imposed by the total momentum sum rule. The singlet analysis has been repeated ((b) in Table 5.6.1) using R_{QCD}, F_2 in the interval $0.03 < x < 0.7$, $\bar{q}$ in the interval, $0.3 < x < 0.7$, correcting the data by using the slow rescaling model for the charm threshold with $m_c = 1.5$ GeV. The results of the fit are compared in Figs. 5.6.5a, 5.6.5b, and 5.6.5c with the experimental points for F_2, xF_3, and $\bar{q}$, respectively.

The CHARM collaboration performed the nonsinglet analysis at leading [BER 83] and next-to-leading [BER 85] order, using the data with $x > 0.03$. The singlet analysis uses F_2 and $\bar{q}$ simultaneously, and fixes the valence distribution as it is

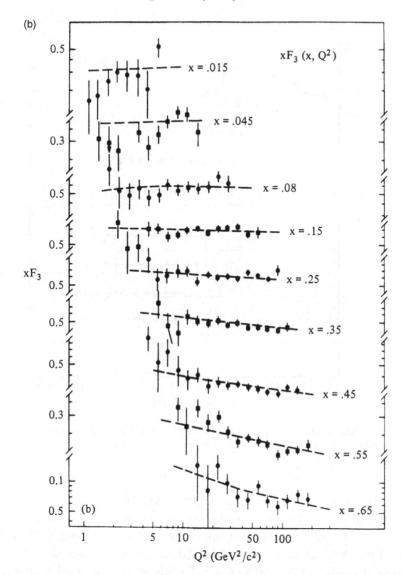

(b)

$xF_3 (x, Q^2)$

x = .015

x = .045

x = .08

x = .15

x = .25

x = .35

x = .45

x = .55

x = .65

xF_3

(b)

Q^2 (GeV^2/c^2)

Fig. 5.6.5(b) Structure functions versus Q^2 for different bins of x, as measured by the CDHS collaboration: xF_3. The solid lines are the result of a leading-order QCD fit to the data.

extracted from the nonsinglet analysis. The antiquark distribution $\bar{q}$, following the experimental observation, is assumed to vanish for $x > 0.5$. The total momentum sum rule is not imposed a priori. The 68 percent confidence contours of the x-dependence for the valence, the sea quarks, and the gluon, resulting from the fit à la Furmanski–Petronzio, are shown in Fig. 5.6.6 for different values of Q^2.

The analysis of CCFR collaboration [SEL 97] uses data with $W^2 > 10 \, \text{GeV}^2$, $Q^2 > 5 \, \text{GeV}^2$ and $x < 0.7$. The next-to-leading order fit is done with a modified

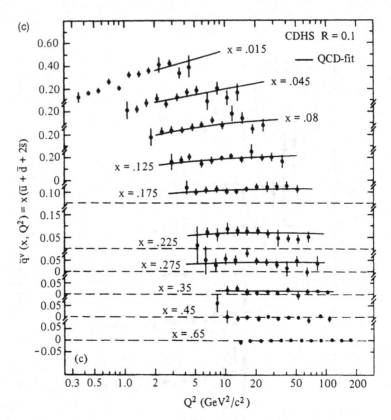

Fig. 5.6.5(c) Structure functions versus Q^2 for different bins of x, as measured by the CDHS collaboration: $x\bar{q}^p$. The solid lines are the result of a leading-order QCD fit to the data.

version of the Duke and Owens program [DUK 84]. The parameters of the gluon distribution at $Q_0^2 = 5\,\mathrm{GeV}^2$ are obtained simultaneously. The effect of target mass is included in the fit by applying one-half of the correction proposed in [DAS 96].

The BEBC-WA25 collaboration [ALL 85] determine Λ assuming $N_f = 3$ through a nonsinglet analysis at the leading order. To improve the statistical accuracy, the data obtained for F_2, assuming $R = 0$, are used in the range $x > 0.4$ instead of those for xF_3.

5.6.5 Systematic effects in the measurements of Λ

Systematic effects on Λ are estimated by different collaborations. They can be gathered in three main classes:

a effects of normalization, cross sections, etc.
b effects of the physical corrections to the data
c effects of the assumptions in the fitting procedure

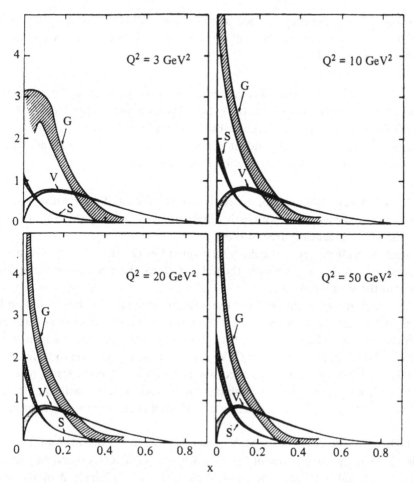

Fig. 5.6.6 The 68 percent confidence contour of the x dependence of valence quark (V), sea quark (S), and gluon (G) distributions as determined by CHARM Collaboration for different values of Q^2.

a   The knowledge of the neutrino fluxes (or, equivalently, of the neutrino cross sections) underlies the absolute values and the relative shapes of the measured structure functions. A global-scale uncertainty pervades all the measured values, as discussed in Section 5.3.1, but it should not greatly affect the determination of Λ. The associated uncertainty on Λ is estimated to be 60 MeV by CHARM. The effect on Λ of an uncertainty in the knowledge of the beam shape has been estimated by CHARM, whose WBB analysis relied on external assumptions on the neutrino cross sections; this contribution to Λ uncertainty is found to be 30 MeV. No explicit estimation of systematics effects is given in CCFR analysis [SEL 97]. In a previous paper [QUI 93] the same collaboration assigned a 20 MeV uncertainty on Λ_{QCD} because of the uncertainty on the cross section and 40 to 50 MeV uncertainty due to possible energy miscalibration of the detector. As a matter of fact, the last value of

Λ_{QCD} differs from the previous one by more than 120 MeV and the difference is attributed in [SEL 97] to the effects of a new energy calibration. The two results are reported in Fig. 5.6.4.

b Radiative corrections can be applied adopting different theoretical prescriptions; however, their effect on Λ is small (switching them off, CHARM finds a 24 MeV decrease of Λ), and the contribution to the global uncertainty should then be negligible. Up to its Q^2-dependent terms, the Fermi motion should not change the evolution and consequently cannot sensitively affect Λ.

c Assumptions and/or approximations introduced when fitting the data refer mainly to the use of F_2, both in the singlet analysis and in the high x nonsinglet analysis; xF_3 is practically free of them.

Target mass effects are particularly relevant at low Q^2. High Q^2 experiments are expected to be rather insensitive to them: For CDHS, these corrections contribute 5 MeV to the value of Λ.

The contribution from higher twist terms can be controlled by suitable cuts in W^2 and Q^2; CHARM finds a decrease of 30 MeV with respect to the central value of Λ (with the corresponding loss of statistical accuracy) by applying a cut in W^2 of 5 GeV^2; CDHS applied a cut at 11 GeV^2 and made the assumption to be free from this effect. CCFR correct for this effect and attribute a systematic error, repeating the fit with no correction and doubling the correction itself, with an effect of ± 13 MeV on $\Lambda_{\overline{MS}}$. CDHS finds a decrease of 65 MeV neglecting the sea in the high x analysis.

CDHS finds a decrease of 100 MeV assuming a constant value of $R = 0.1$ instead of the QCD predicted behavior of F_L. A slightly smaller effect is observed by BEBC-WA25, who find a decrease of 20 MeV assuming $R = 0.1$ instead of the exact Callan–Gross relation.

Another source of uncertainty comes from the assumption on the gluon shape, that is strongly correlated with Λ. In CCFR and in CHARM analysis the gluon shape is not fixed a priori and the Λ value is therefore unaffected.

Different fitting procedures [FUR 82; ODO 81] in the nonsinglet xF_3 analysis are applied to the same data sample by BEBC-WA25. The results show a difference of ~ 40 MeV, depending which of the two methods is used.

5.6.6 Comparisons of different determinations of Λ

Low Q^2 versus high Q^2 A clear dependence of the value of Λ on the lower Q^2 cut is observed in low Q^2 experiments: in contrast, CHARM finds negligible changes increasing the cut from 3 to 10 GeV^2. These results are in agreement with the guess that nonperturbative effects are confined to values of Q^2 below a few GeV^2.

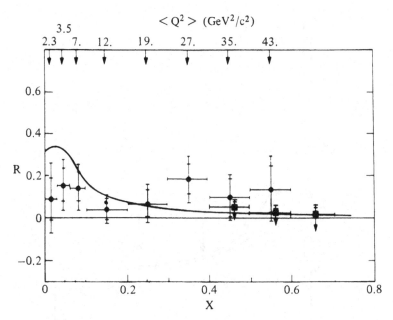

Fig. 5.6.7 R as a function of x as measured by CDHS collaboration compared with the leading-order QCD prediction. Also shown is the averaged value of Q^2 for each bin in x. The data points with an arrow are upper limits.

Nonsinglet versus singlet analysis   Here the choice is between statistical and systematical accuracy; the global level of uncertainties is more or less the same for any experiment.

Finally, if we consider high-statistics, high-Q^2 experiments only, we find quite a satisfactory agreement, within the quoted errors and in spite of the different, not always minor, details of the analysis. The favorite $\overline{MS}$ value for Λ is 300 ± 150 MeV for CDHS, 310 ± 156 MeV for CHARM, and 337 ± 28 (stat.) MeV for CCFR. These values are in good agreement with a recent combined analysis [VIR 92] using SLAC and BCDMS data, giving $\Lambda_{\overline{MS}} = 263 \pm 42$ (stat.) ± 55 (syst.) MeV. This value is reported for comparison with neutrino data in Fig. 5.6.4.

5.6.7 The longitudinal structure function

In principle, the longitudinal structure function can provide a precise test of QCD, yielding a direct estimate of α_s and a different insight into the gluon distribution. Unfortunately, the measurement of F_L is very difficult and it must be considered at most as an important consistency check.

The neutrino results on this subject are in accordance with the expected contributions that can be computed at order α_s. Within the large statistical errors, CDHS seems to be even below the pure perturbative QCD prediction, at least at low x (Fig. 5.6.7). CHARM results leave some room for nonperturbative contributions,

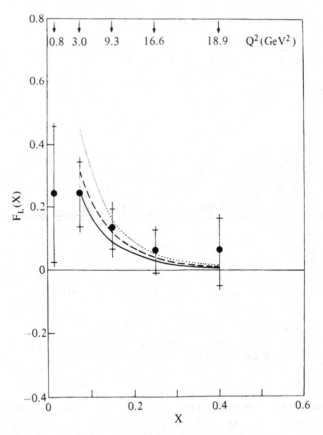

Fig. 5.6.8 x-dependence of the longitudinal structure function F_L observed by CHARM collaboration. The central value of Q^2 for each x bin is indicated on the top. The continuous line is the perturbative QCD prediction; the dotted (dashed) line includes nonperturbative effects following the recipe of [ELL 83a,b] for $k_T = 300$ MeV/c (500 MeV/c), respectively.

to F_L. These contributions are expected to be nonnegligible at the measured values of Q^2. F_L, taken by combining perturbation theory with the structure functions determined by the same experiment, is compared with the experimental data in Fig. 5.6.8. Nonperturbative corrections of order Q^{-2} are also estimated following the recipe of [ELL 83a,b] in terms of the intrinsic parton transverse momentum k_T. The theoretical predictions for two different values of k_T are shown in Fig. 5.6.8.

All data present an increase of F_L at low x showing a statistically significant difference from zero, as expected from QCD.

5.7 The gluon structure function

The gluon density, $G(x)$, is probably the single most important parton density for applications to present and future hadron collider physics. Its detailed knowledge is

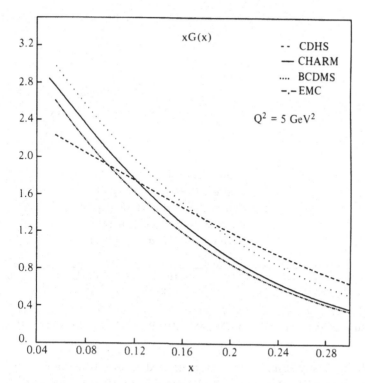

Fig. 5.7.1 The gluon distributions as determined by neutrino and muon experiments at $Q^2 = 5\,\text{GeV}^2$.

an essential ingredient in the evaluation of several key processes. For example, small-x gluons are responsible for the production of b quarks, while large-x gluons give a fundamental contribution to the production of t quarks, and of high-E_T jets. The gluon density is, however, very hard to determine in lepton scattering, as the gluon does not couple directly to electroweak probes. This means that $G(x)$ can only be extracted, assuming the validity of QCD, by looking at the variation of the singlet parton combinations, explicity related to the gluon by the Altarelli–Parisi equations. In any case, an unavoidable correlation remains between the value of Λ and the shape of the gluon density.

The leading order determinations of the gluon density from deep inelastic lepton scattering prior to the start of HERA are shown in Fig. 5.7.1, for $Q^2 = 5\,\text{GeV}^2$. The clear discrepancies among the fits of the different experiments are mostly due to different assumptions on the behavior of the gluon density at small x. This situation improved dramatically since the beginning of HERA, where the large increase of F_2 at small x is entirely driven by the gluon content of the proton, and allows a precise determination of the $x \to 0$ behavior of $G(x)$. The HERA data (see [AID 96; DER 96] for the most recent results), allow us to probe the partonic structure of the proton down to values of x of O (few $\times\, 10^{-5}$) in a Q^2 range ($Q^2 \gtrsim 1.5\,\text{GeV}^2$) where a perturbative QCD analysis is believed to be usable.

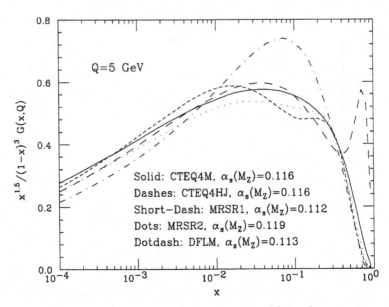

Fig. 5.7.2(a)   The gluon density at $Q = 5\,\text{GeV}$, scaled by $x^{-1.5}(1-x)^3$, for different parton-density parameterizations: DFLM [DIE 88], MRSR [MAR 96] and CTEQ4 [LAI 97].

The gluon density at small x, as parameterized in various sets of parton densities, is shown in Fig. 5.7.2(a) (for $Q = 5\,\text{GeV}$) and 5.7.2(b) (for $Q = 100\,\text{GeV}$). There we show the results of pre-HERA fits (DFLM, [DIE 88]) as well as the most recent fits using the HERA data as inputs (CTEQ4 [LAI 97] and MRSR [MAR 96]).

For x values above 0.1, the most direct measurements of $G(x)$ come instead from fixed-target prompt-photon hadroproduction [BON 88; SOZ 93]. In this case, however, several theoretical uncertainties remain [HUS 95], which leave these determinations rather loose. This is reflected in the differences between various fits of $G(x)$ in the region $x \sim 0.1$, shown in Fig. 5.7.2(a). Notice nevertheless that, with the exception of the older DFLM sets, most of the differences between various fits of $G(x)$ are significantly reduced already at $Q = 100\,\text{GeV}$, Fig. 5.7.2(b).

More recently, information on the gluon density in the range $x \gtrsim 0.1$ has come from the measurements of high-E_T jets at the Tevatron [ABE 96a; ABB 97]. The parton-density set CTEQ4HJ [LAI 97], also shown in Fig. 5.7.2, includes the Tevatron data, with an artificially enhanced weight, in the global fit of $G(x)$. The excess of jets in the region $E_T > 300\,\text{GeV}$ reported by CDF [ABE 96a] causes the peculiar change in shape of $G(x)$ for $x \gtrsim 0.3$. This is a clear indication that the determination of the large-x gluon density is still not very robust.

5.8 QCD-based extrapolation of parton densities

QCD describes the Q^2 dependence of the parton densities through the solution of the Altarelli–Parisi equations. Its predictive power is given by the universality of these

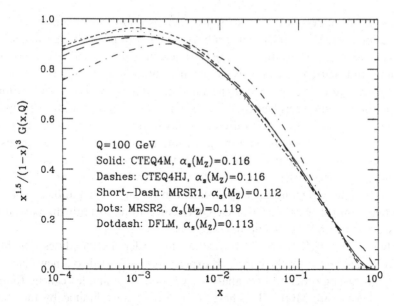

Fig. 5.7.2(b) The gluon density at $Q = 100\,\text{GeV}$, scaled by $x^{-1.5}(1-x)^3$, for different parton-density parameterizations: DFLM [DIE 88], MRSR [MAR 96] and CTEQ4 [LAI 97].

densities: they can be extracted from a given process at some scale and used to perform calculations for different reactions in a wide range of Q^2 provided that the elementary cross sections at the parton level are known.

In this section, different determinations of parton distributions are discussed together with the associated uncertainties. The results are compared to experimental data in hadron–hadron collisions where highly nontrivial dynamical situations occur. This comparison is therefore a severe test of the validity of perturbative QCD over a Q^2 range that could never be spanned by a single experiment.

5.8.1 Representations of parton densities

High-statistics experiments on deep inelastic scattering supply a wide amount of data on structure functions from which it is possible to extract quark and gluon densities to a high degree of accuracy. These processes are the cleanest from both a theoretical and an experimental point of view in order to derive a picture of the proton in terms of its elementary constituents.

Deep inelastic scattering is not the only source of information on the parton distributions. Data coming from different reactions such as Drell–Yan processes and the longitudinal momentum distribution in the J/Ψ hadroproduction can also be used. The interpretation of the latter results is, however, much more model dependent.

In the sector of deep inelastic scattering there is also the problem of combining information from different leptonic probes (e, μ, ν). As already discussed, neutrinos are unique among leptons for studying the nucleon structure and its evolution in a complete and self-consistent way. Certainly, muons give useful information, especially because of the very high statistics attainable. The combination of data samples from several probes entails mixing completely different kinds of systematics. Uncertainties from differences in various data samples must be taken into account in order to estimate the precision attainable on the predictions.

The program of combining all this information in global fits of the parton distributions has been carried out over the past several years by many groups, with constant updates reflecting the improvement in the experimental data and in the theoretical control over higher-order QCD corrections to both the parton-level matrix elements and the evolution equations.

Early studies [DUK 84; EIC 84] used leading-order (LO) analyses. The higher accuracies demanded by the improved experimental data led to a new generation of next-to-leading-order (NLO) analyses, whose first examples can be found in [DIE 88; MAR 88; MOR 91]. These early studies were limited by the lack of experimental data which could constrain (i) the parton distributions at small x $(x < 10^{-2})$ and (ii) the independent contribution of $\bar{u}$ and $\bar{d}$ densities. Furthermore, early data could only provide limited information on the precise ratio of the valence u_v and d_v distributions, as well as on the shape of the strange and charm quark distributions. Recent data from a whole class of fixed target and collider experiments have dramatically improved our control over the fine details of the partonic structure of the nucleon. They have become the standard input for the most recent global QCD analyses of parton distributions, and will be briefly summarized here.

Early indications on the isospin breaking in the light-quark sea from the Goddfried sum rule [AMA 91a; ARN 94a] have been strengthened by the NA51 measurement of the DY production asymmetry in pp vs. pn collisions [BAL 94]. The asymmetry distribution as a function of the DY-pair mass directly constrains the relative shape of the $\bar{u}(x)$ and $\bar{d}(x)$ densities. As already mentioned in Section 5.5.3, recent CCFR data on dimuon production in deep inelastic neutrino scattering [BAZ 95] have improved our knowledge of the strange sea, providing information not only on the global strange content of the proton, but on its shape as well. A precise knowledge of the $s(x)$ density is important for the applications to hadronic collisions, as several leading order processes for the associated production of W and charm plus additional jets, backgrounds to the W plus heavy flavor final states of top production, are directly proportional to the strange content of the proton. Furthermore, the precise knowledge of $s(x)$ is important for accurate estimates of the inclusive W cross section. The d/u ratio in the region $x \sim 0.1$ is today strongly constrained by the CDF data on the lepton-charge rapidity-asymmetry in $W^{\pm}$ decays [ABE 95], shown in Fig. 5.8.1. This information is of particular phenomenological relevance, as the d/u ratio determines the W rapidity and transverse-mass spectra in hadronic collisions, thereby affecting the measurement of the W mass.

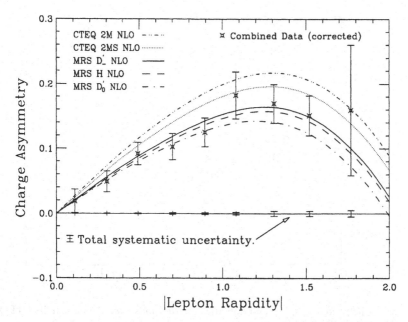

Fig. 5.8.1 The lepton-charge asymmetry in W decays as measured by the CDF Collaboration [ABE 95] vs. the predictions of several sets of parton densities.

Current sets of parton densities resulting from global fits to all data reviewed above are given in terms of a fixed functional form, input at a given Q^2 scale Q_0^2 and then evolved to larger Q^2 values according to the NLO Altarelli–Parisi equations. The densities as a function of Q^2 are then presented either as interpolations from a table of values evaluated at fixed points in x and Q^2, or by fitting the Q^2 evolution of the parameters defining the input functional form.

The most recent global fits available are the MRSR fits [MAR 96] and the CTEQ4 fits [LAI 97]. These fits mostly differ in the choice of some of the input data sets; for example, the CTEQ4 fits include the CDF medium-E_T jet data, to better constrain the gluon density. Other differences lie in the scheme employed to incorporate threshold effects in the evolution of the heavy quark densities, and in the choice of Q_0: the MRSR fits use $Q_0 = 1$ GeV (but only fit to data with $Q^2 > 1.5$ GeV^2), the CTEQ4 fits use $Q_0 = 1.6$ GeV.

For both sets, the input functional forms for u_v, d_v, g and $S = 2(\bar{u} + \bar{d} + \bar{s})$ depend on five parameters, given by:

$$xf_i(x, Q_0^2) = A_i x^{-\lambda_i}(1 - x)^{\eta_i} P_i(x), \quad i = u_v, d_v, S, g, \tag{5.8.1}$$

with $P_i(x) = 1 + \varepsilon_i \sqrt{x} + \gamma_i x$ for the MRS sets, and $P_i(x) = 1 + C_i x^{D_i}$ for the CTEQ sets. Three out of four of the A_i constants are determined from momentum and flavor sum rules. Notice that, contrary to early fits, the small-x shape of the gluon and sea distributions are kept independent, since the HERA data have become

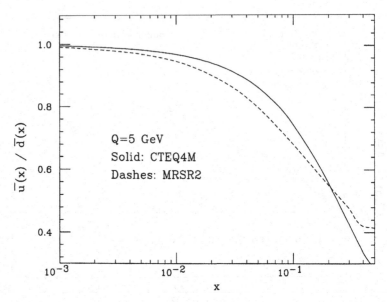

Fig. 5.8.2   The ratio u_s/d_s of up and down sea distributions from the parton fits CTEQ4M [LAI 97] and MRSR2 [MAR 96].

accurate enough to allow their separation. The flavor structure of the sea is taken for both sets to be:

$$2\bar{u} = 0.4S - \Delta$$
$$2\bar{d} = 0.4S + \Delta \qquad (5.8.2)$$
$$2\bar{s} = 0.2S,$$

with

$$x\Delta = x(\bar{d} - \bar{u}) = A_\Delta x^{\lambda_s}(1 - x)^{NS}(1 + \gamma_\Delta x). \qquad (5.8.3)$$

The assumption that the light-quark and the strange sea distributions have the same shape (up to the small isospin breaking effects described by Δ), as well as the choice of the relative suppression factor, are justified by the recent CCFR data [BAZ 95].

As an illustration, we show in Figs. 5.8.2–5.8.5 some results of the CTEQ4 and MRSR fits. Figure 5.8.2 shows the ratio $\bar{u}(x)/\bar{d}(x)$. Figure 5.8.3 shows the strange and charm content of the proton. The ratio $d_v(x)/u_v(x)$ is shown in Fig. 5.8.4, and the total valence content of the proton is shown in Fig. 5.8.5.

As already remarked, the extraction of the parton densities is strongly correlated with the value of α_s. The accuracy of the data and of the theoretical analyses is such that the extraction of α_s from the small-x F_2 measurements at HERA (via a simultaneous fit of the gluon densities and of α_s [BAL 95; BAL 97]) compete today in precision with the extractions based on F_2 scaling violations at large x or on sum

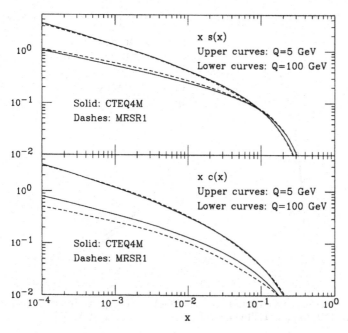

Fig. 5.8.3   The strange and charm distributions, at $Q = 5$ and $100\,\text{GeV}$, from the parton fits CTEQ4M [LAI 97] and MRSRI [MAR 96].

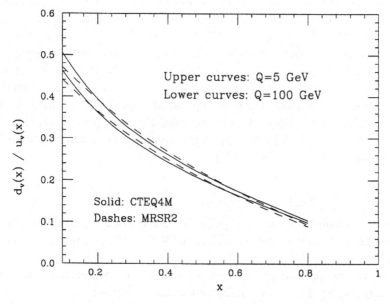

Fig. 5.8.4   The ratio d_v/u_v from the parton fits CTEQ4M [LAI 97] and MRSR2 [MAR 96].

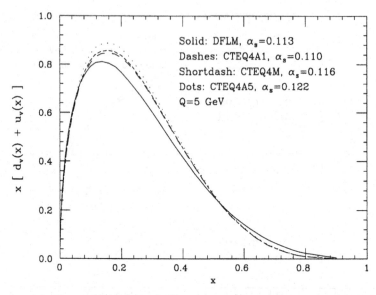

Fig. 5.8.5   The valence distributions at $Q = 5\,\text{GeV}$ are compared for several parton-density fits: DFLM [DIE 88] and CTEQ4 for different values of α_s [LAI 97].

rules (for a recent view, see [KAT 96]). The most recent CCFR [SEL 97] data on F_2 scaling violations give $\alpha_s(M_Z) = 0.119 \pm 0.002(\text{exp.}) \pm 0.004(\text{th.})$. This value is perfectly consistent with the small-x HERA values [BAL 95; BAL 97; MAR 95], with the measurements of LEP/SLC and with the rate of medium-E_T jets at the Tevatron [ABE 96a; ABB 97; GIE 96]. Similar indications for a value of α_s in the 0.12 range come from the final NMC analysis of $F_2^{\mu,(p,d)}$ [ARN 97]. The residual uncertainty on the value of α_s is dealt with by providing different fits for various values of $\alpha_s(M_Z)$.

The effect of the correlations between the value of α_s and the value of the parton densities is clearly shown in Fig. 5.8.6, which shows the NLO QCD predictions for the $t\bar{t}$ production cross section in $p\bar{p}$ collisions at $\sqrt{s} = 1.8\,\text{TeV}$ as a function of $\alpha_s(M_Z)$. The dotted line uses a fixed set of parton densities and α_s is varied only in the matrix elements. The solid line is instead obtained by using parton densities refitted for each value of $\alpha_s(M_Z)$ [MAR 95]. In this case the cross section depends much more weakly on the value of $\alpha_s(M_Z)$.

5.8.2 Comparison with experimental results

The parton densities derived from global fits can be used to make predictions for other processes testing the parton model and QCD. A few examples are reported in the following.

Drell–Yan processes in proton–nucleon and proton–proton collisions are very interesting as the theoretical predictions crucially depend on the shape of the antiquark content of the nucleon. In fact, the main contribution to the

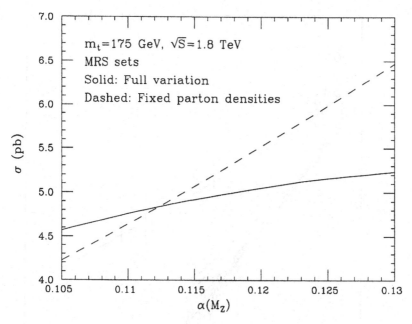

Fig. 5.8.6 Top cross-section as a function of $\alpha_s(M_z)$ [CAT 96]. The dotted line does not include the variation of the parton densities due to the change in α_s. The solid line uses the α_s-dependent MRS fits of [MAR 95].

cross section comes from the annihilation of valence quarks with antiquarks. In Fig. 5.8.7 experimental data at two different values of the center-of-mass energy $\sqrt{s} = 27.4\,\mathrm{GeV}$ [ITO 81] and $\sqrt{s} = 62\,\mathrm{GeV}$ [KOU 80; ANG 84] are compared to theoretical predictions. The cross section is computed including the correction of order α_s [ALT 85].

Results on proton structure functions at $\langle Q^2 \rangle \sim 2\,000\,\mathrm{GeV}^2$ have been published by the UA1 [ARN 84b] and UA2 [BAG 84] collaborations. They are determined from the two-jet differential cross section in proton–antiproton interactions measured at $\sqrt{s} = 540\,\mathrm{GeV}$, using the expression

$$\frac{d^3\sigma}{dx_1 dx_2 d\cos\theta} = \frac{F(x_1)}{x_1}\frac{F(x_2)}{x_2}\frac{d\sigma}{d\cos\theta},\qquad(5.8.4)$$

where, under some simplifying assumptions, $F(x)$ can be written as [HAL 83]:

$$F(x) = xG(x) + \frac{4}{9}x(\bar{q}(x) + q(x)),\qquad(5.8.5)$$

where the factor $\frac{4}{9}$ is due of the different color coupling of gluons and quarks. The experimental results for $F(x)$ as defined in Eq. (5.8.1), together with the extrapolation at $Q^2 = 2\,000\,\mathrm{GeV}^2$ of the parton densities in the combination of Eq. (5.8.2),

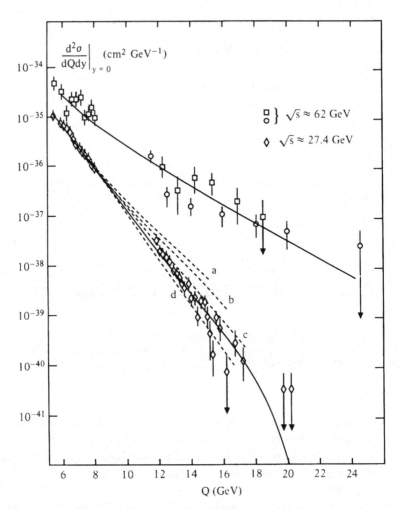

Fig. 5.8.7 Lepton pair production at fixed target energies ($\sqrt{s} = 27.4\,\text{GeV}$) [ITO 81] and ISR energies ($\sqrt{s} = 62\,\text{GeV}$) [KOU 80, ANG 84] are compared with DFLM predictions (solid curves). Dashed curves *a, b, c, d* are the predictions obtained by assigning to the power of the term $(1-x)$ of $x\bar{q}$ respectively the values 6, 6.6, 7.5, 9.5.

are presented in Fig. 5.8.8. The prediction and the experimental results are in substantial agreement. The singlet structure function alone ($\frac{4}{9}F_2(x)$) is also shown in the figure, demonstrating the relevance of the gluon contribution to the total $F(x)$.

More accurate tests using high-energy jets have since become possible, thanks to the improved energy reach and accuracy of the Tevatron experiments [ABE 96a; ABB 97]. The comparison of data and theoretical calculations using different sets of recent parton densities is shown on a linear scale in Fig. 5.8.9.

Another process where accurate experimental data exist is the direct-photon production at high transverse momentum in hadron–hadron collisions. This reaction is particularly sensitive to the gluon density. The elementary processes

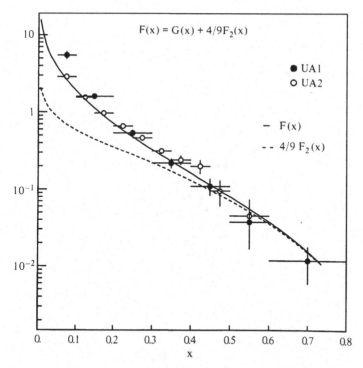

Fig. 5.8.8 $F(x, Q^2)$ defined in Eq. (5.8.1) as measured by the UA1 (full dots) and UA2 (open dots) collaborations in $p\bar{p}$ collisions ($\sqrt{s} = 540\,\text{GeV}$) at a scale $Q^2 \sim 2\,000\,\text{GeV}^2$. The solid line has been derived using DFLM parton densities evolved at $Q^2 = 2\,000\,\text{GeV}^2$ (Eq. (5.8.2)). The dashed line is the quark contribution alone.

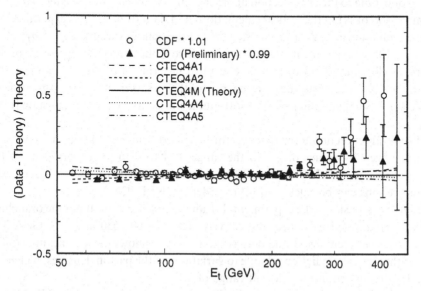

Fig. 5.8.9   Comparison of recent CDF and D0 jet E_T distributions with predictions based on different sets of parton densities.

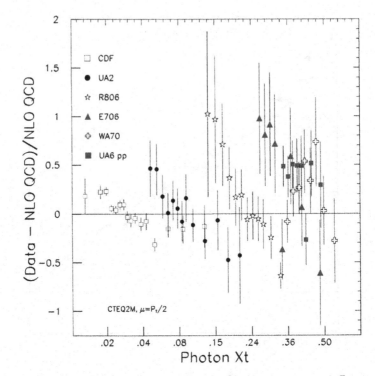

Fig. 5.8.10   The quantity (Data–Theory)/Theory plotted vs. $x_T = 2p_T/\sqrt{s}$ for a number of direct-photon experiments. Theory corresponds to NLO QCD with CTEQ2M parton densities.

that contribute to the cross section at order α_s are in fact $q\bar{q} \rightarrow g\gamma$ and $qg \rightarrow q\gamma$. The comparison of NLO theoretical calculations and data from various experiments is shown, on a linear scale and as a function of the scaling variable $x_T = 2p_T/\sqrt{s}$ in Fig. 5.8.10. All experiments show a slope relative to the theory. The presence of this slope at all values of x_T cannot be explained in terms of parton densities, and is nowadays interpreted as due to a smearing of the p_T spectra induced by intrinsic k_T effects [HUS 95] and higher-order soft-gluon emission from the initial state [BAE 96].

The production of intermediate vector bosons W and Z is a Drell–Yan process, characterized by values of $\sqrt{\tau}$ in the range 0.04 to 0.16 for experiments at the Tevatron ($\sqrt{s} = 1.8\,\text{TeV}$) and CERN $Sp\bar{p}S(\sqrt{s} = 540, 630\,\text{GeV})$ colliders. The cross sections can be calculated up to order α_s^2 [NEE 92].

Figure 5.8.11 shows the experimental W and Z cross-sections times the branching ratio in $\ell\nu$ and $\ell^+\ell^-(\ell = \mu, e)$ respectively, at $\sqrt{s} = 540, 630$ and $1800\,\text{GeV}$. The agreement between theory and data is excellent, and shows the significant stability of the theoretical prediction relative to variations of the parton densities, which are highly constrained in the relevant range of x.

To conclude, we show in Fig. 5.8.12 a comparison between the NLO QCD predictions for the $t\bar{t}$ production rate at $\sqrt{s} = 1.8\text{TeV}$ and the data from the CDF

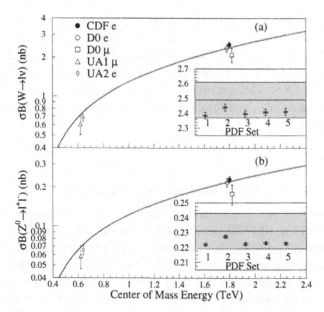

Fig. 5.8.11   Comparison of measured (a) $\sigma \cdot B(W \to e\nu)$ and (b) $\sigma \cdot B(Z \to e^+e^-)$ to theoretical predictions using NNLO QCD result of [NEE 92] and MRSA parton densities [MAR 94]. The shaded area in the inset shows 1σ region of the CDF measurement; the stars show the prediction using various parton density sets of the MRS and CTEQ groups [ABE 96]. The theoretical points include a common uncertainty in the predictions from the choice of renormalization scale μ ($M_W/2 < \mu < 2M_W$).

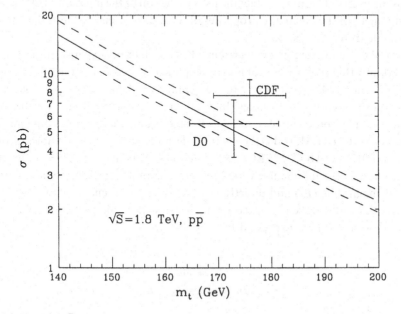

Fig. 5.8.12   The $t\bar{t}$ cross section measured by CDF and D0, compared to the NLO QCD theoretical predictions. The band in the theoretical prediction corresponds to the uncertainty due to the choice of parton densities and of renormalization and factorization scales.

[ABE 95a] and D0 [ABA 97] experiments. At this energy the $t\bar{t}$ production is dominated by quark–antiquark annihilation, with typical x values of order 0.2 and Q of order 350 GeV. At these values of x and Q, the dominant uncertainty in the parton densities comes from the impact of the value of α_s on the evolution, described earlier in Fig. 5.8.6. The convolution of the overall theoretical uncertainties, including the value of α_s and the choice of renormalization scale, is shown by the band in Fig. 5.8.12.

5.9 Neutral-current structure functions

Since the discovery of neutrino-induced reactions without a charged lepton in the final state [HAS 73a; BEN 74], considerable interest has developed in the properties of weak neutral currents. Assuming the validity of the Standard Model, the necessity of tests in the neutral sector is clear. On the other hand, until the discovery of Z^0, neutrino neutral currents were the main source of information.

One of the most significant experiments was the study of the nucleon partonic structure obtained in the neutral-current–neutrino interactions. The resultant picture was in fact expected to be the same as in charged-current interactions, and deviations from the expectations would have revealed peculiar properties of the neutral intermediate boson.

The experimental study of the neutral-current–neutrino events is a task of the utmost difficulty. The aim is in fact the reconstruction of the interaction through the measurement of the kinematic properties of the hadronic energy flow, the only one observable in the final state.

Neutral currents were first detected in a bubble chamber experiment [HAS 73b], and certainly this kind of detector is in principle suited for these studies. However, as seen in Section 5.4, statistics is a fundamental ingredient of any partonic study. Although some pioneering studies [BAL 80] and some later [BAL 84] works were performed in bubble chambers, statistically significant results came from fine-grain calorimeters (CHARM, FMMR) especially designed to match the complexity of the problem [DID 80; DOR 87; BOG 82]. Such detectors allow the measurement of the position of the interaction vertex to determine the neutrino interaction point and the magnitude (E_h) and direction (θ_h) of the hadron energy flow.

These quantities are required in order to measure the scaling variable x, which in a neutral-current event is expressed by

$$x = \frac{E_\nu E_h \sin^2 \theta_h}{2m_p(E_\nu \cos^2 \theta_h - E_h)}. \tag{5.9.1}$$

In this formula the neutrino energy is derived from the energy–radius relation existing in a narrow-band beam (up to the π, K ambiguity).

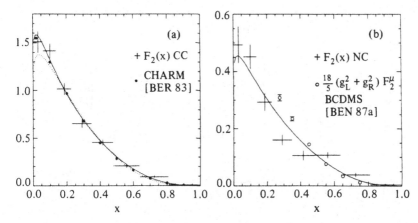

Fig. 5.9.1 (a) The structure functions $F_2(x)$ at $Q_0^2 = 10\,\mathrm{GeV}^2$ derived from CC (where the muon was ignored (+)), with a fit superimposed (continuous line). The dotted curve represents a compilation of charged-current data [PDG 88] and the full points results of the CHARM collaboration on charged-current data [BER 83]. (b) The structure functions $F_2(x)$ at $Q_0^2 = 10\,\mathrm{GeV}^2$ derived from the NC iteractions (+), with a fit superimposed (continuous line). The open points ($\bigcirc$) represents $\frac{18}{5}(g_L^2 + g_R^2)F_2^\mu$ with F_2^μ derived from deep inelastic muon scattering on carbon.

The basic formulas for the NC neutrino cross sections are given in Section 5.2.4. In the framework of the quark parton model, structure functions are expected to be equal for NC and CC reactions up to trivial differences in the quark couplings and different contributions from s and c quarks to F_2. Fixing the quark couplings with $\sin^2\theta_W = 0.236$ [ALL 87a], one finds:

$$F_2^{\mathrm{NC}} = 0.326x(q + \bar{q}) + 0.07x(s - c) \tag{5.9.2}$$

$$xF_3^{\mathrm{NC}} = 0.264x(q - \bar{q}). \tag{5.9.3}$$

The ambiguity existing in the neutrino energy makes it impossible to determine x for a single event, so that a procedure of statistical unfolding is needed [JON 83], unless only pion neutrinos are retained with a great loss in statistics [BOG 85]. In the first case, one assumes that the measured distribution is the result of the physical distribution folded with the spectrum of the incoming beam and the experimental acceptance and resolution. This procedure is tested on charged-current events where the results can be compared to those obtained using muon measurement (Figs. 5.9.1a, 5.9.2a, 5.9.3a) [ALL 88b].

The results of CHARM on $F_2^{\mathrm{NC}}(x)$, $xF_3^{\mathrm{NC}}(x)$ and $x\bar{q}(x)$ are shown in Figs. 5.9.1b, 5.9.2b, 5.9.3b. They summarize the analysis of 36 000 ν and 2000 $\bar{\nu}$ NC interactions. A simple parameterization of the valence and sea quark distributions was fitted to the measured functions (both NC and CC) in order to compare them. The functions chosen were

$$xq_{val}(x) = N_{val}\frac{\Gamma(\alpha + \beta + 1)}{\Gamma(\alpha)\Gamma(\beta + 1)}x^\alpha(1 - x)^\beta \tag{5.9.4}$$

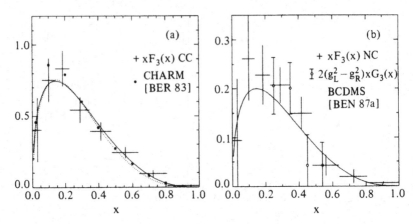

Fig. 5.9.2   (a) The structure functions $xF_3(x)$ at $Q_0^2 = 10\,\mathrm{GeV}^2$ derived from CC (where the muon was ignored (+)), with a fit superimposed (continuous line). The dotted curve represents a compilation of charged-current data [PDG 88] and the full points results of the CHARM collaboration on charged-current data [BER 83]. (b) The structure functions $F_2(x)$ at $Q_0^2 = 10\,\mathrm{GeV}^2$ derived from the NC iteractions (+), with a fit superimposed (continuous line). The open points (○) represent $2(g_L^2 - g_R^2)xG_3$ with xG_3 determined from the charge asymmetry of deep inelastic muon scattering.

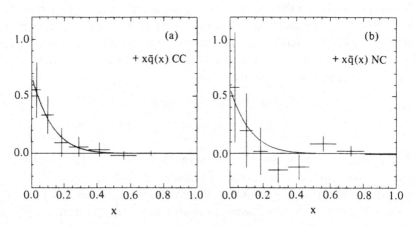

Fig. 5.9.3   (a) The structure function of the nonstrange sea at $Q_0^2 = 10\,\mathrm{GeV}^2$ derived from CC (where the muon was ignored (+)), with a fit superimposed (solid line). The dotted curve represents a compilation of charged-current data [PDG 88]. (b) The structure function of the nonstrange sea at $Q_0^2 = 10\,\mathrm{GeV}^2$ derived from NC interactions (+), with a fit superimposed (solid line).

and the results of the fit for CC and NC are shown in Table 5.9.1. They are, as expected, in good agreement. The statistics is the limiting factor concerning

$$xq_{sea}(x) = \tfrac{1}{2}C(\gamma+1)(1-x)^\gamma \qquad (5.9.5)$$

Table 5.9.1. *Fit parameters for CC and NC distributions*

	N_{val}	α	β	C	γ
CC	3.05 ± 0.13	0.455 ± 0.018	2.77 ± 0.08	0.155 ± 0.006	7.58 ± 0.61
NC	3.0	0.476 ± 0.030	2.75 ± 0.15	0.126 ± 0.008	8.0

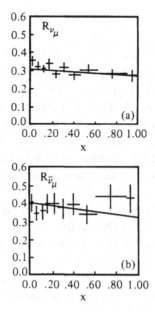

Fig. 5.9.4 (a) NC/CC ratio as a function of x for neutrinos. The solid line indicates the results of a fit. (b) NC/CC ratio as a function of x for antineutrinos. The solid line indicates the results of a fit.

the sea quark distribution analysis. A direct measurement of $s(x)-c(x)$ was therefore impossible.

Results in agreement with those just described have been obtained by the FMMR collaboration, although with meager statistics [BOG 85] (2300 neutrino and 740 antineutrino events). They are shown in Figs. 5.9.4a and 5.9.4b where the ratio of NC/CC events is plotted as a function of x for neutrinos and antineutrinos.

As a further test of the universality, the nucleon densities as probed by neutral current in neutrino interactions can be compared to those seen by electromagnetic current in muon scattering. In Fig. 5.9.1b $F_2^{NC}(x)$ (CHARM [ALL 88b]) is reported together with $\frac{18}{5}(g_L^2 + g_R^2)F_2^\mu(x)$ (BCDMS carbon [BEN 87a]) at a fixed $Q^2 = 10\,\text{GeV}^2$. The ratio of the two functions is expected to be one in the QPM framework and the measured value is $R_{F_2} = 1.13 \pm 0.10$.

The valence quark structure function xF_3^{NC} [ALL 88b] can be also compared in the same fashion to the so-called electroweak interference structure function xG_3

[BEN 87a] (Fig. 5.9.2b). The measured ratio of the functions, again expected to be one, is $R_{xF_3} = 0.98 \pm 0.19$.

In conclusion, all the existing data are compatible with the universality of nucleon structure functions as probed by weak charged and neutral current and by the electromagnetic current.

5.10 Conclusions

Many years of intense experimental activity in neutrino deep inelastic scattering have resulted in a detailed knowledge of the nucleon structure and of the weak and strong interactions of its constituents. It is a general fact that all the neutrino experiments agree with each other within the errors and are consistent with the electroproduction experiments. Within the stated precisions, all the results can be interpreted in the framework of the QCD improved parton model.

More recently, this picture of the nucleon has been complemented by the data collected at HERA, the DESY electron–proton collider. Particularly important is the contribution to the low x region, tightly constraining the previously unravelled low x behavior of the gluon density.

The structure of the nucleon is now explored in an extremely large kinematic range, from 10^{-1} to $10^4 \, \mathrm{GeV}^2$ for Q^2 and from 10^{-5} to 1 for x, with an impressive consistency among completely different classes of experiments and in complete agreement with the evolution predicted by perturbative QCD. Deep inelastic lepton–nucleon scattering, with the relevant contributions from neutrino experiments, still remains one of the fundamental probes of the theory of strong interactions.

References

[ABA 97]	D0 Collab., S. Abachi *et al.*, *Phys. Rev. Lett.* 79 (1997) 1203.
[ABB 80]	L. F. Abbott and R. M. Barnett, *Ann. Phys.* 125 (1980) 276.
[ABB 97]	D0 Collab., B. Abbott *et al.*, *FERMILAB-CONF-97-269-E.*
[ABE 95]	CDF Collab., F. Abe *et al.*, *Phys. Rev. Lett.* 74 (1995) 850.
[ABE 95a]	CDF Collab., F. Abe *et al.*, *Phys. Rev. Lett.* 74 (1995) 2626.
[ABE 96]	CDF Collab., F. Abe *et al.*, *Phys. Rev. Lett.* 76 (1996) 3070.
[ABE 96a]	CDF Collab., F. Abe *et al.*, *Phys. Rev. Lett.* 77 (1996) 438.
[ABR 82a]	CDHS Collab., H. Abramowicz *et al.*, *Z. Phys.* C12 (1982) 289.
[ABR 82b]	CDHS Collab., H. Abramowicz *et al.*, *Z. Phys.* C15 (1982) 19.
[ABR 82c]	CDHS Collab., H. Abramowicz *et al.*, *Z. Phys.* C13 (1982) 199.
[ABR 83]	CDHS Collab., H. Abramowicz *et al.*, *Z. Phys.* C17 (1983) 237.
[ABR 84]	CDHS Collab., H. Abramowicz *et al.*, *Z. Phys.* C25 (1984) 29.
[ABR 86]	CDHS Collab., H. Abramowicz *et al.*, *Phys. Rev. Lett.* 57 (1986) 298.
[ADA 95]	M. R. Adams *et al.*, *Z. Phys.* C67 (1995) 403.
[ADE 86]	M. Aderholz *et al.*, *Phys. Lett.* B173 (1986) 211.
[AID 96]	H1 Collab., S. Aid *et al.*, *Nucl. Phys.* B470 (1996) 3.
[ALL 81a]	D. Allasia *et al.*, *Phys. Lett.* B102 (1981) 374.
[ALL 81b]	P. Allen *et al.*, *Phys. Lett.* B103 (1981) 71.
[ALL 84]	D. Allasia *et al.*, *Phys. Lett.* B135 (1984) 231.

[ALL 85]	D. Allasia *et al.*, *Z. Phys.* C28 (1985) 321.
[ALL 87a]	CHARM Collab., J. Allaby *et al.*, *Z. Phys.* C36 (1987) 611.
[ALL 87b]	CHARM Collab., J. Allaby *et al.*, *Z. Phys.* B197 (1987) 281.
[ALL 88a]	CHARM Collab., J. Allaby *et al.*, *Z. Phys.* C38 (1988) 403.
[ALL 88b]	CHARM Collab., J. Allaby *et al.*, *Z. Phys.* B213 (1988) 554.
[ALL 89]	CHARM Collab., J. Allaby *et al.*, *Phys. Lett.* B231 (1989) 317.
[ALP 89]	P. P. Allport *et al.*, *Phys. Lett.* B232 (1989) 417.
[ALT 77]	G. Altarelli and G. Parisi, *Nucl. Phys.* B126 (1977) 298.
[ALT 78a]	G. Altarelli, R. K. Ellis and G. Martinelli, *Nucl. Phys.* B143 (1978) 521.
[ALT 78b]	G. Altarelli, R. K. Ellis and G. Martinelli, *Nucl. Phys.* B146 (1978) 544.
[ALT 78c]	G. Altarelli and G. Martinelli, *Phys. Lett.* B76 (1978) 89.
[ALT 79a]	G. Altarelli *et al.*, *Nucl. Phys.* B160 (1979) 301.
[ALT 79b]	G. Altarelli, R. K. Ellis and G. Martinelli, *Nucl. Phys.* B157 (1979) 461.
[ALT 84]	G. Altarelli, R. K. Ellis, M. Greco and G. Martinelli, *Nucl. Phys.* B246 (1984) 12.
[ALT 85]	G. Altarelli, R. K. Ellis and G. Martinelli, *Phys. Lett.* B151 (1985) 457.
[ALT 88]	G. Altarelli, M. Diemoz, G. Martinelli and P. Nason, *Nucl. Phys.* B308 (1988) 724.
[AMA 78a]	D. Amati, R. Petronzio and G. Veneziano, *Nucl. Phys.* B140 (1978) 54.
[AMA 78b]	D. Amati, R. Petronzio and G. Veneziano, *Nucl. Phys.* B146 (1978) 492.
[AMA 87]	U. Amaldi *et al.*, *Phys. Rev.* D36 (1987) 1385.
[AMA 91a]	NMC Collab., P. Amaudruz *et al.*, *Phys. Rev. Lett.* 66 (1991) 2712.
[AMA 91]	NMC Collab., P. Amaudruz *et al.*, *Z. Phys.* C51 (1991) 387.
[AMA 92a]	NMC Collab., P. Amaudruz *et al.*, *Phys. Lett.* B295 (1992) 159.
[AMA 92]	NMC Collab., P. Amaudruz *et al.*, *Z. Phys.* C53 (1992) 73.
[AMA 95]	NMC Collab., P. Amaudruz *et al.*, *Nucl. Phys.* B441 (1995) 3.
[AMM 87]	V. V. Ammosov *et al.*, *IFVE/87–82*, 1987.
[ANG 84]	A. L. S. Angelis *et al.*, *Phys. Lett.* B147 (1984) 472.
[APP 86]	J. A. Appel *et al.*, *Phys. Lett.* B176 (1986) 239.
[ARM 81]	N. Armenise *et al.*, *Phys. Lett.* B102 (1981) 374.
[ARN 84a]	R. G. Arnold *et al.*, *Phys. Rev. Lett.* 52 (1984) 727.
[ARN 84b]	G. Arnison *et al.*, *Phys. Lett.* B136 (1984) 294.
[ARN 94a]	NMC Collab., M. Arneodo *et al.*, *Phys. Rev.* D50 (1994) 1.
[ARN 94b]	M. Arneodo, *Phys. Rep.* 240 (1994) 301.
[ARN 97]	NMC Collab., M. Arneodo *et al.*, *Nucl. Phys.* B483 (1997) 3.
[ARR 94]	CCFR Collab., C. G. Arroyo *et al.*, *Phys. Rev. Lett.* 72 (1994) 3452.
[AUB 83]	EMC Collab., J. J. Aubert *et al.*, *Phys. Lett.* B123 (1983) 275.
[AUB 85]	EMC Collab., J. J. Aubert *et al.*, *Nucl. Phys.* B259 (1985) 189.
[AUB 87]	EMC Collab., J. J. Aubert *et al.*, *Nucl. Phys.* B293 (1987) 740.
[AUR 84]	P. Aurenche *et al.*, *Phys. Lett.* B140 (1984) 87.
[AVE 88]	F. Aversa *et al.*, *Phys. Lett.* B210 (1988) 225.
[BAE 96]	H. Baer and M. H. Reno, *Phys. Rev.* D54 (1996) 2917.
[BAG 84]	P. Bagnala *et al.*, *Phys. Lett.* B144 (1984) 283.
[BAK 82]	N. J. Baker *et al.*, *Phys. Rev.* D25 (1982) 617.
[BAK 83]	N. J. Baker *et al.*, *Phys. Rev. Lett.* 51 (1983) 735.
[BAL 77]	C. Baltay *et al.*, *Phys. Rev. Lett.* 39 (1977) 62.
[BAL 80]	C. Baltay *et al.*, *Phys. Rev. Lett.* 44 (1980) 916.
[BAL 84]	C. Baltay *et al.*, *Phys. Rev. Lett.* 52 (1984) 1948.
[BAL 94]	NA51 Collab., A. Baldit *et al.*, *Phys. Lett.* B332 (1994) 244.
[BAL 95]	R. D. Ball and S. Forte, *Phys. Lett.* B358 (1995) 365.
[BAL 97]	R. D. Ball and S. Forte, in *Deep Inelastic Scattering and Related Phenomena (DIS-96): Proceedings*, eds. G. D'Agostini and A. Nigro. World Scientific, 1997, p. 208 (hep-ph/9607289).
[BAR 76]	B. C. Barish *et al.*, *Phys. Rev. Lett.* 36 (1976) 939.
[BAR 77]	B. C. Barish *et al.*, *Phys. Rev. Lett.* 39 (1977) 981.
[BAR 79a]	D. S. Baranov *et al.*, *Phys. Lett.* B81 (1979) 255.
[BAR 79b]	S. J. Barish *et al.*, *Phys. Rev.* D19 (1979) 2521.

[BAR 85]   BCDMS Collab., G. Bari *et al.*, *Phys. Lett.* B163 (1985) 282.
[BAZ 95]   CCFR Collab., A. O. Bazarko *et al.*, *Z. Phys.* C65 (1995) 189.
[BEL 64]   J. S. Bell, *Phys. Rev. Lett.* 13 (1964) 57.
[BEN 74]   A. Benvenuti *et al.*, *Phys. Rev. Lett.* 32 (1984) 800.
[BEN 75a]   A. Benvenuti *et al.*, *Phys. Rev. Lett.* 34 (1975) 419.
[BEN 75b]   A. Benvenuti *et al.*, *Phys. Rev. Lett.* 35 (1975) 1199.
[BEN 78]   A. Benvenuti *et al.*, *Phys. Rev. Lett.* 41 (1978) 1204.
[BEN 87a]   BCDMS Collab., A. Benvenuti *et al.*, *Phys. Rev. Lett.* B198 (1987) 483.
[BEN 87b]   BCDMS Collab., A. Benvenuti *et al.*, *Phys. Lett.* B195 (1987) 91.
[BEN 89]   BCDMS Collab., A. Benvenuti *et al.*, *CERN–EP/89–06*.
[BER 83]   CHARM Collab., F. Bergsma *et al.*, *Phys. Lett.* B123 (1983) 269.
[BER 84]   CHARM Collab., F. Bergsma *et al.*, *Phys. Lett.* B141 (1984) 129.
[BER 85]   CHARM Collab., F. Bergsma *et al.*, *Phys. Lett.* B153 (1985) 111.
[BER 87]   CDHS Collab., P. Berge *et al.*, *Z. Phys.* C35 (1987) 443.
[BER 91]   CDHSW Collab., J. P. Berge *et al.*, *Z. Phys.* C49 (1991) 187.
[BJO 68]   J. D. Bjorken, *Proc. Inst. School of Phys. 'Enrico Fermi'*, Course 41, Academie
              Press, NY (1968) 55.
[BLA 83]   R. Blair *et al.*, *Phys. Rev. Lett.* 51 (1983) 739.
[BLI 75]   J. Blietschau *et al.*, *Phys. Lett.* B58 (1975) 361.
[BLI 76]   J. Blietschau *et al.*, *Phys. Lett.* B60 (1976) 207.
[BLO 69]   E. D. Bloom *et al.*, *Phys. Rev. Lett.* 23 (1969) 930.
[BOD 79]   A. Bodek *et al.*, *Phys. Rev.* D20 (1979) 1471.
[BOG 82]   FMM Collab., D. Bogert *et al.*, *IEEE Trans. Nucl. Sci.* 29 (1982) 363.
[BOG 85]   FMM Collab., D. Bogert *et al.*, *Phys. Rev. Lett.* 55 (1985) 1969.
[BON 88]   WA70 Collab., M. Bonesini *et al.*, *Z. Phys.* C38 (1988) 371.
[BOO 78]   C. de Boor, *A practical guide to splines*, Springer (1978).
[BOS 77]   B. C. Bosetti *et al.*, *Phys. Rev. Lett.* 38 (1977) 1248.
[BOS 78a]   P. Bosetti *et al.*, *Nucl. Phys.* B142 (1978) 1.
[BOS 78b]   B. C. Bosetti *et al.*, *Phys. Lett.* B73 (1978) 380.
[BOS 82]   P. Bosetti *et al.*, *Nucl. Phys.* B203 (1982) 362.
[BRE 69]   M. L. Breidenbach *et al.*, *Phys. Rev. Lett.* 23 (1969) 935.
[BRN 76]   R. M. Barnett, *Phys. Rev.* D14 (1976) 70.
[CAL 75]   C. Callan and D. J. Gross, *Phys. Rev. Lett.* 34 (1975) 419.
[CAT 96]   S. Catani, M. L. Mangano, P. Nason and L. Trentadue, *Phys. Lett.* B378 (1996) 329.
[CHA 75]   C. Chang *et al.*, *Phys. Rev. Lett.* 35 (1975) 901.
[CIA 79]   S. Ciampolillo *et al.*, *Phys. Lett.* 84B (1979) 281.
[CLO 73]   F. E. Close, *Phys. Lett.* B43 (1973) 422.
[COL 79]   D. C. Colley *et al.*, *Z. Phys.* C2 (1979) 187.
[COL 86a]   J. C. Collins, in *Proc. SSC Workshop*, UCLA (1986).
[COL 86b]   J. C. Collins and W. K. Tung, *Nucl. Phys.* B278 (1986) 934.
[COO 84]   A. M. Cooper *et al.*, *Phys. Lett.* B141 (1984) 133.
[COS 87]   G. Costa *et al.*, *Nucl. Phys.* B297 (1988) 244.
[CUR 80]   G. Curci, W. Furmanski and R. Petronzio, *Nucl. Phys.* B175 (1980) 27.
[DAS 96]   M. Dasgupta and B. R. Webber, *Phys. Lett.* B382 (1996) 273.
[DER 96]   ZEUS Collab., M. Derrick *et al.*, *Z. Phys.* C69 (1996) 607, *ibid* C72 (1996) 399.
[DEV 83]   A. DeVeto *et al.*, *Phys. Rev.* D27 (1983) 508.
[DID 80]   CHARM Collab., A. N. Diddens *et al.*, *Nucl. Instrum. Meth.* 178 (1980) 27.
[DIE 86]   M. Diemoz, F. Ferroni and E. Longo, *Phys. Rep.* 130 (1986) 293.
[DIE 88]   M. Diemoz, F. Ferroni, E. Longo and G. Martinelli, *Z. Phys.* C39 (1988) 21.
[DOR 87]   J. Dorenbosch *et al.*, CHARM Collab., *Nucl. Instrum. Meth.* A253 (1987) 203.
[DUK 84]   D. W. Duke and J. F. Owens, *Phys. Rev.* D30 (1984) 49.
[EFR 79]   V. I. Efremenko *et al.*, *Phys. Lett.* B84 (1979) 511.
[EIC 84]   E. Eichten, I. Hinchliffe, K. Lane and C. Quigg, *Rev. Mod. Phys.* 56 (1984).
[ELL 78]   R. K. Ellis *et al.*, *Phys. Lett.* B78 (1978) 281; *Nucl. Phys.* B152 (1979) 285.
[ELL 83a]   R. K. Ellis, W. Furmanski and R. Petronzio, *Nucl. Phys.* B207 (1983) 1.

[ELL 83b]	R. K. Ellis, W. Furmanski and R. Petronzio, *Nucl. Phys.* B212 (1983) 29.
[ELL 85]	R. K. Ellis and J. C. Sexton, *Nucl. Phys.* B282 (1987) 624.
[ELL 88a]	S. D. Ellis, Z. Kunszt and D. E. Soper, Oregon preprint OITS 395 (1988); OITS 396 (1988).
[ERR 79]	O. Erriquez *et al.*, *Phys. Lett.* B80 (1979) 309.
[FAR 75]	G. Farrar and J. D. Jackson, *Phys. Rev. Lett.* 35 (1975) 1416.
[FIE 77]	R. Field and R. P. Feynman, *Phys. Rev.* D15 (1977) 2590.
[FLO 77a]	E. G. Floratos, D. A. Ross and C. T. Sacharjda, *Nucl. Phys.* B129 (1977) 66; B139 (1978) 545.
[FLO 79]	E. G. Floratos, D. A. Ross and C. T. Sacharjda, *Nucl. Phys.* B152 (1979) 493.
[FLO 81]	E. G. Floratos *et al.*, *Nucl. Phys.* B192 (1981) 417.
[FUR 80]	W. Furmanski and R. Petronzio, *Phys. Lett.* B97 (1980) 437.
[FUR 82]	W. Furmanski and R. Petronzio, *Z. Phys.* C11 (1982) 293.
[GEO 76]	H. Georgi and H. D. Politzer, *Phys. Rev.* D14 (1976) 1829.
[GIE 96]	W. T. Giele, E. W. N. Glover and J. Yu, *Phys. Rev.* D53 (1996) 120.
[GOM 94]	J. Gomez *et al.*, *Phys. Rev.* D49 (1994) 4348.
[GON 79]	A. Gonzales-Arroyo, C. López and F. Ynduráin, *Nucl. Phys.* B153 (1979) 161.
[GRO 73a]	D. J. Gross and F. Wilczek, *Phys. Rev. Lett.* 30 (1973) 1343; *Phys. Rev.* D8 (1973) 3633.
[GRO 79]	CDHS Collab., J. C. H. de Groot *et al.*, *Z. Phys.* C1 (1979) 143.
[HAL 83]	F. Halzen and P. Hoyer, *Phys. Lett.* B130 (1983) 326.
[HAN 80]	J. Hanlon *et al.*, *Phys. Rev. Lett.* 45 (1980) 1817.
[HAS 73a]	F. J. Hasert *et al.*, *Phys. Lett.* B46 (1973) 138.
[HAS 73b]	F. J. Hasert *et al.*, *Phys. Lett.* B46 (1973) 121.
[HEA 81]	S. M. Heagy *et al.*, *Phys. Rev.* D23 (1981) 1045.
[HOL 77]	M. Holder *et al.*, *Phys. Lett.* B69 (1977) 377.
[HUS 95]	CTEQ Collab., J. Huston *et al.*, *Phys. Rev.* D51 (1995) 6139.
[ITO 81]	A. S. Ito *et al.*, *Phys. Rev.* D23 (1981) 604.
[JAF 81]	R. L. Jaffe and M. Soldate, *Phys. Lett.* B105 (1981) 467.
[JAR 82]	T. Jarosziewicz, *Phys. Lett.* B116 (1982) 291.
[JON 81a]	CHARM Collab., M. Jonker *et al.*, *Phys. Lett.* B102 (1981) 67.
[JON 81b]	CHARM Collab., M. Jonker *et al.*, *Phys. Lett.* B107 (1981) 241.
[JON 82]	CHARM Collab., M. Jonker *et al.*, *Phys. Lett.* B109 (1982) 133.
[JON 83]	CHARM Collab., M. Jonker *et al.*, *Phys. Lett.* B128 (1983) 117.
[KAP 76]	J. Kaplan and F. Martin, *Nucl. Phys.* B115 (1976) 333.
[KAT 96]	A. L. Kataev, *Proc. of the 1996 Annual Divisional Meeting* (DPF 96) of the Division of Particles and Fields of the American Physical Society, Minneapolis, August 1996, hep-ph/9609398.
[KIT 82]	T. Kitagaki *et al.*, *Phys. Rev. Lett.* 49 (1982) 98.
[KOU 80]	C. Kourkoumelis *et al.*, *Phys. Lett.* B91 (1980) 475.
[KRO 76]	J. Von Krogh *et al.*, *Phys. Rev. Lett.* 36 (1976) 710.
[KUB 79]	J. Kubar-Andre and F. E. Paige, *Phys. Rev.* D19 (1979) 221.
[LAI 97]	CTEQ Collab., H. L. Lai *et al.*, *Phys. Rev.* D55 (1997) 1280.
[LEU 93]	CCFR Collab., W. C. Leung *et al.*, *Phys. Lett.* B317 (1993) 655.
[LUT 81]	S. P. Luttrell, S. Wada and B. R. Webber, *Nucl. Phys.* B188 (1981) 219.
[MAC 84]	D. B. Mac Farlane *et al.*, *Z. Phys.* C26 (1984) 1.
[MAR 88]	A. D. Martin, R. G. Roberts and W. J. Stirling, *Phys. Rev.* D37 (1988) 1161.
[MAR 94]	A. D. Martin, R. G. Roberts and W. J. Stirling, *Phys. Rev.* D50 (1994) 6734.
[MAR 95]	A. D. Martin, R. G. Roberts and W. J. Stirling, *Phys. Lett.* B356 (1995) 89.
[MAR 96]	A. D. Martin, R. G. Roberts and W. J. Stirling, *Phys. Lett.* B387 (1996) 419.
[MOR 81]	J. Morfin *et al.*, *Phys. Lett.* B104 (1981) 235.
[MOR 91]	J. G. Morfin and W.-K. Tung, *Z. Phys.* C52 (1991) 13.
[MUL 78]	A. H. Muller, *Phys. Rev.* D18 (1978) 3705.
[NEE 92]	W. L. van Neerven and E. B. Zijlstra, *Nucl. Phys.* B382 (1992) 1.
[ODO 81]	R. Odorico, *Phys. Lett.* B102 (1981) 341.

[PAN 68]   W. K. H. Panofsky, *Proc. Fourteenth Int. Conf. HEP*, Vienna (1968) 23.
[PAR 79]   A. Para and C. T. Sacharajda, *Phys. Lett.* B86 (1979) 331.
[PDG 88]   Particle Data Group, *Phys. Lett.* B204 (1988) 1.
[PDG 96]   Particle Data Group, R. M. Barnett *et al.*, *Phys. Rev.* D54 (1996) 1.
[PER 75]   D. Perkins, in *Proc. Int. Symp. on Lepton and Photon Interactions at High Energies*, Univ. California, Stanford (1975) 571.
[POL 73]   H. D. Politzer, *Phys. Rev. Lett.* 30 (1973) 1346.
[POL 80]   H. D. Politzer, *Nucl. Phys.* B172 (1980) 349.
[QUI 93]   CCFR Collab., P. Z. Quintas *et al.*, *Phys. Rev. Lett.* 71 (1993) 1307.
[REU 85]   CCCFRR Collab., P. G. Reutens *et al.*, *Phys. Lett.* B152 (1985) 404.
[RIO 75]   E. M. Riordan *et al.*, *SLAC-PUB* 634 (1975).
[RUJ 77]   A. De Rújula, H. Georgi and H. D. Politzer, *Ann. Phys.* 103 (1977) 315.
[RUJ 79]   A. De Rújula, R. Petronzio and A. Savoy-Navarro, *Nucl. Phys.* B154 (1979) 394.
[SAL 64]   A. Salam and J. C. Ward, *Phys. Lett.* 13 (1964) 168.
[SAL 68]   A. Salam, *Proc. Eighth Nobel Symp.*, ed. N. Svartholm, Almqvist and Wiksell, Stockholm (1968).
[SCH 46]   I. J. Schoenberg, *Quart. App. Maths*, 4 (1946) 45, 112.
[SEL 97]   CCFR Collab., W. G. Seligman *et al.*, *Phys. Rev. Lett.* 79 (1997) 1213.
[SHU 82a]   E. V. Shuryak and A. I. Vainshtein, *Nucl. Phys.* B199 (1982) 451.
[SHU 82b]   E. V. Shuryak and A. I. Vainshtein, *Nucl. Phys.* B201 (1982) 141.
[SMI 93]   CCFR Collab., W. H. Smith *et al.*, *Nucl. Phys. Proc. Suppl.* 31 (1993) 262.
[SOP 89]   D. E. Soper, *Proc. Twenty-Fourth Int. Conf. on High Energy Physics*, Munich, ed. R. Kolthaus and J. H. Kuhn, Springer-Verlag (1989).
[SOZ 93]   UA6 Collab., G. Sozzi *et al.*, *Phys. Lett.* B317 (1993) 243.
[STE 75]   S. Stein *et al.*, *Phys. Rev.* D12 (1975) 1884.
[t'HO 71]   G. 'tHooft, *Nucl. Phys.* B33 (1971) 173.
[TAY 69]   R. E. Taylor, *Proc. Fourth Symp. on Electron and Photon Interactions*, Liverpool, Daresbury Nucl. Phys. Lab. (1969) 251.
[TAY 83]   G. N. Taylor *et al.*, *Phys. Rev. Lett.* 51 (1983) 739.
[VAR 87]   K. Varvell *et al.*, *Z. Phys.* C36 (1987) 1.
[VIR 92]   M. Virchaux and A. Milzstein, *Phys. Lett.* B274 (1992) 221.
[VOV 79]   A. S. Vovenko *et al.*, *Sov. J. Nucl. Phys.* 30 (1979) 527.
[WEI 67]   S. Weinberg, *Phys. Rev. Lett.* 19 (1967) 1264.
[WEI 71]   S. Weinberg, *Phys. Rev. Lett.* 27 (1971) 1688.

6

Neutrinos in astrophysics and cosmology

6.1 Solar neutrinos

6.1.1 Detection of solar neutrinos*

6.1.1.1 Introduction

During the sun's lifetime on the main sequence, hydrogen is burnt into helium within the solar core. The fusion chains involve weak interactions like

$$p + p \rightarrow d + e^+ + \nu_e \qquad (6.1.1)$$

$$^7\text{Be} + e^- \rightarrow {}^7\text{Li} + \nu_e \qquad (6.1.2)$$

$$^8\text{B} \rightarrow {}^8\text{Be} + e^+ + \nu_e \qquad (6.1.3)$$

The low-energy ($\sim$ MeV) neutrinos produced in these reactions escape almost freely from the solar interior. Consequently, their detection can probe the physical state of the innermost sun very directly. By comparing measured neutrino fluxes and spectra with theoretical predictions, stellar structure models, in particular the standard solar model (SSM), can be experimentally tested. In addition, solar neutrino observations have the potential to unravel a nonvanishing neutrino restmass by virtue of neutrino oscillations (see Section 6.1.2). For this application, the absolute strength of the neutrino source "sun" must be unambiguously known. This is best fulfilled for the "pp neutrinos" from reaction 6.1.1. In each completed chain $4\text{H} \rightarrow {}^4\text{He}$, 2 pp neutrinos and 26.73 MeV energy are generated. On average, 0.59 MeV escape with neutrinos; 26.14 MeV or 13.07 MeV per pp neutrino make the sun luminous. Dividing the solar luminosity by 13.07 MeV yields the total pp-neutrino flux. At the earth, 150 million km away from the source, this flux is $6 \cdot 10^{10}$ cm^{-2}s^{-1}.

In spite of this high flux, solar neutrino detection experiments are very difficult because of the extremely low interaction cross sections of $\sim$ (sub)MeV neutrinos with matter (favorable cases range from 10^{-46}–10^{-42} cm^2). Typical event rates are illustrated by the magnitude of the appropriately defined solar neutrino unit:

$$1 \text{ SNU} = 1 \text{ event per } 10^{36} \text{ target atoms per second.} \qquad (6.1.4)$$

---

* T. Kirsten, Max-Planck-Institut für Kernphysik, Heidelberg, Germany.

499

As an example, for pp-neutrinos reacting with ^{71}Ga, $< \Phi_{pp} \cdot \sigma \geq 5.9 \cdot 10^{10}\,\text{cm}^{-2}$ $\text{s}^{-1} \cdot 1.18 \cdot 10^{-45}\,\text{cm}^2 = 69.7 \cdot 10^{-36}\,\text{s}^{-1} = 69.7\,\text{SNU}$. In any case, tens to thousands of tons of target material are typically required to obtain just one neutrino-induced reaction per day. Rigorous control of competing side reactions and extremely sensitive low background detectors are clearly required for any solar neutrino experiment. In fact, without the pioneering work of R. Davis, Jr., in running the famous Homestake chlorine experiment, solar neutrino detection could well be considered impossible even today. Instead, five detectors have recorded solar neutrinos, and more are coming up in the near future.

6.1.1.2 Standard Solar Model and neutrino flux predictions

Given its mass and chemical composition, the structure and evolution of a star is uniquely defined. The standard ingredients of stellar models are

> hydrostatic equilibrium
> the ideal gas equation of state
> thermal equilibrium (energy production equals luminosity)
> radiation dominated energy transport in the dense interior
> secular energy production by fusion

For the sun, known observables are its mass and today's luminosity, radius, and surface temperature. The age of the sun ($4.6 \cdot 10^9$ years) is also known. Concerning the chemical composition, the SSM assumes homogeneous accretion; that is, the present-day surface abundances (in particular, of hydrogen and helium) are set equal to the initial composition throughout the sun.

The principles being straightforward, the details of actual solar model calculations are intricate. Required, for example, is explicit knowledge of the absolute nuclear reaction cross sections (S factors) for the relevant fusion reactions and of the opacities that control the internal temperature distribution. The most explicit solar model calculations have been performed and continuously updated by Bahcall during the past 20 years [BAH 82a, 88, 89, 95]. Some results are compiled in Table 6.1.1. The nuclear fusion chains responsible for the sun's energy generation are shown in Fig. 6.1.1.

Given the central temperature of 15.84 million K, the dominant cycle for the sun is the PP chain, whereas the CNO cycle contributes only marginally to the energy production. In the context of this discussion, the theoretical neutrino fluxes arising from these reaction chains are of particular interest. The corresponding neutrino spectrum is shown in Fig. 6.1.2. By far the most abundant solar neutrinos ($5.9 \cdot 10^{10}\,\text{cm}^{-2}\,\text{s}^{-1}$) are those produced in the initial reaction (6.1.1) (pp neutrinos; continuous spectrum; maximum energy 420 KeV only). After deuterium is burnt into ^{3}He, neutrinos arise either in the PPII branch from electron capture of the

Table 6.1.1. *Properties of the sun according to the SSM*

Property	Present sun $t = 4.6 \times 10^9$ years	Initial sun $(t = 0)$
Luminosity $L_\odot$	$= 1$	0.67
Radius $R_\odot$	$696\,000$ km $= 1$	0.87
Surface temperature T_S	5773 K	5665 K
Central temperature T_C	15.84×10^6 K	
Central density ρ_C	156 g/cm^3	
X (hydrogen)	33.3% by mass	70.25% by mass
Y (helium)	64.6% by mass	27.75% by mass
Z ($Z > 2$)	2.0% by mass	2.0% (input)

Note: Ninety-five percent of total luminosity $L_\odot$ produced within $\leq .21 \times R_\odot$ where $T(.21\ R_\odot) = 9 \times 10^6$ K; $\rho(.21\ R_\odot) = 32$ g/cm^3.
Source: [BAH 88, 95].

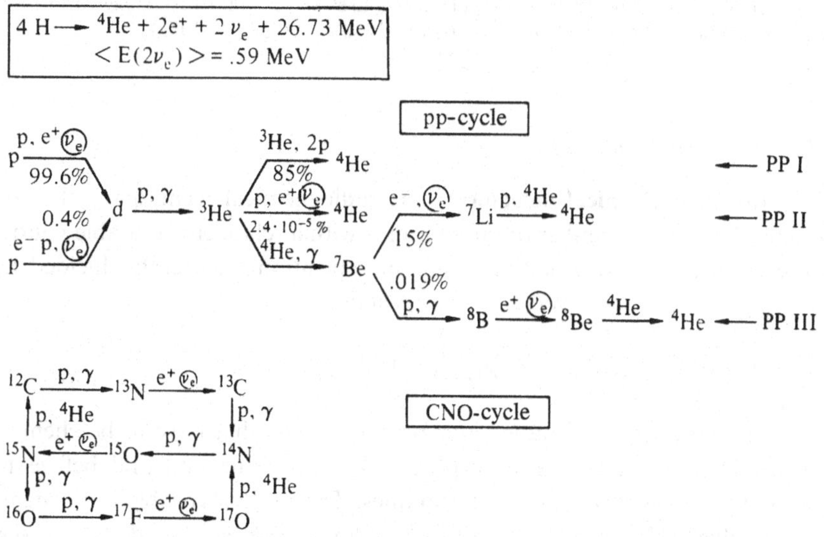

Fig. 6.1.1   The nuclear fusion chains responsible for the energy generation in the sun.

intermittently produced ^{7}Be nuclei (^{7}Be neutrinos; line spectrum) or, in the very rare PPIII chain, from the e^+ decay of ^{8}B (^{8}B neutrinos, $6.6 \cdot 10^6$ cm^{-2}s^{-1}, continuous spectrum, maximum energy 14 MeV).

Among all neutrinos sources, the ^{8}B neutrinos are those whose production depends most strongly on temperature. Consequently, they are produced closest to the center, their production peaks at $\sim 0.04\ R_\odot$. At $.11 R_\odot$, where the differential pp-neutrino production rate is maximal, the ^{8}B-ν production has nearly ceased.

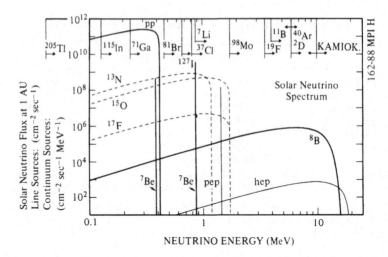

Fig. 6.1.2   Solar neutrino spectrum as calculated from the standard solar model [BAH 88, 95]. Energy thresholds for various neutrino detection schemes are shown on top. Absolute fluxes in $10^6 \, \text{cm}^{-2} \, \text{s}^{-1}$ and their 1σ uncertainties are: *pp:* 59 100 ($\pm 1\%$); *pep:* 140 ($\pm 2\%$); ^{7}Be: 5150 ($\pm 6\%$); ^{8}B: 6.6 ($\pm 15\%$); ^{13}N: 618 ($\pm 20\%$); ^{15}O: 545 ($\pm 20\%$).

6.1.1.3 Detection techniques

(a) Radiochemical detectors   The radiochemical technique is based on exposing a very large number of target atoms with a relatively favorable neutrino-capture cross section to solar neutrinos in a position sufficiently shielded from cosmic ray muons. The inverse β-decay reaction

$$^A Z(\nu_e, e^-)^A (Z+1) \tag{6.1.5}$$

leads to the production of radioactive product nuclei that have to be chemically separated from the target after exposure to near saturation and before their detection with low-level counting techniques. The more favorable radiochemical detection schemes into which at least some serious work has already been invested are listed in Table 6.1.2. It includes the prototype radiochemical solar neutrino detector, based on ^{37}Cl-^{37}Ar. The following considerations influence the prospective merits of potential solar neutrino detection schemes.

Threshold energy. This determines the spectral response of the detector (see Fig. 6.1.2); *pp* neutrino detectors require $E_{tr} < 420 \, \text{keV}$; ^{7}Be neutrinos can be detected if $E_{tr} \leq 862 \, \text{keV}$, and ^{8}B neutrinos are accessible as long as $E_{tr} \leq 14 \, \text{MeV}$.

Value and uncertainty of the expected production rate. This requires a knowledge of the energy-dependent neutrino-capture cross sections for the various neutrino sources. In general, such cross sections for inverse β-decay can be deduced rather reliably from the ft values resulting from the β-decay characteristics of the

Table 6.1.2. *Radiochemical and geochemical solar neutrino detection schemes*

Reaction	$E_{thresh.}$ (keV)	Half-life	Major contributing ν-type	SSM prediction (SNU)[#] 1σ-error	Target, form and size	Detection method	Collaboration	Project status
Radiochemical experiments								
^{7}Li(ν_e, e$^-$)^{7}Be	862	53.3 d	^{8}B-ν	≈60 ± 10	Li-metal, ≈10 t (9 t ^{7}Li)	Low-T calorimeter [GAL 97]	Moscow/Genova [KOP 97]	Pilot phase
^{37}Cl(ν_e, e$^-$)^{37}Ar	814	35 d	^{8}B-ν	$9.3^{+1.2}_{-1.4}$[&]	615 t C$_2$Cl$_4$ (133 t ^{37}cc)	Gas proportional counter	HOMESTAKE [CLE 97a]	Running since 1970
^{71}Ga(ν_e, e$^-$)^{71}Ge	233	11.43 d	pp-ν	137^{+8}_{-7}[&]	100 t GaCl$_3$-solut. (12 t ^{71}Ga)	Gas proportional counter	GALLEX/GNO [HAM 96]	Running 1991–97/ 1998–
^{71}Ga(ν_e, e$^-$)^{71}Ge	233	11.43 d	pp-ν	137^{+8}_{-7}[&]	up to 56t Ga-metal (22.3 t ^{71}Ga)	Gas proportional counter	SAGE [ABD 97]	Running since 1990
^{127}I(ν_e, e$^-$)^{127}Xe*	789	36.4 d	^{7}Be-ν(?)	≈50 (?)	235 t NaI-solution (100 t ^{127}I)	Prop. counter + NaI γ-coin.	U. Pennsylvania [CLE 97b]	Starting 1997
^{131}Xe(ν_e, e$^-$)^{131}Cs	352	9.7 d	^{7}Be + ^{8}B-ν	45^{+12}_{-6}	1000 t liquid Xe (210 t ^{131}Xe)$^\alpha$	Si ionization semiconductor	Kiev [GEO 97]	Very early ideas
Geochemical experiments§								
^{81}Br(ν_e, e$^-$)^{81}Kr	471	2.3 × 10^5 a	^{8}B-ν	31^{+20}_{-14}[&]	10 Kt old K-salts (50 t ^{81}Br)$^\alpha$	Res. Ioniz. Mass Sp. [HUR 84]	MPI Heidelberg [KIR 78]	Parked after pilot studies
^{205}Tl(ν_e, e$^-$)^{205}Pb	43	1.5 × 10^7 a	pp-ν	≈270 (?)[&]	3 t lorandite (TlAs$_2$S$_2$)$^\alpha$ (1 t ^{205}Tl)	Accel. Mass Spectr. (AMS)	LOREX [PAV 88]	Some preparations

& [BAH 95] (explicit, or implicit through update of [BAH 89]).
$^\alpha$ envisioned target size, so far without funded substance.
SNU = Solar Neutrino Unit = 1 ν-capture per 10^{36} target atoms per second.
§ work on ^{98}Mo(ν_e, e$^-$)^{98}Tc is not listed since the scheme seems not feasible. Respective attempts have failed.

product nucleus. However, this applies to ground state transitions only and excited state contributions to the production rate may cause large uncertainties. They may be significantly reduced if the Gamov–Teller strength of such transitions is inferred from (p, n) forward-angle scattering experiments [RAP 85]; however, there remain both principle and experimental problems, and uncertainties may be large in individual cases (see Table 6.1.2).

Availability and affordability of multiton quantities of the relevant target isotope.

Radiochemical purity of the target to avoid interfering side reactions. This refers in particular to U, Th, and their decay chains since fast neutrons (from (α, n) and from U fission) produce, via (n, p), MeV protons, which, in turn, cause (p, n) reactions on the target, leading to the same product nuclei as neutrino capture. *Suitable experiment location.* The dangerous protons can also be produced as secondaries from U, Th in the experimental environment (target tank, rock walls), and cosmic rays. This is why all solar neutrino experiments must be performed deep underground to depress the muon-induced production rate to an acceptable or negligible level.

Feasibility of chemical separation of the few product nuclei from the target. A common way is to purge the target if the reaction product is in a volatile chemical form. For extraction yield determination, a measurable quantity of inactive carrier can be added to the target.

Existence of a feasible detection technique for the electron-capture decay of the few product nuclei. Commonly, Auger electrons and/or X-rays are detected in proportional counters containing the product in the counting gas. Low temperature calorimeters may become an alternative.

Lifetime of the product nuclide. Optimal are weeks to months. Too short half-lives conflict with the time required for the chemical extraction; longer lifetimes reduce the specific activity in the product detection.

(b) Geochemical detectors Here, one accepts even very long lifetimes of neutrino-capture products (6.1.5) but measures the product nuclei themselves rather than their decay. This technique is applied to natural minerals from deeply buried ore deposits that are rich in suitable target isotopes. They must be sufficiently shielded from cosmic rays over geological times exceeding the mean life τ of the neutrino-capture product. In saturation, the number of product atoms N_p becomes $N_p = \bar{P}\tau$ where $\bar{P}$ is the mean production rate over the last $\sim 3\tau$ years. This opens up the possibility of extracting information about the mean neutrino flux in the past millions of years with relatively moderate target quantities (kilograms).

One depends, however, on favorable ore deposits that by virtue of nature are free of U, Th and are deeply shielded. Ways to detect the long-lived product nuclei are various forms of advanced mass spectrometry (MS, AMS [accelerator MS], RIMS [resonance ionization MS]). The more promising cases for geochemical solar neutrino detection are listed in Table 6.1.2.

(c) Real-time detectors While radiochemical detectors can only measure a rate, real-time detectors yield time, energy, and eventually also the direction of individual events. The principal reaction modes for real-time detection of solar neutrinos are

charged-current reaction:

$$^AZ(\nu_e, e^-)^A(Z+1)^* \xrightarrow{\gamma} {}^A(Z+1) \ [e^-, \gamma \text{ coincidence}] \tag{6.1.6}$$

$$(\text{or } ^2\text{H}(\nu_e, e^-)pp) \tag{6.1.6a}$$

neutral-current reaction:

$$^AZ(\nu_x, \nu_x)^AZ^* \xrightarrow{\gamma} {}^AZ \tag{6.1.7}$$

$$(\text{or } ^2\text{H}(\nu_x, \nu_x)pn; \text{ detection e.g. via } {}^{35}\text{Cl} + n \rightarrow {}^{36}\text{Cl} + \gamma \tag{6.1.7a}$$

neutrino–electron scattering (directional):

$$e^-(\nu_x, \nu_x)e^- \qquad (\sigma(\nu_e) \sim 6 \cdot \sigma(\nu_\mu)). \tag{6.1.8}$$

In principle, Čerenkov detectors, scintillation detectors, or fine-grained imaging drift chambers could all be used. The principal limitation of such devices is the high background rate as one shifts the event acceptance threshold to lower and lower energies. In 1987, the Kamiokande group [HIR 87] demonstrated for the first time that it is indeed possible to reduce the intrinsic contamination of a real-time (Čerenkov) detector to a level at which it becomes feasible to observe neutrino-induced recoil electrons at energies as low as ~ 8 MeV, that is, within the solar ^{8}B-neutrino energy range. Table 6.1.3 lists the real-time detectors existing at present or under consideration.

(d) Cryogenic and bolometric detectors In the long run, completely new techniques may become available for ultimate solar neutrino spectroscopy using cryogenic and bolometric detectors with extreme energy resolution. The principle is to measure the energy of recoiling electrons or even of recoiling nuclei from coherent neutrino scattering. Small energy deposits lead to phonon excitation of solid granules or crystals in the super-conducting state or to increasing temperature in dielectric crystals at very low temperature. Various ways to read out such events are under study (see 6.1.1.6.2).

6.1.1.4 Solar neutrino observations

So far, five experiments have actually measured solar neutrinos: The radiochemical Homestake chlorine detector [CLE 97a]; the real-time Kamiokande and

Table 6.1.3 *Real-time solar neutrino detection schemes*

Reaction	Effective ν-energy threshold [MeV]	Major contributing ν-type	Target form	Detection method	Size (fiducial) [tons]	Rate# (approx.) [events/d]	Collaboration or proposing institute	Time of operation or status of project
$e^-(\nu_x, \nu_x)e^-$	7.0	^{8}B-ν	Water	Cerenkov	680	0.4 (actual)	KAMIOKANDE [SUZ 97a]	1986–1995
	6.5 → 5.0	^{8}B-ν	Water	Cerenkov	22 500	15–30 (actual)	SUPER-KAMIOK [SUZ 97b]	1996–
	0.4	^{7}Be-ν	Pseudocumene	Liquid scintillator	100	50 (SSM)	BOREXINO [FEI 97]	2000
	0.2	pp-ν	He gas (5 atm, 77 K)	Time projection chamber	8ᵃ	7 (SSM)	HELLAZ [TAO 97]	≫ 2002
	0.2	pp-ν	CF$_4$ gas (1 atm)	Time projection chamber	7ᵃ	10 (SSM)	SUPER-MuNu [BRO 97]	≫ 2002
	0.2	pp-ν	Liquid He	Cryogenic roton detection	10ᵃ	20 (SSM)	HERON-[BAN 95]	> 2000
^{2}H(ν_x, e^-)pp	6.5	^{8}B-ν	Heavy water	Cerenkov + n-detectors (^{3}He-counters or ^{35}Cl-capture)	1000	25 (SSM)	SNO [MEI 97]	1998–
^{2}H(ν_x, ν_x)pn	2.2					5 (SSM)		

All the following reactions can in principle also exploit: $e^-(\nu_x, \nu_x)e^-$ and neutral-current excitations of type $T(\nu_x, \nu_x)T^* \to T + \gamma$ (T = target nuclide)

Reaction			Substance	Detector type	Size	Rate	Lab [ref]	Status
$^7Li(\nu_e, e^-)^7Be^{\&}$	≈0.5	pep-ν	LiF crystals	Cryogenic thermal bolometer	4^α	0.1 (SSM)	AT&T Bell Lab [RAG 93] Univ. of Maryland [CHA 94]	≫2000 Just ideas
$^{11}B(\nu_e, e^-)^{11}C^*$	6	^{8}B-ν	LiI(Eu) crystals Trimethylborate	Crystals scintillator Liquid scintillator	240	1.6 (SSM)	BOREX [RAG 88]	Parked
$^{19}F(\nu_e, e^-)^{19}Ne^*$	≈5	^{8}B-ν	Hexafluorobenzene	Liquid scintillator	600^α	10 (SSM)	INR [BAR 94]	Just ideas
$^{40}Ar(\nu_e, e^-)^{40}K^*$	≈5	^{8}B-ν	Liquid Ar	Time projection chamber	$600\cdot x^\alpha$	$1\cdot x \ (x=1\dots10)$	ICARUS [RUB 96]	1999–
$^{71}Ga(\nu_e, e^-)^{71}Ge^{\&}$	0.23	$pp\nu$	GaAs crystals Ga single crystals	Ionization semicond. (20°C) cryogenic thermal detector	125^α 60^α	2 2	INR + LANL [BOW 96] Oxford, others	Just ideas Just ideas
$^{81}Br(\nu_e, e^-)^{81}Kr^*$	0.47	^{7}Be-ν	NaBr or CsBr crystals	Cryogenic thermal bolometer	100^α	0.7	Milano University [ALE 95]	≫2002
$^{115}In(\nu_e, e^-)^{115}Sn^*$	0.12	pp-ν	InSb crystals	Superconducting tunnel junct.	4^α	1 (SSM)	Oxford [SWI 94]	≫2002
$^{176}Yb(\nu_e, e^-)^{176}Lu^*$	0.45	^{7}Be-ν	Dissolved in scintill.	Liquid scintillator	10^α	0.5 (SSM)	AT&T Bell Lab [RAG 97]	Just ideas

#SSM = Standard Solar Model. Obviously, actual rates may differ. The latter are quoted where experimental results already exist.

&no coincidence scheme available.

αenvisioned detector size, so far without funded substance.

Table 6.1.4a. *Neutrino capture rates in ^{37}Cl and ^{71}Ga [SNU]*

ν Source	Prediction						Total expected [BAH 95]	Measured	Ref.
	pp	pep	^{7}Be	^{8}B	^{13}N	^{15}O			
^{37}Cl	—	0.22	1.24	7.36	0.11	0.37	$9.3^{+1.2}_{-1.4}$	2.56 ± 0.22	[CLE 97a]
^{17}Ga	69.7	3.0	37.7	16.1	3.8	6.3	137^{+8}_{-7}	76 ± 8	[KIR 97]

Note: The error quoted by [BAH 95] for Cl-capture rate is $\approx 15\%$. However, Standard Solar Model calculations by others predict substantially lower ^{8}B neutrino fluxes and, therefore, lower production rates (e.g., 6.4 SNU [TUR 93], 7.6 SNU [SAC 90]). Hence, the systematic uncertainty for the Cl-rate prediction is at least 30% (note that this does not apply to gallium).

Super-Kamiokande water Čerenkov detectors [HIR 87; SUZ 97a]; and the GALLEX [KIR 90; 97] and SAGE [ABD 97] radiochemical gallium detectors.

Chlorine experiment: (^{37}Cl (v_e, e^-)^{37}Ar) In the Homestake gold mine, 380 000 liters of perchlorethylene (C_2Cl_4) are exposed in a large tank below 4 100 m.w.e. of shielding. Every few months, the radioactive ^{37}Ar (half-life 35 days) produced via ^{37}Cl(v_e, e^-)^{37}Ar is purged out of the target, collected on a charcoal trap, purified, and admixed to the counting gas of small gas-proportional counters. Measured are the Auger electrons resulting from the e^--capture decay of ^{37}Ar.

The energy threshold for neutrino capture on ^{37}Cl is 814 keV (see Fig. 6.1.2). Consequently, the Cl detector is blind for the most abundant (91 percent) pp neutrinos (maximum energy 420 keV), but it can detect ^{8}B and ^{7}Be neutrinos from the PPIII and PPII branches.

Since only a few decays can be expected per Ar extraction, the experiment requires ultimate refinement in the extraction and counting techniques and in the reduction, recognition, or exclusion of backgrounds. To compare the measured rates with expectation, the neutrino fluxes must first be converted into predictions for the capture rate in ^{37}Cl.

In the case of ^{37}Cl, the capture rate is (fortunately) dominated by the transitions to the isobaric analog state near 5 MeV. Therefore, the transition strength can be deduced from the ^{37}Cl-β-decay properties. Overall, the error of the capture rate due to uncertainties in the ^{37}Cl-capture cross section is < 10 percent [BAH 88]. The resulting production rates for the various neutrino sources are given in Table 6.1.4a.

We note that by far the largest contribution (79 percent) is due to ^{8}B neutrinos. Nevertheless, we must bear in mind that owing to the nonnegligible contribution from ^{7}Be neutrinos, the SNU values measured in the Cl experiment cannot be directly converted into ^{8}B neutrino fluxes. The total expected capture rate for the chlorine detector is $9.3 \pm^{1.2}_{1.4}$ SNU (1σ) whereby 90 percent of the total error is associated with ^{8}B neutrinos. The Homestake data record covers the periods 11/ 1970 to 1985 and, after an interruption and transfer of custody from Brookhaven

National Laboratory to Pennsylvania University (Davis and Lande), from 10/1986 to 1994, altogether 108 runs.

The data must be evaluated recognizing the statistics of small numbers. In this respect, the overall distribution of the results from the individual runs is consistent with the assumption of a production rate constant in time, justifying the deduction of a mean value. This yields an ^{37}Ar-production rate of 0.484 ± 0.042 atoms/d (1σ). This corresponds to 2.56 ± 0.22 SNU (1σ). The difference between this figure and the SSM prediction of 9.3 ± 1.3 SNU (1σ) constitutes the first solar neutrino problem (SNP) (see Table 6.1.4b). It signals a significant deficit of ^{8}B neutrinos relative to expectation.

The consistency of the data with a time-constant neutrino flux does, of course, not per se rule out another behavior. Frequently, fluctuations likely to be of statistical nature have been suggested to indicate physical effects. For instance, the earlier data had suggested an apparent anti-correlation of the neutrino flux with the solar activity (see e.g. [AKH 97]). If significant, this could imply a modification of the solar neutrino flux in passing through the outer sun, or a completely different origin of the measured rate in the complete absence of ^{8}B neutrinos. Okun [VOL 86] has suggested a mechanism that, at least in principle, could establish a causal connection for what seems totally mysterious at first glance, namely, the interaction of time-variable magnetic fields in the outer convective zone with a hypothetical magnetic moment of the neutrino in the order of $10^{-10}\,\mu$Bohr. Yet, existing limits on the magnitude of the neutrino magnetic moment from the SN 1987A supernova neutrino observation already seem to rule out this mechanism.

Kamiokande and Super-Kamiokande Čerenkov detectors   These real-time detectors are based on the registration of the Čerenkov light cones produced by recoil electrons resulting from neutrino–electron scattering in water. Kamiokande had originally been installed, below 2700 m.w.e. of shielding in the Kamioka mine in Japan, in order to observe higher energy phenomena such as proton decay ($\approx$ GeV). Even though the Kamioka group succeeded in reducing the intrinsic detector contamination (U, Th series) quite dramatically, the task of recognizing directional solar neutrino-induced events remains formidable and requires extremely good track recognition criteria and timing. This may be illustrated by the fact that the total trigger rate exceeds the signal by factors of 10^5 to 10^6. After ten years of successful solar neutrino recording, Kamiokande was in 1996 replaced by the dedicated Super-Kamiokande detector, having a 33-fold fiducial volume (see Table 6.1.3). It consists of 50 kilotons of water from which the innermost 22.5 kt are used as fiducial volume. The water tank is lined with a dense network of large photomultiplier (PM) tubes. The energy and the vertex of an event are reconstructed from the number, orientation, and relative time of the PM cells hit by the Čerenkov light cone. An auxiliary LINAC is used for energy calibration. The

Table 6.1.4b. *Observed solar neutrino deficits δ (in percent)*

Experiment	Deficit $\delta = (1 - \text{signal/prediction}) \cdot 100\%$		R
HOMESTAKE	$(72 \pm 12)\%$	(2σ)	3.6
SUPER-KAMIOKANDE	$(63 \pm 14)\%$	(2σ)	2.7
GALLEX	$(44 \pm 14)\%$	(2σ)	1.8

The reduction factor $R = \text{prediction/signal} = 100/(100 - \delta)$.
[$\sigma_{\text{theor.}}$ and $\sigma_{\text{experim.}}$ are added in quadrature].

spectrum of the recoil electrons reflects the initial neutrino spectrum in a well-understood fashion.

The results from Kamiokande and Super-Kamiokande confirm the deficit of ^{8}B-neutrinos as indicated by the Cl-experiment. The measured ^{8}B-ν flux is $(2.44 \pm ^{0.26}_{0.11}) \cdot 10^6 \, \text{cm}^{-2} \text{s}^{-1}$, only 37% of the SSM-predicted flux [SUZ 97b]. The reduction is significant, but less than in the Homestake experiment.

This difference has been called "the second solar neutrino problem".

Gallium experiments (^{71}Ga(ν_e, e^-)^{71}Ge) Radiochemical gallium experiments are of outstanding importance and significance because of the low energy threshold of 233 keV. Consequently, the expected neutrino-capture rate on ^{71}Ga is dominated by *pp* neutrinos (see Table 6.1.4a). For many years to come this will remain the only realistic possibility for measuring the crucially important *pp*-neutrino flux. The SSM prediction for *pp* and *pep* neutrinos on Ga has the highest reliability and the smallest error ($< 2\%$) among all neutrino detection schemes. This branch is the major contributor to the total expected SSM production rate of 137 ± 8 SNU [1σ]. Two experiments have produced data since 1990, the GALLEX experiment [KIR 97] in the INFN Gran Sasso Underground Laboratory (Italy) and the SAGE experiment [ABD 97] in the INR Baksan Valley Neutrino Laboratory (Caucasus).

The **GALLEX** collaboration uses 30.3 tons of gallium in an ultra-pure aqueous 8-normal galliuchloride solution. This facilitates Ge extraction as volatile $GeCl_4$ by gas purge of the target [HEN 97]. Crucial experimental steps such as extraction, conversion of $GeCl_4$ into GeH_4 (suitable as counting gas), and ultra low-level counting with full pulse shape analysis for counter background rejection are controlled and verified:

→ regularly in each run:
  by determination of the recovery of ≈ 1 mg inactive Ge-carrier isotope (routinely $> 97\%$ recovery).
→ in special experiments:
  (a) neutrino-induced ^{71}Ge production by means of a man-made Megacurie ^{51}Cr low energy neutrino source ($[93 \pm 8]\%$ of expected signal) [HAM 96].

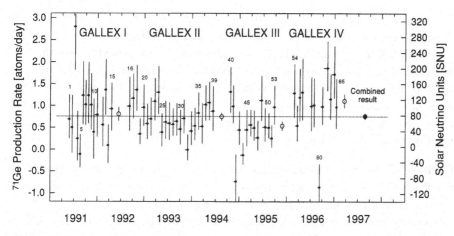

Fig. 6.1.3. Results of the individual runs of the GALLEX detector [KIR 97], [HAM 98]. The overall result is 76 ± 8 SNU (1σ).

(b) in-situ ^{71}Ge production by β-decay of ^{71}As spikes added to the gallium target ($[100 \pm 3]\%$ of expected signal) [KIR 97].

These successful performance tests lend credit to the measured solar neutrino induced production rates. In six years of operation, 65 runs have been performed and a production rate of 76 ± 8 SNU has been obtained (Fig. 6.1.3). This is 56% of the SSM prediction (see Table 6.1.4b). In spite of the obvious deficit ("third solar neutrino problem", see below), the GALLEX data constituted the first observational proof of hydrogen fusion in the solar interior [ANS 92].

The **SAGE** experiment is performed with up to 56 tons of metallic gallium, using a different Ge-extraction scheme but otherwise techniques similar to those used in GALLEX. After early reports of a $\approx$ zero signal, the SAGE result has now also approached the 70–80 SNU level. The latest value of 74 ± 12 SNU [ABD 97] is in good agreement with the GALLEX result.

6.1.1.5 Status of the Solar Neutrino Problem

Because of the far-reaching consequences, the SNP has stimulated exceptional theoretical efforts to provide an adequate explanation. One possible solution of the Solar Neutrino Problem could be neutrino oscillations (see Section 6.1.2).

If this were the cause of the SNP, it would imply that a nonzero mass of the neutrino has been positively detected, with all its consequences for lepton–quark symmetry and grand unification. Before one draws such far reaching conclusions, astrophysical solutions to the SNP must be definitely excluded.

Before the solar neutrino enigma, there was little doubt that the SSM properly describes the sun, a typical main sequence star. In particular, the ^{8}B-neutrino deficit does not seem to be dramatic since the ^{8}B-neutrino flux is proportional to the 18th power of the central temperature. Hence, all that is needed is a reduction of T_C from 15.8 million K to ~ 15 million K. However, as it turns out, in the frame of the SSM this cannot be achieved without severe interference with the fundamentals. As a consequence, many nonstandard models have been proposed to explain the SNP. Most nonstandard models are especially tailored just to reduce the ^{8}B-neutrino flux by various mechanisms lowering the central temperature. For illustration, we mention some of the ad hoc possibilities:

(a) Replace thermal pressure by magnetic or centrifugal pressure [BAH 82b].
(b) Replenish fresh hydrogen into the solar core by turbulent diffusion to maintain the luminosity at low T_C [SCH 85].
(c) Assume inhomogeneous accretion of the sun as cause for lower heavy element abundances (e.g., C, N, O, Fe) in the interior as compared to the surface. This lowers the opacity and therefore also T_C [BAH 71].
(d) Modify the energy transport by means of weakly interacting massive particles (WIMPS) [FAU 85].

To decide between the two principal alternatives for explaining the SNP, experiments measuring the pp- (and ^{7}Be) neutrino flux and the *spectral shape* of ^{8}B neutrinos are most diagnostic. The reason is that astrophysical modifications would hardly alter the pp-flux or the spectral shape of individual neutrino sources, quite distinct from the potential effects of neutrino oscillations.

With the establishment of the ^{7}Be neutrino deficit in GALLEX and with the increasing quality of helioseismological data [PAT 97], many nonstandard models can already be ruled out, only "helioseismologically constrained solar models" (HCSM) should be considered in the future.

From the data summarized in Table 6.1.4, we note that

– *all* solar neutrino experiments observe solar neutrinos above zero signal
– *all* solar neutrino experiments observe fewer solar neutrinos than predicted from stellar theory.

The depression factors are variable, apparently related to the different energy thresholds of the experiments. For ^{8}B-dominated experiments, deficits are high, but theoretical uncertainties are large. For the pp-dominated gallium experiments, the deficit is smaller, while the theoretical prediction is rather precise.

With shrinking errors in GALLEX, the third solar neutrino experiment has also created the third Solar Neutrino Problem [ANS 95; KIR 95]. It consists of the apparent absence of most or all ^{7}Be neutrinos, the second largest expected contributor to the Ga-signal. The measured rate in GALLEX is almost exactly what is expected from the PPI-cycle alone ("minimal model": pp-neutrinos just to account for the solar luminosity), hence there is little space for the sizeable

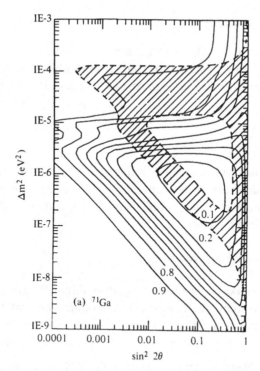

Fig. 6.1.4 Neutrino mixing parameters and domains of sensitivity for neutrinos (antineutrinos) from accelerators, reactors, cosmic ray–atmosphere interactions, and from the sun. The combined results for pp and ^{7}Be-neutrinos (GALLEX, SAGE) and for ^{8}B-neutrinos (HOMESTAKE, KAMIOKANDE+SUPER-KAMIOKANDE) can be consistently explained by the "MSW-solutions" as a consequence of energy dependent ν_e-reduction factors due to flavor conversions.

contribution expected from ^{7}Be. ^{8}B neutrinos are not distinct in the Ga signal, their direct contribution is small. However, ^{8}B neutrinos are made from ^{7}Be, hence their (partial) presence as seen in Kamiokande requires the precursor ^{7}Be to also be present, at least in that proportion, but this is not seen. ^{7}Be is more reduced than its offspring, ^{8}B.

Contrary to PPIII (^{8}B), the 14% PPII (^{7}Be) branch is much more robust and there are not many "astrophysical" possibilities to significantly affect it. This is why it becomes increasingly probable and eventually at some point even inescapable to conclude ν_e-disappearance for reasons other than those related to the sun.

Figure 6.1.4 depicts in a plot of parameter pairs (Δm^2, $\sin^2 2\theta$) the domains of sensitivity for different kinds of neutrinos. Taking all solar neutrino data together, the most favored MSW-solution ("small angle solution"; see Section 6.1.2) corresponds to $\Delta m^2 \approx 7 \cdot 10^{-6}$ and $\sin^2 2\theta \approx 5 \cdot 10^{-3}$, a much smaller range than otherwise accessible.

6.1.1.6 Future experiments[1]

The main goals for future solar neutrino experiments are to measure the full solar neutrino spectrum, to solve the SNP, and, if applicable, to determine the actual values of Δm_i^2, θ or to exclude a large part of the presently unexplored parameter space.

6.1.1.6.1 Near future

The years 2000–2005 will be a rather active period of solar neutrino exploration. The presently *running* experiments will continue to operate, or even be upgraded. In addition, four *new* detectors will definitely go into operation, each of them with at least some new specific quality.

Continuation of running experiments

– **Chlorine at Homestake** has by far the longest continuous solar neutrino flux record and will continue to take data. One envisioned modification would be to collect ^{37}Ar produced at day and at night separately. This is to look for the MSW flavor reconversion of neutrinos penetrating the earth before they reach the detector.

– **GNO** (Gallium Neutrino Observatory) at Gran Sasso is the follow-up project of **GALLEX** to provide a long time record of low-energy neutrino observations from 1998 onwards. It is intended to reduce the total error to 5% for a close examination of the time constancy of the *pp* neutrino flux during a whole solar cycle. Upscaling up to 100 tons of gallium is envisioned. Also, a new cryogenic calorimeter method for ^{71}Ge detection with superior efficiency is being developed at Munich [ALT 97]. Ge is deposited as metal between two absorber crystals connected with superconducting phase transition thermometer (SPT) readouts. The combination of more gallium and higher efficiency would further reduce the errors. Under favorable circumstances, ν_e-disappearance could be proved, assumption free, from just this one experiment.

– **SAGE** is planned to operate for a long time. The rather unstable general situation in Russia has required heroic efforts to keep the experiment running.

– **SUPER KAMIOKANDE** has succeeded **KAMIOKANDE** in 1996 (Table 6.1.3) [SUZ 97b]. As time goes on, it will become the first high statistics solar neutrino experiment in the years to come. Daily, seasonal, and overall variations can be recorded and a statistically significant Kurie-plot of the ^{8}B-neutrino spectrum should be obtained. This will be the earliest test of the small angle MSW solution of the solar neutrino problems.

---

[1] To comprehend this section, consultation of Tables 6.1.2 and 6.1.3 is essential since repetitions from the tables in the text are minimized.

Experiments in an advanced state

- **Iodine at Homestake** (Table 6.1.2) started to take data in 1997 with the first 100 tons of iodine [CLE 97b]. In many aspects this radiochemical experiment resembles the Chlorine Experiment. It has a much higher, but not well known, production rate and benefits from an EC/γ-coincidence in the ^{127}Xe decay. From the beginning the experiment is operating in the alternating day/night extraction mode.

- **SNO** (Table 6.1.3) is a 1 kt heavy water real-time Čerenkov detector for ^{8}B neutrinos. It is under construction at the Creighton mine in Sudbury, Canada. After the crucial background problems were in principle solved, data taking started in 1998 [MEI 97]. The main goal of SNO is to measure not only the spectral shape of ^{8}B neutrinos > 6.5 MeV in the charged-current reaction, but also the ratio $\Phi\nu_e/\Phi\nu_{\text{all}}$, that is, to search for neutrino oscillations not only through disappearance of ν_e but in effect also through appearance of ν_μ and ν_τ. This is possible since the disintegration of the deuteron into its nucleons is a neutral-current reaction and occurs irrespective of neutrino flavor. In this case, the event signature is a 6.25 MeV γ followed by the detection of the neutron, either by capture on added ^{35}Cl or by means of ^{3}He counters.

- **BOREXINO** (Table 6.1.3) is a 100 ton liquid scintillation detector devoted specifically to the detection of ^{7}Be-neutrinos through neutrino–electron scattering. There is no boron in **BOREXINO**, the name is historical since the project developed from the earlier planning on BOREX (see below). Electron–neutrino scattering is dominated by ν_e since the cross section for the charged-current reaction is about six times that for neutral-current scattering of ν_x. Scintillation detectors have superior light output and better energy resolution compared to Čerenkov detectors. They can operate at much lower energy if and only if the internal background can be reduced to acceptable levels. This requires extreme radiopurity, especially for ^{14}C and nuclides from the U, Th decay series. In a pilot experiment (CTF = counting test facility) it has been shown that backgrounds can be kept at the required level down to a neutrino energy threshold of ≈ 0.5 MeV, sufficient to detect the flux of 862 keV ^{7}Be-neutrinos (Fig. 6.1.2) [FEI 97]. This flux would be rather high if the SSM applies. If it were found to be strongly reduced or absent, this would confirm the ^{7}Be-problem encountered in the gallium experiments (see Section 6.1.1.5). The detector is now under construction and planned to go into operation in the year 2000.

- **ICARUS** (Table 6.1.3) is a fine-grained multi-purpose liquid argon drift chamber which can also be used to detect solar ^{8}B-neutrinos above ≈ 6 MeV. It is sensitive to both c.c. and n.c. reactions and to $e-\nu$ scattering. A first 600 ton module is in preparation at Gran Sasso [RUB 96], the final goal is a 6 kt detector. It should register the ^{8}B-neutrino spectrum with high resolution in a relatively short time of running.

6.1.1.6.2 Far future

In the following we describe interesting conceptions and ideas of more or less promising new solar neutrino experiments. Intense work of many years is involved in some of them, others are at present just 'good ideas'. Common to all of them is that at this time one can not judge if or when they could eventually become real solar data taking experiments, since either the physics (mostly background), the technology (large scales), the funding, or all of them are unsolved. Notwithstanding, some of these proposals may sometime later play an important role, and the development of at least one real time *pp*-neutrino detector is an absolute must for the future of solar neutrino spectroscopy.

Radiochemical **Lithium** experiment (Table 6.1.2) [KOP 97]. Early studies of the system ^{7}Li (ν_e, e^-) ^{7}Be by Davis were aimed at the organic solvent extraction of Be from liquid Li metal but the unsolved problem was the detection of ^{7}Be. The only detectable radiation emitted from ^{7}Be is 43 eV Auger electrons. Now the Genoa group has made remarkable progress with a low temperature micro-calorimeter [GAL 97] and joined with the INR group to rejuvenate the project. A detector as small as 10 tons of Li could be sufficient. With respect to energy, the expected signal would be rather unspecific since none of the ν-sources above the threshold of 0.86 MeV is clearly dominating the production rate.

Radiochemical liquid **Xenon** experiment (Table 6.1.2) [GEO 97]. The idea is to extract by ion collection techniques ^{131}Cs from ^{131}Xe$(\nu_e, e^-)^{131}$Cs out of 1 kt of liquid xenon and to detect it with semiconductors. ^{7}Be and ^{8}B neutrinos would contribute in about similar quantities.

We may remark here that for ^{8}B neutrino detection the radiochemical method has been overtaken by present 'state of the art' real-time detector technology. In view of the experimental errors, it is an artificial argument that data from radiochemical detectors (sensitive only to c.c.) can serve to distinguish from signals obtained with real-time detectors (sensitive to c.c. and n.c.).

With respect to *geochemical* experiments (Section 6.1.1.3) there have been some discouraging experiences (Br–Kr, Mo–Tc, see Table 6.1.2). At present, little activity exists in this field. The only exception might be:

LOREX (^{205}Tl $(\nu_e, e^-)^{205}$Pb) (Table 6.1.2). Here one tries to exploit the potential of this very low threshold reaction using the mineral lorandite (TlAs$_2$S$_2$) from the unique Alchar mine in Macedonia by means of AMS-measurements of ^{205}Pb/^{204}Pb ratios of $\approx 10^{-14}$ [ERN 84]. This could in principle yield the average *pp*-neutrino flux during the past few million years, a very attractive goal. Unfortunately, however, the theoretical production rate prediction is very uncertain. The Tl ore (worldwide the only sizable Tl-ore deposit) is insufficiently shielded from cosmic rays (≈ 300 m.w.e. only); hence cosmic ray produced ^{205}Pb competes with neutrino-produced ^{205}Pb. Crucial for the future of this experiment is whether the lorandite mineralization continues to larger shielding depths. For this, the (abondoned) mine must be reopened and explored by drilling.

The **HELLAZ** project aims for real time *pp*-neutrino detection by ν–e scattering (Table 6.1.3) using a high pressure, low temperature He time projection chamber [TAO 97]. Extreme radiopurity is required for an effective threshold energy as low as 200 keV, helium is favorable in this respect. Some lab-scale testing has been performed, but the realization of the envisioned 10 ton detector is rather far away.

An approach quite similar to HELLAZ but using some advantages of gaseous tetrafluorocarbon instead of helium is investigated in the **Super-MuNu** project (Table 6.1.3) [BRO 97]. Such a TPC is primarily developed to determine the magnetic moment of the neutrino, but it has also the potential for solar *pp*-neutrino detection. As in all these ambitious low-threshold experiments, success depends on the practical solution of the background problems in the full scale detector. Upscaling from small models by more than a factor of five in one step is highly questionable.

BOREXINO is the only actively pursued **liquid scintillation detector**, but the attractive feature of being able to simultaneously detect charged-current reactions (often with convenient coincidences), neutral-current reactions, and ν–e scattering has led to the consideration of quite a few potential experiments, either with the target isotope being part of the scintillator (e.g. hexafluorobenzene in the **Fluorine** experiment ($^{19}F(\nu_e, e^-)^{19}Ne^*$, Table 6.1.3, [BAR 94]) or by dissolving the target in the scintillator. In this way, with some knowledge of chemistry, almost any desired nuclide becomes feasible for investigation. The first such attempt was the **Indium Scintillator** experiment ($^{115}In\,(\nu_e, e^-)^{115}Sn^*$, [RAG 76]). The initial hope, to benefit from the low threshold of 128 keV to detect *pp*-neutrinos, failed because of the natural activity of ^{115}In, but with the electronic threshold raised to about 600 keV, In could still make a very good detector for ^7Be-neutrinos. However, such a project is not presently pursued since BOREXINO is more advanced for ^7Be-neutrino detection. BOREXINO (see above) is the stripped down version of the initial **BOREX** proposal (with ^{11}B in trimethylborate dissolved in the scintillator, Table 6.1.3, [RAG 88]) by settling for ν–e scattering only. Another recently proposed scheme, also with low threshold for ^7Be-neutrino detection, is the ^{176}**ytterbium** loaded scintillator (Table 6.1.3) [RAG 97], and the list is by no means complete. Even solid scintillator crystals such as **LiI(Eu)** [CHA 94] or ionization semiconductors such as **gallium arsenide** crystals (Table 6.1.3) [BOW 96] have been considered. At least some of these experiments seem feasible in principle, but whether any will be carried out is open. All solar neutrino experiments are expensive and time consuming projects. Assuming all other problems to be solved, only after full scale installation will one know for sure whether the backgrounds in a particular experiment are manageable.

For the far future, real technological breakthroughs can be hoped for from **cryogenic particle and quasi-particle detectors**.

The importance of spectral measurements of *pp*-neutrinos has motivated the **HERON** project (Table 6.1.3) [BAN 95]. This is planned as a ballistic roton detector using 10 tons of superfluid helium at 20 mK. Excitons from ν–e scattering have long ranges at very low temperature. Their action is transmitted to the surface of the extended liquid helium target. At the surface, rotons evaporate He atoms, which are detected by means of micro-calorimeters (Si wafer plus thermistor).

Bolometers: Owing to the very small specific heat at very low temperature, dielectric crystals (e.g. Si, Ge, Al_2O_3 [sapphire single crystals], alkali halogenides) experience a sizeable temperature rise upon energy deposition. This can be measured with a thermistor or a SPT. Promising advances are reported by the Milano group on **Sodium bromide** for neutrino capture on ^{81}Br (Table 6.1.3) [ALE 95]. Earlier, **Lithium fluoride** cryogenic bolometers had been proposed for inverse beta decay on ^{7}Li, in particular by *pep*-neutrinos (Table 6.1.3) [RAG 93], and even **Gallium single crystals** at low temperature may have some advantages over the above mentioned idea of gallium arsenide semiconductors.

Finally we mention the possibility of using real **superconductivity** for solar neutrino detection. Here one records with tunnel junctions the charge created by Cooper pair breaking due to (quasi)particle excitation in the superconductor. The best known project of this type is *pp*-neutrino detection in **superconducting Indium antimonide** (Table 6.1.3) [SWI 94]. Four tons of superconducting crystals are required for a capture rate of just one per day, obviously there is a long way to go for this type of detector.

6.1.2 Neutrino oscillations in the sun*

Three principal sources of neutrinos are expected to emerge from the central region of the sun as shown [BAH 88] in Table 6.1.5. Experiments up till now have been mainly sensitive to the high energy but rare ^{8}B ν, but future experiments will be sensitive to all three sources and thus to an effective range of energies from about 200 kev to 12 Mev.

If neutrino oscillations exist in nature, there is the possibility that the neutrinos will arrive at the detector as ν_μ or ν_τ rather than ν_e. Detectors based on inverse β-decay, such as the Davis ^{37}Cl detector, cannot detect ν_μ or ν_τ. Detectors based on neutrino–electron scattering, such as the Kamiokande H_2O Čerenkov detector, are about seven times less sensitive to ν_μ or ν_τ than to ν_e. Thus neutrino oscillations could have a drastic effect on solar neutrino experiments. Because of the large size and internal density of the sun and the large distance between the earth and the sun,

Table 6.1.5. *Principal sources of solar neutrinos*

Name	Reaction	Energy spectrum	Relative flux
p ν	$p + p \rightarrow d + e^+ + \nu_e$	Continuous to 420 kev	1
^{7}Be ν	$e^- + {}^7\text{Be} \rightarrow {}^7\text{Li} + \nu_e$	Line mainly 860 kev	0.08
^{8}Be ν	$^8\text{B} \rightarrow {}^8\text{Be} + e^+ + \nu_e$	Continuous to 14 Mev	10^{-4}

---

* L. Wolfenstein, Carnegie Mellon University, Physics Department, Pittsburgh, PA 15213.

solar neutrino experiments are sensitive to a range of neutrino masses and mixings that cannot be explored in laboratory experiments.

Neutrino propagation in vacuum for two generations is described by

$$i\frac{d}{dt}\begin{pmatrix} \nu_e \\ \nu_\mu \end{pmatrix} = \frac{1}{2E}\begin{pmatrix} M_{ee}^2 & M_{e\mu}^2 \\ M_{e\mu}^2 & M_{\mu\mu}^2 \end{pmatrix}\begin{pmatrix} \nu_e \\ \nu_\mu \end{pmatrix} \tag{6.1.10}$$

where E is the neutrino energy and the matrix M^2 is given by

$$M_{ee}^2 = \frac{1}{2}(\mu^2 - \Delta m^2 \cos 2\theta_v) \tag{6.1.11a}$$

$$M_{\mu\mu}^2 = \frac{1}{2}(\mu^2 + \Delta m^2 \cos 2\theta_v) \tag{6.1.11b}$$

$$M_{e\mu}^2 = \frac{1}{2}\Delta m^2 \sin 2\theta_v \tag{6.1.11c}$$

where θ_v is the vacuum mixing angle, $\mu^2 = m_1^2 + m_2^2$, and $\Delta m^2 = m_2^2 - m_1^2$. The probability that a beam originally ν_e remains ν_e after going a distance x (in our extreme relativistic approximation we use x and t interchangeably) is given by

$$|\langle \nu_e|\nu_e(x)\rangle|^2 = 1 - \sin^2 2\theta \sin^2\left(\frac{\pi x}{l}\right) \tag{6.1.12}$$

where $\theta = \theta_v$ and

$$l = l_v \approx 4\pi E/\Delta m^2 = \frac{2.5E(\text{MeV})}{\Delta m^2(\text{eV}^2)}\text{meters}. \tag{6.1.13}$$

The neutrinos coming from the sun form a continuous spectrum (except for the ^{7}Be neutrinos) and detectors typically average over a significant energy interval. As a result, considering only vacuum oscillations, if l_v is much less than the distance to the sun, the result of neutrino oscillations will be a reduction in the detected flux by a factor $(1 - \frac{1}{2}\sin^2 2\theta)$. This will be the case if $\Delta m^2 > 10^{-9}$ eV^2. If l_v just happens to be of the same order of magnitude as the earth–sun distance, then there could be a somewhat larger reduction factor [BOU 86]. It is also possible to get a larger reduction factor if there is a large amount of mixing among three types of neutrinos.

The most popular models [GRO 88] of neutrino mass are based on grand unified theories (GUTs) in which leptons and quarks are unified into a single grand fermion field [LAN 81]. It is natural in these models for the mixing of neutrinos to be similar to that of quarks. On that basis we do not expect θ_v much larger than the Cabibbo angle and so we do not expect $\sin^2 2\theta_v$ to be larger than 0.2. In this case vacuum oscillations will not produce any large changes in the solar neutrino flux arriving at the earth.

The picture of oscillations can be dramatically changed, however, when the effects of the material medium are taken into account [WOL 78]. For the case of ν_e

from the sun, large effects can occur if $m_2 > m_1$; that is, $M_{\mu\mu}^2 > M_{ee}^2$. This is the natural expectation in GUTs corresponding to ν_e being the lightest neutrino. It is now necessary to take into account in Eq. (6.1.10) the change in phase associated with the index of refraction of the neutrinos moving through matter. The index of refraction n is given by the optical theorem

$$k(n-1) = 2\pi N f(0)/k \qquad (6.1.14)$$

where $k \simeq E$ is the neutrino momentum, N is the density of scatterers, and $f(0)$ is the forward scattering amplitude. n and $f(0)$ are in general matrices in flavor space and the right-hand side of Eq. (6.1.14) should be added to the propagation matrix in Eq. (6.1.10). Assuming the neutrino scattering is described by the usual weak interaction theory, $f(0)$ is diagonal. Furthermore, an overall phase factor is of no interest for neutrino oscillations so that we need only consider terms in $f(0)$ that distinguish ν_e and ν_μ. These are only the charged-current terms corresponding to elastic ν_e scattering from electrons

$$2\pi N f(0)/k = -\sqrt{2}GN_e \qquad (6.1.15)$$

where N_e is the number density of electrons in the medium. (The elastic neutral-current scattering is flavor-independent if we neglect small higher-order corrections.) Thus in a medium we can use Eq. (6.1.10) provided we replace Eq. (6.1.11a) by

$$M_{ee}^2 = \tfrac{1}{2}(\mu^2 - \Delta m^2 \cos 2\theta_v) + 2\sqrt{2}GN_eE. \qquad (6.1.16)$$

For a fixed value of N_e, oscillations are described by Eq. (6.1.12) with θ, ℓ replaced by θ_m, l_m:

$$\tan 2\theta_m = \sin 2\theta_v/(\cos 2\theta_v - l_v/l_o) \qquad (6.1.17)$$

$$l_m = l_v[1 - 2(l_v/l_o)\cos 2\theta_v + (l_v/l_o)^2]^{-\frac{1}{2}} \qquad (6.1.18)$$

$$l_o = 2\pi/\sqrt{2}GN_e = (1.6 \times 10^7/\rho_e)\text{meters} \qquad (6.1.19)$$

where ρ_e is the electron number density in units of Avogadro's number. The value l_o defines a characteristic length in matter; for normal matter it is of the order of the earth's radius. For $l_v \ll l_o$, matter effects can be ignored; for $l_v \gg l_o$, oscillations are highly suppressed; of particular interest in the case in which l_v and l_o are comparable. In particular, for

$$l_v = l_o \cos 2\theta_v \qquad (6.1.20)$$

θ_m becomes 45°. The importance of this case was discovered by Mikhaeyev and Smirnov [MIK 86] who refer to it as the resonant amplification of neutrino oscillations by matter. For the case of $\bar\nu_e$, $\bar\nu_\mu$ the sign of the right-hand side of Eq. (6.1.15) is reversed, in effect changing the sign of l_o. Thus the resonance condition can be satisfied for $\bar{\nu}_e$ if $m(\nu_e) < m(\nu_\mu)$ but not for ν_e. For small values of θ_ν, Eq. (6.1.20) gives

$$E_\nu(MeV)\rho_e = 6.5 \times 10^6 \Delta m^2(eV^2). \qquad (6.1.21)$$

For the range of solar neutrino energies and values of ρ_e in the sun this can be satisfied for values of Δm^2 in the range 10^{-4} to 10^{-8} eV^2.

In fact, the solar density varies as the neutrinos move from the center to the edge of the sun so that Eq. (6.1.12) supplemented by Eqs. (6.1.17) and (6.1.18) is not appropriate. Of particular interest is the case in which the condition (6.1.21) holds in the solar interior away from the center or the edge. The ν_e is born approximately in the upper eigenstate of M^2 because the second term in M^2_{ee} in Eq. (6.1.16) dominates the matrix owing to the large value of N_e at the solar center. If θ_ν (and thus $M^2_{e\mu}$) is not too small, the adiabatic approximation can be used and ν_e will remain in this upper state throughout its journey through the sun. It will thus emerge at the surface $(x = R)$ in the state ν_2 which is mainly ν_μ so that

$$|\langle \nu_e | \nu_e(R) \rangle|^2 = \sin^2 \theta_\nu = \tfrac{1}{2}(1 - \cos 2\theta_\nu). \qquad (6.1.22)$$

For small values of θ_ν, the ν_e is almost completely converted to ν_μ. Note that in this approximation the neutrino emerges in a vacuum mass eigenstate so that no further oscillations take place on the way to the earth. It is possible, however, for a narrow range of Δm^2, that the resonance condition holds inside the earth so that significant additional oscillations may affect the flux detected at night [CRI 86; BAL 87].

In order for the adiabatic condition to hold, it is necessary that over the oscillation length l_m there be only a small change in θ_m. The rate of change of θ_m depends on the distance R_o over which there is a sizable density change; for most of the sun

$$R_o = \left[\frac{1}{\rho}\frac{d\rho}{dr}\right]^{-1} \approx 7 \times 10^7 \text{ meters.} \qquad (6.1.23)$$

The distance Δx over which there is large change in θ_m is found by differentiating Eq. (6.1.17)

$$\Delta x = R_o \tan 2\theta_\nu.$$

From Eq. (6.1.18) at resonance $l_m = l_\nu / \sin 2\theta_\nu$ so that the adiabatic condition $l_m \ll \Delta x$ yields

$$l_\nu \ll R_o \sin^2 2\theta_\nu / \cos 2\theta_\nu \qquad (6.1.24)$$

or, using the resonance condition Eq. (6.1.20),

$$l_o \ll R_o \tan^2 2\theta_\nu.$$

When the adiabatic condition fails, it is often possible [HAX 86; PAR 86a] to replace Eq. (6.1.22) by the Landau–Zener approximation

$$|\langle \nu_e | \nu_e(R) \rangle|^2 = \frac{1}{2} - \left(\frac{1}{2} - T \right) \cos 2\theta_v \cos 2\delta \qquad (6.1.25)$$

$$T = \exp \left[- \frac{\pi^2 \sin^2 2\theta_v}{2} \frac{R_o}{\cos 2\theta_v} \frac{R_o}{l_v} \right].$$

Here $((\pi/2) - \delta)$ is the value of θ_m at the point of production; if this point has a much higher density than needed for resonance, θ_m will be close to $\pi/2$ as assumed before; that is, ν_e is essentially in the upper state and $\delta \approx 0$. If, in addition, the adiabatic condition of Eq. (6.1.24) holds, $T \simeq 0$ and Eq. (6.1.25) reduces to the simple result (6.1.22).

Results of numerical calculations are presented in many papers [ROS 86; BOU 86; PAR 86b]. A typical result [BOU 86] for the ^{37}Cl detector which is mainly sensitive to ^{8}B ν is shown in Fig. 6.1.5. It is seen that for values of Δm^2 between 10^{-4} and 10^{-7} eV^2 there is a sizable reduction of the flux. The following features of the curve may be noted:

(1) As Δm^2 grows above 10^{-4} eV^2, the matter effect becomes unimportant because Eq. (6.1.21) cannot be satisfied even for the highest-energy neutrinos and the highest value of ρ_e.
(2) As Δm^2 falls below 10^{-7} eV^2, the adiabatic condition of Eq. (6.1.24) fails and the transition factor T of Eq. (6.1.25) grows.
(3) For a small set of values of Δm^2 between 10^{-5} and 10^{-6} eV^2 there is an increased detection probability of ν_e because the ν_2 reaching the earth oscillate back to ν_e as they pass through the earth at night. The result shown is an average over 24 hours.

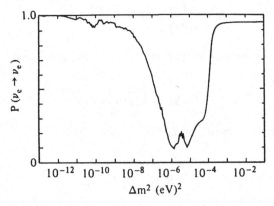

Fig. 6.1.5 Probability $P(\nu_e \to \nu_e)$, the probability that a ν_e produced in the sun arrives as a ν_e, for ^{37}Cl experiment as a function of Δm^2 for $\sin^2 \theta = 0.1$ [BOU 86].

Similar results hold for the ^{71}Ga detector (which is sensitive to the lower-energy *pp* ν), but the large suppression holds for Δm^2 between 10^{-5} and 10^{-8} eV^2.

While we have used $\nu_e - \nu_\mu$ oscillations in our discussion so far, all the results apply equally for $\nu_e - \nu_\tau$ oscillations. The case in which both types of oscillations must be considered together is covered in many papers [KUO 87]. If we assume $m(\nu_\tau)$ and $m(\nu_\mu)$ are much larger than $m(\nu_e)$, then solar neutrino studies probe $m(\nu_\mu)$ and $m(\nu_\tau)$ between 10^{-2} and 10^{-4} eV if the mixing angles are not too small.

An observation of a solar neutrino flux below theoretical explanations may be explained either by a modification of the model of the sun used to calculate the flux or by neutrino oscillations. An important distinction between these is that neutrino oscillations modify the spectrum because of the dependence of l_ν on E. For example, consider a reduction factor of the order of 2 in Fig. 6.1.5. For $\Delta m^2 \sim 10^{-4}$ eV^2 this occurs because the resonance condition (6.1.21) can be satisfied for values of E_ν in the top half of the spectrum and the adiabatic condition works well. As a result, the high-energy part of the spectrum is very suppressed whereas the lower-energy neutrinos get through because ρ_e is not large enough to satisfy Eq. (6.1.21). A reduction factor of 2 can also occur for $\Delta m^2 \sim 10^{-7}$ eV^2. In this case, the resonance condition is satisfied only in the low-density edge of the sun where the adiabatic condition begins to fail. This failure is somewhat worse for the high-energy neutrinos and so somewhat more of these get through. The possibility of detecting the spectral changes in the ^{8}B spectrum via neutrino–electron scattering is discussed by Bahcall *et al.* [BAH 87b].

A way of searching for neutrino oscillations is to use a detection method that is sensitive only to neutral-current interactions. Such a detector would be equally sensitive to ν_e, ν_μ, or ν_τ, and so should detect the total neutrino flux independent of the oscillations. Comparing this result to the flux of ν_e would directly test the oscillation scenario. One possibility is the D$_2$O detector (see Section 6.1.1) using the reaction

$$\nu + d \rightarrow p + n + \nu.$$

The signature of the event is only the final neutron, so this is much more difficult than detecting ν_e by the charged-current reaction. Another project [RAG 86] uses the neutral-current excitation of ^{11}B with the detection of the de-excitation gamma rays.

The study of solar neutrinos provides a unique way of studying neutrino oscillations for values of Δm^2 between 10^{-4} and 10^{-8} eV^2. The only other possibility might be the observation of the neutronization burst from a nearby supernova. In order to prove the existence of oscillations, if they do occur, a variety of solar neutrino experiments will be necessary.

6.2 Supernovae and neutrinos

6.2.1 Experimental observations*

6.2.1.1 Brief history

On February 25, 1987, a sheet of telefax came to us from S. A. Bludman, saying "Supernova went off in Large Magellanic Clouds. Can you see it? This is what we have been waiting 350 years for!" In a few hours, more information arrived. But it was still too early to definitely identify the supernova as type I or type II. The type I supernova is an explosion of a complete star due to uncontrolled nuclear fusion, while the type II supernova is triggered by gravitational collapse of the Fe core of a massive star (≥ 8 solar mass). It is this type II supernova that would leave a neutron star or a black hole after the liberation of an enormous amount of energy (3×10^{53} erg) in the form of neutrinos. Therefore only the type II supernova is a relevant place to look for neutrino signals.

It was also frustrating that the time when the stellar collapse actually took place was not definitely determined, because it was believed that the supernova brightened up about a day after the collapse and there was an ambiguity in a time lag of the optical observation. There was a possibility that it had happened well before February 24.

Kamiokande was happily running continuously on February 21–4. We conveyed the supernova news to the shift members in Kamioka on February 25, and asked them to ship the magnetic tapes for data up to February 25 to Tokyo as soon as possible (at that time, all the analyses were being made at the Faculty of Science, University of Tokyo). It never occurred to us that the time of the supernova occurrence would be important. Remember that Kamiokande was searching for proton decay with a sensitivity of 10^{32} years! Who cared about a 1-minute inaccuracy in the clock timing? Even worse, before we realized its importance, our primary power was shut down for a short period on February 26. We lost a chance to recalibrate our computer clock.

At any rate, the tapes arrived on February 27. It was straightforward to analyze the data quickly. The Kamiokande detector had been upgraded to observe solar neutrinos; anticounter layers with water thickness of ≥ 1.2 m was installed, the water purifier had a new ion exchange column to remove radioactive elements (^{238}U or ^{226}Ra) in the water, and the detector was made airtight to avoid contamination of ^{222}Rn from the air. As a result of these efforts, the trigger threshold was lowered to 7.6 MeV (50 percent efficiency). Software work was also in progress. Programs were ready for space reconstruction and the determination of low-energy events. Energy calibration had already been performed with a gamma ray source from the Ni(n, γ)Ni reaction. We were in quite a good position [NAK 87].

* Y. Totsuka, Institute for Cosmic Ray Research, University of Tokyo, 3-2-1 Midori-cho, Tanashi, 188 Tokyo, Japan.

Analysis of the data between February 21 and February 24 started from the evening of February 27 and ended the next morning. We simply printed out a two-dimensional plot of time versus energy for each event. It was easy to pick up a cluster of 11 events that had lasted for about 10 seconds from 500 pages of printouts. The events were space-reconstructed and their energies determined. A computer display showed that the events were clear electron signals. A correlation of the cluster with preceding energetic cosmic ray muons was checked. Muons occasionally produce multiple pions by a violent interaction with oxygen nuclei in water. The pions then break up other ^{16}O and produce a number of radioactive nuclei, ^{12}N, ^{12}B, ^{8}B, ^{8}Li, and so on, which β-decay in about $10 \sim 20$ sec, and fake a neutrino burst. It was found that there were no preceding muons that had interacted violently in the water. Therefore the cluster was most probably produced by an enormous number of neutrinos. The background was completely negligible.

It was quite natural to associate the cluster with the supernova SN 1987A, because we could not imagine any other physical phenomenon that might emit such an enormous number of neutrinos. This work was completed on February 28. A few days later we learned that the Mt. Blanc UNO experiment also observed a cluster of five events in 7 sec, but strangely enough the time of occurrence was about 5 hours earlier than Kamioka's observation [IAU 87]. We were so confident in our data that we simply considered the UNO data as spurious. The UNO people, however, are still confident in their observation. This difference of opinion has not been resolved yet, and will not be resolved until the next supernova in our galaxy.

It took us another 5 to 6 days to evaluate errors in energy and angle determinations of the 11 events. The paper was ready on March 6, and our result was then delivered to the world.

The IMB experiment, which is another large water Čerenkov detector, searched for a burst of low-energy events after learning about the Kamioka and Mt. Blanc times. They quickly found a cluster of 8 events in 6 sec at the Kamioka time, but none at the Mt. Blanc time. Thus they confirmed Kamioka's result nicely, but the Mt. Blanc anomaly remains a mystery. That the supernova went off in the Large Magellanic Clouds (LMC) has positive as well as negative aspects. It was quite fortunate that the supernova was observed by an optical means, which is not always the case in our galaxy, because the galactic center is invisible owing to thick layers of dust. The data of about 10 events alone would hardly be taken as evidence of the stellar collapse if there were no optical help. The distance to the LMC is known, which enabled us to calculate the neutrino luminosity. The negative aspect is that the LMC is a bit too far, 50 kpc, so that the number of burst events was marginal, and moreover we did not observe a definite signal of a short pulse (~ 10 ms) expected from the neutronization.

We will not need the optical help to observe neutrinos from the next supernova, because we now know a basic property of the neutrino burst. The next supernova will most probably go off in our galaxy and will provide much more information on stellar collapse and neutrino properties. The supernova rate is believed to be $\sim \frac{1}{10}$ yr (see Section 6.2.2 and [BAH 83; SOF 88]). We can only hope that much better detectors will be ready by that time.

6.2.1.2 Detector

Supernova neutrinos have an average energy of $10 \sim 20 \, \text{MeV}$. Cross sections are listed for reactions that can be utilized for detecting such low-energy neutrinos [ARA 87a]:

$$\sigma(\bar{\nu}_e p \rightarrow e^+ n) = 9.77 \times 10^{-42} \quad (E/10 \, \text{MeV})^2$$

$$\sigma(\bar{\nu}_e e \rightarrow \bar{\nu}_e e) = 0.388 \times 10^{-43} \quad (E/10 \, \text{MeV})$$

$$\sigma(\nu_e n \rightarrow e^- p) = 9.8 \times 10^{-42} \quad (E/10 \, \text{MeV})^2$$

$$\sigma(\nu_e e \rightarrow \nu_e e) = 0.933 \times 10^{-43} \quad (E/10 \, \text{MeV})$$

$$\sigma(\nu_\mu e \rightarrow \nu_\mu e) = 0.159 \times 10^{-43} \quad (E/10 \, \text{MeV})$$

$$\sigma(\nu_\mu e \rightarrow \bar{\nu}_\mu e) = 0.130 \times 10^{-43} \quad (E/10 \, \text{MeV})$$

$$\sigma(\nu_e {}^{16}\text{O} \rightarrow e^- {}^{16}\text{F}) = 1.1 \times 10^{-44} \quad (E(\text{MeV}) - 13)^2$$

where the unit is cm^2. Note $\sigma(\bar{\nu}_e p)$ is larger by two orders of magnitude than $\sigma(\nu_e e)$. Taking into account the relative numbers of protons and electrons, the ratio of event rates, $N(\bar{\nu}_e p)/N(\nu_e e) = 20 : 1$ ($25 : 1$) for water (liquid scintillator). Thus, $\bar{\nu}_e$ signals dominate over ν_e signals. However, scattering with electrons has an advantage; a strong directionality of scattered electrons enables us to point back the position of supernovae. The deuterium target that provides almost free neutrons is interesting; $\sigma(\nu_e n)$ is large and one is able to detect ν_e's from the neutronization.

In view of these factors, one may employ either a water or liquid scintillator as a detector material. Remember that one needs a large amount ($\sim \text{kt}$) of material to observe a reasonable number of interactions. Therefore expensive material like a plastic scintillator is not suitable. Table 6.2.1 summarizes the materials and their

Table 6.2.1. *Materials suitable for detection of supernova neutrinos and their properties*

Material	Detection method	$p/n/e$	Directionality	Energy resolution (%)[a]	Angular resolution[a]
H_2O	Čerenkov	1/0/5	○[b]	22[c]	28[c,d]
D_2O	Čerenkov	1/1/5	○[b]	22[c]	28[c,d]
$(CH_2)_n$	Scintillation	1/0/4	×	25[e]	

[a] At 10 MeV.
[b] For electron target.
[c] Parameters of Kamiokande II.
[d] Parameters of Kamiokande II, dominated by multiple Coulomb scattering.
[e] Parameter of UNO.

properties. It is interesting that the energy resolution of the water Čerenkov detector (Kamiokande) is as good as that of the liquid scintillation detector (UNO), although the Čerenkov light yield is 100 times less than the scintillation.

Table 6.2.2 shows the existing four detectors that are capable of detecting supernova neutrinos. Water Čerenkov detectors are in general much larger than liquid scintillator detectors, for a good reason. All the groups listed in Table 6.2.2 claim that they have observed the neutrino burst from SN 1987A. Background rates are, however, quite different from detector to detector, $0.7/\text{day} \sim 1.2 \times 10^{-8}/\text{yr}$. Obviously the Baksan group cannot claim their observation without the help of Kamioka or IMB. The background rate of the Mt. Blanc group is rather small (0.7/yr), but as stated before, their observation time is a problem. The two experiments, Kamioka and IMB, are completely free from background, and their observation times are consistent within quoted errors. Both detectors happen to be water Čerenkov detectors. It is understandable because they are at least 10 times more massive than the scintillation detectors, as seen in Table 6.2.2.

We now describe a typical water Čerenkov detector, namely, the Kamiokande detector. It is located about 300 km west of Tokyo. The rock overburden is 2700 m.w.e. (meter water equivalent). It is essential to go underground in order to reduce cosmic ray-induced background. Figure 6.2.1 shows a schematic diagram of the Kamiokande detector. It consists of an inner and outer part. The outer part is a water Čerenkov counter consisting of water at least 1.2 m thick and completely surrounds the inner one. It serves as a veto counter against incoming cosmic ray muons ($\sim 0.3\,\mu/\text{sec}$) and also as an absorber of gamma rays and neutrons coming from the rock. The inner detector contains 2140 t of pure water viewed by 948 20 inch diameter photomultipliers. Photomultiplier tubes (PMT) are uniformly placed ($1\,\text{PMT/m}^2$) and their gains are adjusted with an rms spread of better than 8 percent. The gain spread of 8 percent does not cause any problems. About 26 PMTs are hit for a 10-MeV electron, so that the energy uncertainty caused by the gain spread is only $8\%/\sqrt{26} = 1.6\%$ at 10 MeV, which is negligible compared to the energy resolution of 22% ($\approx 1/\sqrt{26}$). Water is continuously purified with ion exchangers and a series of filters. Purification keeps the light attenuation length long (≥ 50 m) and eliminates radioactive elements (^{238}U, ^{226}Ra) (contamination $\leq 10^{-13}$); ^{222}Rn are reduced to a minimum level (we have not yet succeeded in completely eliminating them) by making the inner detector airtight.

The detector is triggered by at least 20 PMT discriminators firing within 100 nsec. The trigger dead time is approximately 50 nsec for nonfired PMTs by preceding trigger. Charge and time information for each channel above threshold is recorded for each trigger. The trigger efficiency is shown in Fig. 6.2.2 in which those for IMB, UNO, and Baksan are also shown [ALE 88; BIO 88; O. Saavedra, pers. commun.]. The trigger efficiency saturated at 95 percent, not 100 percent because some of the events that were produced near the PMT wall and directed toward the wall were not accepted. The raw trigger rate was 0.60 Hz, of which 0.37 Hz was cosmic ray muons.

Table 6.2.2. *Existing detectors capable of detecting supernova neutrinos*

	Depth (m.w.e.)	Material	Technique	Mass (t)	Threshold[a] energy (MeV)	Observation time (UT)	Number of events	Background rate	References
Kamioka	2700	H_2O	Čerenkov	2140	7.5	7:35:35	11	1.2×10^{-8}/yr	[HIR 87, 88]
IMB	1570	H_2O	Čerenkov	6800	30	7:35:41	8	5×10^{-3}/yr[b]	[BLO 88]
Mt. Blanc	4000	$(CH_2)_n$	Scintillation	90	5.5	2:52:37	5	0.7/yr	[AGL 87]
Baksan	850	$(CH_2)_n$	Scintillation	200	10	7:36:12	5	0.7/d	[ALE 88]

[a] At 50 percent efficiency.
[b] The rate contains edge-clipping muons. Nonmuon background would be similar to Kamioka.

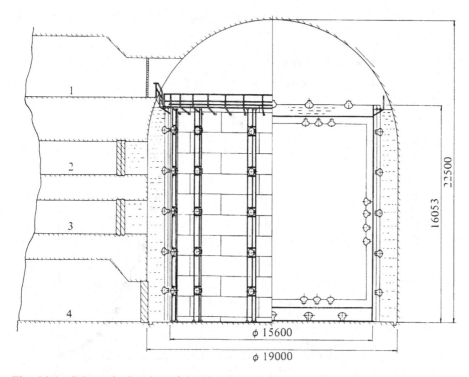

Fig. 6.2.1   Schematic drawing of the Kamiokande detector. The volume inside the PMT wall contains 2140 t of pure water.

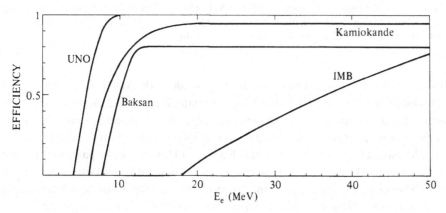

Fig. 6.2.2   Trigger efficiency as a function of electron energy, which is compared with the IMB, UNO, and Baksan trigger efficiencies.

The remaining 0.23 Hz was largely due to β-rays from ^{214}Bi, which is a daughter element of ^{222}Rn or ^{226}Ra.

Energy calibration is performed experimentally with gamma rays of about 9 MeV from the reaction Ni(n, γ)Ni. Figure 6.2.3 shows the spectrum of gamma rays after

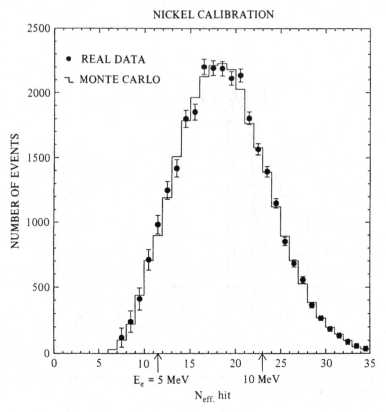

Fig. 6.2.3   Spectrum of the number of hit PMTs ($N_{eff.hit}$) produced by gamma rays from Ni(n, γ)Ni, suitably corrected for water transparency. $N_{eff.hit}$ is slightly different from the number of hit PMTs, N_{hit}, used for the supernova analysis.

background subtraction. The absolute energy scale of the Monte Carlo program is adjusted to obtain the best fit to the Ni(n, γ)Ni data. Figure 6.2.3 shows that not only the peak but also the spectrum shape is well reproduced. Another source of energy calibration is electrons from μ-decays. About 300 muons stop in the detector every day. We have a large number of decay electrons. Figure 6.2.4 compares the electron spectrum with Monte Carlo, whose energy scale was fixed by the Ni(n, γ)Ni gamma rays. The agreement is better than 3 percent; that is, the absolute normalization error is about 3 percent or better. The number of hit PMT, N_{hit}, turns out to be a good energy indicator. Figure 6.2.5 shows the correlation between N_{hit} and electron energy for the observed supernova events [HIR 87, 88]. The energy of each event is determined in the following way. Sets of Monte Carlo events of fixed energy differing by a small amount inferred from $N_{hit}/2.6(MeV)$ are generated with event vertexes in each set varied within the vertex resolution. The estimated energy for the event is chosen to be that energy for which the mean N_{hit} of the Monte Carlo data reproduces the N_{hit} of the event. One sees a small variation of the data points about

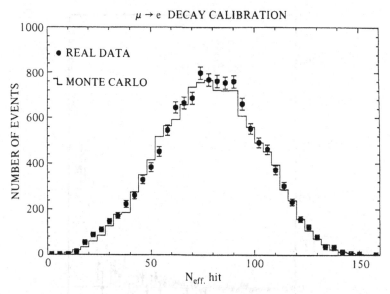

Fig. 6.2.4 Spectrum of $N_{\text{eff.hit}}$ for electrons produced by μ-decays. The histogram is a prediction by Monte Carlo.

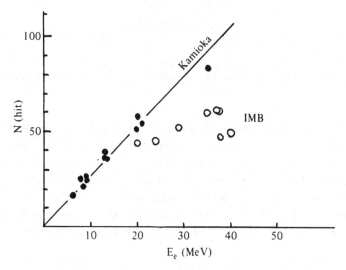

Fig. 6.2.5 N_{hit} vs. electron energy (E_e) for Kamiokande and IMB. IMB needs much larger corrections to estimate E_e than Kamiokande. The line corresponds to the relation $N_{\text{hit}} = 2.6\,E_e$, E_e in MeV.

the straight line $N_{\text{hit}} = 2.6\,E_e$, which indicates a relatively small correction applied to each event. The same figure also shows the IMB data [BIO 88], which have much larger corrections.

Space reconstruction of events is performed with timing information of each hit PMT. The time resolution of the 20 inch diameter PMT is not bad, 5.6 nsec(rms), at

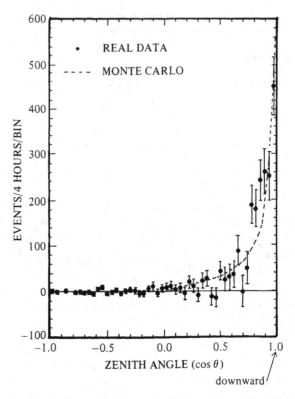

Fig. 6.2.6 Angular distribution of gamma rays that are injected in the detector vertically from Ni(n, γ)Ni. The dashed line is the expected resolution from Monte Carlo. Agreement is good.

the pulse height of single photoelectrons, which rapidly decreases as the pulse becomes larger. The error of the space reconstruction is 1.7 m for a 10-MeV electron. The angular resolution, 28 deg for a 10-MeV electron, however, is dominated by multiple Coulomb scattering, not by the spatial resolution. The angular resolution can also be determined experimentally. Gamma rays from Ni(n, γ)Ni are injected downward vertically into the detector. Figure 6.2.6 shows a peak in the downward direction. Our Monte Carlo nicely reproduces the peak, confirming its reliability [NAK 87].

6.2.1.3 Background

Supernova neutrinos come in a burst of $\sim$ 10-sec duration. In general, constant background does not cause a serious problem, although it is not negligible. We are interested in the energy range from a few MeV to a few tens of MeV. What kind of constant background exists in this energy range? Figure 6.2.7 shows a scatter plot of low-energy events in visible energy versus angle measured from the direction of the sun, observed by Kamiokande. Several events are seen in 20–50 MeV, which are

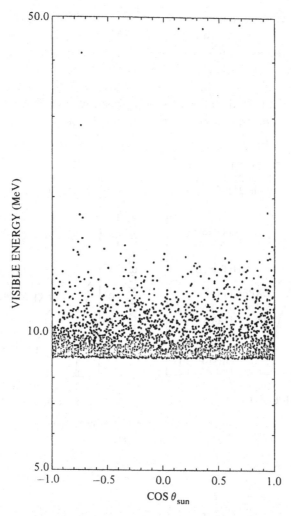

Fig. 6.2.7 Visible energy and $\cos\theta_{sun}$ of low-energy events observed by Kamiokande, where θ_{sun} is the angle measured from the direction from the sun. They are events still remaining after background rejection. Fiducial volume is 680 t in this case. Observation time is 450 days.

presumably electrons from decays of very low energy muons produced by ν_μ (below Čerenkov threshold). But they are too few to become a serious background for supernova neutrinos. Therefore the IMB detector is essential free from background (threshold energy = 30 MeV). Below 20 MeV, however, background increases quite rapidly. They consist of three types; β-rays of ^{214}Bi in the water, γ-rays unabsorbed in the outer detector from the rock, and β-rays of spallation products by cosmic ray muons like ^{12}N, ^{12}B, ^{8}B, and ^{8}Li. In principle, all of them can be eliminated. ^{222}Rn, which is a parent element of ^{214}Bi, can be taken out of the water. A thicker outer detector can absorb external γ-rays. Spallation products are negligible if the detector is located deep underground. However, existing detectors do not have all the requirements. A trigger threshold of 7.6 MeV for Kamiokande is in fact set owing to ^{214}Bi and external γ-rays.

Figures 6.2.8a and 6.2.8b show the distributions of the number of events in 10 sec for threshold energies of 7.5 MeV and 11.5 MeV for Kamiokande. Figure 6.2.8c corresponds to the IMB data [BIO 88]. They just follow the Poisson distribution. Note the points corresponding to the neutrino burst. They are far above the

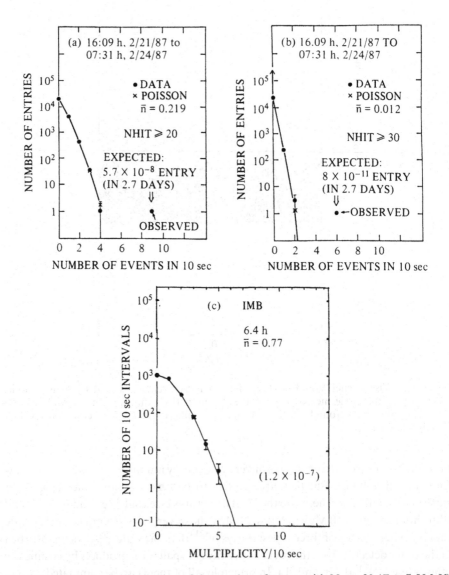

Fig. 6.2.8　(a) Poisson distribution of the number of events with $N_{hit} \geq 20$ ($E_e \geq 7.5$ MeV) per 10 sec in a 2.7-day period. (b) Same for $N_{hit} \geq 30$ ($E_e \geq 11.5$ MeV). (c) Same for the IMB data. The point denoted as observed corresponds to the neutrino burst.

expected entries. Therefore Kamiokande and IMB are still free from background for supernova neutrinos, although the trigger thresholds are limited by the background rate.

One must worry about two kinds of burstlike backgrounds. One is electric noise coming from electric sparks. It produces a train of pulses and would be a serious problem unless some electronic rejection is made. Kamiokande has a flash ADC (FADC) to directly record pulse waveforms. Electric noises produce quite distinct waveforms and are easily distinguished from genuine pulses. Moreover, the timing and topological information of PMT signals completely eliminates electric noises. The other source of burstlike events is multiple spallation products produced by a single muon. In fact, Kamiokande observed a burst of 26 low-energy events in 20 sec on January 23, 1987, besides the supernova burst. This event was an obvious spallation event because a muon with a violent interaction in the detector was found less than 4 msec before the first low-energy event. The Kamiokande location is deep enough to eliminate such multiple spallation background. Detectors at much shallower depth will have a serious trouble on the spallation background owing to a much larger muon flux.

6.2.1.4 Observation

Once the detector is well understood, observing a neutrino burst is quite an easy task. One first eliminates electric noises that happen occasionally, and space-reconstructs each low-energy event. Then one searches for a cluster of events, say, in 10 sec. When such a cluster is found, a check is made for muon activity preceding the cluster by up to 20 sec. One can readily eliminate electrons from multiple $\pi \to \mu \to e$ decays produced by an energetic muon, because the time spread of the cluster is less than about 5 μsec, which is completely different from the supernova time scale. One then estimates the probability that the cluster is of the spallation origin. Some properties of multiple spallation background are as follows: (1) It exhibits an exponential time structure that reflects the known lifetimes of the radioactive fragments from ^{16}O, specifically, an 18 ± 1.2 msec component from ^{12}N and ^{12}B, and also a component with a longer exponential time structure of 1.2 ± 0.5 sec from ^{8}B and ^{8}Li, with relative rates 2 : 1, respectively; (2) the resultant β-decay electrons with observed energies above 15 MeV occur with less than 4 percent probability; (3) the relative total rate of spallation leading to one or more low-energy electrons is less than 10^{-3} per incident muon; (4) the measured multiplicity distribution of low-energy electron events following an incident muon in time yields a probability of multiplicity ≥ 3 of 3×10^{-3}. For Kamiokande, the rate of muons incident in the detector is 0.3 μ/sec. So, clusters with multiplicity ≥ 10 occur very rarely (≤ 1/ month) and are readily recognized as spallation products. This kind of background is in any case negligible for detectors located even deeper underground.

Figure 6.2.9 shows the time sequence of all low-energy events (solid lines) and all cosmic ray muon events (dashed line) in a 45-sec interval centered on 7:35:35 UT,

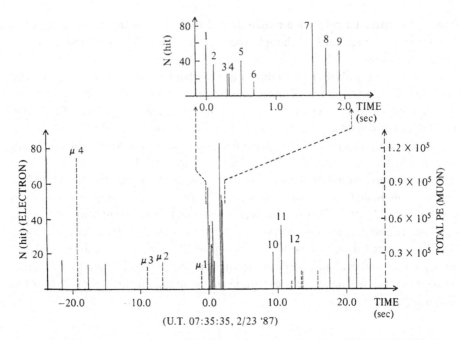

Fig. 6.2.9   Time sequence of events in a 45-sec interval centered on 7:35:35 UT, February 23, 1987. Solid lines represent low-energy electron events in unit of the number of hit PMTs, N_{hit} (left-hand scale). Dashed lines represent muon events in unit of the number of photoelectrons (P.E., right-hand scale). The shower energy is roughly given by (P.E. − 12 000)/3.6 (MeV). The upper right figure is the 0–2 sec time interval on an expanded scale.

February 23, 1987, [HIR 88]. The event sequence during 0 to 2 sec is shown expanded in the upper right corner. The left-hand scale, N_{hit}, indicates the event energy ($E_e \approx N_{hit}/2.6$ MeV. See Fig. 6.2.5), while the right-hand scale is the pulse height (in unit of photoelectrons) of the muons. The threshold energy is set to $N_{hit} = 20$ ($E_e = 7.5$ MeV) at which the detection efficiency drops to 50 percent. The burst consists of 12 events. Event number 6 is below threshold and consistent with background. There is a large gap between event numbers 9 and 10. The three events 10, 11, and 12, most probably belong to the burst. The probability that they are background is 4×10^{-5}. Four muons have passed within 20 sec prior to the burst; $\mu1$, $\mu2$, and $\mu3$ are normal through-going muons with little activity in the detector; $\mu4$ has produced a shower ($E_e \sim 30$ GeV) but cannot be a progenitor of the burst, as it does not exhibit the features of spallation described above. The burst cannot be due to a statistical fluctuation of constant background, as already shown in Fig. 6.2.8. All the events have a typical Čerenkov pattern produced by electrons. Therefore we conclude that the burst has been produced by neutrinos. It is natural to associate the burst with the supernova SN 1987 A, which became optically visible about 3 hours later.

The IMB group has observed a burst of 8 events in 6 sec on 7:35:41 UT, February 23, 1987 [BIO 88]. Since their threshold energy is 30 MeV (at 50 percent efficiency),

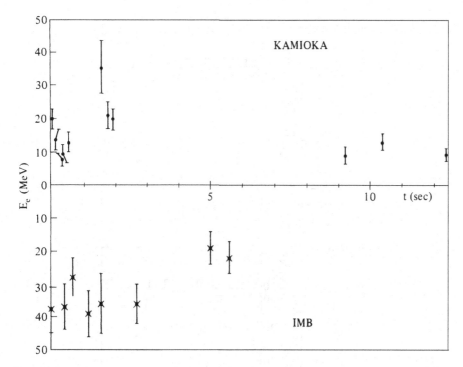

Fig. 6.2.10   Time sequence and energy of the burst events observed by Kamiokande and IMB.

spallation is not a problem. The constant background rate in the IMB detector has been shown in Fig. 6.2.8c. The burst cannot be a statistical fluctuation of the general background.

The time sequence and energy of the burst events are shown in Fig. 6.2.10 for both Kamiokande and IMB. Events observed by IMB of course have higher energies owing to the higher energy threshold. Note that a 7-sec gap in the Kamiokande data is nicely filled by the two events of IMB. One should not take the gap seriously. The angular distributions for the two data samples are shown in Figs. 6.2.11a and 6.2.11b [BIO 88; HIR 87, 88]. The Kamiokande data are consistent with isotropy, while the IMB events are concentrated in the forward direction. This is not due to a possible detector bias [BIO 88], but the probability that the distribution is consistent with expectation from $\bar{\nu}_e p \rightarrow e^+ n$ is 4.5 percent, which is not very small, although not very large [BIO 88]. A simple statistical fluctuation is not excluded to explain the IMB angular distribution.

6.2.1.5 Future outlook

When will the next supernova occur? This question has been asked many times, especially after SN 1987 A. The supernova rate has been estimated by several authors ranging from 1 every 100 years to 1 every 10 years. These days, many tend to

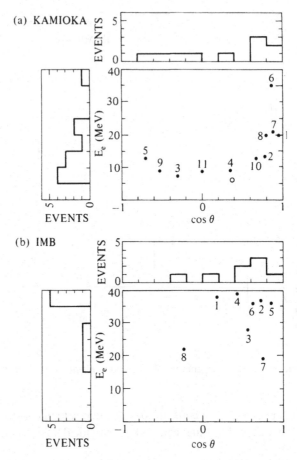

Fig. 6.2.11 Angular and energy distributions of the observed electrons. (a) Kamiokande and (b) IMB. The point denoted by an open circle in (a) is below analysis threshold and should not be taken seriously. The number attached to each event is a sequential number in chronological order.

believe the optimistic rate of about 1 every 10 years [BAH 83; SOF 88]. Thus one expects the next supernova some time around the beginning of the century. It will most probably occur near the center of our galaxy (8.5 kpc from the earth), and will not be seen optically owing to thick layers of dust covering the galactic center. Neutrinos have no problem punching through them and reaching the earth. After about a year from the neutrino burst the supernova will become visible by X-rays and gamma rays.

Since optical help is not expected, the direction of the supernova must be obtained by neutrino data alone. This will encourage satellites to search for X-ray or gamma-ray signals afterwards. It is in principle possible to determine the supernova direction by neutrinos. One detects elastic scattering $\nu e \rightarrow \nu e$. The scattered electron has a strong directionality that points back to the supernova accurately. One expects the scattering rate of about 7/kt for a supernova at 10 kpc away [SAT 87]. The angular resolution of a water Čerenkov detector is about $28°/\sqrt{E_e/10\,\text{MeV}}$ (E_e is the electron energy in MeV). Since the average ν_e and $\nu_{\mu,\tau}$ energies are 10 and 20 MeV, respectively, the accuracy of the direction is better than $28°/\sqrt{7W} = 11°/\sqrt{W}$, where W is the detector mass in kt. For example, the Super-Kamiokande detector [TOT 86] will have a fiducial volume of 22 kt, which leads to the directional accuracy of 2°. Another way of determining the direction is to use at least three detectors separated widely and to determine the accurate timing of the neutronization burst. Theory tells us that the pulse width of the neutronization burst is about 10 msec or less. If the separation of the detectors is 10^4 km, then the directional accuracy is about $3 \times 10^5 \times 10^{-2}/10^4 = 0.3\,\text{rad} = 17°$. This is in fact the worst case, because each detector would be able to detect more than one event, say, N events. Then the directional accuracy would become $17°/N$. In this case, one needs large water or scintillation detectors (≥ 10 kt, expected yield ≈ 0.4/kt), or D_2O detectors of kt size.

The observed rate of $\bar{\nu}_e p \rightarrow e^+ n$ is about 130/kt for a supernova 10 kpc away. For example, the Super-Kamiokande detector will observe 2.800 $\bar{\nu}_e$ events, which are certainly sufficient for a detailed study of stellar collapse, prompt or delayed explosion, and so on.

Another important observation on the next supernova will be ν_μ and ν_τ elastic scatterings $\nu_{\mu,\tau} e \rightarrow \nu_{\mu,\tau} e$. The average energy of $\nu_{\mu,\tau}$ is expected to be 20 MeV, which is higher than the average energy of $\nu_e \sim 10$ MeV. Note that $\nu_{\mu,\tau}$ signals are well separated from $\bar{\nu}_e$ by their sharp angular distribution. The only confusion comes from $\nu_e e \rightarrow \nu_e e$ scattering, but a factor of two difference in the average energies will be able to pick up the $\nu_{\mu,\tau}$ sample about 2.5 events/kt, from the ν_e background (4.5 events/kt). For example, the Super-Kamiokande detector will observe 100 ν_e and 55 $\nu_{\mu,\tau}$ events, which are probably large enough to find a two-step energy distribution and to estimate the $\nu_{\mu,\tau}$ flux as well as $\nu_{\mu,\tau}$ energies. Observation of $\nu_{\mu,\tau}$ will have a significant impact on neutrino physics. One could set the $\nu_{\mu,\tau}$ mass limit as low as 50 eV if the $\nu_{\mu,\tau}$ signals are observed in 10 sec. This value is better by a factor of 10^3–10^6 than the present laboratory limits, and will settle the question of whether the relic $\nu_{\mu,\tau}$ can close the universe.

Table 6.2.3 shows four detectors being built or planned. All of them are sufficiently large to study the neutrino burst in detail.

Table 6.2.3. *Large underground detectors being build or planned*

	Material	Mass (t)	Threshold energy (Mev)	Directionality
Super-Kamiokande	H_2O	22 000	~ 5	O
SNO	D_2O	1000	~ 5	O
LVD	$(CH_2)_n$	1340	~ 5	×
ICARUS	Argon	2000	~ 5	O

6.2.1.6 Conclusions

A neutrino burst from the supernova SN 1987A was clearly observed coincidentally by Kamiokande and IMB. The detection itself was easy and straight forward. The total number of observed neutrino events was 11 and 8 for Kamiokande and IMB, respectively. The small data samples are due to the long distance to the LMC (50 kpc) and small size of the detectors ($2 \sim 6$ kt). Nevertheless, they provided important information on stellar collapse and neutrino properties.

The supernova watch must continue, but with better detectors. The principal goal of the observation of the next supernova should be to provide a clear answer to the stellar collapse theory and to set limits on ν_e and $\nu_{\mu,\tau}$ masses. It is essential to have a large detector to observe enough numbers of ν_e from the neutronization and of $\nu_{\mu,\tau}$. Let us hope the next supernova will occur in the near future.

6.2.2 Supernovae and neutrino production*

For over 20 years, it has been known that the gravitational collapse events, thought to be associated with Type II supernovae and neutron star or black hole formation, are copious producers of neutrinos. In fact, the main form of energy transport in these objects comes from neutrino interactions. It has long been predicted that the neutrino fluxes produced by these events would be high enough that if an event occurred within the galaxy, it could be detected. The following discussion of the neutrinos borrows heavily from Schramm [SCH 87a].

It has been well established in the models of Arnett [ARN 73a] and Weaver *et al.* [WEA 83] that massive stars with $M \gtrsim 8\,M_\odot$ evolve to an onion-skin configuration with a dense central iron core of about the Chandrasekhar mass, surrounded by burning layers of silicon, oxygen, neon, carbon, helium, and hydrogen. Collapse inevitably occurs when no further nuclear energy can be generated in the core. The collapsing iron core mass is always about $1.4 \pm 0.2\,M_\odot$. (Some authors emphasize the variation about $1.4\,M_\odot$; however, for general understanding the key is a close convergence to the Chandrasekhar mass, not the variance.)

Bethe and Brown [BET 85] and Baron *et al.* [BAR 85] have argued that, provided the equation of state of matter above nuclear density is very soft, stars in the mass range $10 \le M \lesssim 16\,M_\odot$ with cores slightly below the $1.4\,M_\odot$ Chandrasekhar mass may explode owing to the prompt exit of the shock wave formed after the core

* D. N. Schramm, The University of Chicago, Chicago, IL 60637, and NASA/Fermilab Astrophysics Center, Fermi National Accelerator Laboratory, Box 500, Batavia, IL 60510-0500.

I would like to acknowledge useful conversations with David Arnett, Gene Beier, Adam Burrows, John Ellis, Josh Frieman, Bruce Fryxel, Wolfgang Hillebrand, Rocky Kolb, Toshi Koshiba, Al Mann, Ron Mayle, Keith Olive, Bitt Press, Olga Radnyaskaya, Alvaro De Rujala, Richard Schaefer, David Spergel, Albert Stebbins, Leo Stodolsky, Michael Turner, Jim Truran, Jack Vandervelde, Terry Walker, Joe Wampler, and Jim Wilson. This work was supported in part by the National Science Foundation and by the Department of Energy at the University of Chicago and by the National Aeronautics and Space Administration at Fermilab.

bounces upon reaching supranuclear density. For stars with $16 \lesssim M \leq 80 \; M_\odot$, the shock wave stalls on its exit from the core and becomes an accretion shock. Wilson *et al.* [WIL 75] have shown that such stars will eventually (~ 1 second later) eject their envelope as a result of neutrino heating in the region above the neutrinosphere and below the shock. (The delayed ejection can also occur in the lower-mass collapsion if the initial bounce does not produce an explosion.) The success of such delayed ejection seems to depend critically on the calculational details. Wilson [WIL 88] has shown that small variations in the numerical treatment of the neutrino transport can make or break the possibility of ejection. Obviously the above scenarios are sensitive to the stiffness of the core equation of state, which is still poorly known at and above nuclear mass densities. As was first emphasized by Arnett and Schramm [ARN 73b], the ejecta have a composition that fits well with the observed "cosmic" abundances for the bulk of the heavy elements.

Regardless of the details of collapse, bounce, and explosion, it is clear that to form a neutron star the binding energy, $\epsilon_B \approx 2 \times 10^{53}$ erg must be released. The total light and kinetic energy of a supernova outburst is about 10^{51} erg. Thus, the difference must come out in some invisible form, either neutrinos or gravitational waves. It has been shown [SHA 78] that gravitational radiation can at most carry out 1 percent of the binding energy for reasonable collapses because neutrino radiation damps out the nonsphericity of the collapse (see [KAZ 76, 77]). Thus, the bulk ($\gtrsim 99$ percent) of the binding energy comes off in the form of neutrinos.

It is also well established [FRE 77] that for densities greater than about 2×10^{11} g/cm^3, the core is no longer transparent to neutrinos. Thus, as Mazurek and Sato [MAZ 76] first established, the inner core has its neutrinos degenerate and in equilibrium with the matter. Because of the trapping of neutrinos, the neutrino emission time and thus the collapse time scale is governed by neutrino diffusion times of the order of seconds rather than hydrodynamic free-fall times, which would be of the order of milliseconds.

For electron neutrinos, the "neutrinosphere" has a temperature such that the average neutrino energy is around 10 MeV. This was established once it was realized that the collapsing iron core mass is $1.4 \; M_\odot$, owing to the role of the Chandrasekhar mass in the pre-supernova evolution fixing the scale. Since the μ and τ neutrinos and their antiparticles only interact at these temperatures via neutral- rather than charged-current weak interaction, their neutrinosphere is deeper within the core. Therefore, their spectra are hotter than that of the electron neutrinos. The electron antineutrino opacity will initially be dominated by charged-current scattering off protons but as the protons disappear, it will shift to neutral-current domination. Thus the effective temperature for $\bar{\nu}_e$'s changes from that for ν_e's to that for ν_μ and ν_τ's.

The average emitted neutrino energy is actually quite well determined [SCH 74, 76, 87a] for the peak of the neutrino distribution and is very insensitive to model parameters. The peak emission occurs at the highest temperature for which neutrinos can still free-stream out of the star. The Fermi–Dirac (F-D) temperature of this peak varies as the $\sim \frac{1}{5}$ power of the model-dependent parameters, thus yielding a well-determined value [SCH 87a] regardless of the input.

For ν_e this yields

$$T_{\bar{\nu}_e} \simeq 3.5\,\text{MeV}$$

or an average energy

$$\langle E_{\nu_e} \rangle \approx 10\,\text{MeV}.$$

This is in good agreement with detailed numerical results. As mentioned above, for $\bar{\nu}_e$'s the average energy increases with time. The time averaged value is about 15 MeV.

It should also be noted that since the interaction cross sections in the star are proportional to the square of the neutrino energy, the lower-energy neutrinos can escape from deeper in the star. In addition, as time goes on, the core evolves, so some higher-energy neutrinos are able to get out from deeper inside. Thus, the energy distribution of the emitted neutrinos is not a pure thermal distribution at the temperature of the neutrinosphere. Also, particularly for the $\bar{\nu}_e$ where T changes with time, the time integrated distribution is a superposition of many temperatures, so its shape will not be purely Fermi–Dirac. In fact, Mayle *et al.* [MAY 87a] argue that the high-energy tail of the distribution is above the thermal tail of a distribution that fits the peak.

While the general scenario for collapse events is well established, the detailed mechanism for the ejection of the outer envelope in a supernova as the core collapses to form a dense remnant continues to be hotly debated. Therefore, most theorists working on collapse prior to SN 1987A have focused on these details in an attempt to solve the mass-ejection problem. As a result, most of the pre-1987 papers in the literature are concerned with the role played by neutrinos internal to the stellar core, rather than the nature of the fluxes that might be observed by a neutrino detector on earth. In particular, while it has been known since the early 1970s [FRE 77; SCH 74, 76] that the average energy of the emitted neutrinos was about 10 MeV, with neutrino luminosities of a few 10^{52} erg/sec, the detailed nature of the emitted spectra was only recently explored by Mayle, Wilson, and Schramm [MAY 87; WIL 81]. Their calculation emphasized the high-energy neutrinos that are easier to detect. The diffusion approximation used in most collapse calculations does not treat the high-energy tail of the spectrum accurately.

In addition to the basic energetic arguments, there is the neutronization argument (see [SCH 87a] and references therein). The collapsing core has $\sim 10^{57}$ protons that are converted to neutrons via

$$p + e^- \rightarrow n + \nu_e$$

to form a neutron star. (This process is also called deleptonization by some authors.) Each ν_e, so emitted from the core, carries away on the average 10 MeV; thus around

1.3×10^{52} erg are emitted by neutronization ν_e's. This is $\lesssim 10$ percent of the binding energy. The remainder of the neutrinos come from pair processes such as

$$e^+ + e^- \rightarrow \nu_i + \bar{\nu}_i,$$

where $i = e, \mu,$ or τ, with ν_μ and ν_τ production occurring via neutral currents, and ν_e via both charged and neutral currents. (See review by [FRE 77].) As an aside, it is curious to note that the neutral-current role focused on in the 1970s was the coherent scattering off heavy nuclei and its possible role in the ejection. These early papers (see [SCR 74]) also recognized the neutral-current emission of all species but did not emphasize it owing to the preoccupation with ejection mechanisms and presumed unlikelihood of ever expecting to see a neutrino burst.

Some fraction ($\lesssim 50$ percent) of neutronization occurs as the initial shock hits the neutrinosphere (the remainder occurs on a neutrino diffusion time scale). The pair ν's always come from the "thermally" radiating core on a diffusion time scale. The time scale for an initial neutronization ν_e burst will be much less ($\lesssim 10^{-2}$ sec) than the diffusion time ($\sim$ seconds) that governs the emission of the bulk of the flux. Some so-called advection/convection models increase the initial ν_e's burst by convecting high-T, degenerate core material out. These models have higher-energy ν_e's with larger fluxes, and suppress the $\bar{\nu}_e$ fluxes.

As to the time scale for the bulk neutrino emission, in the "detailed" explosion models, more than half of the thermal neutrino emission comes out in the first one or two seconds for the delayed ejection models owing to the high neutrino luminosity during the accretion phase. For prompt models, the exponentially falling Kelvin–Helmholtz cooling starts immediately. Thus the bulk of the emission also occurs early. But for prompt models the neutrino luminosity does not stay uniformly high for as long as it does in accretion models. In both models, once the ejection has occurred the remainder of the neutrinos come out over many seconds as the hot, newborn, neutron star cools down via Kelvin–Helmholtz neutrino cooling to become a standard "cold" neutron star. Figure 6.2.12 is a schematic summarizing this. Burrows and Lattimer [BUR 85] carried out detailed cooling calculations prior to SN 1987 A. Most other authors cut their calculations off after the bulk of the neutrino emission occurred and mass ejection was established. Detailed models for the bulk of the neutrino emission (see [May 87b]) seem to find that the pair processes yield an approximate equipartition of energy in the different species. The ν_μ and ν_e's have a higher energy per ν; thus their flux is down to preserve this equipartition.

Despite the explosive mechanism, for stars in the mass range $10 \lesssim M \lesssim 16\, M_\odot$ the most distinctive structure in the neutrino signal is the initial neutronization burst. However, in the delayed explosions seen by Wilson *et al.* [WIL 81], for stars with $M \geq 16\, M_\odot$, besides the burst, the neutrino luminosity shows an oscillatory behavior superimposed on an almost constant neutrino luminosity during the postbounce pre-ejection accretion phase. The oscillations in luminosity are related to oscillations in the mass accretion rate onto the protoneutron star. The physical nature of the instability that is responsible for the oscillations in luminosity and

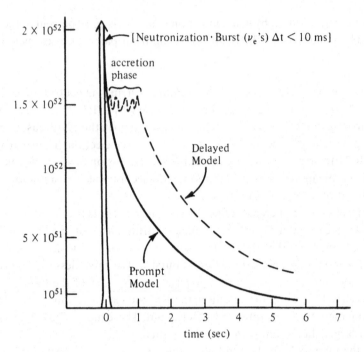

Fig. 6.2.12 A schematic showing the neutrino emission time profile for prompt and delayed models.

mass-accretion rate is described in Wilson *et al.* [WIL 75], and in more detail in Mayle [MAY 87a]. After the envelope is ejected, the luminosity will smoothly decrease as the remaining binding energy is emitted during the Kelvin–Helmholtz cooling. Models without the accretion phase go directly from the neutronization burst to Kelvin–Helmholtz cooling. Those models thus have the $\bar{\nu}_e$ emission fall-off with a single characteristic cooling time. However, models with an accretion phase have a high average emission rate for a second or so after the neutronization burst before the mass ejection and onset of the cooling phase with its dropping emission.

It is important to remember that the mean neutrino energy and total emitted energy depend only on the initial iron-core mass and are otherwise independent of the explosive mechanism. Because the opacity is less for the ν_μ and ν_τ's, they are emitted from deeper in the core where temperature is higher. Thus, they have a higher average energy. Several calculations [WIL 81; MAY 87a; BUR 88] find $E_{\nu_\mu} \simeq E_{\nu_\tau} \approx 2E_{\nu_e}$. As mentioned above, the easier-to-observe $\bar{\nu}_e$'s start out with energy comparable to ν_e's and gradually shift over to the ν_μ–ν_τ energy as their emission continues from progressively deeper in the core [MAY 87a].

By using simple, model-independent arguments, one obtains a crude estimated $\bar{\nu}_e$ counting rate for an H_2O detector

$$n = \frac{(1-f_n)\epsilon_B}{2N_\nu\langle E_\nu\rangle}\frac{\langle\sigma\rangle}{4\pi r^2}\frac{2}{18}\frac{M_D}{m_p} \tag{6.2.1}$$

where f_n is the fraction radiated in the neutronization burst, $\langle E_\nu \rangle$ is the average neutrino energy, $\langle \sigma \rangle$ is the average cross section above threshold (note that the cross section goes as $p_e E_e$, not E_ν^2, as discussed in the appendix to [SCH 87a]); however, this effect can be treated as an additional detector sensitivity factor), r is the distance to the source (~ 50 kpc for the LMC), M_D is the mass of the detector, m_p is the proton mass, and N_ν is the number of neutrino flavors. (For the Mt. Blanc liquid-scintillator detector, one should multiply by 1.39 for the average number of free protons in $H_{2+2n}C_n$.) Using F-D statistics yields

$$\langle \sigma \rangle \approx \frac{\int_{E_e}^{\infty} \bar{\sigma} E^4 \, dE / (1 + e^{E/T})}{\int_{E_e}^{\infty} E^4 \, dE / (1 + e^{E/T})} \qquad (6.2.2)$$

where E_c is the low-energy cutoff and $\bar{\sigma} \equiv \sigma / E_\nu^2$.

Plugging in values for the LMC and SN 1987 A yields the expected number of counts n of

$$n = 5.2 \left(\frac{T}{4 \, \text{MeV}} \right) \left(\frac{\epsilon_B}{2 \times 10^{53} \, \text{erg}} \right) \left(\frac{1 - f_n}{0.9} \right) \frac{1}{(N_\nu/3)} \left(\frac{M_D}{\text{ktons}} \right) \left(\frac{50 \, \text{kpc}}{r} \right)^2 . \qquad (6.2.3)$$

For the 2.14-kt Kamioka detector, this yields about 10 counts. Similarly, for the Mt. Blanc detector with 0.09 kt, times 1.39 extra free protons in the scintillator, a simple prediction is ~ 0.05 counts. IMB is a little more difficult because its threshold is not below the peak $\bar{\nu}_e$ counting rate. In addition, it is totally dominated by the high T tail where a constant T may not be an ideal approximation. However, we can crudely estimate from integrating the Mayle *et al.* spectra with the IMB efficiencies and 6 kt of detector that IMB should see ~ 7 counts from SN 1987A [MAY 87a].

To estimate the expected number of electron scattering events one must do a bit more if threshold effects are to be included. Electron scattering yields a very flat energy distribution. When such a flat energy distribution is combined with a finite temperature F-D distribution for the initial neutrinos, one finds an expected energy distribution for the scattered electrons that is quite peaked at low energies. If pure constant temperature F-D distributions are assumed for the neutrinos, the total number of scatterings is expected to be $\lesssim 0.5$ for $10 \, \bar{\nu}_e$ capture events. If the high-energy tails are supressed by absorption as assumed by Imshennik and Nadyoshen in [CHE 77], then the expected scattering rate is even lower. However, if the high-energy super-thermal tails of Mayle *et al.* [MAY 87a] are included, one finds that for every $10 \, \bar{\nu}_e$ absorptions, one expects about 0.7 to $1 \, \nu_e$ scattering and about $0.7 \, \nu_x e$ scattering, where ν_x is either ν_μ, $\bar{\nu}_\mu$, ν_τ, $\bar{\nu}_\tau$, or $\bar{\nu}_e$. We can understand why the scattering rate is $\sim \frac{1}{15}$, even though the cross-sectional ratio at 10 MeV is ~ 80, by remembering that there are five electrons for each free proton in an H_2O target. In addition, at a given energy from one cross section table in [SCH 87a]

$$(\sigma_{\nu_\tau e} + \sigma_{\nu_\mu e} + \sigma_{\bar{\nu}_\tau e} + \sigma_{\bar{\nu}_\mu e} + \sigma_{\bar{\nu}_e e})/\sigma_{\nu_e e} \simeq 1. \tag{6.2.4}$$

Thus, if fluxes are equal, the rate is doubled. Actually, the average energy of other species is about twice that of ν_e, but fluxes are reduced accordingly to roughly maintain equipartition of energy per neutrino species, thus keeping scattering constant. The difference in expected number of scatterings is an important probe of the high-energy tail.

For the 615-t C_2Cl_4 Homestake there are 2.2×10^{30} ^{37}Cl atoms. As seen from the appendix to [SCH 87a], the cross section is not a simple integer power of E_ν, however; it seems to fall roughly between an E^3 and E^4 relationship for $E_\nu \lesssim 30$ MeV. For temperatures above 5 MeV, the peak contribution to the thermal average would be coming from energies above 30 MeV where the cross section no longer rises as rapidly and the expected counting rate no longer continues to rise with temperature. In the standard case, one expects about half a count above the solar background. However, for advection models, one might expect several ^{37}Cl events. As in the solar case, ^{37}Cl is once again a potentially sensitive thermometer.

All the predictions described above assume a simple, spherically symmetric collapse. If large amounts of rotation or magnetic fields were present (with energies comparable to the binding energy) then the Standard Model would be altered with different time scales and different core masses and binding energies, since such conditions would alter the initial core mass as well as the dynamics. As we shall see, the Kamioka/IMB neutrino burst fits the standard assumptions well and the collapse that created that burst did not have significant rotation or magnetic fields.

Before SN 1987 A, it was also obvious that a supernova, if detected by its neutrinos, would constrain neutrino properties. In particular, if the neutrinos got here, we would have a lifetime limit. If the time pulse was not too spread out, that would mean a mass limit on those neutrino types that were clearly identified. Also, from the number of $\bar{\nu}_e$ counts, one could constrain N_ν since if N_ν was large, the fraction of thermally produced $\bar{\nu}_e$'s would go down. In addition, neutrino mixing could be constrained by detecting different types and comparing with Mikheyev–Smirnov [MIK 86] matter mixing, as parameterized to solve the solar neutrino problem, $\nu_e \rightarrow \nu_\mu$ (or ν_τ), and ν_μ (or ν_τ) $\rightarrow \nu_e$, but nothing happens in the antineutrino sector. Such mixing would eliminate seeing the initial ν_e burst, but give higher energies to the later, thermal ν_e's since they would be mixed with ν_μ's [WAL 87]. Of course, nonsolar Mikheyev–Smirnov can be used if antineutrino mixing is seen. All of these effects will be examined with the data from SN 1987 A. Future collapse will have much better statistics and stronger statements might be made.

6.2.2.1 Neutrino observations

On February 23, 1987, neutrino detections were reported at two separate times 2 h 52 m U.T. in the Mt. Blanc detector and 7 h 35 m at Kamioka [BUR 87] and IMB

[ARN 87] detectors (the Baksan and Mt. Blanc detectors also reported excess events at this latter time but their observations are consistent with background and would not have been reported if it were not for the Kamioka/IMB events at the same time). It is important to note that there clearly was a detection on February 23 near 7 h 35 m U.T. at the IMB and Kamioka detectors. Thus, unquestionably, *extra solar system neutrino astronomy has been born*!

As to the initial Mt. Blanc report at 2:52, it has been argued [SCH 88a] that such a burst would violate conservation of energy if it originated in the LMC and triggered the small Mt. Blanc detector without causing large numbers of events at the larger Kamioka detector. To have such a low temperature emission as to be below the Kamioka threshold and still excite Mt. Blanc requires more than the rest mass of the progenitor star to be converted into neutrinos. Including possible coincident gravitational wave detections with room temperature detectors makes energy conservation matters worse! An alternative is that this event was not in the LMC but was *much* closer, thus reducing the energy requirements but requiring a remarkable timing coincidence. (Background noise in the Mt. Blanc neutrino detector produces similar events every few hundred days.) Given all these problems, we quote Eddington: "Observations should not be believed until confirmed by theory". It should be remembered that the Mt. Blanc detector, unlike Kamioka and IMB, was actually constructed to look for collapse neutrinos; unfortunately, it was optimized for collapses within 10 kpc.

Let us now turn our attention to the well-established Kamioka/IMB burst. For the sparse data of Kamioka and IMB, four observables may be useful:

1 The total number of events, which can be used to estimate the emitted luminosity.
2 The average energy of the events, which, when deconvoluted with detector efficiencies, can be used to estimate an F-D temperature assuming a constant emission temperature.
3 The duration of the burst, which gives hints about diffusion time scales, softness, or stiffness of equations of state and possibly gives hints regarding prompt versus delayed models.
4 Angular distributions, which give hints about electron scattering events versus proton absorption, which in turn may tell something about high-energy supra-thermal distribution tails or other emission physics.

Figure 6.2.13 is a plot showing the energy and timing of the Kamioka and IMB events. (Kamioka's event no. 6 is ignored as being below their criteria for a definitive event.) Note that almost all the counts are concentrated in the first few seconds, as one expects in collapse models. A reasonable tail, as predicted by theory (see Fig. 6.2.12), yields low but finite rates after 10 sec. Such rates following the bulk early emission from an accretion phase have little difficulty in producing apparent gaps in counts due to the problems of small-number statistics [BAH

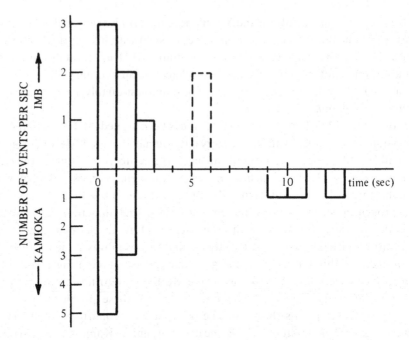

Fig. 6.2.13   The ν counting rates for IMB/Kamioka.

87a; MAY 87b]. Note also that the IMB late counts (dashed lines) nicely fill in the 6-sec gap in the Kamioka data.

To examine consistency, let us use the number of counts and mean energies measured in the experiments to determine the implied temperature and energy emitted in $\bar{\nu}_e$'s. Such estimates require detailed consideration of efficiency and threshold effects.

To convert a mean neutrino energy to an effective temperature, one must assume that the emitted ν spectrum was well described by F-D statistics. Mayle *et al.* argue that this is a reasonable assumption, but, as already mentioned, they did find that their models had a higher tail at high energies than a simple, single-temperature model would yield [MAY 87a]. Thus, one might expect the IMB temperature to be slightly higher than the Kamioka temperature because of its weighting on the high-energy events. Schramm [SCH 87a] and others have shown how one can deconvolute an F-D distribution with thresholds and efficiencies to derive an effective temperature for the emitted neutrinos from the measured energies of the events assuming a single-temperature model.

The inferred emitted energy, $\epsilon_{\bar{\nu}_e}$ in $\bar{\nu}_e$'s is obtained (see [SCH 87a]) by integrating the $\bar{\nu}_e + p \rightarrow n + e^+$ cross section over the F-D distribution with appropriate efficiencies and comparing it to the observed number of counts. The total emitted energy, ϵ_T, can be related to $\epsilon_{\bar{\nu}_e}$ by

$$\epsilon_T \approx \frac{2N_\nu \epsilon_{\bar{\nu}_e}}{(1 - f_n)}.$$

The numbers in Fig. 6.2.13 are calculated assuming $N_\nu = 3$ and $f_n = 0.1$, with Kamioka having $M_D = 2.14$ kt and IMB having $M_D = 5$ kt. IMB now argues their effective mass is over 6 kt, although all their counts were in the internal 5 kt. This revision would shift the IMB curve down by ~ 20 percent and enlarge the overlap region somewhat. Figure 6.2.14 shows the energy radiated versus $T_{\bar{\nu}_e}$. The boundaries of the region come from 1 σ errors in counts as well as the range of reasonable assumptions one might make about cutoff energies and stated experimental errors in energy.

Although one might expect (from [MAY 87a]) IMB to measure a slightly higher T, it is interesting that there is nevertheless a region of overlap where both data sets yield the same T_{ν_e} and $\epsilon_{\bar{\nu}_e}$. It is particularly satisfying that this region of overlap is exactly where one might have expected a standard gravitational collapse event to plot, namely, $\epsilon_T \sim 2 \times 10^{53}$ erg, $T \sim 4.5$ MeV. Similar conclusions were reached by Sato and Suzuki [SAT 87] and Bahcall *et al.* [BAH 87a] using a different treatment than has been applied here. Once T and ϵ_T are determined, one can use the luminosity–temperature relationship to solve for the radius, R, of the neutrino-sphere and obtain, in our case, a few tens of kilometers, in reasonable agreement with the Standard Models, whether or not the first one or two or the last three events from Kamioka are included. Note that the above analysis is very crude.

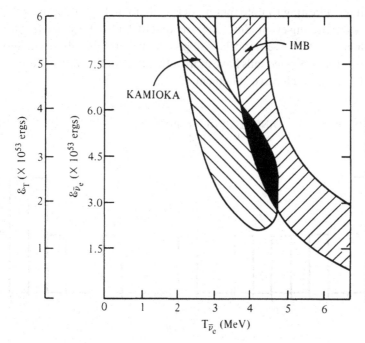

Fig. 6.2.14   Emitted energy, $\epsilon_{\bar{\nu}_e}$ in $\bar{\nu}_e$ and total emitted energy, (assuming $N_\nu = 3$) versus temperature of Kamioka and IMB data, allowing for statistical errors as well as systematic shifts due to possible electron scattering events and variations in threshold and efficiency assumptions. Note overlap region is a good fit to the Standard Model.

Kolb *et al.* [KOL 87] have pointed out that simple converting of the emitted positron energy E_e to $E_\nu - Q$, as was done here, is inaccurate, although it does not affect these conclusions. Also note that the boundaries used in Fig. 6.2.14 do not have a quantitative statistical meaning since systematic as well as statistical uncertainties were mixed in obtaining them. Nonetheless, the results are suggestive, and more detailed analyses seem to yield similar conclusions [BAY 87a; MAY 87b; SAT 87].

The angular distribution for Kamioka is presented in Fig. 6.2.15. It appears to show an isotropic distribution with a possible slight excess in the direction of LMC. From the isotropic rate background and the angular resolution, the number of excess directed events (note, Kamioka now only explicitly claims one probable scattering, but considering resolution, etc., we feel that our estimate is reasonable) is ~ 2. Since $\bar{\nu}_e + p$ would yield an isotropic distribution, the number of directed electron scattering events should be relatively small, as might be expected by the ratio of cross sections. Mayle *et al.* [MAY 87b] expect ~ 1.5 for $12\,M_\odot$ or 2 for their $15\,M_\odot$ model in reasonable agreement with the observations. One also expects that ~ 50 percent of these scattering events are higher-energy ν_μ, ν_τ, $\bar{\nu}_\mu$, $\bar{\nu}_\tau$ or $\bar{\nu}_e$ events. This also fits well since the highest-energy Kamioka events have $\cos\theta > 0.7$. It is also intriguing, although not statistically significant, that the first event had $\cos\theta$ closest to unity. Remember that the initial 0.01-sec neutrino burst is expected to be ν_e's with no $\bar{\nu}_e$'s. It is interesting to note that models with no high-energy tail would predict less than $\frac{1}{2}$ a scattering event. Since the data hint at 2 or so, they lean toward models with high-energy tails over models with pure constant T distributions, and models with absorption supressed tails have more difficulty, but the statistics are not strong enough to make a definitive statement.

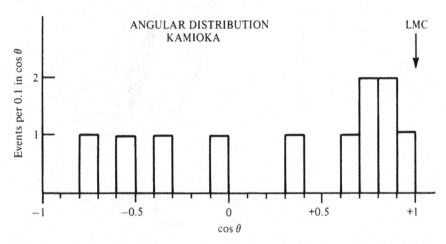

Fig. 6.2.15 Angular distribution of Kamioka data. If level of isotropic events is chosen from directions away from LMC, then there appears to be a couple of excess counts in the direction of the LMC, presumably due to electron scattering. The model of Mayle *et al.* [MAY 87a] predicts ~ 1.5 to 2; pure F-D with constant T predicts < 0.5.

The angular distribution for IMB is more problematic [VAN 88]. Initially, the failure of one of the four power supplies was thought to bias the data, but subsequent analysis showed that the effect was not significant. The IMB distribution peaks at $\sim 45°$, with most of the events forward and no significant backward scattered events. It clearly is not fitted by an isotropic source, however, if it is recognized that at high energy and with the particular detector $\bar{\nu}_e + p$ should yield $\sim 1 + 0.2 \cos \theta$, not isotropic, and with a high-energy tail giving ~ 1 e-scattering, then the distribution is at a $\lesssim 10$ percent probability, so it is not too ($\lesssim 2\sigma$) unlikely (the $1 + 0.2 \cos \theta$ distribution by itself is at $\lesssim 4$ percent probability level).

The ^{37}Cl experiment of Davis was operating at the time of the supernova, and counting began shortly after the light was observed. This experiment is only sensitive to ν_e's. After 45 days of counting, Davis saw one count, completely consistent with his normal solar counting rate [TAM 77; R. Davis, pers. commun.]. As mentioned earlier, for a standard collapse one expects from the LMC event ~ 0.5 events in the Homestake chlorine detector. However, if one interprets the Kamioka data as implying a large excess [ARA 87] of ν_e's, then one might have expected several ^{37}Cl counts. The lack of observed Cl counts argues that the ν_e flux is not in disagreement with standard predictions of $\sim 2 \times 10^{52}$ erg of neutronization ν_e's, plus 3×10^{52} erg of thermal ν_e's, all at $E_\nu \sim 10$ MeV ($T_{\bar{\nu}_e} \sim 3.5$ MeV). This constrains models [BUR 87; ARN 87] with "advection" producing excessively large high-energy ν_e fluxes and reducing the $\bar{\nu}_e$ fluxes. Such models can predict at most about 5 ^{37}Cl counts. While extreme models with $T_{\nu_e} \gtrsim 5$ MeV and fraction of energy in ν_e's, $f_{\nu_e} \sim 1$ may be in difficulty, intermediate models with $T_{\nu_e} \lesssim 4$ MeV and/or $f_{\nu_e} \lesssim 0.5$ are still allowed.

Another constraint on ν_e's comes from interactions with ^{16}O, which would be backward peaked at high energy. There is no evidence for this.

The total time spread of the IMB/Kamioka events (see Fig. 6.2.13) shows that ν-emission (or at least detection) lasted for ~ 10 sec. The duration of neutrino emission varies in different collapse models due to the equation of state and the total mass of the collapsing core (is it slightly greater or less than $1.4 M_\odot$?) and the dynamics (prompt vs. accretion). Longer time scales favor soft equations of state and higher core masses ($1.4-1.6$ vs. $1.2-1.3 M_\odot$) and thus favor accretion versus prompt mechanisms. However, until we have a collapse in our galaxy with a more detailed time evolution of the ν-signal, it will be hard to make detailed statements on the collapse mechanism.

6.2.2.2 Constraints on neutrino physics

The detection of neutrinos from a reasonably well-understood source 170 000 light-years away makes for useful constraints on neutrinos and other weakly interacting particles that might have been emitted. Unfortunately, the statistics of so few counts have limited the accuracy of some of these arguments so that some of the more interesting ones such as neutrino mass or number of flavors do no better than

Table 6.2.4. *Physics constraints from SN 1987 A*

Physics	Limit	Comment
ν_e	$\gamma\tau \gtrsim 1.6 \times 10^5$ yr	Best limit – rules out ν_e decay as solar ν solution
ν_e, ν_μ, ν_τ, photo decay limits	$\tau/m_\nu \gtrsim 10^{16}$ sec/eV	Best limit – comes from lack of SMM γ coincidence
e^+e^- branch limits	Comparable	Best limit from ionization of surrounding media
ν_e mass	$m_{\nu_e} \lesssim 25$ eV	Comparable to the 1987 experiments (but not better)
Number of flavors	$N_\nu \lesssim 7$	Comparable to 1987 accelerator limits; not as strong as big bang nucleosynthesis argument
ν charge	$\lesssim 10^{-17}|e|$	Best limit
Axion coupling	$f_a \gtrsim 10^{11}$ GeV	Best lower bound
Weak equivalence	Fermions and bosons affected equivalently by gravitational fields to $\lesssim 10^{-5}$	Best limit
ν mixing	Current supernova does not give strong constraints owing to poor statistics	

current terrestrial experiments. However, a future collapse in our galaxy with 1000's of counts will be able to make significant improvements and may be our only way of getting limits on m_{ν_μ} and m_{ν_τ} that are of cosmological significance.

The detailed arguments on constraining neutrino properties from astrophysical arguments (including SN 1987 A) are given in Section 2.7. Hence, here, only a brief summary is given in Table 6.2.4.

6.2.2.3 Collapse rates

Over the past 1000 yr, there have been only five visual supernovae in the Milky Way galaxy, implying at first glance a rate of 1/200 yr. However, if we look at galaxies like our own, that is, standard evolved Sb and Sc galaxies, we find [TAM 77] in other galaxies rates of 1/15 to 1/40 yr. Obviously our galaxy's observed low rate is probably the result of most of our galaxy being obscured from view by dust in the disk. In fact, the five historical supernovae were all in our sector of the galaxy, which implies a *minimal* enhancement of a factor of 5 to 1/50 yr to include the entire disk volume. Now that we can detect collapses by neutrinos alone, we do not need to worry about the obscuration of our disk, so traditional astronomers argue that the rates in other galaxies where we sample their entire disk might be more relevant.

However, with neutrino detectors we only see Type II supernovae. Thus the rates quoted may, at first glance, be on the high side, since these include all types ("neutrinoless" Type I's account for about one-third to one-half of the supernovae by such direct counting of supernovae in these galaxies). Such direct counting of supernovae is fraught with uncertainties. For example, SN 1987 A would probably not have been included since it was so underluminous owing to its progenitor being a blue rather than red star. If the fraction of blue stars collapsing is only minimally related to metallicity, then SN 1987A types could enhance the supernova rate for the high metallicity disk populations. It may even be that metallicity enhances the blue progenitor fraction as high mass loss rates might move more stars from red to blue prior to collapse. (It is now known that SN 1987A was once a red giant and lost its red envelope by mass loss prior to exploding as a more compact blue star. For a detailed discussion of SN 1987 A, see [SCH 88b]. Of course, if the blue progenitors occur only in metal-poor populations, SN 1987 A would not alter the statistics for the Milky Way. Similarly, other underluminous collapses, such as Cassioppe A would not be detected in extragalactic surveys. Remember the popular bias toward type II's being red stars is caused primarily by the fact that red stars are more luminous when they explode and thus dominate the statistics in distant galaxies or in bright massive galaxies where dimmer blue star explosions would go undetected. People have also tried to use pulsar formation rates and supernova remnant statistics, but these are plagued with uncertainties such as distances, beaming fractions, and remnant lifetimes and can at most give order-of-magnitude estimates. Similarly, arguments on X-ray heating against high SN rates can also be circumvented if much hot SN ejecta leaves the disk of the galaxy via SN-generated "chimneys" before cooling and falling back in.

An alternative approach to direct searches is to do statistics on stellar types. Arnett, Schramm, and Truran [ARA 87b] have argued that we do not need Type I's to make iron or any other major element. Thus apparent large numbers of Type I's may be a selection effect due to their greater brightness or a recently high rate having little to do with the integral rate over the history of the universe, since otherwise Type I's would have produced too much iron. Bahcall and Piran [BAH 85] have shown that the rate of formation of all stars $\gtrsim 8 M_\odot$ is $\sim 1/8$ yr using a Salpeter mass function and a constant star formation rate. All such stars presumably undergo collapse. Of course, the Salpeter mass function is probably most uncertain for these more massive stars, and the assumption of a constant rate can be argued.

We do know from the 2 percent heavy element content of our galaxy and the assumption that $\gtrsim 1 M_\odot$ of heavies is ejected per collapse that the $10^{11} M_\odot$ disk requires $\lesssim 2 \times 10^9$ ejections over the 15×10^9 year history of the galaxy. Thus our average Type II rate [SCH 87b] is $\lesssim 1/7$ yr. Since our current rate of explosion is $\lesssim$ the average, this is certainly a good limit. Since some galactic evolution models seem to have roughly constant nucleosynthesis rates [HAI 76; SCA 88], this limit is also not a bad estimate and is in good agreement with the Salpeter estimate. Of course, other galactic evolution models argue for high initial supernova rates that would yield very low present collapse rates to fit an average of $\sim 1/10$ yr; now that we know that SN 1987 A was changed from red to blue by mass loss, it is not unreasonable to think that many SN are missed in external surveys. Thus the nucleosynthesis rates and massive star formation rates implying SN rates of $\sim 1/10$ yr may be reasonable, or at least not excludable.

6.2.2.4 Summary

The recent supernova SN 1987 A has confirmed that our basic understanding of gravitational collapse and neutrino emission is on firm ground. Given that we now know what a neutrino burst looks like, we should have confidence that if a collapse occurs anywhere in our galaxy, regardless of the visibility of the SN, we should observe it and be able to extract many new physics constraints.

6.3 Massive neutrinos and cosmology*

6.3.1 Introduction

The importance of the consequences of a nonzero neutrino rest mass on cosmology was, perhaps, first recognized by Gershtein and Zeldovich [GER 66], after the discovery of the 3-K microwave background radiation (hereafter MBR). They argued that since the number density of MBR photons and neutrinos of each kind is roughly the same (small difference arises due to both different statistics and input of energy in radiation when electron–positron pairs annihilate), the mass of muonic neutrinos must be less than about 400 eV. Otherwise the total mean mass density of the universe would be too high, and its age too young.

Since the first works on the primordial synthesis of ^{4}He [HOY 64], it has been known that additional neutrino species increase the rate of expansion of the universe during the epoch of the primordial nucleosynthesis, which increases the yield of ^{4}He. Combining the results of the theory with astronomical measurements of the ^{4}He abundance and the estimate of the mass density of MBR, Shvartsman [SHV 69] suggested the upper limit on the mass density of all relativistic matter at that epoch: $\rho_{\text{rel}} \lesssim 5\rho_{\text{MBR}}$ which eventually became the upper limit for the number of neutrino species: $N_\nu \lesssim 7$ [STE 77].

At that time, the constraints based on cosmological arguments were much stronger than ones based on laboratory experiments.

Further progress occurred when it was suggested that massive (i.e., with $m_\nu \neq 0$) neutrinos could be excellent candidates for solving the so-called missing mass problem. About the time that Fermi coined the term "neutrino," Zwicky [ZWI 33] discovered that the masses of clusters of galaxies (the largest gravitationally bound

* Sergei F. Shandarin, Institute for Physical Problems, Moscow. To Ya. B. Zeldovich, my teacher. This work was done during my visits to the Department of Physics, Johns Hopkins University and Institute of Astronomy Cambridge University. I express my gratitude to A. Szalay and M. Rees for helping to arrange these visits.

systems in the universe known so far and having masses roughly $10^{15} M_\odot$ where $M_\odot \approx 2 \times 10^{33}$g is the mass of the sun) had to be much greater than one could infer from counting galaxies in the clusters. Measuring the velocities of galaxies, he found that the gravitational potential wells determined by the total mass of the clusters were deeper than they would be if the masses of clusters were equal to the sum of galaxy masses. Later, a similar phenomenon was found in large spiral galaxies.

It was natural to speculate that the missing mass both in galaxies and galaxy clusters *could consist of neutrinos* if neutrinos are massive [MAR 72; COW 72, 73; SZA 76]. A decade or so earlier, Markov [MAR 64] had shown that massive neutrinos could form "superstars," that is, gravitationally bound objects. Thus it became clear that neutrinos can determine the potential wells of galaxy clusters as well as form massive halos around galaxies. In this case, most of the masses of galaxies and clusters are "dark," since neutrinos neither emit nor absorb electromagnetic radiation. Subsequently, Schramm and Steigman [SCH 81] came to a more definite conclusion that the dark mass of rich clusters of galaxies *could not be all baryonic, and perhaps massive neutrinos would be required.*

In attempting to determine whether some kind of weakly interacting particles can explain the dark mass of a particular kind of astronomical objects, one should keep in mind an important restriction identified by Tremaine and Gunn [TRE 79]. Based on phase density restriction arguments, it relates the size r_c and velocity dispersion σ in the object, on the one hand, and the mass of elementary particles m_ν comprising the halo of the object, on the other hand:

$$m_\nu > 100 \left(\frac{100 \, \text{km/s}}{\sigma} \right)^{\frac{1}{4}} \left(\frac{1 \, \text{kpc}}{r_c} \right)^{\frac{1}{2}} g_\nu^{-\frac{1}{4}} \, \text{eV}. \qquad (6.3.1)$$

Here the normalizations of σ and r_c are chosen to illustrate typical galaxy velocity dispersions and coradii (1 kpc $\approx 3 \times 10^{21}$ cm); g_ν is a statistical weight. However, one should also note that the size of the halo may be larger than the visible size of the galaxy.

In 1980 the ITEP (Institute of Theoretical and Experimental Physics, Moscow) group [LUB 80] announced that their study of tritium decay possibly indicated the existence of a nonzero mass for the electronic neutrino. Whether or not their interpretation is confirmed, they have started a new phase in cosmology. The efforts and the number of papers per year on neutrino and other "ino"-dominated cosmological models increased by perhaps two orders of magnitude after that. One result of this work was the "standard" model of neutrino-dominated universe [DOR 81; SHA 83d]. In this model it was assumed that most of the mass of the universe consists of stable neutrinos with a mass of about 30 eV. It was demonstrated [KLY 83; SHA 83b] that this model can also explain (at least qualitatively) a peculiar distribution of galaxies resembling giant layers or filaments at scales of several tens of megaparcecs (1 Mpc $\approx 3 \times 10^{24}$cm) separated by giant voids where galaxies have not been found. These structures were found to form a connected system called the large-scale structure of the universe [OOR 83]. However, after a couple of years of euphoria, the Standard Model was found to possess severe if not fatal flaws [WHI 83, 84; MEL 85]. As we shall see, the problem with the standard neutrino-dominated model is not that independent confirmation of the results of the ITEP group is lacking, but that it fails to satisfy purely cosmological arguments.

This section examines possible interrelations between cosmology and neutrino physics, but not to prove or disprove a particular cosmological scenario. However, I will use the hot big bang cosmological model as a basis for the following discussion. Ya. B. Zeldovich liked to say that the hot big bang has been established as firmly as Copernicus' system. By that he did not mean that all cosmological problems are solved by this model, but he believed that the principal features like Hubble expansion explaining galaxy and quasar red-shifts; the high homogeneity and isotropy of the universe on the horizon scale; high temperature at early stages resulting in thermodynamical equilibrium between photons, baryons, and leptons; primordial formation of light elements (D, ^{3}He, ^{4}He, ^{7}Li); and so on have been established definitively.

6.3.2 Neutrinos in the homogeneous universe
6.3.2.1 Neutrino mass, mean density, and the age of the universe

We begin with the estimation of mean neutrino mass density in the universe. Measurements of the spectrum of the MBR have shown that it has a Planckian shape within an accuracy of a few percent. However, to show how other parameters might depend upon the uncertainty of the temperature of the MBR, the parameter θ is usually used:

$$T_\gamma = 2.7\,\theta\,\mathrm{K}. \tag{6.3.2}$$

The thermal spectrum of the MBR assumes that the present number density of photons is

$$n_\gamma = 400\,\theta^3\,\mathrm{cm}^{-3}. \tag{6.3.3}$$

Photons of the MBR are the most abundant kind of particles detected so far in the present universe.

Baryons are many orders of magnitude less populous than the photons

$$\bar{n}_b = 10^{-10}\,\eta_{-10}n_\gamma \tag{6.3.4}$$

where η_{-10} is a parameter describing the uncertainty caused primarily by the uncertainty in astronomical measurements of the mean density of baryons $\bar{n}_b$.

In the early universe, at temperatures higher than about 1 MeV ($t < 1$s), all kinds of light neutrinos (i.e., with masses less than 1 MeV) were in thermal equilibrium with photons. Later, they decoupled and experienced adiabatic expansion owing to the expansion of the universe. This resulted in a decrease in both the density and the thermal velocity dispersion of neutrinos. However, the form of the distribution function remained unchanged:

$$\bar{n}_\nu(t) \propto a(t)^{-3}$$

$$n_\nu(p, t) = [1 + \exp(cp/kT_\nu(t))]^{-1} \tag{6.3.5}$$

with $T_\nu(t) \propto a(t)^{-1}$, where $\bar{n}_\nu(t)$ is the space number density, $n_\nu(p, t)$ is the distribution function in phase-space, p is a momentum, k is the Boltzmann constant, $T_\nu(t)$ is the "temperature," and $a(t)$ is a scale factor describing the expansion of the universe, which is related to the redshift z as follows:

$$a(t) = (1 + z)^{-1} \tag{6.3.6}$$

Let us recall that the redshift z is defined by the ratio of the observed to the emitted wavelength

$$1 + z \equiv \frac{\lambda_{obs}}{\lambda_{em}}. \tag{6.3.7}$$

Hereafter $a(t)$ is assumed to be normalized at the present epoch: $a(z = 0) = 1$.

The ratio of neutrino density to that of the photons remains constant unless either neutrinos or photons are produced or destroyed in amounts comparable to the number of MBR photons, Eq. (6.3.3). So far, no physical processes except e^+e^- annihilation have been found that could do this provided that the spectrum of the MBR remains thermal with the observed accuracy. Thus, on the basis of these arguments, one can easily calculate the mean density of every light neutrino species in the present universe:

$$\bar{n} = \left(\tfrac{3}{4}\right)\left(\tfrac{4}{11}\right)n_\gamma \approx 109\,\theta^3 \text{ cm}^{-3}. \tag{6.3.8}$$

Here the first multiplier $\left(\tfrac{3}{4}\right)$ reflects the difference between densities of fermions (neutrinos) and bosons (photons) at equilibrium, and the second one $\left(\tfrac{4}{11}\right)$ describes the increase of the photon number due to e^+e^- annihilation after the decoupling of neutrinos from photons and electrons.

If neutrinos have masses greater than $m_\nu \sim 3 \cdot 10^{-4}$ eV, then at present they are nonrelativistic. In order to find the mean neutrino mass density in this case, one should simply sum products of the neutrino number density and their masses over all neutrino species. However, one does not know exactly either the neutrino masses or the number of neutrino species. Therefore it is more convenient to put this in the form of an upper limit on the sum of neutrino masses

$$\sum m_{\nu,i} \leq \bar{\rho}_t/\bar{n}_\nu \leq 100\,\Omega_t\,h^2\theta^{-3} \text{ eV} \tag{6.3.9}$$

where all light neutrino species are included in the sum and $\bar{\rho}_t$ is the total mean density in the universe.

Two essential dimensionless parameters h and Ω_t were introduced in (6.3.9), so let me remind the reader what they mean in cosmology. Parameter h is called a dimensionless Hubble constant that is usually expressed in terms of $100 \, \text{km s}^{-1} \, \text{Mpc}^{-1}$. Thus, the Hubble constant equals

$$H_0 = 100 \, h \, \text{km s}^{-1} \, \text{Mpc}^{-1}. \tag{6.3.10}$$

Parameter h reflects the uncertainty of astronomical measurements of the Hubble constant H_0, which characterizes the expansion rate of the universe at present

$$u = H_0 \cdot r \tag{6.3.11}$$

where u is the velocity and r is the distance of an object, assuming that $u \ll c$ and the universe is homogeneous and isotropic, that is, neglecting peculiar velocities of objects. At present it is generally accepted that $0.5 \leq h \leq 1$ [SAN 84].

Parameter Ω_t relates the total mean density of the universe $\bar{\rho}_t$ to the critical density ρ_{cr}:

$$\bar{\rho}_t = \Omega_t \rho_{cr} \tag{6.3.12}$$

where

$$\rho_{cr} = 3H_0^2/8\pi G \approx 2 \times 10^{-29} h^2 \, \text{g cm}^{-3} \tag{6.3.13}$$

where G is the gravitational constant, and condition $\Omega_t = 1$ separates closed Friedmann cosmological models with $\Omega_t > 1$ from open ones with $\Omega_t < 1$. Available astronomical data restrict Ω_t in the range of $0.03 \leq \Omega_t \leq 2$.

The Hubble constant H_0 together with the total mean density Ω_t determine the age of the universe

$$t_0 = H_0^{-1} f(\Omega_t) \tag{6.3.14}$$

where $f(\Omega_t)$ is a known but rather complicated function of the order of unity in the range mentioned above. However, for our purpose it is sufficient to have a few values of t_0 at different Ω_t. Table 6.3.1 demonstrates that the greater Ω_t, the shorter the age of the universe. If neutrinos are massive, they can dominate the mean density of the universe (i.e., $\Omega_\nu \approx \Omega_t$). This can constrain the mass of light neutrinos if the age of the universe is known independently.

The most severe low limit on the age of the universe comes from the ages of globular star clusters: $t_{gc} \geq 16.3 \times 10^9$ yr [SAN 84]. It is clear that the universe must be older. Table 6.3.1 also shows that only at $\Omega_t \leq 0.5$ and $h \leq 0.5$ can the age of globular clusters and the age of the universe be reconciled.

Table 6.3.1. *The age of the matter-dominated universe*

Ω_t	0.0	0.5	1.0	2.0
t_0 (10^9 yr)	$9.8\,h^{-1}$	$7.3\,h^{-1}$	$6.5\,h^{-1}$	$5.6\,h^{-1}$

In the above discussion, we assumed that both relativistic particles and the cosmological term are unimportant in evaluating mean density. The age of the universe decreases slightly with an increase in the density of relativistic particles. A positive Λ-term increases the age of the universe, but if it is negative the age becomes shorter. Of course, in order to affect the dynamics of the universe significantly, the density of relativistic particles or of the cosmological constant must be comparable with the critical density.

Assuming $h \approx 0.5$, $\Omega_t \approx 0.5$, and $\theta = T_\gamma/2.7K \approx 1$ in Eq. (6.3.9), one gets the upper limit on the sum of the masses of all neutrino species: $\sum m_{\nu,i} \leq 12.5$ eV, which is very severe. Unfortunately, the age of globular clusters t_{gc} as well as Ω_t and h can be estimated only by means of very indirect procedures, which involve many steps and phenomenologically established laws. Parameter θ is known better but it comes into (6.3.9) raised to a rather high power. Altogether this must relax the upper limit (6.3.9), perhaps within a factor of 2 or 3. Thus, it is probably safe to accept a less severe limit

$$\sum m_{\nu,i} \leq 25 \, \text{eV}. \tag{6.3.15}$$

Now let us look at the problem from another point of view. Let us ask, what are the inputs of known forms of matter into the mean density of the universe?

As we have seen, Eq. (6.3.3), the MBR photons are very populous in the universe. But at temperature $T_\gamma \approx 2.7$ they are extremely "light." Therefore their input into the mean mass density is very small: $\Omega_\gamma = \rho_\gamma/\rho_{cr} \sim 10^{-4}$. Baryons, in contrast, are very heavy but sparse. The theory of the primordial nucleosynthesis of light elements gives the following upper limit on the baryon density in all forms of matter [YAN 84]:

$$\Omega_b \equiv \bar{\rho}_b/\rho_{cr} \leq 0.0035 \, h^{-2}\theta^3\eta_{-10}. \tag{6.3.16}$$

Again assuming $h \approx 0.5$, $\theta \approx 1$ and $\eta_{-10} \leq 10$, one gets $\Omega_b \leq 0.14$. Thus, if one believes that the universe is baryon-dominated, then it must be open. In this case at present, practically, there is no problem with the age of the universe. Generally speaking, if the universe was really homogeneous then it could be baryon-dominated, but one has also to explain the structure of the universe, that is, galaxies, clusters, and superclusters of galaxies. As we shall see from the discussion in the next section, the problem of structural formation is much more difficult to explain in the baryon-dominated universe.

Using Eq. (6.3.9) together with (6.3.16) one can obtain two useful estimates. At

$$\sum m_{\nu,i} \geq 0.35 \, \eta_{-10} \, \text{eV} \tag{6.3.17}$$

the universe becomes neutrino-dominated (i.e., $\Omega_\nu \geq \Omega_b$) and at

$$\sum m_{\nu,i} \geq 100 \, h^2\theta^{-3} \, \text{eV} \tag{6.3.18}$$

the neutrino-dominated universe becomes closed (i.e., $\Omega_\nu \geq 1$).

The last comment in this section concerns heavy stable neutrinos. Neutrinos with masses in excess of about $1 \div 10\,\text{MeV}$ become nonrelativistic, remaining in equilibrium. Therefore, their number density must be estimated as freeze out abundance [LEE 77; DIC 77]. In this case, the heavier the neutrinos the less abundant they are. However, the increase in mass does not compensate the decrease in the number density, with the result that the Ω_ν decreases with the growth of the neutrino mass. Heavy neutrinos dominate the mean density at

$$m_\nu \leq 8\,\text{GeV} \tag{6.3.19}$$

and at

$$m_\nu \leq 1.5\,\text{GeV} \tag{6.3.20}$$

the mean neutrino density exceeds the critical value. However, this range is excluded by the experimental upper limits on the neutrino masses: $m_{\nu_\mu} < 250\,\text{keV}$ and $m_{\nu_\tau} < 70\text{MeV}$ [PRI 86], and therefore it will not be discussed any further.

6.3.2.2 The number of neutrino species and the primordial abundance of ^4He

The theory of primordial nucleosynthesis is a cornerstone of the big bang cosmological model. It explains with reasonable accuracy the observed abundances of the light elements: D, ^{3}He, ^{4}He, ^{7}Li. However, the problem is not at all simple. Perhaps the most difficult question to answer is how much the primordial abundances differ from the ones we observe now. Stars burn hydrogen to helium and helium to heavier nuclei; some of the ^{3}He is destroyed in hot central regions of stars but can be produced in cooler layers of some stars (see [BOE 85; MAT 88]).

The primordial abundance of ^{4}He, the most abundant element in the universe except hydrogen, depends primarily on three parameters: (1) the ratio of baryon to photon abundances η_{-10}, Eq. (6.3.4), (2) neutron half-life time $\tau_{\frac{1}{2}}$, and (3) the number of relativistic (at the time of nucleosynthesis) neutrino species N_ν. For the interesting range in baryon abundance ($1.5 < \eta_{-10} < 10$), the primordial mass helium abundance

$$Y \equiv \frac{\text{(helium mass density)}}{\text{(total baryon mass density)}} \tag{6.3.21}$$

fits well with the following expression:

$$Y \approx 0.230 + 0.011 \ln \eta_{-10} + 0.013(N_\nu - 3) + 0.014 \tag{6.3.22}$$
$$(\tau_{\frac{1}{2}} - 10.6\,\text{min})$$

[YAN 84; BOE 85]. Most observational data on ^{4}He abundance are grouped close to

$$Y = 0.245 \pm 0.003 \tag{6.3.23}$$

[KUN 83]. Thus $N_\nu = 3$ gives a good agreement of the theory of primordial nucleosynthesis with observations.

6.3.3 The formation of structure in the neutrino-dominated universe

6.3.3.1 Origin of the primordial fluctuations

If neutrinos have a nonzero mass and dominate the mean mass density of the universe, then, perhaps, the hypothesis that might best explain the observable structures (galaxies, clusters, and superclusters of galaxies) is the gravitational instability scenario [PEE 80; DOR 81; PRI 87; SHA 83d, 89]. It is based on the conclusion that if the mass density distribution is not exactly homogeneous then under definite conditions gravitation can amplify density fluctuations. This question is discussed in the following sections. Now it is a well-known point that if the universe was exactly homogeneous it would remain homogeneous forever. More precisely, this means that the statistical fluctuations arising during the Friedmann stage (which is believed to follow the inflationary stage after the average energy of particles falls below roughly 10^{14} GeV [LIN 84; PRI 87]) and amplified by the gravitational instability cannot explain the existence of the observable structure since they have too small an amplitude [LIF 46]. Thus we need to have primordial fluctuations of some kind originating at early stages, with an amplitude significantly exceeding the statistical fluctuations.

At present, the idea that primeval perturbations originated during the inflationary stage is widely accepted (e.g., see reviews in [LIN 84; PRI 87]). It is assumed that they arise as null quantum fluctuations of a scalar field or metric (an idea originally suggested by Sakharov [SAK 66], however, in the framework of a quite different cosmological model). From a theoretical point of view, the most attractive idea seems to be the hypothesis of adiabatic density perturbations with a scale invariant spectrum (so-called Harrison–Zeldovich spectrum, [YUP 69; HAR 70; ZEL 72]). In this case the perturbations of the metric have approximately equal amplitudes on all scales. The typical dimensionless amplitude needed for structure formation is of the order of 10^{-4}.

6.3.3.2 Linear stage: $\delta\rho/\rho \ll 1$

At the Friedmann stage perturbations evolve for a very long time, remaining quite small: $\delta\rho/\rho < 1$. This stage of their evolution is mathematically described by a linear theory, which at present is well developed (see, e.g., [PEE 80]).

Schematically, the typical scenario of the linear evolution of density fluctuations in the universe dominated by weakly interacting particles can be divided into four stages.

In the first stage the universe is dominated by relativistic particles, and weakly interacting massive particles are relativistic and the baryonic component is fully ionized.

The main feature of this stage is that all perturbations with scales less than the horizon size are erased because of the free streaming of relativistic, collisionless, weakly interacting particles. This process is similar to Landau damping in plasma.

The second stage begins when the weakly interacting massive particles become *nonrelativistic*. The universe can be still dominated by *relativistic* particles (photons), but it depends on the parameters of the model. The baryonic component remains fully *ionized*.

In the case of light neutrinos, the transition to this stage happens at

$$1 + z_\nu \approx 6 \cdot 10^4 \, (m_\nu/30 \, \text{eV}) \qquad (6.3.24)$$

where z_ν is redshift, Eq. (6.3.7). After this transition, the originally scale-free spectrum of density perturbations acquires a characteristic scale because the perturbations do not damp in the nonrelativistic medium of weakly interacting particles. The scale is approximately equal to the scale of the horizon at the time of the transition, and in terms of mass is as follows:

$$M_\nu \approx M_h(z_\nu) \sim 2 \cdot 10^{15} (m_\nu/30 \, \text{eV})^{-2} M_\odot. \qquad (6.3.25)$$

It can be also expressed in terms of fundamental constant [BIS 80]:

$$M_\nu \sim m_{Pl} \cdot (m_\nu/m_{Pl})^{-2} \qquad (6.3.26)$$

(where $m_{Pl} = (ch/G)^{\frac{1}{2}}$ is the Planck's mass), which coincides with the mass of the Markov's neutrino "superstars" [MAR 64].

The perturbations entering the horizon at this stage neither damp nor grow because relativistic particles dominate the mean density. As a result, the slope of this part of the spectrum changes by roughly -4 [PEE 80].

The third stage begins when the *nonrelativistic* weakly interacting particles come to dominate the mean density of the universe. This happens because the mean density of relativistic particles in the expanding universe decreases faster than that of nonrelativistic particles: $\rho_{rel} \propto a^{-4}$ and $\rho_{nonrel} \propto a^{-3}$. In the neutrino-dominated model the transition takes place at

$$1 + z_{eq} \approx 4 \cdot 10^4 \, \Omega_t h^2 (1 + 0.68 N_\nu) \qquad (6.3.27)$$

where N_ν the number of relativistic neutrino species. After the transition, the other scale is imprinted into the spectrum of density perturbations. Now it is equal to the horizon scale at that time:

$$M_{eq} \approx M_h(z_{eq}) \sim 10^{15} (\Omega_t h^2)^{-2}. \qquad (6.3.28)$$

In the universe dominated by stable light neutrinos with a mass of about 30 eV, the second stage does not exist. In this case, neutrinos become nonrelativistic roughly at the time when they come to dominate the mean density of the universe: $z_\nu \sim z_{eq}$. So the masses M_ν and M_{eq} are also close. This results in a very simple spectrum of density perturbations after the epoch of equality, Eq. (6.3.27). Assuming the primordial spectrum was of the Harrison–Zeldovich type, it becomes roughly

$$\delta_k^2 \propto \begin{cases} k, & \text{if } k < k_\nu \\ 0, & \text{otherwise} \end{cases} \tag{6.3.29}$$

where k is a comoving wave number that is related to the wavelength: $k = 2\pi a(t)/\lambda$.

Accurate numerical calculations of the linear evolution of density perturbations in the neutrino-dominated universe were done by Bond and Szalay [BON 83]. Again assuming the primordial spectrum of the Harrison–Zeldovich type, the results of the numerical calculations can be well fitted by

$$\delta_k^2 \propto k \cdot 10^{-2}(k/k_\nu)^{1.5} \tag{6.3.30}$$

where $k_\nu \approx 0.49\Omega_t h^2 \theta^{-2} \text{Mpc}^{-1}$. From that time until the beginning of the nonlinear stage, the spectrum keeps its shape within the accuracy of the linear approximations.

The fourth stage begins when the baryonic component becomes *neutral*. Earlier the universe was hot enough to keep hydrogen fully ionized. For this reason, baryons were tightly coupled with the radiation that prevented their peculiar motions. However, in the course of time the universe becomes cooler. Finally, at

$$1 + z_{\text{dec}} \approx 1500 \tag{6.3.31}$$

when $t_r = 4500$ K electrons recombine with protons into neutral hydrogen. Helium recombines somewhat earlier. Recombination is an extremely important cosmological event as from that time the baryonic component of matter decouples from the radiation and becomes involved in the process of gravitational instability.

The amplitude of density perturbations in the neutrino component begins to grow as $\delta\rho/\rho \propto a(t) = (1 + z)^{-1}$ after the epoch of equality, Eq. (6.3.27). This is a considerable advantage of the neutrino-dominated universe over the baryon-dominated universe because in the latter perturbations begin to grow only after decoupling. As a result, the amplitude of the primordial fluctuations in the neutrino-dominated universe can be at about $(1 + z_{\text{eq}})/(1 + z_{\text{dec}}) \approx 20$ times smaller than that in the baryon-dominated universe (it was assumed that $\Omega_t = 1$; $h = 0.5$; $N_\nu = 3$, and $z_{\text{dec}} \approx 1500$). The reduction of the density perturbation amplitude in the neutrino-dominated universe results in an equal reduction in the amplitude of the angular fluctuations of the MBR because before recombination baryons are distributed much more smoothly than neutrinos. This rescues the hypothesis of adiabatic primordial perturbations from the severe constraints on their amplitude imposed by the observational upper limits on the angular fluctuations of the microwave background radiation [DOR 81].

After decoupling, baryons quickly fill the potential wells formed by neutrino density inhomogeneities. Afterward there is quite a long stage of linear gravitational instability, when perturbations in baryons and neutrinos grow in amplitude as a whole, keeping similar spatial structure except for stretching caused by the expansion of the universe.

6.3.3.3 Nonlinear stage: $\delta\rho/\rho \geq 1$

When the amplitude of density perturbations reaches a value of the order unity ($\delta\rho/\rho \approx 1$), the nonlinear stage begins. It is usually assumed that at this stage real objects (stars, galaxies, clusters, and superclusters of galaxies) begin to form. This is justified by the fact that all these objects are evidently nonlinear concentrations of mass.

The physics of star and galaxy formation is very complicated and is still poorly understood. It should include fully nonlinear gasdynamic and thermal processes fed by the release of huge amounts of nuclear energy in stars. Therefore we restrict the present discussion to a description of the evolution of density inhomogeneities, thereby leaving unanswered the question concerning possible connections between mass and galaxy distribution. However, in modern cosmology this is a subject of exciting speculations. Perhaps it is worth mentioning that the simplest assumption that the galaxy distribution is linearly related to the total mass distribution (i.e., $\rho_{gal} \propto \rho_{mass}$) is no longer considered valid. Instead the biasing hypothesis has become widely accepted (e.g., [PRI 87]). The notion of biasing is based on the assumption that the distribution of galaxies in space does not completely replicate the total mass distribution. A particular example of biasing assumes that galaxies form only in the regions where the density perturbation is above some threshold.

The physics of cluster and supercluster formation is perhaps simpler because the larger the mass of the object, the stronger is the influence of pure gravitational processes on its evolution and structure. However, if one takes into account nonlinearity of the process, the question becomes far from simple.

The gravitational instability at the nonlinear stage in the neutrino-dominated universe is well understood. The theory is based on completely classical equations of motion and gravity, which, however, have been modified to take into account the expansion of the universe [PEE 80]. One can gain a good qualitative understanding of the process in the framework of the approximate solution suggested by Zeldovich [ZEL 70] to describe the nonlinear evolution of density perturbations in a dustlike medium in the expanding universe. This solution is given in terms of the equation explicitly relating the Lagrangian and Eulerian positions of every particle:

$$\mathbf{r}(\mathbf{q}, t) = a(t)(\mathbf{q} - b(t)\nabla_q \Phi(\mathbf{q})) \tag{6.3.32}$$

where $\mathbf{r}$ is the Eulerian position of the particle at the time t; $\mathbf{q}$ is the Lagrangian coordinate of the particle that would coincide with its position at the present epoch ($a(t_0) = 1$) if the universe was homogeneous; $a(t)$ is as usual the scale factor describing the homogeneous and isotropic expansion of the universe, Eq. (6.3.6); $b(t)$ is a growing function of time describing the growth of perturbations (in the universe with a closure mean density $b(t) \propto a(t) = (1 + z)^{-1}$. However, in the open universe $b(t)$ grows slower, and in the closed one faster than $a(t)$). The scalar function $\Phi(\mathbf{q})$ is proportional to the perturbation of the gravitational field in the growing mode at the linear stage and in fact contains the full information concerning the development of the structure.

At the linear stage, Eq. (6.3.32) describes the growing mode of the linear theory of gravitational instability. However, it also works surprisingly well at the beginning of the nonlinear stage. It is surprising because there is no formal reason for this approximation to be good at the nonlinear stage. One can provide some arguments for it [SHA 83b], but the decisive argument has come from a comparison of the particle distribution computed according to this solution with one obtained in direct three-dimensional N-body simulations with the same initial conditions [EFS 83]. Unfortunately, the approximation cannot be used for a long time after the formation of the first nonlinear structures.

Using the approximation (6.3.32), Zeldovich has predicted that the first nonlinear structures must be highly anisotropic. They resemble thin pancakes rather than spherical concentrations of mass. This conclusion has been undoubtedly confirmed by N-body simulations of various kinds [DOR 80; KLY 83; CEN 83; SHA 83b; DEK 83]. This is a nonlinear effect because one cannot find very much in the way of asymmetric structures from studying the shapes of the inhomogeneities at the linear stage when they are assumed to be random field of Gaussian type. At the linear stage the level surfaces of density perturbations in the vicinities of peaks are smooth and approximately ellipsoidal. The ratio of axes typically is not larger than 2 or 3.

If the universe was homogeneous and neutrinos had masses of about 30 eV, then at present the neutrino velocity dispersion would be about $\sigma_\nu = 6$ km/s [DOR 81]. This means that typical comoving sizes of the large-scale structure ($\sim 30 \ h^{-1}$ Mpc) are several orders of magnitude larger than the Jeans length $l_{\text{Jeans}} \sim \sigma_\nu t_0 \sim 0.04 \ h^{-1}$ Mpc. Therefore in the first approximation one can consider the medium to be cold and suppose that $\sigma_\nu = 0$. In this case, the pancakes arise as singularities in the density distribution [ZEL 70]. At the very beginning a pancake is a point singularity, having resulted, however, from extremely anisotropic (one-dimensional) contraction. At the next moment of time it becomes a very flattened region of three stream flows resembling a pancake. The pancake boundary is a caustic surface, where the density is singular. The phenomenon is very similar to the formation of caustic faces in geometrical optics [ZEL 83]. Since it is assumed that initial perturbations are of generic type, then the pancakes and other structures to form at the beginning of the nonlinear stage are generic singularities having quite specific geometrical shapes [ARN 82]. But, even small thermal velocity dispersion in the neutrino medium (as well as actual microscopic discreteness of the neutrino distribution) destroys the singularities, reducing the density to the values of several tens $\rho_{\text{max}} \sim 20 \div 40$ assuming the standard neutrino model with $m_\nu \approx 30$ eV [ZEL 82c]. Thus these singularities are not physical but just convenient abstractions that have resulted from the simplified approach to the problem.

In this scenario, it is usually assumed that galaxies form later inside pancakes as well as in other nonlinear structures by means of fragmentation [SHA 83d]. Of course, gasdynamic and thermal processes in the baryonic component must play an extremely important role on the scale of galaxies. Galaxies cannot form in the space between pancakes because the density outside the pancakes is even lower than the mean density in the universe and there are no perturbations of galaxy size in the initial spectrum. Thus the space between pancakes is devoid of galaxies, but is not completely empty because neutrinos and diffuse baryonic gas of primordial chemical composition can be present with a density about a tenth of the mean matter density in the universe [ZEL 82b]. Thus the pancake model provides the natural biasing for galaxy formation and distribution.

According to the pancake model, the whole large-scale structure represents a connected system of flattened and elongated superclusters with large clusters in the nodes of the structure. It possesses a particular percolation [ZEL 82a; SHA 83a] and topological properties [SHA 83c; GOT 87].

The above picture of structural formation is on the whole in quite good qualitative agreement with astronomical observations, but suffers from severe quantitative disagreements (WHI 83; 84). In my opinion, the most difficult problems it faces are the following:

1 At present, the model cannot reconcile the formation of the galaxies at least at $z_g \approx 3$ or even earlier (which is necessary to explain the observation of quasars and galaxies with $z > 3$) with the correlation scale $r_g \approx 5 \, h^{-1}$Mpc calculated for galaxy distributions (the correlation scale is the distance where the two point dimensionless correlation function of galaxies equals 1 [PEE 80]). The model predicts either the correct correlation scale but a very late ($z_g < 1$) epoch for galaxy formation, or early galaxy formation (say at $z_g \approx 3$) but too large a value for r_g [WHI 83; KLY 83].

2 In the latter case, the model also predicts the existence of very massive "clusters" of galaxies that are impossible to hide since they must be very strong X-ray sources [WHI 84].

3 There is evidence for the existence of dark matter halos in dwarf spheroidal galaxies [AAR 83; FAB 83]. Since the gravitational potential is rather low in such galaxies they give the most severe lower limit on the possible mass of the particles, Eq. (6.3.1), which is in conflict with the upper limit based on the age of the universe, Eq. (6.3.15).

Despite the fact that until now we have not understood galaxy formation quite well, one can speculate that the future theory of galaxy formation will explain all difficulties; this "solution" does not look very promising.

One possible way out infers the hypothesis that "neutrinos" (of any kind) having masses in the range roughly $m_\nu \sim 30 \div 100 \, \text{eV}$ must be unstable with a half-life time of $\tau_\nu \sim 1109 \, \text{yr}$ [DOR 84, 86]. This model has also to assume rather specific decay channels producing few photons as well as the existence of stable neutrinos to explain the low level of the UV-background and dark matter in galaxies and clusters. The main advantage of the model is that at the time when pancakes form, the particle decay slows down further evolution. This allows the epoch of galaxy formation to push to higher redshifts without any conflict with present epoch. Unfortunately, it assumes fine tuning between apparently unrelated parameters: the half-life time of the particles and the time of the beginning of the nonlinear stage have to be equal with a rather high precision. This model also does not help to solve the problem of the dark matter in dwarf galaxies.

6.3.4 Summary

The hypothesis that neutrinos of any kind have a nonzero mass has had rather a strong influence in cosmology. The age of the universe and primordial abundances of ^{4}He, restrict the mass and the number of neutrino species. The model of the cosmological nucleosynthesis together with observations of the ^{4}He abundance restrict the number of neutrino species practically to the number of known kinds: $N_\nu = 3$. As we have seen from the discussion in Section (6.3.2), in order to avoid the conflict with the age of the universe the mass of neutrinos must be rather small: $m_\nu < 25\,\mathrm{eV}$, Eq. (6.3.15). However, this restriction produces very serious difficulties for understanding the formation of observational large-scale structures of the universe. Present understanding of this process perhaps excludes the hypothesis that neutrinos of *any kind* possessing a mass in this range can solve *alone* the problem of the dark matter in the universe.

References

[AAR 83]   M. Aaronson, *Astrophys. J. Lett.* 266 (1983) L11.
[ABD 97]   J. N. Abdurashitov *et al.*, *Proc. Fourth Int. Solar Neutrino Conf.*, ed. W. Hampel, Max-Planck-Inst. f. Kernphysik, Heidelberg (1997) 109.
[AGL 87]   M. Aglietta *et al.*, *Europhys. Lett.* 3 (1987) 1315.
[AKH 97]   E. K. Akhmedov, *Proc. Fourth Int. Solar Neutrino Conf.*, ed. W. Hampel, Max-Planck-Inst. f. Kernphysik, Heidelberg (1997) 388.
[ALE 88]   E. N. Alexeyev *et al.*, *Phys. Lett.* B205 (1988) 209.
[ALE 95]   A. Alessandrello *et al.*, *Astroparticle Phys.* 3 (1995) 239.
[ALT 97]   M. Altmann, *Proc. Fourth Int. Solar Neutrino Conf.*, ed. W. Hampel, Max-Planck-Inst. f. Kernphysik, Heidelberg (1997) 183.
[ANS 92]   P. Anselmann *et al.*, (GALLEX-Collaboration), *Phys. Lett.* B285 (1992) 376.
[ANS 95]   P. Anselmann *et al.*, (GALLEX-Collaboration), *Phys. Lett.* B357 (1995) 237.
[ARA 87a] J. Arafune and M. Fukugita, *Phys. Rev. Lett.* 59 (1987) 367.
[ARA 87b] J. Arafune, M. Fukugita *et al.*, *Phys. Rev. Lett.* 59 (1987) 1864; W. D. Arnett, J. Truran and D. Schramm, 1988, University of Chicago preprint.
[ARN 73a] W. D. Arnett, *Explosive Nucleosynthesis*, University of Texas Press, Austin (1973).
[ARN 73b] W. D. Arnett and D. N. Schramm, *Astrophys. J. Lett.* 184 (1983) L47; S. Woolsey et al., *Astrophys. J. Lett.*, submitted (1987).
[ARN 82]   V. I. Arnold, S. F. Shandarin and Ya. B. Zeldovich, *Geophys. Astrophy. Fluid Dynamics* 20 (1982) 111.
[ARN 87]   W. D. Arnett, University of Chicago preprint (1987); R. M. Bionta *et al.*, *Phys. Rev. Lett.* 58 (1987) 1494.
[BAH 71]   J. N. Bahcall and R. K. Ulrich, *Astrophys. J.* 170 (1971) 593.
[BAH 82a] J. N. Bahcall, W. F. Hübner, S. H. Lubov, P. D. Parker and R. K. Ulrich, *Rev. Mod. Phys.* 54 (1982) 767.

[BAH 82b]   J. N. Bahcall and R. Davis, *Essays in Nuclear Astrophysics*, Cambridge Univ. Press (1982) 242.
[BAH 83]   J. N. Bahcall and T. Piran, *Ap. J.* 267 (1983) L 77.
[BAH 87a]   J. Bahcall, T. Piran, W. Press and D. Spergel, IAS preprint (1987).
[BAH 87b]   J. N. Bahcall, J. Gelb and S. P. Rosen, *Phys. Rev.*, D35 (1987) 2976.
[BAH 88]   J. N. Bahcall and R. K. Ulrich, *Rev. Mod. Phys.* 60 (1988) 297.
[BAH 89]   J. N. Bahcall, *Neutrino Astrophysics*, Cambridge Univ. Press (1989).
[BAH 95]   J. N. Bahcall and M. H. Pinsonneault, *Rev. Mod. Phys.* 67 (1995) 781.
[BAL 87]   A. J. Baltz and J. Weneser, *Phys. Rev.* D35 (1987) 528.
[BAN 95]   S. R. Bandler *et al.*, *Phys. Rev. Lett.* 74 (1995) 3169.
[BAR 85]   E. Baron *et al.*, *Phys. Rev. Lett.* 55 (1985) 126.
[BAR 94]   I. R. Barabanov *et al.*, *Nucl. Phys. B (Proc. Suppl.)* 35 (1994) 461.
[BET 85]   H. Bethe and G. Browne, *Scientific American* 252 (1985) 60.
[BIO 88]   R. M. Bionta *et al.*, *Phys. Rev. Lett.* 58 (1987) 1494; C. B. Bratton *et al.*, *Phys. Rev.* D37 (1988) 3361.
[BIS 80]   G. S. Bisnovatyi-Kogan and I. D. Novikov, *Sov. Astron.* 24 (1980) 516.
[BOE 85]   A. M. Boesgaard and G. Steigman, *Ann. Rev. Astron. Astrophys.* 23 (1985) 319.
[BON 83]   J. R. Bond and A. S. Szalay, *Astrophys. J.* 274 (1983) 443.
[BOO 87]   N. E. Booth, *Sci. Prog., Oxford* 71 (1987) 563.
[BOU 86]   J. Bouchez *et al.*, *Z. Phys.* C32 (1986) 499.
[BOW 96]   T. Bowles and V. N. Gavrin, *Seventh Int. Workshop on Neutrino Telescopes*, ed. M. Baldo-Ceolin, INFN Venezia (1996) 253.
[BRO 97]   C. Broggini, *Nucl. Phys. B (Proc. Suppl.)* 35 (1994) 441 and *ibid.* (1998), *Proc. TAUP 97*, Gran Sasso, to appear.
[BUR 85]   A. Burrows and J. Lattimer, *The Birth of Neutron Stars*, preprint (1985).
[BUR 87]   A. Burrows, University of Arizona preprint (1987); K. S. Hirata *et al.*, *Phys. Rev. Lett.* 58 (1987) 1490.
[BUR 88]   A. Burrows, Univ. of Arizona preprint, 1988.
[CEN 83]   J. M. Centrella and A. L. Melott, *Nature* 304 (1983) 196.
[CHA 94]   C. C. Chang *et al.*, *Nucl. Phys. B (Proc. Suppl.)* 35 (1994) 464.
[CHE 87]   M. L. Cherry and K. Lande, *Phys. Rev.* D36 (1987) 3571.
[CHO 00]   CHORUS Collab. *Phys. lett. B* to be published.
[CHO 00b]   CHORUS Collab. *Phys. lett. B* to be published.
[CIO 88]   A. Ciocio, *Proc. 13th Int. Conf. Neutrino Physics and Astrophysics*, ed. J. Schneps, World Scientific, Singapore (1988) 280.
[CLE 97a]   B. T. Cleveland *et al.*, *Proc. Fourth Int. Solar Neutrino Conf.*, ed. W. Hampel, Max-Planck-Inst. f. Kernphysik, Heidelberg (1997) 85.
[CLE 97b]   B. T. Cleveland *et al.*, *Proc. Fourth Int. Solar Neutrino Conf.*, ed. W. Hampel, Max-Planck-Inst. f. Kernphysik, Heidelberg (1997) 228.
[COW 72]   R. Cowsik and J. McClelland, *Phys. Rev. Lett.* 29 (1972) 669.
[COW 73]   R. Cowsik and J. McClelland, *Astrophys. J.* 180 (1973) 7.
[COW 82]   G. A. Cowan and W. C. Haxton, *Science* 216 (1982) 51.
[CRI 86]   M. Cribier *et al.*, *Phys. Lett.* 182 (1986) 891.
[DAV 88]   R. Davis, NSF Proposal, 1988.
[DEK 83]   A. Dekel, *Astrophys. L.* 264 (1983) 373.
[DIC 77]   D. A. Dicus, E. W. Kolb and V. Teplitz, *Phys. Rev. Lett.* 39 (1977) 168.
[DOR 80]   A. G. Doroshkevich, E. V. Kotok, I. D. Novkov, A. N. Polyudov, S. F. Shandarin and Yu. S. Sigov, *Mon. Not. R. Astr. Soc.* 192 (1980) 321.
[DOR 81]   A. G. Doroshkevich, M. Yu. Khlopov, R. A. Sunyaev, A. S. Szalay and Ya. B. Zeldovich, *Ann. N. Y. Acad. Sci.* 375 (1981) 32.
[DOR 84]   A. G. Doroshkevich and M. Yu. Khlopov, *Mon. Not. R. Astr. Soc.* 211 (1984) 277.
[DOR 86]   A. G. Doroshkevich, A. A. Klypin and E. V. Kotok, *Sov. Astron.* 30 (1986) 251.
[EPS 83]   G. Efstathiou and J. Silk, *Fund. Cosm. Phys.* 9 (1983) 1.
[EL 85]   A. L. Melott, *Astrophys. J.* 289 (1985) 2.
[ERN 84]   H. Ernst *et al.*, *Nucl. Instr. Meth.* B5 (1984) 426.

[EWA 87]   G. T. Ewan *et al.*, SNO-87-12 (Proposal to Canadian Research Council) (1987).
[FAB 83]   S. M. Faber and D. N. C. Lin, *Astrophys. J.* 266 (1983) L17.
[FAU 85]   J. Faulkner and R. L. Gilliland, *Astrophys. J.* 299 (1985) 994.
[FEI 97]   F. v. Feilitzsch, *Proc. Fourth Int. Solar Neutrino Conf.*, ed. W. Hampel, Max-Planck-Inst. f. Kernphysik, Heidelberg (1997) 192.
[FIO 88]   E. Fiorini, *Proc. 13th Int. Conf. Neutrino Physics and Astrophysics*, ed. J. Schneps, World Scientific, Singapore (1988) 344.
[FRE 77]   D. Z. Freedman, D. N. Schramm and D. Tubbs, *Ann. Rev. Nucl. Sci.* 27 (1977) 167.
[GAL 97]   M. Galeazzi *et al.*, *Phys. Lett.* B398 (1997) 187.
[GAV 88]   V. A. Gavrin, *Proc. 13th Int. Conf. Neutrino Physics and Astrophysics*, ed. J. Schneps, World Scientific, Singapore (1988) 317.
[GEO 97]   A. Sh. Georgadze *et al.*, *Proc. Fourth Int. Solar Neutrino Conf.*, ed. W. Hampel, Max-Planck-Int. f. Kernphysik, Heidelberg (1997) 283.
[GER 66]   S. S. Gershtein and Ya. B. Zeldovich, *Sov. Phys. JETP Lett.* 4 (1966) 174.
[GOT 87]   J. R. Gott, D. H. Weinberg and A. L. Melott, *Astrophys. J.* 319 (1987) 1.
[GRO 88]   M. Gronau *et al.*, *Phys. Rev.* D37 (1988) 2597.
[HAI 76]   K. Hainebach and D. Schramm, *Astrophys. J.* (1976).
[HAM 88]   W. Hampel, *Neutrino 88*, ed. J. Schneps, World Scientific, (1989) 311.
[HAM 96]   W. Hampel *et al.*, (GALLEX-Collaboration), *Phys. Lett.* B388 (1996) 384.
[HAM 98]   W. Hampel *et al.*, (GALLEX-Collaboration), *Phys. Lett.* B420 (1998) 114.
[HAR 70]   E. R. Harrison, *Phys. Rev.* D1 (1970) 2726.
[HAX 86]   W. C. Haxton, *Phys. Rev. Lett.* 57 (1986) 1271.
[HAX 87]   W. C. Haxton, *Phys. Rev.* D35 (1987) 2352.
[HAX 88]   W. C. Haxton, Los Alamos National Laboratory 40048-46-IV7 (1988).
[HEN 97]   E. Henrich *et al.*, *Proc. Fourth Int. Solar Neutrino Conf.*, ed. W. Hampel, Max-Planck-Inst. f. Kernphysik, Heidelberg (1997) 151.
[HIR 87]   K. S. Hirata *et al.*, *Phys. Rev. Lett.* 58 (1987) 1490.
[HIR 88]   K. S. Hirata *et al.*, *Phys. Rev.* D38 (1988) 48.
[HIR 89]   K. Hirata *et al.*, *Phys. Rev. Lett.* 63 (1989) 16.
[HOY 64]   F. Hoyle and R. J. Tayler, *Nature* 203 (1964) 1108.
[HUR 84]   G. S. Hurst *et al.*, *Phys. Rev. Lett.* 53 (1984) 1116.
[IAU 87]   International Astronomical Union Circular No. 4323 (1987).
[KAZ 76]   D. Kazanas and D. N. Schramm, *Nature* 262 (1976) 671.
[KAZ 77]   D. Kazanas and D. N. Schramm, *Astrophys. J.* 214 (1977) 819.
[KIR 78]   T. Kirsten, *Proc. Inform. Conf. on Status and Future of Solar Neutrino Research*, Brookhaven BNL 50879, 1 (1978) 305.
[KIR 84]   T. Kirsten, *Inst. Phys. Conf. Ser.* 71 (1984) 251.
[KIR 90]   T. Kirsten, *Inside the Sun, Proc. IAU Coloq.*, Versailles (1989) 121; ed. G. Berthomieu and M. Cribier (1990) 187.
[KIR 95]   T. Kirsten, *Ann. N. Y. Acad. Sci.*, 759 (1995) 1.
[KIR 97]   T. Kirsten, *Proc. Fourth Int. Solar Neutrino Conf.*, ed. W. Hampel, Max-Planck-Inst. f. Kernphysik, Heidelberg (1997) 138.
[KIR 99]   T. Kirsten, Gallex Collab., *Nucl. Phys. B (Proc. Suppl.)* 77 (1999) 26.
[KLY 83]   A. A. Klypin and S. F. Shandarin, *Mon. Not. R. Astr. Soc.* 204 (1983) 891.
[KOL 87]   E. W. Kolb, A. Stebbins and M. S. Turner, *Phys. Rev.* D35 (1987) 3598, 3820(E).
[KOP 97]   A. V. Kopylov, *Proc. Fourth Int. Solar Neutrino Conf.*, ed. W. Hampel, Max-Planck-Inst. f. Kernphysik, Heidelberg (1997) 263.
[KUN 83]   D. Kunth and W. L. Sargent, *Astrophys. J.* 273 (1983) 81.
[KUO 87]   T. K. Kuo and J. Pantaleone, *Phys. Rev.* D35 (1987) 3432.
[LAN 81]   P. Langacker, *Phys. Rep.* C72 (1981) 185.
[LAN 87]   R. E. Lanou *et al.*, DOE Proposal, Brown Univ. (1987).
[LEE 77]   B. W. Lee and S. Weinberg, *Phys. Rev. Lett.* 39 (1977) 165.
[LIF 46]   E. M. Lifshitz, *J. Phys. USSR Acad. Sci.* 10 (1946) 116.
[LIN 84]   A. D. Line, *Rep. Prog. Phys.* 47 (1984) 925.

[LUB 80]   V. A. Lubimov, E. G. Novkov, V. Z. Nozik, E. F. Tretyakov and V. S. Kozik, *Phys. Lett.* B94 (1980) 266.

[MAR 64]   M. A. Markov, *Phys. Lett.* 10 (1964) 122.

[MAR 72]   G. Marx and A. S. Szalay, *Proc. Neutrino 72*, Technoinform, Budapest (1972) 123.

[MAT 88]   G. J. Mathews, ed., *Origin and Distribution of the Elements*, World Scientific, Singapore, 1988.

[MAY n.d]   R. Mayle, Ph.D. diss., University of California, Berkeley (available as Lawrence Livermore preprint UCRL 53713).

[MAY 87a]   R. Mayle, J. Wilson and D. N. Schramm, *Astrophys. J.* 318 (1987) 288.

[MAY 87b]   R. Mayle and J. Wilson, Livermore preprint (1987), unpublished.

[MAZ 76]   T. Mazurek, *Astrophys. J. Lett.* 207 (1976) L 87.

[MEI 97]   R. Meijer Drees, *Proc. Fourth Int. Solar Neutrino Conf.*, ed. W. Hampel, Max-Planck-Inst. f. Kernphysik, Heidelberg (1997) 210.

[MIK 86]   S. P. Mikhaeyev and A. Yu Smirnov, *Nuovo Cimento* 9C (1986) 17.

[NAK 87]   M. Nakahata, Ph.D. diss., UT-ICEPP-88-01, Faculty of Science, University of Tokyo.

[OOR 83]   J. H. Oort, *Ann. Rev. Astron. Astrophys.* 21 (1983) 373.

[PAR 86a]   S. J. Parke, *Phys. Rev. Lett.* 57 (1986) 1275.

[PAR 86b]   S. J. Parke and T. P. Walker, *Phys. Rev. Lett.* 57 (1986) 2322.

[PAT 97]   L. Paterno, *Proc. Fourth Int. Solar Neutrino Conf.*, ed. W. Hampel, Max-Planck-Inst. f. Kernphysik, Heidelberg (1997) 54.

[PAV 88]   M. K. Pavicevic, *Nucl. Instr. Meth.* A271 (1988) 287.

[PEE 80]   P. J. E. Peebles, *The Large-Scale Structure of the Universe*, Princeton University Press (1980).

[PER 88]   D. Perret-Gallix, *Neutrino 88*, ed. J. Schneps, World Scientific (1989) 337.

[PET 88]   S. T. Petcov, *Phys. Lett.* B200 (1988) 373.

[PIZ 87]   P. Pizzochero, *Phys. Rev.* D36 (1987) 2293.

[PRE 87]   K. Pretzl, N. Schmitz and L. Stodolsky, eds., *Low Temperature Detectors for Neutrinos and Dark Matter*, Springer (1987).

[PRI 86]   J. R. Primack, in *Cosmology, Astronomy, and Fundamental Physics*, ed. G. Setti and L. Van Hove, ESO, Garching bei Munchen (1986) 183.

[PRI 87]   J. R. Primach, in *Proc. Int. School of Physics 'Enrico Fermi'*, Course 92, ed. N. Cabibbo, North-Holland (1987) 137.

[RAG 76]   R. S. Raghavan, *Phys. Lett.* 37 (1976) 259.

[RAG 86]   R. S. Raghavan, S. Pakvasa and B. A. Brown, *Phys. Rev. Lett.* 57 (1986) 1801.

[RAG 88]   R. S. Raghavan, *Neutrino 88*, ed. J. Schneps, World Scientific (1989) 305.

[RAG 93]   R. S. Raghavan *et al.*, *Phys. Rev. Lett.* 71 (1993) 4295.

[RAG 97]   R. S. Raghavan, *Proc. Fourth Int. Solar Neutrino Conf.*, ed. W. Hampel, Max-Planck-Inst. f. Kernphysik, Heidelberg (1997) 248.

[RAP 85]   J. Rapaport *et al.*, *Phys. Rev. Lett.* 54 (1985) 2325.

[ROS 86]   S. P. Rosen and J. M. Gelb, *Phys. Rev.* D34 (1986) 969.

[ROW 85]   J. K. Rowley, B. T. Cleveland and R. Davis, *AIP Conf. Proc.* 126 (1985) 1.

[RUB 96]   C. Rubbia, *Nucl. Phys. B (Proc. Suppl.)* 48 (1996) 172.

[SAC 90]   J. Sackman *et al.*, *Astrophys. J.* 360 (1990) 727.

[SAK 66]   A. D. Sakharov, *Sov. Phys. JETP* 22 (1966) 241.

[SAN 84]   A. Sandage and G. A. Tammann, in *First ESO-CERN Symp: Large Scale Structure of the Universe, Cosmology and Fundamental Physics*, ed. G. Setti and L. Van Hove, CERN, Geneva (1984) 127.

[SAT 87]   K. Sato and H. Suzuki, *Phys. Rev. Lett.* 58 (1987) 2722.

[SCH 74]   D. Schramm and W. D. Arnett, *Astrophys. J.* 198 (1974) 629.

[SCH 76]   D. N. Schramm, *Proc. 2nd Dumand Symposium* (1976) 87.

[SCH 81]   D. N. Schramm and G. Steigman, *Astrophys. J.* 243 (1981).

[SCH 85]   E. Schatzman, *AIP Conf. Proc.* 126 (1985) 69.

[SCH 87a]   D. Schramm, *Comm. Nucl. Part. Phys.* (1987).

[SCH 87b]   D. Schramm, *Proc. SLAC Summer School* (1987).

[SCH 88a]   D. Schramm, *Proc. Rencontre de Moriond Workshop on Exotic Phenomena* (1988).

[SCH 88b]   D. Schramm and J. Truran, *Phys. Rep.* 189 (1990) 89.

[SHA 78]   S. Shapiro, in *Gravitational Radiation*, ed. L. Smarr, Oxford University Press, (1978).

[SHA 83a]   S. F. Shandarin, *Sov. Astron. Lett.* 9 (1983) 104.

[SHA 83b]   S. F. Shandarin, in *The Origin and Evolution of Galaxies*, ed. B. J. T. Jones and J. E. Jones, Reidel, Dordrecht (1983) 171.

[SHA 83c]   S. F. Shandarin and Ya. B. Zeldovich, *Comm. Astrophys.* 10 (1983) 33.

[SHA 83d]   S. F. Shandarin, A. G. Doroshkevich and Ya. B. Zeldovich, *Sov. Phys. Usp.* 26 (1983) 46.

[SHA 89]   S. F. Shandarin and Ya. B. Zeldovich, *Rev. Mod. Phys.* 61 (1989) No. 2.

[SHV 69]   V. F. Shvartsman, *Sov. Phys. JETP Lett.* 9 (1969) 184.

[SOF 88]   Y. Sofue, *Proc. Second Workshop on Elementary-Particle Picture of the Universe*, KEK, Japan (1988).

[STE 77]   G. Steigman, D. N. Schramm and J. Gunn, *Phys. Lett.* B66 (1977) 202.

[SUZ 97a]   Y. Suzuki, *Proc. Fourth Int. Solar Neutrino Conf.*, ed. W. Hampel, Max-Planck-Inst. f. Kernphysik, Heidelberg (1997) 163.

[SUZ 97b]   Y. Suzuki, *Nucl. Phys. B (Proc. Suppl.)* (1998), *Proc. TAUP 97*, Gran Sasso, *Nucl. Phys. B (Proc. Suppl.)* 35 (1999) 35.

[SUZ 99]   Y. Suzuki, *Proc. Neutrino 98*, Takay-ama 1998, *Nucl. Phys. B (Proc. Suppl.)* 77 (1999) 35.

[SWI 94]   A. M. Swift *et al.*, *Nucl. Phys. B (Proc. Suppl.)* 35 (1994) 405.

[SZA 76]   A. S. Szalay and G. Marx, *Astron. Astrophys.* 49 (1976) 437.

[TAO 97]   C. Tao, *Proc. Fourth Int. Solar Neutrino Conf.*, ed. W. Hampel, Max-Planck-Inst. F. Kernphysik, Heidelberg (1997) 238.

[TAM 77]   G. Tamman, in *Supernovae*, ed. D. Schramm, Reidel (1977).

[TOS 87]   S. Toshev, *Phys. Lett.* B198 (1987) 551.

[TOT 86]   Y. Totsuka, UT-ICEPP-86-06, Faculty of Science, University of Tokyo.

[TRE 79]   S. Tremaine and J. E. Gunn, *Phys. Rev. Lett.* 42 (1979) 407.

[TUR 93]   S. Turck-Chieze *et al.*, *Astrophys. J.* 408 (1993) 347.

[VAN 88]   J. Vandervelde, 'One Year Later', *Proc. Aosta Symp.*, SN 1987a (1988).

[VOL 86]   M. B. Voloshin, M. I. Vysotsky and L. B. Okun, *Sov. J. Nucl. Phys.* 44 (1986) 440.

[WAL 87]   T. Walker and D. N. Schramm, *Phys. Lett. B* 195 (1987) 331; R. Mayle, J. Wilson and D. Schramm, *Ap. J.* 318 (1987) 288.

[WEA 83]   T. Weaver and S. Woolsey *Proc. Wilson Symp.*, ed. J. Centrella (1983).

[WHI 83]   S. D. M. White, C. S. Frenk and M. Davis, *Astrophys. J. Lett.* 274 (1983) L1.

[WHI 84]   S. D. M. White, M. Davis and C. S. Frenk, *Mon. Not. R. Astro. Soc.* 209 (1984) 27.

[WIL 75]   J. Wilson *et al.*, *Ann. N. Y. Acad. Sci.* 262 (1975) 54.

[WIL 81]   J. Wilson, *Astrophys. J.* (1981); D. N. Schramm, J. Wilson and R. Mayle, in *Proc. First Int. Conf. on Underground Physics*, in *Nuovo Cimento* 9C (1986) 443.

[WIL 88]   J. Wilson, 1988, presented at Aspen Winter Meeting on SN 1987A.

[WOL 78]   L. Wolfenstein, *Phys. Rev.* D17 (1978) 2369.

[YAN 84]   J. Yang, M. S. Turner, G. Steigman, D. N. Schramm and K. A. Olive, *Astrophys. J.* 281 (1984) 493.

[YUP 69]   J. T. Yu and P. J. E. Peebles, *Astrophys. J.* 158 (1969) 103.

[ZEL 70]   Ya. B. Zeldovich, *Astron. Astrophys.* 5 (1970) 84.

[ZEL 72]   Ya. B. Zeldovich, *Mon. Not. R. Astr. Soc.* 160 (1972) 1.

[ZEL 82a]   Ya. B. Zeldovich, *Sov. Astron. Lett.* 8 (1982) 102.

[ZEL 82b]   Ya. B. Zeldovich and S. F. Shandarin, *Sov. Astron. Lett.* 8 (1982) 139.

[ZEL 83]   Ya. B. Zeldovich, A. Mamaev and S. F. Shandarin, *Sov. Phys. Usp.* 26 (1983) 77.

[ZWI 33]   F. Zwicky, *Helv. Phys. Acta.* 6 (1933) 110.

Index